CORRELATIVE LEARNING

BICENTENNIAL

1807 WILEY 2007

BICENTENNIAL

THE WILEY BICENTENNIAL–KNOWLEDGE FOR GENERATIONS

Each generation has its unique needs and aspirations. When Charles Wiley first opened his small printing shop in lower Manhattan in 1807, it was a generation of boundless potential searching for an identity. And we were there, helping to define a new American literary tradition. Over half a century later, in the midst of the Second Industrial Revolution, it was a generation focused on building the future. Once again, we were there, supplying the critical scientific, technical, and engineering knowledge that helped frame the world. Throughout the 20th Century, and into the new millennium, nations began to reach out beyond their own borders and a new international community was born. Wiley was there, expanding its operations around the world to enable a global exchange of ideas, opinions, and know-how.

For 200 years, Wiley has been an integral part of each generation's journey, enabling the flow of information and understanding necessary to meet their needs and fulfill their aspirations. Today, bold new technologies are changing the way we live and learn. Wiley will be there, providing you the must-have knowledge you need to imagine new worlds, new possibilities, and new opportunities.

Generations come and go, but you can always count on Wiley to provide you the knowledge you need, when and where you need it!

William J. Pesce
PRESIDENT AND CHIEF EXECUTIVE OFFICER

Peter Booth Wiley
CHAIRMAN OF THE BOARD

CORRELATIVE LEARNING

A Basis for Brain and Adaptive Systems

Zhe Chen
RIKEN Brain Science Institute

Simon Haykin
McMaster University

Jos J. Eggermont
University of Calgary

Suzanna Becker
McMaster University

BICENTENNIAL 1807 WILEY 2007 BICENTENNIAL

WILEY-INTERSCIENCE

A JOHN WILEY & SONS, INC., PUBLICATION

Published by John Wiley & Sons, Inc., Hoboken, New Jersey.
Published simultaneously in Canada.

Wiley Bicentennial Logo: Richard J. Pacifico

Library of Congress Cataloging-in-Publication Data:

Correlative learning : a basis for brain and adaptive systems / Zhe Chen . . .
[et al.].
p. ; cm. – (Wiley series on adaptive and learning systems for signal processing, communications, and control)
Includes bibliographical references and index.
ISBN 978-0-470-04488-9 (cloth)
1. Learning–Physiological aspects. 2. Brain–Physiology. 3. Artificial intelligence. 4. Computer simulation. 5. Correlation (Statistics) I. Chen, Zhe, 1976- II. Series: Adaptive and learning systems for signal processing, communications, and control.
[DNLM: 1. Brain–Physiology. 2. Artificial Intelligence. 3. Computer Simulation. 4. Learning–Physiology. WL 300 C824 2007]
QP408.C67 2007
612.8′2–dc22

2007006012

10 9 8 7 6 5 4 3 2 1

To Spring

CONTENTS

FOREWORD

The world we live in is complex, but that complexity is not so obscure that it is undecipherable. In fact, the laws of physics and chemistry that have governed the universe since the big bang are the same laws providing order to our seemingly chaotic world and have enabled life to evolve. Even the human brain, while being a highly complex and enormously organized system, coheres with the laws of the universe. We seek to understand how these first principles structure our minds and our external world.

We attempt to unlock the tangled secrets of our world and minds by finding correlations that are the result of the highly organized structures that exist, the same structures that provide us with the means to survive. The brain is no exception. It, too, learns and organizes itself according to its interactions with and in the world. Design principles also use correlations to guide the development of sophisticated engineering systems.

Correlation is not merely the co-occurrence of two events. Correlation between two events implies deeper relationships within space and in time. When two or more events have temporal, spatial, and higher order correlations, there is a relevant relationship between the events—whether these are linear or nonlinear structures.

This monograph focuses on how efforts to understand the mechanisms of learning in the brain and in engineering systems use generalized concepts of correlation. The neurons in the brain form complex networks and our understanding of these networks is increasingly used to develop sophisticated engineering systems. So, while they appear to be vastly different structures based on unrelated principles, a look under the surface reveals surprising similarities.

Unlike scientific and technological pursuits in the last century that were strictly divided between disciplines, multidisciplinary approaches are increasingly more essential and useful in these pursuits in the twenty-first century. In this volume, efforts to reveal the mysterious working of the brain are incorporated into the designs of sophisticated and intelligent engineering information systems. This is a good example of interdisciplinary collaboration to understand intelligence.

The present monograph broadly covers the latest output in brain science and engineering learning systems as it introduces the learning mechanisms of the brain as well as approaches to adaptive signal processing and intelligent information systems. Since the histories of these three disciplines are long and not easily accessible, it is attempted to demonstrate their common intrinsic structures. The results should prove intriguing.

Such a book cannot be written without close collaboration between active researchers—young and old—whose combined interests include brain science, cognitive science, and signal processing. This highly correlated effort has produced a wonderful, engaging book that touches on aspects of learning from a unified perspective.

I stand in admiration of their accomplishment and I am pleased to be able to recommend this book to researchers and students working in diverse areas of science and engineering.

SHUN-ICHI AMARI
Director of RIKEN Brain Science Institute
Professor-Emeritus at the University of Tokyo

PREFACE

> Learning without thought is useless, thought without learning is dangerous.
> —Confucius

> Cogito Ergo Sum (I think, therefore I am).
> —René Descartes

MOTIVATION

Computational neuroscience, according to Terrence Sejnowski and Tomaso Poggio, is an approach to understanding the information content of neural signals by modeling the nervous system at many different structural scales, including the biophysical, the circuit, and the system levels. Therefore, an essential goal of computational neuroscience is to build a computational model, paradigm, or theory for understanding the brain's functions. With its intrinsic interdisciplinary nature that invokes many disciplines such as neuroscience, biology, physiology, psychology, computer science, physics, mathematics, and engineering, the past decades have witnessed significant gains in approaching the goal of understanding the human brain. Many of us are fascinated by the fact that numerous ideas in different disciplines have been cross-fertilized; in particular, the horizons of neuroscience research have been greatly expanded by the ever-developing statistical and computational modeling paradigms. It is our belief that developing powerful computational tools would provide an accessible means of modeling and comprehending the functions of the brain; in so doing, an emerging understanding of the nature of the brain would be beneficial and insightful. Challenges certainly still remain, but that is why we are motivated and where our work shall start.

The human brain, being a highly sophisticated and complex system, has provided us with many insights for designing adaptive learning systems. In turn, developing intelligent adaptive systems has also deepened our understanding of the human brain's function. For many years, developing brain-style signal processing or machine learning algorithms has been the Holy Grail of artificial intelligence research. Unraveling the mysteries of the brain has attracted many sharp minds from a wide range of disciplines.

This research monograph represents an effort to bridge the communication gap between neuroscientists and engineers. For many years, it has been our feeling that signal processing researchers and neuroscientists do not share a common langauge that could help engineers to understand and appreciate this highly sophisticated biosystem—the human brain—although this is vitally important for engineers whose aim is to build complex, reliable (robust), adaptive systems in practice. It is this belief that brought out the writing of this research monograph, coauthored by four researchers with varying backgrounds from signal processing, neuroscience, psychology, and computer science. It is our hope that this monograph might be helpful as a step forward to approaching this goal.

ROAD MAP

Correlations are arguably ubiquitous phenomena that occur in the human brain. According to [241], correlation is believed to occur at many timescales and also to exist at both macroscopic and microscopic levels, which are useful for adapting the synaptic strengths, for sensory perception, for learning and memory, as well as for high-level cognition. Correlation is important not only for brain function but also for building adaptive systems in practical engineering applications, such as spectrum analysis, signal detection, statistical analysis, as well as optimization.

This research monograph is aimed at providing a bridge between two distinct disciplines: computational neuroscience/neural computation and signal processing. To do so, we first try to lay down the necessary neuroscience background for engineers. In particular, the first part (Chapters 1 and 2) of the monograph presents an overview of the role of correlation in the human brain as well as in signal processing. The next part (Chapters 3–5) of the monograph is intended to unify many well-established synaptic adaptation (learning) rules within the correlation-based learning framework. Specifically, Chapter 6 focuses on a particular correlative learning paradigm known as ALOPEX. The final part (Chapter 7) presents some case studies that illustrate how to use computational tools for either helping us understand brain functions or fitting specific engineering applications.

ORGANIZATION

This monograph is structured in three major parts that include an introduction and eight other chapters:

- The introduction presents a general account of why correlation is important and its omnipresent role in the brain; it also discusses the important notion of learning that functions as the backbone of this monograph.
- Chapter 1 addresses the correlative brain, which highlights the key role that correlation plays in many aspects of the human brain, ranging from synaptic

plasticity, neocortical receptive fields, population synchrony coding, hippocampal coding of episodic memory, synchrony in feature binding and attention, sensory coding, and motor control. The aim of this chapter is to provide a general neuroscience background as well as to underscore the breadth of ways in which correlation is a vital concept for understanding brain function. The neuroscience material in Chapter 1, combined with the signal processing material in Chapter 2, should provide a reader with a general science background with a sufficient foundation for understanding the algorithms described in the remainder of the book.

- Chapter 2 discusses the role of correlation in statistical and adaptive signal processing. This is a chapter that takes an engineering perspective. Starting with the roots of modern signal processing, we discuss in detail the correlation functions for developing the relevant concepts in spectrum analysis, Wiener filtering, least-mean-square (LMS) filters, recursive least-squared (RLS) filters, matched filters, correlated detectors, and statistical data analysis.
- Chapter 3 is devoted to a general overview of correlation-based learning rules and correlation-based computational neural models. In this relatively lengthy chapter, it is shown that many statistical learning rules, despite their varying motivations, can be traced back and unified within the framework of generalized Hebbian learning; this is done by reinterpreting the pre- and postsynaptic terms of Hebb's original rule.
- Chapter 4 is devoted to correlation-based kernel learning. The kernel is a natural tool for extending correlation-based similarity measures from linear spaces to nonlinear feature spaces; many correlation-based statistical kernel methods will be developed by employing the "kernel trick" in reproducing kernel Hilbert space (RKHS).
- Chapter 5 extends the correlation concept to the complex-valued domain and naturally defines various second-order and higher order statistics for complex-valued random variables. In a similar vein, we also extend our discussions to complex-valued generalized Hebbian learning, which has many engineering applications in communications and array signal processing, such as blind channel equalization, blind separation and blind deconvolution, and beamforming.
- Chapter 6 discusses a special correlation-based learning paradigm—ALOPEX, short for *ALgorithm Of Pattern EXtraction*. While being a correlative learning rule, ALOPEX distinguishes itself from Hebb's rule in many different ways, especially in the use of feedback. We will present the canonical version and several sophisticated variants of the ALOPEX that were developed by the authors and many others.
- Chapter 7 presents a few case studies of applying the notion of correlative learning to various applications in computational neuroscience (auditory and visual modeling) and engineering (human–machine interface design and training artificial neural networks).
- Chapter 8 concludes the book with a discussion on future perspectives.

While most chapters stand by themselves, they are also intrinsically related by their contents. Nevertheless, maximum gain can be anticipated for reading while following the given chapter order. At the end, some mathematical backgrounds are presented in the appendices for completeness.

PRODUCTION

Writing a book involves a huge time commitment and coordinated efforts while considering the fact that the current four coauthors are geographically separated and overloaded with busy schedules. The main coordination job was conducted by the first author, who often solicited inputs from the others while sending back the updated versions. The back-and-forth process went on mainly via email communications. This sometimes also caused inconvenience to achieve a harmony when preparing some materials. I owe my deep gratitude to my coauthors for their patience for revising and correcting many versions of the printout. The efficiency of the production of this monograph is partly due to the inventors of LaTeX, Donald Knuth and Leslie Lamport, without whom this job would have been extremely painful. The majority of the editing job was done by the first author, for which he shall take full responsibility and blame for any unnoticed mistakes that occur in the text.

It is noted that some of research results reported in this book were partially published earlier in some journal articles, for which the copyright shall be borne by the associated publishers (Elsevier, IEEE, Wiley, MIT Press, the American Physiological Society, and the Society for Neuroscience). We are very grateful to the publishers for their kind permissions to reproduce the research results here.

FURTHER READING

This research monograph is by no means comprehensive; rather, sometimes it is our intent to ignore the details when describing specific contents. No claim is made that our coverage of the materials is exhaustive or that our bibliography of the literature is complete. Instead, we intend to provide the reader a concise yet clear picture while directing the reader to other archives for more detailed accounts. It is our hope that such a treatment would help to accelerate the circulation of the idea for the general audience.

The scope of the readership of this research monograph is intended for audiences from a wide range of disciplines, including neuroscientists, signal processing researchers, computer scientists, graduate students, and people who have a general interest in understanding the brain and building adaptive learning systems.

As the complementary references that might catch the attention of the reader of this book, the following bibliographical resources are highly recommended by the current authors:

- For the *correlative brain*, see J. J. Eggermont's insightful monograph, *The Correlative Brain: Theory and Experiment in Neural Interaction* (Springer, New York, 1990).
- For the *cerebral cortex*, see V. B. Mountcastle's encyclopedic book, *Perceptual Neuroscience: The Cerebral Cortex* (Harvard University Press, Cambridge, MA, 1998).
- For *neuron models and Hebbian synaptic plasticity*, see the book by W. Gerstner and W. M. Kistler, *Spiking Neuron Models: Single Neurons, Populations, Plasticity* (Cambridge University Press, Cambridge, 2002).
- For a general account of *computational neuroscience and the brain*, see the classic book by P. Churchland and T. J. Sejnowski, *The Computational Brain* (MIT Press, Cambridge, MA, 1992).
- For a sophisticated and detailed textbook treatment of *computational neuroscience*, see the excellent book by P. Dayan and L. F. Abbott, *Theoretical Neuroscience: Computational and Mathematical Modeling of Neural Systems* (MIT Press, Cambridge, MA, 2001).
- For *correlation-based engineering applications*, see the book by B. Vijayakumar, A. Mahalanobis, and R. D. Juday, *Correlation Pattern Recognition* (Cambridge University Press, Cambridge, 2005).
- For the *ALOPEX algorithms*, see the edited volume by one of its early developers, E. M. Tzanakou, *Supervised and Unsupervised Pattern Recognition: Feature Extraction and Computational Intelligence* (CRC Press, Boca Raton, FL, 2000).

ABOUT THE COVER ILLUSTRATION

The cover illustration was designed and created by the first author using a computer software (http://www.andreaplanet.com) for generating the photo mosaic. The original image, as an illustration of the human brain, is used to generate the mosaic image. The mosaic image consists of 2000 tiles, each of which consists of a patch of the image sampled and flipped among a collection of a few hundreds human face images. The careful observer may pick out some familiar faces from the neural computation communities. Some faces belong to those who have made great contributions to the computing machinery and neural network literature (including those late, great minds of Alan Turing, John von Neumann, Claude Shannon, Warren McCulloch, and Donald Hebb). For design reasons, we must apologize in advance for using these face images without the direct consent of the individual persons that appear here. The symbol of this image is to show that numerous researchers (mathematicians, physicists, neuroscientists, computer scientists, engineers) are working together to unveil the mysteries of the neural networks, either biological or artificial.

Upon appropriate scaling and compression, the correlation coefficient between the original image and the resultant mosaic image is about 0.87, reaching a high degree of positive correlation.

ZHE CHEN
Tokyo, Japan

ACKNOWLEDGMENTS

This monograph initially stems from some of research that I did in my Ph.D. thesis at McMaster University, Canada. I am greatly indebted to my thesis advisor, Professor Simon Haykin, for giving me the freedom and support to pursue my research interests and for his confidence and encouragement of my work. The privilege of working with Simon has been an enjoyable and productive journey in my scientific career. I would also like to express my deep gratitude to Dr. Sue Becker for serving as my Ph.D. supervision committee member. Sue's insightful discussions and critical comments throughout the supervision are deeply helpful and appreciated.

The majority of the book was written during my stay at Japan. I am deeply grateful for the support and advice from Professors Shun-ichi Amari and Andrzej Cichocki at the Brain Science Institute of RIKEN (The Institute of Physical and Chemical Research). Professor Cichocki has given me many opportunities to pursue brain-related research at the Laboratory for Advanced Brain Signal Processing. For many years, Professor Amari has been a personal hero to me for his pioneering contributions in the field of neural networks and information geometry; his ever-increasing enthusiasm for pursuing scientific knowledge as well as his incisive view of mathematical neuroscience has had a great impact on the people surrounding him. I also owe Amari a deep gratitude for the effort and time that he dedicated to provide invaluable constructive suggestions in writing as well as his kind agreement to write the foreword of this book.

The academic atmosphere and freedom at the Brain Science Institute and the excellent research environment at the laboratories have always been source of inspiration to me. Many parts of this monograph have benefited, directly or indirectly, from frequent yet fruitful discussions with my friends and former colleagues at the institute, to name a few, Dr. Sergei Gepshtein, Dr. Jon Hatchett, Dr. Kosuke Hamaguchi, Dr. Kukjin Kang, Dr. Naoki Masuda, and Dr. Taro Toyoizumi. Dr. Hiroyuki Nakahara and Dr. Danilo Mandic also provided me with helpful feedbacks during the writing process. In addition, I would like to thank Dr. K. P. Unnikrishnan for sharing some early valuable feedbacks. In addition, the case studies presented in Chapter 7 are based on a number of earlier publications of the ongoing research work, for which the current four book authors owe their special gratitude to a number of collaborators, including Ian Bruce, Ron Racine, Gaurav Patel, Jeff Bondy, Arnaud J. Noreña, Boris Gourévitch, and Naotaka Aizawa.

I will continue my research journey at the Neuroscience Statistics Research Laboratory, Massachusetts General Hospital/Harvard Medical School, headed by Professor Emery N. Brown, for which I am also grateful for the opportunity. Needless to say, there are a lot of interesting and challenging research problems ahead of me, which, in the meantime, is also very exciting.

I also thank Dr. Christine (Joyce) Boucard, Dr. GuoQiang Bi, Dr. Zhi Ding, Dr. DeLiang Wang, Dr. Miguel Á. Carreira-Perpiñán, and Rong Dong for the courtesy of using some figures for illustration in this book. Special thanks also go to a number of publishers, including MIT Press, Springer, IEEE, Elsevier Science, Marcel Dekker, Nature Publishing Group, Annual Reviews, Society for Neuroscience, and the American Physiology Society, for allowing us to reproduce some results and figures that appeared in their previous publications. In preparing this monograph, I am also indebted to George Telecki, Rachel Witmer, and Christine Punzo from John Wiley & Sons for their patient assistances during the final production process.

Last but not the least, I would like to take this opportunity to thank my parents and my best friend Ying-Chun (Spring) Sun for their persistent and unfailing support. I owe a special gratitude to Spring, who has been sharing my joys and griefs these years whenever and wherever possible.

ZHE CHEN

ACRONYMS

AAF	Anterior auditory field
ACF	Autocorrelation function
AES	Anterior ectosylvan sulcus
ALOPEX	Algorithm of pattern extraction
AM	Amplitude modulation
AMUSE	Algorithm for multiple unknown signals extraction
APEX	Adaptive principal-components extraction
AR	Autoregressive
ARMA	Autoregressive moving average
AWGN	Additive white Gaussian noise
BAM	Bidirectional associative memory
BCI	Brain–computer interface
BCM	Bienenstock–Cooper–Munro
BIC	Bayesian information criterion
BOLD	Blood oxygenation level dependent
BPSK	Binary phase shift keying
BPTT	Backpropagation through time
BSB	Brain state in a box
BSS	Blind source separation
CAM	Content-addressable memory
CASA	Computational auditory scene analysis
CCA	Canonical correlation analysis
CF	Characteristic frequency, climbing fiber
CGHA	Complex generalized Hebbian algorithm
CM	Constant modulus
CMA	Constant-modulus algorithm
CMAC	Cerebellar model articulation controller
CR	Conditioned response
CS	Conditioned stimulus
CSD	Correntropy spectral density
CSP	Common spatial pattern
DCN	Dorsal cochlar nucleus
DOA	Direction of arrival
EC	Entorhinal cortex
EEL	Electroencephalography

EKF	Extended Kalman filter
EM	Expectation–Maximization
EMD	Empirical mode decomposition
EPP	Exploratory projection pursuit
EPSP	Excitatory postsynaptic potential
EVD	Eigenvalue decomposition
FA	Factor analysis
FFT	Fast Fourier transform
FIR	Finite-duration impulse response
FM	Frequency modulation
fMRI	Functional magnetic resonance imaging
FOBI	Fourth-order blind identification
GABA	Gamma-aminobutyric acid
GC	Granule cell
GCC	Generalized cross-correlation
GHA	Generalized Hebbian algorithm
GLM	Generalized linear model
GSD	Gated steepest descent
GSVD	Generalized singular-value decomposition
HHT	Hilbert–Huang transform
HMC	Hybrid Monte Carlo
HOS	Higher order statistics
ICA	Independent-component analysis
ICC	Inferior colliculus
IE	Instantaneous energy
IIR	Infinite-duration impulse response
IPS	Interacting particle systems
IPSP	Inhibitory postsynaptic potential
IT	Inferotemporal
ITD	Interaural time difference
JADE	Joint approximate diagonalization of eigenmatrices
JPSTH	Joint peristimulus time histogram
KCCA	Kernel canonical correlation analysis
KGHA	Kernelized generalized Hebbian algorithm
KGV	Kernel generalized variance
KICA	Kernel independent-component analysis
KL	Kullback–Leibler (divergence)
KPCA	Kernel principal-component analysis
LDA	Linear discriminant analysis
LFP	Local field potential
LGN	Lateral geniculate nucleus
LMF	Least mean fourth
LMS	Least mean square
LPZ	Lesion projection zone
LTD	Long-term depression

LTI	Linear time invariant
LTP	Long-term potentiation
LVQ	Learning vector quantization
MAP	Maximum a posteriori
MCA	Minor-component analysis
MCLMS	Multichannel least mean square
MCMC	Markov chain Monte Carlo
MDL	Minimum description length
MDP	Markov Decision process
MEG	Magnetoencephalography
MF	Mossy fiber
MGB	Medial geniculate body
MGN	Medial geniculate nucleus
MIMO	Multiple input–multiple output
MISO	Multiple input–single output
MLE	Maximum-likelihood estimate
MLP	Multilayer perceptron
MMI	Minimum mutual information
MMN	Mismatch negativity
MMSE	Minimum mean-square error
MSE	Mean-square error
MSF	Matched spatial filter
MTL	Medium temporal lobe
MUA	Multiunit activity
NDEKF	Node-decoupled extended Kalman filter
NMDA	N-Methyl-D-aspartate
NMF	Nonnegative matrix factorization
OD	Ocular dominance
ODE	Ordinary differential equation
OP	Orientation preference
PC	Purkinje cell
PCA	Principal-component analysis
PES	Posterior ectosylvan sulcus
PF	Parallel fiber
PI	Performance index
PLS	Partial least squares
PSD	Power spectral density
PSK	Phase shift keying
PSP	Postsynaptic potential
QAM	Quadrature amplitude modulation
QPSK	Quadrature phase shift keying
RBF	Radial basis function
RBM	Restricted Boltzmann machine
REM	Rapid eye movement
RF	Receptive field

RKHS	Reproducing kernel Hilbert space
RLS	Recursive least squares
RMLP	Recurrent multilayer perceptron
RTRL	Real-time recurrent learning
SDE	Stochastic differential equation
SFA	Slow feature analysis
SIMO	Single input–multiple output
SIR	Sampling–importance–resampling
SIS	Sequential importance sampling
SISO	Single input–single output
SNR	Signal-to-noise ratio
SOBI	Second-order blind identification
SOM	Self-organizing map
SOS	Second-order statistics
SPL	Sound pressure level
SSM	State-space model
STDP	Spike-timing-dependent plasticity
STFT	Short-time Fourier transform
STRF	Spectrotemporal receptive field
SVD	Singular-value decomposition
SVM	Support vector machine
SWS	Slow-wave sleep
TD	Temporal difference
TDE	Time-delay estimation
TSP	Traveling salesman problem
US	Unconditioned stimulus
VCN	Ventral cochlear nuclei
VOT	Voice-onset time
VOR	Vestibular–ocular reflex
VQ	Vector quantization
WTA	Winner take all
WVD	Wigner–Ville distribution
XOR	Exclusive OR

INTRODUCTION

Correlation

CORRELATION, by definition, according to the *Encyclopedia Britannica*, eleventh edition is "a causal, complementary, parallel, or reciprocal relationship, especially a structural, functional, or qualitative correspondence between two comparable entities." More concisely, it is defined as "simultaneous change in value of two numerically valued random variables." Commonly, when we say that two things are correlated, we mean that two things have a causal relationship. However, correlation is not identical to causation, since correlation is a term that describes a "stochastic" behavior that involves random variations. Correlation does not imply a directionality to the relationship, nor does it convey whether the relationship is direct or mediated by a hidden cause. In contrast, causation entails a directional relationship that is not explainable by some additional hidden cause and often implies an almost "deterministic" relationship. In mathematics or statistics, correlation is defined as the degree of association between one, two (or more) random variables, which can be in the form of either autocorrelation or cross-correlation. To evaluate the degree of association, the term *correlation coefficient* was introduced by Sir Francis Galton in 1888 (while examining forearm and height measurements), with the value ranging from -1 to $+1$: with 1 representing the highest degree of association and 0 being totally uncorrelated (see Figure 0.1 for an illustrative example on two correlated Gaussian random variables). Notably, correlation alone does not necessarily imply causality, since correlation is independent of spatial and temporal arrangement of random samples. As seen in Figure 0.1, interchanging the abscissa and ordinate of two variables would not affect their correlation relationship, and we cannot make any inference about the causal relationship between them. On the other hand, causality imposes strong temporal asymmetry between the occurrence of random events.

Quantitatively, correlation serves as a useful statistic for characterizing random variables, although the complete characterization of a random variable is given by its probability distribution function. For continuous random variables, the Gaussian distribution is the most popular distribution that is sufficiently characterized by the first- and second-order moment statistics, which also turns out to be the distribution

Correlative Learning: A Basis for Brain and Adaptive Systems, by Zhe Chen, Simon Haykin, Jos J. Eggermont, and Suzanna Becker

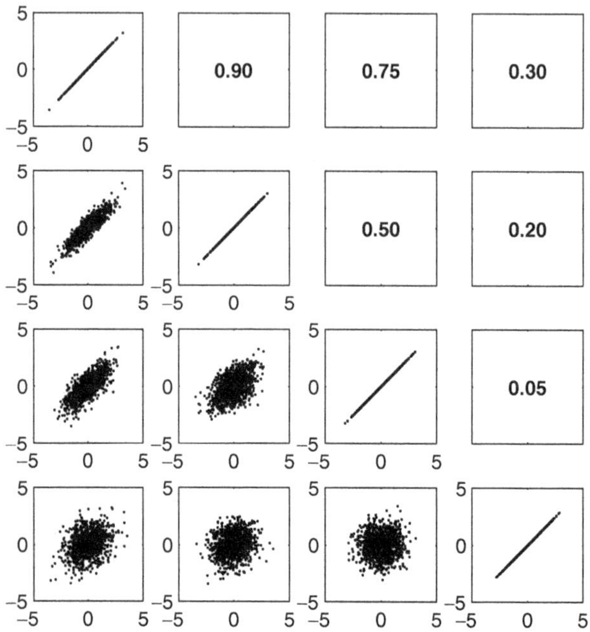

Figure 0.1 Visual illustration of correlation: The scatter plots of 1000 pairs of two-dimensional Gaussian distributed random variables are plotted against each other in the lower diagonal panels, and the corresponding correlation coefficients are shown in the symmetric upper diagonal panels; along the diagonal each set of numbers is plotted against itself, displaying a line with correlation coefficient +1.

that has the maximal entropy given a fixed-variance constraint. A generalized concept for random variable is a random process which involves a number of random variables that are functions of time. A well-studied stochastic process is the so-called Gaussian process. The popularity and ubiquity of the Gaussian distribution and Gaussian process is credited to the *law of large numbers* and the fact that they have finite and easy-to-compute sufficient statistics. Therefore, the correlation statistic or correlation function plays the dominant role in statistical decisions and random data analysis. Autocorrelation and cross-correlation functions are the basic tools for characterizing statistical dependency.

Correlation also serves as a similarity measure. Two things that are similar tend to have higher correlation coefficients. To characterize higher order dependency, a more powerful similarity measure is *mutual information*, which was first introduced by Claude Shannon, the father of information theory, in his landmark 1948 paper "A Mathematical Theory of Communication" [823]. Simply put, mutual information is based on the information-theoretic notion of entropy, which is defined as the expected log probability of a random variable. At an intuitive level, this characterizes the average degree of surprise one would have at observing any particular value of the random variable given the expected distribution. The mutual information between two random variables may be interpreted as the amount of surprise one would have at observing the second variable having observed already the first variable, or in other words, the part of the information that is common to two or more random variables. Generally, things that are correlated have more mutual information, whereas independent random variables have zero mutual information. Throughout this book, we will treat mutual information as a generalized notion of correlation.

Correlative Brain

The brain is a truly extraordinary system that enables animals or humans to conduct tasks varying from low-level perception to high-level cognition. We may view the brain as a computing machine that is amazingly powerful, highly functionally organized, and extremely robust. It is also these properties that highlight the fundamental difference between the brain and a supercomputer. In the past decades, achieving such brain-style computing is the Holy Grail of research in artificial intelligence.

Understanding the brain and its fundamental functions is the central goal of the brain sciences. To fully understand the brain, we need to study the brain mechanisms at the biological, biophysical, physiological, and psychological levels. The brain is also a hierarchical architecture that includes macroscopic and microscopic levels such as cortices, neuronal circuits, neurons, synapses, and molecules. Different parts of the brain cooperate as a seamless machine, and invoke different levels and scales of correlations, in both space and time.

The brain, in a multitude of ways, explores the sensory environment and uses the information obtained to control behavior. In doing so, its primary mechanism to evaluate, control, and learn is that of correlation. Correlation of nervous activity can take many forms: It can be the detection of coincidences in the firing of two neighboring nerve cells (see Figure 0.2 for an illustration) or the detection of the covariation in the firing rates of two nerve cells. It can be the covariation in the activity pattern of neuronal groups, but it can also be the covariation in the postsynaptic currents entering the same cell at distinct dendritic synapses. Neuroscience currently emphasizes spike timing and coincidences between spikes from different neurons as important in learning and plasticity, and our emphasis in this book will be likewise. Looking for coincidences provides a means of making inferences about the environment. In the case that two event-generating processes A and B are independent, the joint probability density of the two series of events, $P_{AB}(t, u)$, is equal to the product of the probability densities of the individual series of events $P_A(t)$ and $P_B(u)$:

$$P_{AB}(t, u) = P_A(t)P_B(u).$$

In the case that two processes A and B are dependent (i.e., whenever coincidences occur more often or less often than expected on the basis of mere chance), there is a correlation between the events generated by these two processes represented in $C_{AB}(t, u)$, which is called the cross-correlation function of the events generated

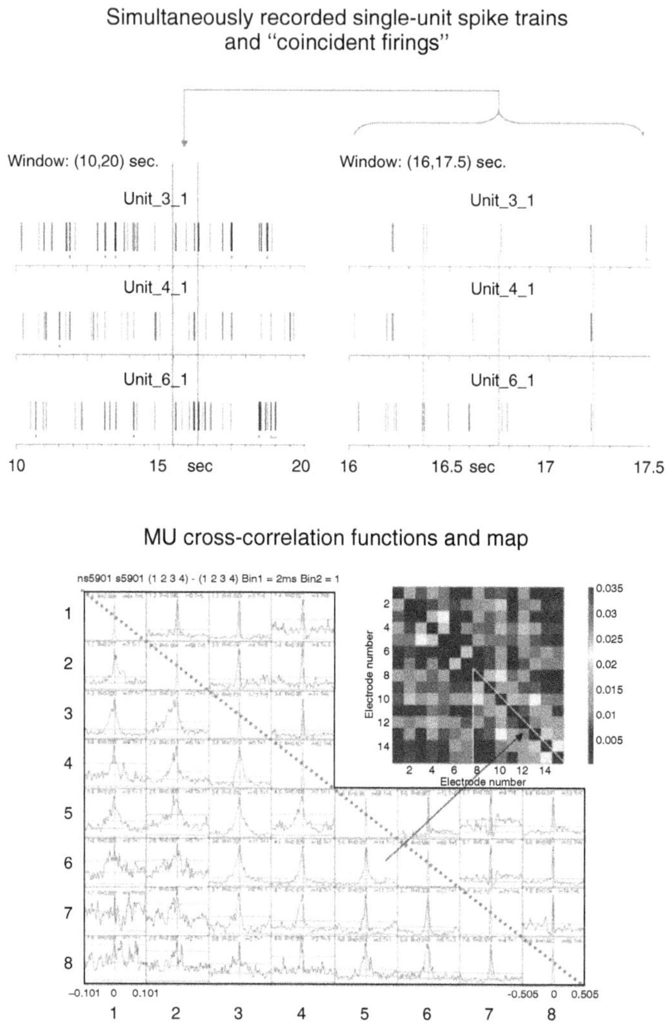

Figure 0.2 Coincident firings are signs of neural interaction or of shared input from a common source. Three spike trains that were simultaneously recorded are shown on both a 10-s timescale and for a selected portion on a 1.5-s timescale. The red lines indicate near coincidences. These can be statistically evaluated from the multiunit (MU) cross-correlograms. The bottom part of the figure shows below the main diagonal the pairwise correlograms between 8 simultaneously recorded units using a bin size of 2 ms. The green lines indicate mean ± 3 SD (standard deviations) and peaks exceeding the upper level are considered to represent correlations that are significantly different from zero (for details, see (242)). The 8-electrode recording was part of a 16-electrode one, and the full pairwise matrix of the peak cross-correlation coefficients is shown in the inset. The lower triangle in the matrix represents the correlograms; the arrow indicates the position of one particular value. The colorbar indicates the peak values between 0 and 0.035 on a linear scale.

by processes A and B. Let τ denote the time difference $t - u$. Then we may write $C_{AB}(t, \tau)$ as the time-dependent cross-correlation function. For stationary processes, we have $C_{AB}(t, \tau) = C_{AB}(\tau)$. The cortex, and most other parts of the brain, may have evolved to detect correlated events.

In addition, it is important to realize the prominent role of correlation within the life span of the brain. It has been reported that the synapses of the visual system in the brains of human infants within the first few months of life undergo rewiring or self-organization by utilizing correlations [84], while their receptive fields may have been already established to some degree in the prenatal stage [416]. On the other hand, correlation-based associative memory will continue to function in a healthy brain right up to its ultimate death. It is also worth pointing out the universal role of correlation in both microscopic and macroscopic levels of brain functions. Indeed, it is widely believed that correlation serves as the basis of synaptic plasticity, learning, association, pattern recognition, and memory recall [241]. In Chapter 1, we will present a detailed overview of the correlative brain.

Learning

Hallmark characteristics of humans are the amazing capability to *learn* and the flexibility to adapt to a dynamic environment. In neurobiological terms, learning is referred to as *synaptic plasticity*. From the time of birth, humans never stop learning across a wide range of domains, including language, vocabulary, reading, and memorizing. Learning new environments requires the brain to adapt in a self-organizing fashion. The adaptation is reflected by the changes in neural firing patterns inside the brain as well as the changes in emergent behavior. In addition, learning is also an essential component of the human's intelligent behavior. By *intelligence*, we mean "the capacity to learn or to profit by experience" and "a biological mechanism by which the effects of a complexity of stimuli are brought together and given a somewhat unified effect in behavior" ([717], p. 6–7). The notion of intelligence is omnipresent in almost every aspect of human activities, such as perception, action, thinking, memory recall, recognition, and so on. Despite significant progress, a full understanding of intelligence is far from complete, and the enigma of the human brain remains elusive. Reported scientific evidence has revealed that the human brain is capable of learning new things from birth to death; the potential of the brain to learn is truly overwhelming and often underestimated. Now, the questions arise: *How does learning occur? What are the underlying neural mechanisms? How can we model the learning process?* This monograph attempts to explore these questions according to what we know so far. In so doing, a central tenet will be the importance of correlation as an underlying organizing principle. This tenet will be discussed throughout this monograph in various aspects, ranging from biological human brains to artificial adaptive systems, along with the design of learning algorithms. Another purpose of this monograph is to convey the message that correlation is omnipresent and important; it is certainly our hope to have convinced the reader of this after finishing this monograph.

Correlation-based theories of learning have a long history in psychology and neurobiology [436, 747]. In retrospect, the notion of correlation-based learning can be traced back to the Greek philosopher Aristotle. The earliest formulation of correlative learning as it relates to brain processes, however, was due to William James [436]. Specifically, he stated ([436], Chapter XVI; see also [39]): "When two elementary brain-processes have been active together or in immediate succession, one of them, on re-occurring, tends to propagate its excitement into the other." Following William James, the formal establishment of correlation-based learning was credited to Donald Hebb, whose postulate is now known as Hebbian learning [377]. Describing a correlative synaptic mechanism, Hebbian learning is a *local* rule, meaning that it requires only information that would be available locally to a neuron, and therefore it is physiologically [855] and biologically plausible [89]. More specifically, the modification of synaptic strength depends on the pre- and postsynaptic firing rates and the present strength of the synapse. In fact, Hebb's profoundly influential idea has not only withstood the test of time in neurobiological circles but also become the starting point and foundation of a wide range of neural learning algorithms.

Correlative learning can be viewed as a generic case of the Hebbian rule and therefore appealing to serve as a neurobiological model of learning. Following the seminal work of Hebb, many researchers have developed numerous correlation-based computational neural models in a wide range of domains, including memory, vision, audition, and synaptic modulation. In modeling synaptic plasticity, various correlative learning rules and computational models have been proposed and developed [93, 342, 818, 961]. Correlation activity was believed to play a critical role in the central nervous system [183], and is arguably the ubiquitous basis for learning, association, pattern recognition, novelty detection, and memory recall [241]. Chapter 3 will be dedicated to elucidating many biologically inspired correlation-based computational neural models that mimic the correlative mechanisms in the brain.

Bearing in mind the goal of building adaptive systems in engineering applications, we also discuss the role of correlation functions in developing statistical signal processing or machine learning algorithms. In the literature, learning has been categorized into three major types according to the nature of the task: supervised learning (learning with teachers), unsupervised learning (learning without teachers), and reinforcement learning (learning with critics). Simply put, they can be formulated as solving different problems:

Supervised learning can be understood as a multivariate function approximation problem [731]; in the statistical jargon, it amounts to regression for a specific parametric, semiparametric, or nonparametric statistical model. Supervised learning includes two instances: regression and classification; and classification can be viewed as a special case of regression.

Unsupervised learning is aimed at learning the structure or regularity of unlabeled data [60, 389]; unsupervised learning exploits the basic information processing principles (e.g., self-organization or maximum entropy) using either *bottom-up* or *top-down* approaches.

Reinforcement learning can be understood as a Markov decision process (MDP) that is aimed at learning proper actions leading to optimal outcomes; it attempts to solve a *temporal credit assignment* problem [85, 868]. Motivated by dynamic programming, reinforcement learning has been extended for several varieties of prediction and control problems.

Despite their seemingly different goals and motivations, the common correlative nature will be emphasized to better understand the principles for developing adaptive learning systems in practical applications. In particular, Chapter 2 discusses the unique role of the correlation function that is used for developing modern signal processing techniques and statistical decision analysis. Chapter 3 discusses the role of correlation in developing various biological (synaptic) and machine learning algorithms as well as computational neural models. It will be shown that various types of statistical learning algorithms can be unified within the correlative learning framework. Chapter 4 introduces the notion of kernel and discusses correlation-based kernel learning. Chapter 5 discusses the correlation concept for complex-valued signals and extends the notion of correlative learning to the complex domain. Finally, Chapters 6 and 7 discuss a few correlation-based learning paradigms and computational models, with several selected applications in modeling perceptual (auditory and visual) systems, time series analysis, and pattern recognition.

1

THE CORRELATIVE BRAIN

The human brain is a hugely complex information processing system. In this chapter, it is our intention neither to review the brain anatomy and structures in detail nor to discuss every aspect of brain functions. Instead, we try to present an overview of the correlative brain at both the microscopic and macroscopic levels. Before discussing various correlative neural mechanisms, we first provide a brief background of some fundamental concepts of the human brain.

1.1 BACKGROUND

1.1.1 Spiking Neurons

The human brain consists of about 10^{11} (a hundred billion) neurons and $10^{15}-10^{16}$ (quadrillion) synapses. Each neuron is connected via synapses to about 1000–10,000 other neurons. It is the vast amounts of neurons and synapses that empower the brain with a high capacity for memory and "computing power" in a way that is quite different from the Turing machine or von Neumann–type computer.

A neuron is the basic functioning unit in the nervous system; it is responsible for receiving, integrating, and transmitting information. Despite the fact that there are many different types of neurons in terms of shape or size, most of them share a similar structure, as illustrated in Figure 1.1. Typically, a single cortical neuron receives thousands of inputs from other connecting neurons and sends its output spikes to about the same number of other neurons.

Correlative Learning: A Basis for Brain and Adaptive Systems, by Zhe Chen, Simon Haykin, Jos J. Eggermont, and Suzanna Becker

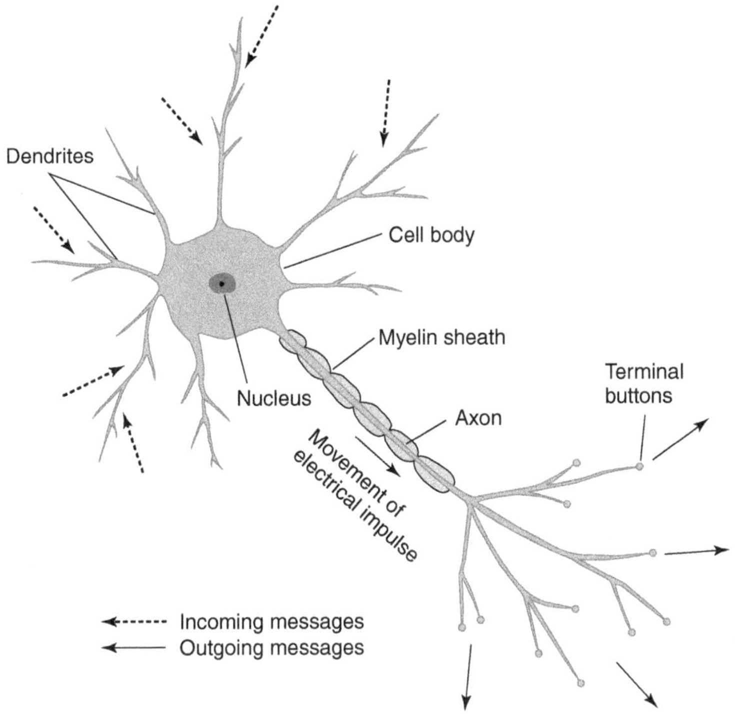

Figure 1.1 Schematic of neuron structure.

In Figure 1.1, there are several distinct components inside or outside the neuron:

Soma (cell body): Soma (Latin, meaning "body") is the cell body of the neuron and contains the nucleus and other structures that support the chemical processing.

Dendrite: Dendrites (Greek, meaning "tree") are the branching fibers that connect the soma; the fibers are the site of the synapses that are responsible for receiving incoming information from other neurons.

Axon: Axon is a singular fiber that carries information away from the soma to the synaptic sites of other neurons (dendrites and somas).

Synapse: Synapse (Greek, meaning "association")[1] is the connection that bridges two neurons or the connection between a neuron and a muscle. The synapse consists of three elements: (i) the presynaptic membrane, which is formed by the terminal button of an axon; (ii) the postsynaptic membrane consisting of a segment of dendrite or soma; and (iii) the space between these two structures, which is called the synaptic cleft.

Terminal buttons (boutons) are the small knobs at the end of an axon that release chemicals called neurotransmitters; the terminal buttons (boutons) form the presynaptic side of the synapse.

Myelin sheath consists of fat-containing cells that insulate the axon from electrical activity and increase the rate of transmission of signals. Axons that carry information over long distances, for example, from the periphery to the brain or between the two hemispheres of the cortex, tend to be myelinated while short-range axons do not.

Synapses are commonly believed to be the initial places where information is gained and stored. The massive number of synapses connecting the neurons across the brain constitutes a distributed memory system for storing the knowledge learned from experience. Depending on their electrical and chemical properties, synapses can be either *excitatory* or *inhibitory*. For the excitatory synapse, the neurotransmitters "depolarize" the postsynaptic membrane, that is, make the inside of the cell less negative with respect to its resting value (about −70 mV). The change in membrane potential due to depolarization (i.e., electrical discharge) is called the *excitatory postsynaptic potential* (EPSP). If the depolarization of the postsynaptic membrane reaches a threshold (about −55 mV), an action potential (i.e., spike) is generated in the postsynaptic neuron. In contrast, at the inhibitory synapse, the neurotransmitters "hyperpolarize" the postsynaptic membrane, that is, make the membrane potential more negative. The change in membrane potential due to hyperpolarization (i.e., electrical charge) is called the *inhibitory postsynaptic potential* (IPSP). The IPSP will make the neuron much less likely to spike when simultaneously receiving excitatory input.

The action potential generated at the postsynaptic neuron is a pulse of electrical activity that is created by a depolarizing current that exceeds the critical threshold level. This occurs because the exchange of ions across the membrane causes more sodium ions to enter the neuron; the spiking process often occurs over a time course of 2–100 ms, depending on the specific neuron. As a function of time, the spike trains can be observed at the location of a specific postsynaptic neuron, and these spike trains produce the spiking neural codes (see Figure 1.2). The spike train sequences can be roughly modeled as a homogenous Poisson process with the average firing rate as a rate parameter [201]. Specifically, let k denote the number of the spikes in the interval $(0, T]$, and let $r = k/T$ denote the average firing rate; by letting k and T approach infinity in the limit while keeping the ratio r constant, it follows that the probability of N spikes falling within an interval of time of bin

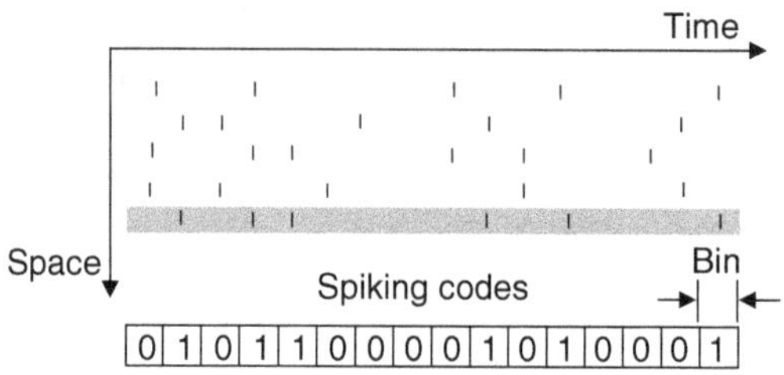

Figure 1.2 Graphical illustration of spiking neural codes.

size Δt is equal to

$$\Pr(N \text{ spikes in } \Delta t) = e^{-r\Delta t}\frac{(r\,\Delta t)^N}{N!}, \tag{1.1}$$

which defines a Poisson probability density function (pdf). Calculating the mean and variance of spike counts with respect to the Poisson probability would yield

$$\langle N \rangle = r\,\Delta t, \qquad \mathrm{var}\,[N] = r\,\Delta t. \tag{1.2}$$

Additionally, given a spike at the present time, the waiting time (denoted by τ) between the current spike and the next spike follows an exponential distribution that has the pdf form

$$p(\tau) = re^{-r\tau}. \tag{1.3}$$

Calculating the mean and variance of τ with respect to $p(\tau)$ would yield

$$\langle \tau \rangle = \frac{1}{r}, \qquad \mathrm{var}[\tau] = \frac{1}{r^2}. \tag{1.4}$$

A graphical illustration of simulated Poisson distributed spike trains is given in Figure 1.3. Figure 1.4 also presents an illustration of measuring firing rate via spike counting.

To understand brain function, we have to look into the "code" that neurons use. Action potentials (or spikes) are the primary way in which neurons communicate with each other; hence neural spikes are the unique "language" used inside the brain. In addition to the rate code (i.e., the number of spikes in a specific time interval), neurons may also use spike *timing* to code information (therefore referred to as temporal code). It appears that spike timing is important, at least in some neural systems such as the auditory regions, in that specific times between action potentials may carry information that is not available from the rate code. Experiments in vivo suggest that firing rates and synchrony are often simultaneously relevant. However, how firing rate and synchrony comodulate and which aspects of inputs are effectively encoded have yet remained elusive.

Functionally, a neuron is often simplified as an *integrate-and-fire* unit: The input x_i to a neuron i is generated by the firing rates x_j of other neurons j subject to a gain function

$$x_i = f\left(\sum_{j \in N_i} \theta_{ij} x_j - b_i\right), \tag{1.5}$$

where θ_{ij} denotes the synaptic efficacy and $f(\cdot)$ is a *gain function* which can be linear, nonlinear, or binary (all or none). Biologically speaking, equation (1.5) has the following interpretation:

- The parameter N_i defines the neighborhood region where neurons are connected to neuron i.

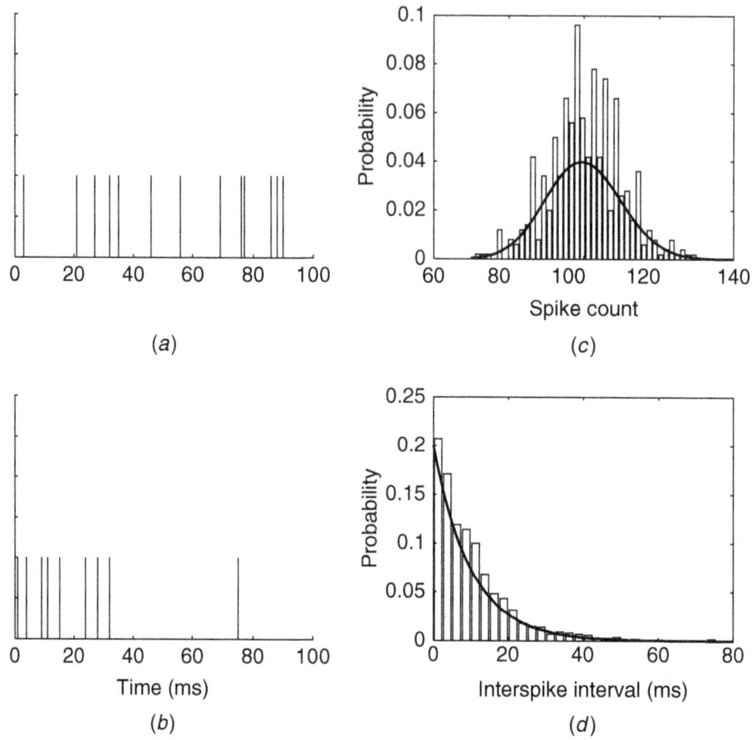

Figure 1.3 A graphical illustration of the Poisson spike trains. (*a, b*) Simulations of two Poisson spike trains with $r = 100$ and $\Delta t = 1$ ms. (*c*) Spike count histogram calculated from 1000 Poisson trains simulated within 1s duration; the solid curve is the Poisson spike count density. (*d*) Interspike interval (waiting time) histogram calculated from the simulations; the solid curve is the exponential interspike interval density.

- The weighted summed current $I_i = \sum_{j \in N_i} \theta_{ij} x_j$ is often called the *postsynaptic potential* (PSP) of neuron i.
- The voltage x_i is viewed as the firing rate of neuron i.
- The threshold bias b_i is viewed as a baseline current.
- The function f can be viewed as an operation that is implemented via dendritic integration.

It is this "integrate-and-fire" mechanism described in (1.5) that motivated Warren McCulloch and Walter Pitts [606] to first develop the computational neuron model. The McCulloch–Pitts neuron is a static model; despite its simplicity, the McCulloch–Pitts neuron model has been widely used in the neural network literature. In the meantime, more biologically accurate neuron models, such as Caianiello's neuron model [132] and the Hodgkin–Huxley model [395], also have been developed to analyze neuronal dynamics.

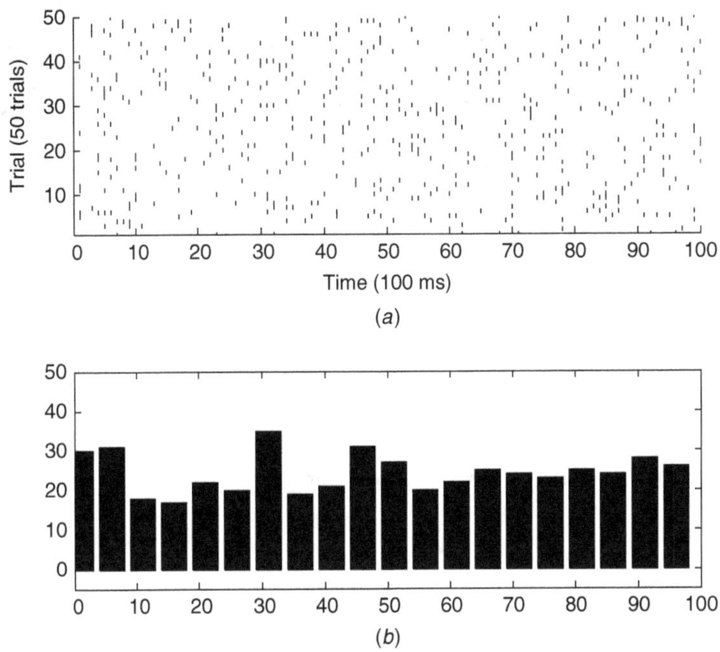

Figure 1.4 *(a)* The spike trains observed within 100 ms over 50 independent trials. *(b)* The total number of spike counts per 5 ms within 50 trials, from which we can calculate the mean firing rate as about 100 spikes/s.

Specifically, Caianiello [132] introduced the *time delay* into the model of a neuron's temporal dynamics,

$$x_i(t) = f\left(\sum_j \sum_k \theta_{ij} x_j(t - k\tau) - b_i(t)\right). \tag{1.6}$$

The above so-called *neuronic equation* essentially states that neuron j can influence the firing of neuron i up to $k\tau$ time steps in the future, and the dynamics can be modeled as a Markov process.[2]

To model the single neuron's firing rate, a simple way is to link the Poisson rate to the membrane potential from a biophysical viewpoint:

$$r(t) \approx \alpha[V(t) - V_{\text{th}}], \tag{1.7}$$

where V_{th} (in millivots) denotes a potential threshold value, α (in spikes per second per millivolt) denotes the slope parameter, and $V(t)$ denotes the instantaneous membrane potential. Taking the time average of (1.7) would yield the mean firing rate expression

$$\langle r(t)\rangle \approx \alpha[V_0(t) - V_{\text{th}}], \tag{1.8}$$

where $V_0(t) = \langle V(t) \rangle$ denotes the time-averaged membrane potential. Nevertheless, the neural firing of a single cell is known to be very noisy. If we measure the firing rate in different trials by presenting the same or correlated stimuli, a significantly different firing pattern can be observed. Such random firing effects can be overcome by averaging an ensemble of neurons or a population of cells; by doing that the firing rate function appears more deterministic. In practice, the firing rate is modeled as a filtered version of a known stimulus signal

$$r(t) = r_0 g\left(\int_{-\infty}^{\infty} d\tau \, f(\tau)s(t-\tau)\right), \tag{1.9}$$

where r_0 denotes the background firing rate when no stimulus occurs (i.e., $s = 0$), $f(t)$ denotes a filter, and $g(\cdot)$ denotes a memoryless nonlinear function whose argument is a *reverse correlation function*. Note that if the stimulus signal $s(t)$ is close in shape to that of the filter $f(t)$, specifically $s(t) = f(-t)$, then the rate function $r(t)$ will increase its value considerably, thereby achieving the maximum modulation.

1.1.2 Neocortex

The brain of vertebrates consists of the forebrain, brainstem, and spinal cord. In the forebrain the most recently evolved component, and the most prominent component in higher vertebrates, is the neocortex. In addition, the forebrain includes phylogenetically older cortical areas (allocortex) such as the olfactory cortex and hippocampus as well as many nuclei important for emotion (e.g., the amygdala), motor control (the basal ganglia), and numerous other functions. The brain is divided into left and right hemispheres. Different sides of the brain are responsible for controlling their *opposite* sides of the body. While the precise role of each hemisphere is still under debate, it is generally agreed that the left hemisphere plays a greater role in language and object recognition while the right plays a greater role in spatial cognition. The hemispheres of the cerebral cortex are also divided into four divisions, or lobes, the *frontal, parietal, occipital*, and *temporal* lobes. The gray matter volume within a given region of the brain often correlates positively with specific skills associated with that region.

In different cortical areas, there are specialized *functional* cortices responsible for specific tasks of sensory perception, cognition, or motor control. The neurons in those specific cortical areas often form specific topographic maps; the neurons within the same cortical region also have similar functional roles and structures. In particular, five important cortices of the neocortex are described here:

Visual cortex is specialized for vision; it is located at the back of brain in the occipital lobe. There are also numerous visual areas within the temporal and parietal lobes. The neurons within the visual cortex receive and process the information from the eyes (namely, their retinae) and complete the visual tasks. In monkeys nearly half of the cerebral cortex is related to visual processing.

Auditory cortex is specialized for audition or hearing; it is located in the temporal lobe. The neurons in the auditory cortex process the information received at the auditory nerves from the inner ear (cochlea) and further propagated through the auditory brainstem and the ascending auditory system.

Somatosensory cortex is mainly specialized for haptic sensations; it is located in the parietal lobe.

Motor cortex is specialized for movement; it is located in the back portion of the frontal lobe.

Association cortex refers to the areas of the lobes that are multimodal, receiving converging inputs from multiple sensory modalities. Different association cortices may be specialized for different functions, such as language comprehension, spatial imagery, memory, or sensorimotor transformations.

Within the motor or sensory cortices, there are also *primary* and *secondary* motor or sensory areas. The primary motor or sensory areas are those where motor or sensory information first arrives at the cortex. These primary areas are responsible for processing the primitive motor command or low-level sensory stimuli. For representing cortical areas of neocortex, Table 1.1 lists some abbreviated terms commonly used in neuroscience.

The neocortex is thought to be a self-organizing system[3] in the sense that a larger degree of order emerges from the system as time progresses. The neocortex is structurally ordered at many levels, including the layered and columnar structure, groupings of columns into hypercolumns, and at a larger scale into topographically organized feature maps. A central and long-standing theme in neuroscience has been to study *why* and *how* these ordered structures and maps are formed in the neocortex. Information arriving at the neocortex, in the form of spatiotemporal spike patterns, is structured, redundant, high dimensional, and somewhat random. In terms of their roles, there are two categories of maps: *functional* and *topographic*. Topographic maps are by definition functionally structured, but functional maps

Table 1.1 Common Terminology for Areas in Sensory and Motor Cortices

Term	Description
V1	Primary visual cortex, striate cortex
V2	Secondary visual cortex
MT	Medial temporal, V5
IT	Inferior temporal
A1	Primary auditory cortex
A2	Secondary auditory cortex
S1	Primary somatosensory cortex
S2	Secondary somatosensory cortex
M1	Primary motor cortex
M2	Secondary motor cortex

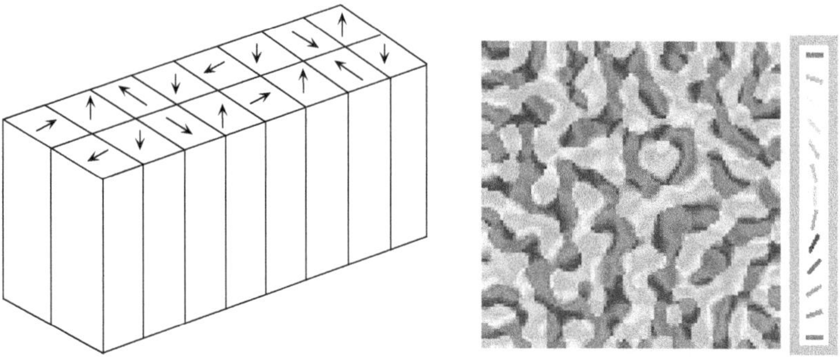

(a) (b)

Figure 1.5 (a) Graphical illustration of three-dimensional columnar structure with two arrays of orientation selective cells. (b) Computer simulation of two-dimensional orientation maps of visual cortex.

might not be topographically organized. Different cortical areas have their own specific functional maps, for example:

- Visual maps can represent the distance to an object, line orientation, movement direction, binocular disparity, and so on.
- Auditory maps can represent the object in terms of azimuth, elevation, and distance by synthesizing the maps of time and intensity disparity.
- Motor maps can, for instance, represent gaze direction; variations in motor commands are represented topographically into spatiotemporal patterns within the motor maps.

Topographic (such as retinotopic, somatotopic, or tonotopic) maps arise as a result of the anatomical structure of the sensory receptor surface and the innervating nerve fibers preserving this orderliness in the fiber tracts and in each interposed nucleus. Although the roles of topographic maps vary, a commonly accepted view is that the maps provide a low-dimensional representation of complex stimuli in the cortices. Topographic map formation has been widely studied using correlation-based neural models and learning rules (to be discussed in Chapter 3). As an example, the orientation-selective columnar cells in the visual cortex are illustrated in Figure 1.5.

1.1.3 Receptive Fields

Another important notion for understanding how neurons process and respond the sensory stimuli is the so-called *receptive field* (RF).[4] Each neuron has its own RF. Although the size and property of different neurons may vary, their common goals are to detect, match, and encode the (primitive or abstract) features of the information flow. By appropriate tuning of the synaptic strengths of inputs within a neuron's RF, that neuron can be viewed as a feature detector whose task is to extract a set of information-bearing features to represent (with maximum information retention) the complex sensory stimuli.

Within the neural maps, neighboring cells often have *similar* and *overlapping* RFs, which enable them to cooperate with each other in processing the incoming stimuli. For instance, the neurons in the visual orientation maps have RFs that cause them to respond only to a small subset of visual stimuli that are strongly localized in the retinal space as well as the orientation angle space. Computationally, Daugman [199] used two-dimensional Gabor filters to model the spatial RFs of simple cells in the visual cortex,

$$\mathrm{RF}(x, y) = \exp\left(-\frac{\tilde{x}^2 + \gamma^2 \tilde{y}^2}{2\sigma^2}\right) \cos\left(2\pi \frac{\tilde{x}}{\lambda} + \varphi\right), \tag{1.10}$$

where

$$\tilde{x} = x \cos \Theta + y \sin \Theta, \qquad \tilde{y} = -x \sin \Theta + y \cos \Theta,$$

where the arguments x and y define the spatial position of the visual RF; parameter γ is the aspect ratio that specifies the support of the Gabor filter; parameter λ defines the *wavelength*, and $1/\lambda$ defines the *spatial frequency*; parameter σ defines the *size* of the RF, and the ratio σ/λ determines the spatial frequency bandwidth of the cells; the angle parameter $\Theta = 2\pi/k$ ($k \in \mathbb{N}$) specifies the *orientation* of the impulse response, and φ is a *phase offset* parameter (when $\varphi = 0$, the RF function is symmetric; when $\varphi = \pi$, the function is antisymmetric). It is well believed that the Gabor filter provides a good approximation of the response properties of visual cells [276]. Figure 1.6 depicts some computer simulations of visual RFs using a Gabor filter with varying parameters $(\gamma, \lambda, \sigma, \Theta, \varphi)$.

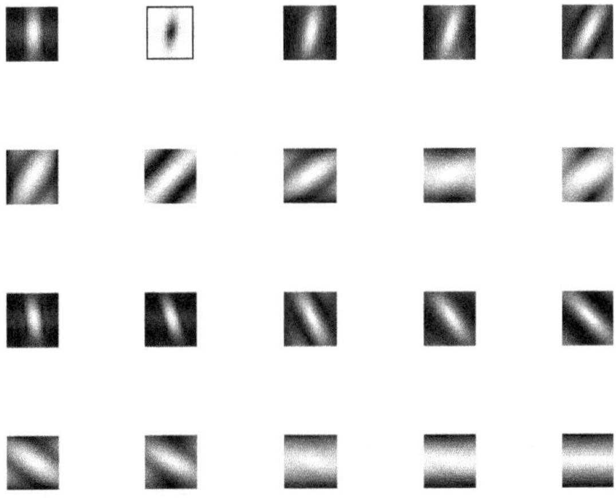

Figure 1.6 Illustration of visual receptive fields. The orientation-selective receptive fields are simulated by two-dimensional Gabor filters. The first two correspond to the "ON-center-OFF-surround" and "OFF-center-ON-surround" cells, respectively.

Likewise, the neurons in the auditory maps have similar and overlapping spectrotemporal receptive fields (STRFs) in terms of either the amplitude (modulation) or frequency (tone) of the sound stimuli. In a similar vein, we can define the STRF with a two-dimensional complex Gabor filter,

$$\mathrm{STRF}(t, f) = \frac{1}{2\pi\sigma_t\sigma_f}\exp\left(-\frac{(t-t_0)^2}{2\sigma_t^2} - \frac{(f-f_0)^2}{2\sigma_f^2}\right) \times \exp\left[j\omega_t(t-t_0) + j\omega_f(f-f_0)\right] \quad (j=\sqrt{-1}) \tag{1.11}$$

which is modeled by the product of a Gaussian envelope and a complex-valued Euler function. The Gaussian envelope is specified by the mean parameters t_0 and f_0 (central frequency) and the standard deviation parameters σ_t and σ_f. The periodicity is defined by the radian frequencies ω_t and ω_f. The scaling factors σ_t and σ_f at time and frequency make the Gabor filter act like a wavelet function for multiresolution analysis. Therefore, the auditory neurons with a waveletlike STRF can tune their auditory responses according to varying auditory stimuli.

1.1.4 Thalamus

Most sensory input to the cortex (including visual, auditory, and somatosensory but not olfactory) project to the cortex primarily via the thalamus, although there are also nonthalamic pathways. Thus the thalamus is the last region in the primary processing chain between sensory receptors and the cortex. Despite its relatively compact volume, its role in information processing is extremely important. It is now widely believed that the thalamus is more than a relay station between the received sensory stimuli and sensory cortices. Indeed, surprisingly it has been found that the number of feedback connections in the corticothalamic loop is about 10 times as many as that of feedforward connections in the thalamocortical loop.[5] In the visual pathway, the thalamic structure is known as the lateral geniculate nucleus (LGN); whereas in the auditory pathway, it is referred to as the medial geniculate nucleus (MGN) or medial geniculate body (MGB). The motor information generated by the cerebellum or basal ganglia also passes through thalamus to motor cortex. The feedback projections are believed to play a crucial role for selective attention, top-down expectation, or prediction (given the contextual prior). See Figure 1.7 for an illustration of thalamocortical and corticothalamic loops in the visual system.

1.1.5 Hippocampus

The hippocampus,[6] an older part of cerebral cortex, is located inside the temporal lobe of the brain. The perforant path constitutes the predominant input pathway to the hippocampus and it projects mainly to the superficial layers of the entorhinal cortex (EC), which in turn projects to the dentate gyrus and CA fields (CA stands for cornu ammonis—so called because the whole structure looks like rams' horns). There are also connections from the dentate gyrus to CA3, from CA3 to CA1, and

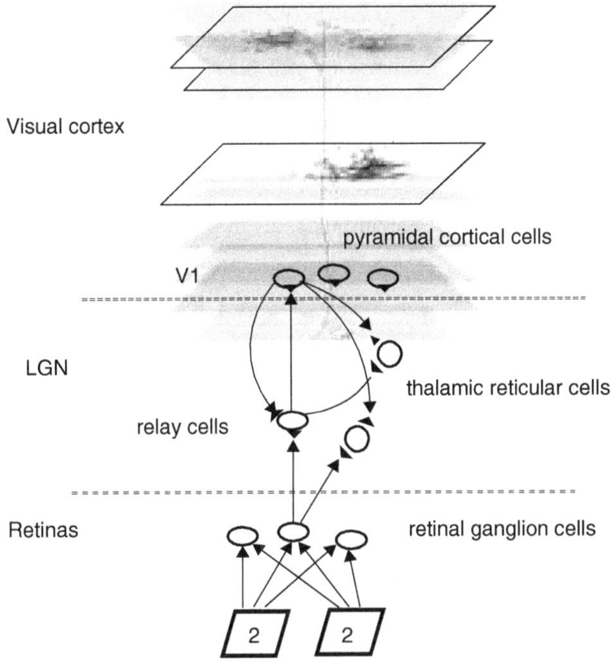

Figure 1.7 Schematic of thalamocortical and corticothalamic loops between the LGN and primary visual cortex (V1).

CA1 back to the EC (as shown later in Figure 1.15). Studies in rats have shown that neurons in the hippocampus have spatial firing fields, for which these cells are known as the *place cells*. The discovery of place cells has led to the idea that the hippocampus might act like a cognitive map [682].

1.2 CORRELATION DETECTION IN SINGLE NEURONS

The most important characteristic of a well-functioning brain is that it learns by experience. Learning starts with modifiable synapses, which are considered more and more as important computational systems of the brain [2]. The idea of synapse involvement in memory and thus implicitly that of modifiable synapses has a rather long history [747].

The Law of Neural Habit and Correlative Synapses. An early idea of the correlative synapse can be traced back to William James. In his classic work on psychology [436] (excerpted in [39]), James proposed the laws of association ([39], p. 225):

> How does a man come, after having the thought of A, to have the thought of B the next moment? or how does he come to think of A and B always together? These

> were the phenomena which Hartley undertook to explain by cerebral physiology. I believe he was in essentially respects, on the right track, and I propose simply to revise his conclusions by the aid of distributions which he did not make.

In James's theory, he claimed that ([39], p. 566; also in [122])

> there is no other elementary causal law of association than the law of neural habit: When two elementary brain-processes have been active together or in immediate succession, one of them, on reoccurring, tends to propagate its excitement into the other.

Essentially, James's *law of neural habit* indicates the basic conditions ("being coactive" and "reoccurring") for the modification of neural synapses, although he did not restrict himself to the synapses; instead, he used the term "elementary brain processes." However, James's theory clearly bears a resemblance with the theory on synaptic plasticity established later.[7]

Herbert Spencer, in *The Principles of Psychology* [844], has also described similar concepts of correlation-based modification of synaptic connections; he also indicated the fundamental connection between nervous changes and psychological states and discussed the psychological aspects of intelligence. In his words ([844], p. 408)

> when any state a occurs, the tendency of some other state d to follow it, must be strong or weak according to the degree of persistence with which A and D (the objects or attributes that produce a and d) occur together in the environment.

Basically, this law of connection states that if two external events occur in a correlative fashion, the associated internal states will also be correlated correspondingly; it is the "strengths of the connection" between the internal states and external events that are important to encode the information or knowledge within the brain [844].

Following the early research studies in psychology, Young [990] also suggested that repeated excitation leads to a permanent facilitation, that is, stronger and more efficacious synapses between neurons. McCulloch and Pitts [606] were among the first to phrase the properties of what later would be called Hebb's synapse in the following words:

> The phenomena of learning, which are of a character persisting over most physiological changes in nervous activity, seem to require the possibility of permanent alterations in the structure of [neural] nets. The simplest such alteration is the formation of new synapses or equivalent local depressions of threshold. We suppose that some axonal termination cannot at first excite the succeeding neuron; but if at any time the neuron fires, and the axonal terminations are simultaneously excited, they become synapses of the ordinary kind, henceforth capable of exciting the neuron. The loss of inhibitory synapses gives an entirely equivalent result.

According to Changeux and Heidmann [155], the first mention of changes in strength or number of connections in neural networks can already be found in

Descartes' *Traité de l'homme* (1677). In this case, we have to convert several aspects of Descartes' concept of a hydraulic nervous system to those fitting the present electrochemical one.

Postulate of Hebbian Learning. The most influential proponent of learning as a correlative process was Donald Hebb, who postulated the following, now referred to as *Hebb's postulate* ([377], p. 62)[8]:

> When an axon of cell A is near enough to excite a cell B and repeatedly or persistently takes part in firing it, some growth process or metabolic changes take place in one or both cells such that A's efficiency as one of the cells firing B, is increased.

The clause "*takes part in firing it*" indicates the causality condition and implies both temporal specificity, that is, the spikes from cell A occur prior to and within a short time window of the firings in cell B, and spatial specificity so that only the synapse involved in firing cell B gets strengthened. Stated mathematically, Hebb's postulate can be formulated as

$$\Delta\theta_{AB}(t) = \eta x_A(t) y_B(t), \tag{1.12}$$

where x_A and y_B represent the pre- and postsynaptic activities (i.e., firing rates), respectively, between the synapse connecting neurons A and B; $\Delta\theta_{AB}$ denotes the change of synaptic strength; and η is a small step-size (also known as learning-rate) parameter. Namely, the change of the synaptic weight $\theta_{AB}(t)$ is proportional to the product of input $x_A(t)$ and output $y_B(t)$. The learning rule is local, since the information for modifying the synapse is easily available at the location of the synapse. Averaged over many time steps, the synaptic weight becomes proportional to the *correlation* between pre- and postsynaptic firing [320].

Although Hebb's postulate became well known in 1949, it is not until nearly a quarter of a century later that physiological experiments first offered the validated evidence of Hebb's proposal. In 1973, Bliss and Lomo [100] published a paper describing a form of activation-induced synaptic modification in the hippocampus of the brain. In their experiments, they applied pulses of electrical simulation to the major pathway entering the hippocampus while recording the synaptically evoked responses, and they reported the *long-term potentiation* (LTP) phenomenon.[9] Long-term potentiation shows a number of associative properties in that there are interaction effects between *coactive* pathways. Specifically, if a weak input that would not normally cause a strong postsynaptic response is paired with a strong input, the weak input can be potentiated. Such an associative property can find its links with Pavlov's conditioning experiments and Hebb's postulate, and it is believed to form the cellular basis of memory.

Hence, the "Hebb-like effect" can be long lasting. In Hebb's original words, this consequence is described as ([377], p. 70)

> any two cells or systems of cells that are repeatedly active at the same time will tend to become associated, so that the activity in one facilitates activity in the other ... such that a reverberation in the structure might be possible.

In the literature, synapses that follow Hebb's postulate, when using the standard LTP protocol described above, are called *Hebbian synapses*. The important features of Hebb's rule include (i) a time-dependent mechanism, (ii) a local mechanism, (iii) an associative mechanism, and (iv) a correlational mechanism for which the Hebbian synapses are often referred to as *correlational synapses* [36].

Nowadays, Hebb's postulate has been widely accepted and supported by numerous neurophysiological data. It is believed that Hebbian correlation between presynaptic and postsynaptic neurons, which leads to synaptic plasticity, is mediated by backpropagating action potentials that are actively or passively transmitted to the synapse.

Experience-Dependent Synaptic Plasticity in Neocortex. The formulation of the Hebb rule $\Delta\theta_{AB}(t) = \eta\langle x_A(t)y_B(t)\rangle$, that is, the change in synaptic weight is proportional to the correlation of presynaptic and postsynaptic activities, appears to lead to untenable predictions [10, 11]. These authors recorded from pairs of neurons that either directly excited or directly inhibited each other in the auditory cortex of behaving monkeys. They found that functional plasticity is a function of the change in correlation (or covariance) and not of correlation or covariance per se. They also found that the size of the plasticity effect was increased approximately sixfold during appropriate behavior. The spike activity of the presynaptic cell was considered as the conditioned stimulus (CS), the response from the postsynaptic cell the conditioned response (CR), and the auditory stimulus the unconditioned stimulus (US) when presented 2–4 ms after a spike of the presynaptic cell. The monkey was trained to respond to the US. Specifically, they suggested a modified Hebbian learning rule as follows:

$$\Delta\theta_{AB}(t) = \eta\left[x_A(t)y_B(t+\tau) - \langle x_A y_B\rangle\right], \tag{1.13}$$

where the time interval τ is only a few tens of milliseconds after the time of a CS spike at time t and the average correlation $\langle x_A y_B\rangle$ is taken over at least several minutes. Thus the changes in synaptic weights are proportional to the changes in correlation. Appropriate behavior increases the modification factor by about a factor 6, as more or less required by *Thorndike's law of effect* [882]. Ahissar and colleagues [10, 11] also suggested that, alternatively, fractional changes in synaptic weights could be proportional to fractional changes in the correlation.

Spike-Timing-Dependent Plasticity. There are two main problems with the classical Hebbian synapse; one is that under the standard formulation the synaptic strength can only increase. Such a system, when linear, is inherently unstable and results in unlimited growth of excitatory synapse strength. The system can be kept stable through nonlinear saturation or by imposing normalization conditions. One could, for instance, keep the total summed weight of all synapses to a given neuron constant, that is, when one synapse increases in strength the others have to decrease collectively by the same amount. This mechanism contradicts with the supposed spatial selectivity of synaptic strengthening or weakening. However, numerous reports about the occurrence of heterosynaptic LTP and LTD have surfaced in recent years [89], so this is a feasible solution. Another problem with the firing-rate-based Hebb synapse is the way the association between the firings of the input and output neuron is supposed to occur. This can be assessed much more effectively on the basis of a spike-timing-based correlation procedure compared to a rate-based one.

Recently several investigators [80, 584, 589] presented evidence that the precise timing difference between pre- and postsynaptic action potentials determines whether LTP or LTD will occur. Long-term potentiation occurs when the presynaptic spikes precede the postsynaptic ones, whereas LTD occurs when the postsynaptic spikes precede the presynaptic ones. The time window for these phenomena is rather short (Figure 1.8), of the order of tens of milliseconds, and the phenomenon is called *spike-timing-dependent plasticity* (STDP). Essentially, STDP imposes a *temporally asymmetric* time window on Hebbian learning [89]; that is, if a presynaptic neuron fires a short time before the postsynaptic neuron, positive Hebbian learning occurs, whereas if the postsynaptic neuron fires a short time before the presynaptic neuron, anti-Hebbian learning occurs. This form of spiking-time-dependent Hebbian learning is more realistic in that it captures the causal relationship that exists between presynaptic and postsynaptic firing [317, 320, 484]. Specifically, the STDP learning rule has several distinct features [195]: (i) the bidirectionality of synaptic modification with approximately balanced LTP and LTD, which helps the neural circuit maintain its net synaptic excitation at a stable level; (ii) the spike sequence dependence of synaptic modification, which allows the circuit to learn sequences and to encode causality of external events; and (iii) the narrow

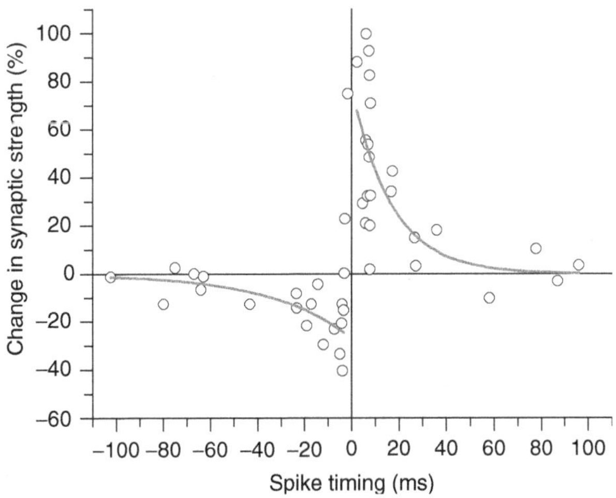

Figure 1.8 Illustration of temporally asymmetric spiking-time-dependent Hebbian synaptic plasticity. The synaptic modifications (LTP or LTD) are induced by correlated pre- and postsynaptic spiking. (Reprinted, with permission, from the *Annual Review of Neuroscience*, Vol. 24. Copyright © 2001 by Annual Reviews.)

temporal window, which allows the system to select inputs based on its response latency with a millisecond precision, thus shaping the temporal dynamics of the circuit.

The biphasic learning window of STDP overcomes the instability problem inherent in the rate-based Hebbian learning rule if there is slightly more depression than potentiation. The temporal window length arises naturally in a model where backpropagation of action potentials from the cell soma, where they are initiated, into the dendrites toward the synapses is considered. This makes the timing of the postsynaptic spikes available at the synapse, and the backpropagated signal functions as an associative signal for synapse modification. The conduction velocity of these backpropagated action potentials is of the order of 0.5 m/s in cortical pyramidal cells [130, 863], and with a typical dendritic length of 0.5 mm this translates in a delay of about 1 ms between the initiation of the action potential and its availability at the dendritic synapse. Recently several investigators [80, 584, 589] presented evidence that the precise timing difference between pre- and postsynaptic action potentials determines whether activity-dependent LTP or LTD will occur.

Depolarization of the postsynaptic membrane (e.g., by a backpropagating action potential) can remove a Mg^{2+} ion from the pore of an NMDA (*N*-methyl-D-aspartate) receptor channel, thereby allowing an influx of Ca^{2+} when the presynaptic terminal releases glutamate. This mechanism allows an NMDA receptor channel to function as a molecular detector of the coincidence of presynaptic activity and postsynaptic depolarization [106]. The resulting influx of Ca^{2+} may lead to synaptic potentiation.

The STDP is dependent not only on the timing interval between pre- and postsynaptic spikes but also on the timing of preceding presynaptic spikes. Such spikes can depress the efficacy of following spikes in producing STDP. Therefore the first spike of a burst in the presynaptic neuron is the dominant one in causing synaptic modification [297]. Recent studies [298] suggest that STDP is also locationdependent; specifically, the activity-dependent synaptic modification depends on dendritic location according to the temporal characteristics of presynaptic spikes.

In experimental studies, STDP was shown to be instrumental in eliciting changes in orientation columns in cat visual cortex, thereby demonstrating the link between *synaptic plasticity* and *representational plasticity*. Schuett et al. [803] paired brief flickering gratings of low spatial frequency and with a particular orientation with one 60-μA electrical pulse in about 300 μm below the cortex surface for 3–4 h. The timing of the pairing was critical; a shift in orientation preference toward the paired orientation occurred at the site of electrical stimulation if cortex was activated first visually and then electrically. A similar result was found by repetitive pairing of two visual stimuli with different orientations for 3–6 min [988]. A shift in orientation tuning of cortical neurons was found with the direction of shift determined by the order of presentation. An effect was found when the time difference of the presentation was about 40 ms. They also demonstrated that this stimulation paradigm in humans produced a shift in perceived orientation, thereby demonstrating a link between synaptic plasticity, representational plasticity, and perception. Song and Abbott [842] in a modeling study demonstrated that the formation of orientation columns during development as well as their remapping in adulthood follows the timescales and biphasic shape of STDP.

1.3 CORRELATION IN ENSEMBLES OF NEURONS: SYNCHRONY AND POPULATION CODING

Correlative Firing. In neuroscience, correlative firing refers to two or more neurons (or ensembles of neurons) that tend to be activated at the same time [786]. According to Cook [183], correlated firing occurs at two levels. In the short term, since few neurons can be driven reliably by a single axon, the relative timing of multiple inputs is crucial to their influence. For a population of neurons, a "window of opportunity" focuses on the moment at which a strong volley of afferent impulses shifts the membrane potential toward the firing threshold; within that window the effect of another input on the neuron's output may be enhanced. In the longer term, for some neurons and synapses, the relative timing of multiple inputs can modulate synaptic efficacy in long-lasting ways and thus change the functional properties of the circuit.

Correlated activities are widely witnessed in various sensory (visual, auditory, olfactory, or somatosensory) systems (e.g., [336, 337, 531, 582, 583]) and motor system (e.g., [501]). Although there remain some distinctions between different systems, the basic functional principles are similar. For example, in the visual system, neighboring neurons, in areas from retina to cortex, tend to fire synchronously more often than would be expected by chance; correlated firing among neural assemblies abounds at cortical and subcortical (e.g., thalamic) levels [16, 833].

For the auditory system, Eggermont [243] reviewed the role of correlation and synchrony in auditory cortex. Specifically, in the auditory brainstem and midbrain, inhibitory interactions between neurons further add to the highly nonlinear nature of the coding of sound whereby the firings of individual cells become highly interdependent and their firing times may become correlated. The way sound is represented at various levels of the auditory system forms the basis for its neural coding. A *neural code* is considered here as a vocabulary of the firings represented at a subcortical and/or cortical level on which perceptual discrimination is based. This vocabulary, an N-dimensional vector (with N the number of participating neurons, i.e., the size of the assembly), contains all the information needed for the perceptual decision process. Examples of such vocabularies are those based on instantaneous firing rates, integrated firing rates, and mean interspike interval duration of a group of specialized neurons [248]. How a neural code is constructed out of neural representations depends on (i) the sensitivity of the neurons to detect changes in the stimulus, (ii) the variability in the individual neurons' responses to the same stimulus, and (iii) the correlation between the responses of the individual neurons. If a neural code was based on firing rate, then independence of the firings in neighboring neurons would allow more information to be transmitted and correlations between the firings of individual neurons would generally diminish the information capacity of a neuronal population [1002]; however, it can improve the accuracy of the neural code [1, 770].

Population Coding in Motor and Sensory Systems. Animals extract information in parallel from an initially unknown, usually time-varying stimulus on the basis of short segments of a large number of spike trains to allow real-time estimation of some aspects of the stimulus [761]. Potential examples of pseudo-real-time estimation procedures are found in the *population vector coding* method applied to motor cortex [315] and the superior colliculus [694]. In these models, assuming independence of neuronal firing, the firing rates of neurons were weighted by their preferred hand-pointing or saccadic eye-movement directions and added up to provide a movement vector that predicted the motor output in strength and direction. If the motor neurons are assumed to be tuned in cosine fashion to a particular angle-of-motion direction (d, in radians), that is, the individual neuronal firing rate r_n that depends on d and achieves its maximum $r_{n,\max}$ in the preferred angle of direction d_n satisfies a cosine tuning function

$$r_n(d) = r_{n,\max} \cos(d - d_n), \tag{1.14}$$

where only positive cosine values are taken into account, then the weight of each individual contribution to the final compound saccade vector will be given by the correlation of its preferred firing direction and the desired direction of motion, which is proportional to the cosine of the angle between the two vectors. The population vector model then states that the direction of motion induced by the population activity is given by

$$d_{\text{pop}} = \frac{1}{N} \sum_{n=1}^{N} \frac{r_n}{r_{n,\max}} d_n., \tag{1.15}$$

where N denotes the total number of motor neurons. This model is equally applicable to encoding of a stimulus direction, for example, orientation of a visual object, but its assumption about independence of individual neuron activity and its sensitivity to noise (i.e., the spontaneous firing activity) make it less than ideal [736, 785].

Place cells in the hippocampus that code the position of the animal in reference to its environment, cells in visual cortical field MT that detect direction of motion, and cells in visual cortical field V1 that are tuned to the orientation of a stimulus are also prime examples of population coding on the basis of firing rate that can produce adequate stimulus reconstruction [203, 206, 735]. Recently the importance of dedicated subgroups of neurons in the hippocampus ("cliques") that can initiate various startle responses has been highlighted [556]. Dedicated subgroups of neurons ("clusters") have been identified for representation of auditory space in the midbrain and forebrain [179]. These clusters are not part of topographic maps because neighboring clusters may be coding for completely different sound location cues.

Examples of population coding in auditory cortex based on the firing rate of (presumably independently firing) neurons are found in the panoramic code of sound location [299, 619], in the population vector model of sound azimuth coding [252], and in the coding of vocalizations [312, 797] or periodic sounds [574].

The sampling of the neuronal population in all these studies was done sequentially, thereby making their activities in fact independent. The coding of the sound direction features by firing rate was much better than those of the vocalizations or periodic sounds. Thus, better representational codes must exist for aspects of sensory stimuli other than those related to direction or location.

Role of Correlated Firing in Neural Coding. Sensory systems often represent distinct features of the environment by spatially distinct sets of neurons. For instance, in the visual system, color, texture, and size are encoded in different visual areas. Thus, a yellow, fuzzy tennis ball and a red, smooth pool ball would be coded in one area as yellow versus red, in another area as fuzzy versus smooth, and in the third one as slightly different sizes. Somehow, the relationships of the properties belonging to the tennis ball and the pool ball need to be tagged to prevent us from seeing a fuzzy, red pool ball. This may require a mechanism, such as enhanced neural synchrony between cortical areas [335], to group the extracted features belonging to a specific object. It is also possible that the common spatial location of [yellow, fuzzy] for the tennis ball is a sufficient tag that could be accomplished by connections of the color and texture areas to the retinal maps in V1.

In the auditory system, important sound features are "components of an auditory scene [that] appear to be perceptually grouped if they are harmonically related, start and end at the same time, share a common rate of amplitude modulation or if they are proximate in time and frequency" [184]. Thus, important sound features allow correlations in the temporal domain and spectral domain that signal sufficient overlap to be grouped into one percept or assigned to one sound source (Figure 1.9). Sounds can be meaningfully decomposed into contours (e.g., temporal envelopes) and texture (e.g., frequency content), as is common for visual images [248]. The most meaningful aspects of speech are likely the sound envelopes as these play a crucial role in speech recognition as demonstrated by replacing the detailed frequency information by octavewide bands of noise without affecting recognition to an appreciable extent [824]. These sound envelopes also produce the largest changes in the correlation of neural activity, compared to a nonstimulus condition, in auditory cortex [248]. The correlated activity across a neural population may emphasize these stimulus contours above their texture, despite the fact that STRF overlap accounts for up to 40% of the variance in pairwise neural correlation [250]. This suggests that the fraction of shared inputs from the auditory thalamus by cortical cells represents those that potentially take part in a correlated neural assembly but the firing times of a neuron are codetermined by the sound envelopes as filtered by the neuron's STRF.

Coding of complex sounds requires a population of neurons. In response to complex sounds, cortical neurons typically show a correlation in their time-varying firing rates and even in their spike-firing times. Thus, the coding mechanism utilized by a cell population to extract stimulus information cannot be inferred from the activities of different neurons recorded at different times. The role of these correlated firings in the coding of complex sound is not fully known. Coincident

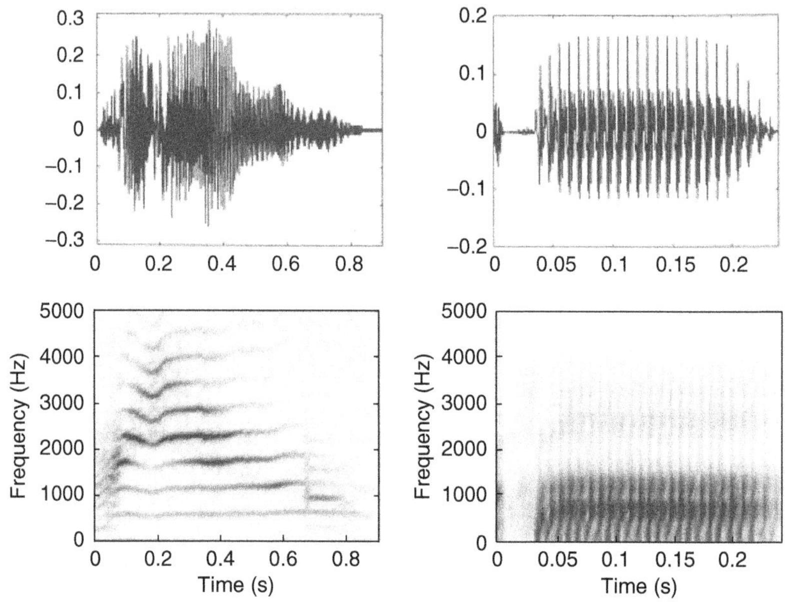

Figure 1.9 Two vocalization sounds that illustrate similarities and differences in binding features. In the left-hand column, the waveform and spectrogram of a kitten meow are presented. The average fundamental frequency is 550 Hz, and the highest frequency component (not shown) is 5.2 kHz. Distinct downward and upward frequency modulations occur simultaneously in all formants between 100 and 200 ms after onset. The meow has a slow amplitude modulation. In the right-hand column, the waveforms of a /pa/ syllable with a 30-ms voice-onset time (VOT) and its spectrogram are shown. The periodicity of the vowel and the VOT are evident from the waveform. The fundamental frequency (i.e., the periodicity) started at 125 Hz and remained at that value for 100 ms and dropped from there to 100 Hz at the end of the vowel. The first formant started at 512 Hz and increased in 25 ms to 700 Hz, the second formant started at 1019 Hz and increased in 25 ms to 1200 Hz, and the third formant changed in the same time span from 2153 to 2600 Hz. The dominant role of the periodicity in binding of frequency components is noted. (Reprinted from *Hearing Research*, Vol. 157, J. J. Eggermont, Between sound and perception: Reviewing the search for a neural code, pp. 1–42. Copyright © 2001, with permission from Elsevier.)

firings that frequently occur without concomitant firing rate changes (such as in the neural response to the steady-state portion of a pure tone, which can show the same firing rate as under silence but with increased neural synchrony between pairs of neurons [205]) can in principle be detected by depressing cortical synapses [819]. These synapses have an initial high probability of transmitter release and act as low-pass filters that are most effective at the onset of presynaptic activity and respond most vigorously to transient stimuli and to slow modulation envelopes. These synapses are responsible for the low-pass properties of temporal modulation transfer functions (Figure 1.10) as measured electrophysiologically in primary auditory cortex (A1) [246, 248].

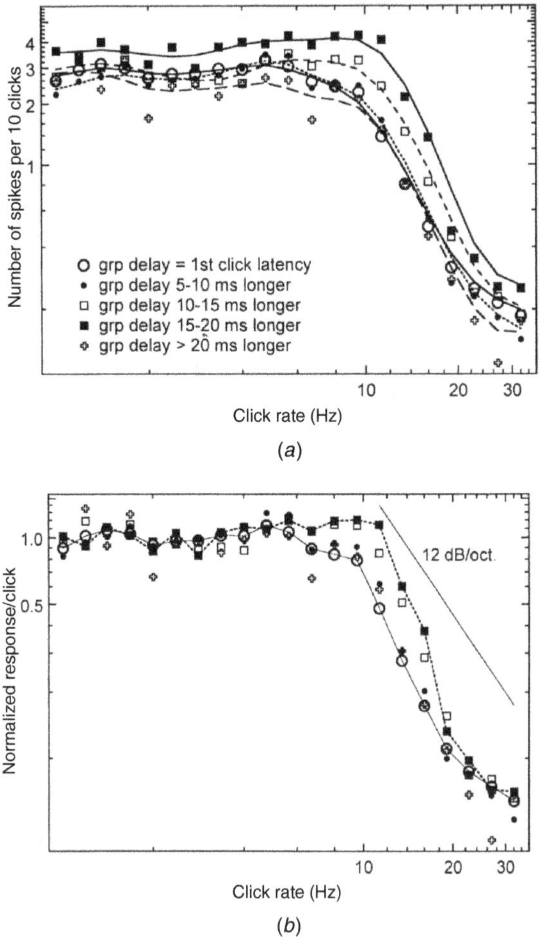

Figure 1.10 Low-pass filtering in auditory cortex neurons. Stimuli presented were 1-s-long periodic click trains and the number of synchronized spikes per click is shown here as a function of the click repetition rate. (*a*) Group averages are distinguished by group delay as determined from the phase repetition rate dependence. This plays only a modest role, except that neurons with large group delays show a slightly higher cutoff rate compared to those with group delays below 15 ms. (*b*) Various curves are normalized to their mean response between 1 and 4 Hz. (Reprinted from [246], with permission. Copyright © 1999, *Journal of Neuroscience*, by the Society for Neuroscience.)

It has been predicted [485], and shown recently in the avian forebrain [481] and in vitro [759], that correlated neural activity is capable of propagating through cortical structures without diminishing in strength and with preserved temporal precision. This would facilitate grouping across distinct cortical fields and the formation of interarea neural codes. This is reminiscent of the theory of synfire chains [4, 920], which require this property.

Observations That Favor Role of Coincident Firings in Neural Coding. In the primary motor cortex (M1) of behaving macaque monkeys, correlated neural firings play a significant role in coding movement direction [601]. The information carried by neural interactions using a simultaneous recording from 12–16 neurons during an arm-reaching task was investigated. Pairs of simultaneously recorded cells revealed significant correlations in firing rate variation when estimated over 600-ms time intervals. This covariation was only weakly related to the preferred directions of the individual M1 neurons estimated from their maximal firing rate. Interelectrode distance had no significant effect either. In some of the cell pairs, the strength of the neural correlation varied with the direction of the arm movement. Prediction of the direction was consistently better when correlations were incorporated as compared to one based on the average firing rate of presumably independent neurons. Thus, neural interactions quantified by correlated activity carried additional information about movement direction beyond that based on the firing rates of the individual neurons. The correlated neural activity was also much higher for a planned sequence of movements compared to the same movements when executed independently by the monkey, although the firing rates were the same in the two conditions [360]. Simultaneously recorded activities of neurons in M1 of monkeys during performance of a delayed-pointing task showed that accurate spike time synchronization occurred in relation to stimuli and movements and was commonly accompanied by discharge rate modulations but without precise time locking of the spikes to these external events [760].

In primary somatosensory cortex (S1) of the anesthetized cat, stimulation of the front paw with an air jet resulted in neuron pair correlograms (see examples in Figure 0.2) with much sharper peaks than observed without stimulation [776]. The incidence and rate of stimulus-induced synchronization decreased with the distance between the recording sites. These results suggest that neuronal synchronization measures may supplement the changes in firing rate that code intensity and other attributes of a tactile stimulus.

The synchronous firing in the secondary somatosensory cortex (S2) of three monkeys trained to switch attention between a visual task and a tactile discrimination task increased in up to 35% of the pairs tested and so did the firing rates, however without a significant correlation between the changes in firing rate and changes in synchrony [854].

Cells in cat primary visual cortex showed enhanced orientation discrimination by including the synchronization of the firings between two to six cells in addition to their firing rates [787, 788].

Pairs of neurons recorded with electrodes in different auditory cortical areas showed a fourfold increase in firing synchrony during stimulation with tones or noise compared to silence combined with modest increases in firing rate [247]. Neural synchrony in rat auditory cortex also increased in a delayed go/no-go task, a task where one stimulus required a behavioral response after some prescribed time and the other one did not, but specifically in the waiting period [916].

Observations That Argue Against Role of Coincident Firings in Cortical Neural Coding. In V1 of the awake monkey, neural synchrony was observed between neurons with distant RFs in response to textured "figure–ground" stimuli. However, there was no difference in synchrony between pairs with both RFs overlapping the "figure" part and pairs in which one or both units had RFs within the "background" part of the stimulus. Thus, no evidence was found for a role of neural synchrony in the binding of those features that lead to texture segregation [521]. In a coherent motion detection task, the neural synchrony in awake monkey visual field MT was actually lower than for noncoherent conditions [879] and thus not likely to play a role in binding of motion by synchrony. Pairwise correlation strength for units recorded on the same electrode in MT of the behaving monkey was independent of the presence of visual stimulation and the behavioral choice of the animal [53]. Rolls et al. [770] and Aggelopoulos et al. [9] also found little gain of stimulus-dependent synchronization on the information available about the stimulus in the neuronal firing rate in inferior temporal visual cortex.

Simultaneously recorded firings from 30–40 neurons from three somatosensory cortical areas were able to predict the type of stimulus regardless of whether the trials were shuffled for each single neuron [659]. This suggests that precise timing information between those neurons was irrelevant. In secondary somatosensory cortex (S2) of anesthetized cats, Alloway et al. [15] found no evidence that synchrony played a role in the coding of the direction of movement of a tactile stimulus. Similarly, in rat barrel cortex, synchronized firing did not contribute to coding the stimulated whiskers [714]; coding was instead solidly based on first-spike latency.

A similar absence of change in correlation strength with increased auditory stimulation level was reported for units recorded on separate electrodes in A1 of the anesthetized cat [247]. Thus neural synchrony likely does not code for stimulus level.

Hence, it appears that in the early stages of motor and sensory cortical processing (M1, S1, V1, A1) neural synchrony may play a greater role than in later stages (S2, MT, IT). We return to this issue later in our discussion of the role of synchrony in feature binding via bottom-up versus top-down attentional processes.

1.4 CORRELATION IS THE BASIS OF NOVELTY DETECTION AND LEARNING

It has been already suggested that coincidence detection by rather broadly tuned neurons may result in sharper tuning or greater specificity for particular stimuli [57]. This can be obtained either by a simple convergence of two neural activity patterns on a coincidence detecting neuron [250, 886] or by strengthening the direct connections between simultaneously active neurons. The latter mechanism has been postulated for the creation of sharply tuned neural assemblies [377], secondary repertoires [238], and synfire chains [3].

Neural Assemblies. Hebb [377] has pointed out that there are two extreme views of neural assembly action. One was called the *switchboard theory*: The cortex is considered as an elaborate kind of telephone exchange with precise connections; the other was called the *field theory*, which regards the cortex as an aggregate of cells forming a statistically homogeneous medium with mostly random connections. An example of a switchboard theory was presented by Ballard [55]. Examples of field theories are those by Beurle [86], Cowan [190], Griffith [339], and Hopfield [399], to mention a few. Hebb's own assembly model was somewhat intermediate in assuming that precise connections existed but with modifiable synapses that could be changed by experience. An elaboration of such an assembly theory was presented by John [444] in what he called a "statistical configuration theory." In this theory, learning and memory are envisioned as the establishment of a representational system of a large number of neurons in different parts of the brain. The activity of these neurons will be affected in a coordinated way by the spatiotemporal characteristics of the stimuli presented during the learning task. This was assumed to initiate a common mode of activity in various brain regions specific for that stimulus. Information about an event is represented by the average behavior of such a responsive neural ensemble. Another event can be represented by the same ensemble but with a different correlation pattern.

A big leap in the concept of neural assemblies was made by von der Malsburg [922] by proposing the following description: "a cell assembly is a set of neurons cross-connected such that the whole set is brought to become simultaneously active upon activation of appropriate subsets which have to be sufficiently similar to the assembly to single it out from overlapping others." Thus, given suitable input, the assembly can be ignited and then acts as a logical unit by going through a spatiotemporal activity pattern characteristic for that assembly. The ignition character of an assembly is also evident in the concept of the synfire chain [3, 4]: "the activity of the neurons that transmit information is organized along a chain of sets of neurons. Each link in the chain is made of a set of neurons that fire in exact synchrony whenever the chain becomes active." The concept of neural assembly also includes the necessarily hierarchical character of the organization and is related to the concept of repertoires [238] defined such that "the main unit of function and selection in the higher brain is a group of cells connected in various ways. Groups of cells build repertoires."

Neural assemblies have more recently been defined as "a group of neurons [that are] at least transiently working together as indicated by correlation of unit activity" [316]. In visual cortex, cells with approximately 0.5-mm separation showed the highest correlation among cells with similar RFs and similar connectivity from the LGN [522]. This suggests that overlapping or shared connectivity is a dominant factor in neural assembly formation. It is common to think about a neural assembly as widely distributed in cortical space, potentially extending over various subdivisions of cortex [838]. For instance, connections over large spatial divisions of auditory cortex are provided by the thalamic cell axonal divergence and convergence, often estimated to be between 2 and 5 mm at the cortical level [536] and intracortically through horizontal fibers [932] that can range up to 8 mm. In visual cortex, the spatially periodic effects of the patchy connections of these horizontal fibers have been shown by cross-correlation [893]. These cortico-cortical connections are for a sizable part heterotopic. In auditory cortex, they connect cell groups with characteristic frequencies (CFs) differing by more than one octave [537]. In visual cortex, the horizontal fibers connect cell groups without spatial RF overlap but with similar orientation tuning.

Neural assembly membership is expected to be stimulus dependent and context specific and may reflect the number and functional strength of its common inputs under different conditions [903]. It is however likely that at any point in time several spatially overlapping neural assemblies are active. In response to external events, a group of neurons forming a dynamical cell assembly may spontaneously organize itself temporarily by correlated firing of their spiking activity.

Neural assemblies, thus defined, may potentially be probed using microelectrode arrays that allow recording from a set of relatively widely spaced neurons. These neurons could participate in one or more neural assemblies. The quantification of the correlation in spiking activity occurring between pairs of such widely spaced neurons thus becomes crucial in defining membership of neural assemblies. The stimulus will be one of the dominant sources of neural correlation because of its common input character. Although it is common to correct for stimulus-induced correlations by using shift predictors or joint peristimulus time histogram (JPSTH) techniques [244], the brain does not have that luxury but may exploit this stimulus-dependent correlation to change the extent and structure of the neural assemblies.

Secondary Repertoires. The selection theory of brain function [238–240] assumes that after ontogeny and early development the brain contains cellular configurations (groups) that can already respond in a discriminatory way to sensory stimuli (e.g., the orientation selectivity in the visual system of newborn monkeys), because of their genetically determined structures or because of epigenetic alterations that have occurred independently of the structure of these sensory signals. This prespecified collection of neuronal groups is called a primary repertoire and consists of a large number of groups (of the order of 10^6), each with a modest number (50–10,000) of cells. The primary repertoire is degenerate, that is, it contains multiple neuronal groups, with different internal structures, that are capable of carrying out the same function. The primary repertoire should contain enough neuronal groups such that sensory signals have a high probability to find matching groups; and finally it must have provisions for amplifying a selective recognition event, probably by synaptic alterations, either through the formation of new synapses or through changes in already existing contacts. All these properties are very much the same as in the classical perceptron [773].

In addition, the neuronal group selection theory requires a secondary repertoire as a collection of different, higher order neuronal groups whose internal and external synaptic connectivity can be altered by selection during experience. This cell group selection can occur in two stages, first by filtering—selecting all groups that react more or less well to the spatiotemporal input pattern—and second by an inhibition process (a threshold mechanism) that eliminates those selected groups from stage

1 that have an insufficient response. An important aspect of the theory is the reentrance of signals at the level of the secondary repertoire. The dominant cell type, the pyramidal cells in cortex receive far more collaterals from other pyramidal cells (>99%) than from specific afferents (<1%). Thus, the cortex is to a large extent a thinking machine working on its own output [110]. For the pyramidal cells there is probably no way to distinguish these inputs. Thus, internally generated signals are reentered as if they were external signals; this reentrance might be able to guarantee a continuity in the neural construct as well as a succession of temporal order of associated memory events. Moreover, reentrant signals in the secondary repertoire might be of a different modality from the more direct input from first repertoire groups. Thereby the secondary repertoire might be able to relate multimodal activity patterns. Reentrance is also a mechanism that allows ongoing cross-correlations between inputs from first repertoire groups and second repertoire groups, thereby providing the possibilities of association and classification [845]. The role of STDP in the neuronal group selection theory has been elucidated in [435].

Learning. Learning is a process that generates a "brain" that is different from that prior to the learning process. The results of learning are memories. Memories are likely laid down in spatial patterns of synaptic connectivity. Novelty detection is related to memory recall; the evidence for the existence of novelty detectors comes among others from evoked potential studies where an "oddball" stimulus creates large activity in the latency region beyond 100 ms following the stimulus, called the *mismatch negativity* (MMN) [646]. A novelty detector is most likely a neuronal group and not an individual neuron. Whenever information is processed through various assemblies, familiar input will activate already formed functional connections in a neuronal group. An acceptable match between incoming information and stored information is based on correlation and will generate either an inhibitory signal to arousal centers or no signal at all. Whenever there is a mismatch, a signal is sent to the arousal centers whose nervous activity, or activity induced by the arousal center in the sensory cortex, may result in detectable evoked potentials. Novelty detection then in this view is the result of a "template-matching" procedure that has to be carried out in parallel for a very large number of potential templates: In principle all information that does not match with the templates has a novelty character.

In monkeys, MMN-like activity corresponded to spikes generated in superficial layers of primary auditory cortex. When NMDA receptors in the auditory cortex were blocked, the MMN disappeared [439]. In analogy, administering the anesthetic ketamine, which is an NMDA blocker, to human volunteers also abolished the MMN [508]. This suggests that the MMN requires NMDA receptors involved in the formation of memory traces used in the novelty detection. This is not surprising as NMDA channels have also been linked to LTP and working memory [564].

Topographic and Functional Brain Maps. Braitenberg [110] came to the conclusion, on anatomical grounds, that the probability of two neighboring pyramidal cells in cortex being interconnected by one synapse was about 10%. Experimentally, this was confirmed much later. Dual intracellular recordings from neighboring pyramidal cells were used to assess the probability of synaptic contact between randomly selected neuron pairs within 250 μm distance from each other [880] as 1 : 11 (layer III to layer V), 1 : 21 (layer III to layer III), 1 : 41 (layer V to layer V), and 1 : 86 (layer V to layer III). Thus, most of the pyramidal cells are not interconnected, and if they are, the connection is very weak (by only one synapse out of the 5000–10,000 located on each pyramidal cell) compared to the estimated firing threshold of a pyramidal cell, which is about 10–30 for time-correlated input and more than 300 for asynchronously arriving inputs [4]. Thus, correlated input activity keeps the brain going. In rat somatosensory cortex, Bruno and Sakmann [124] measured in vivo the excitatory postsynaptic potential evoked by a single thalamocortical synaptic connection and confirmed its low efficacy. They also demonstrated that the thalamocortical inputs are numerous and synchronous, thereby confirming the suggested synchrony mechanism by which thalamic inputs alone can drive cortex.

Most correlations of spiking activity in the cortex are through shared specific afferent (i.e., thalamocortical) input and therefore serve to functionally link neurons with overlapping STRFs. What the experimenter observes as topographic maps (e.g., retinal position, sound frequency, body surface) does not have meaning for the brain itself based on this spatial organization but only through the correlations in the neuronal firings. Thus, activity in one cell can just be spontaneous ("noise") and will be very difficult for other neurons to distinguish from stimulus-induced activity. Correlated activity in two or more neurons that project to another neuron potentially signifies outside world activity if their firings are correlated. Furthermore, as described above, correlated inputs are more likely to fire the receiving cell.

As discussed earlier, two types of neuronal maps are found in the cortex: *topographic maps* and *functional maps*. Topographic maps arise from the anatomical structure of the receptor surface and from the afferent nerve fibers preserving this orderliness in the nerve fiber tracts and in each interposed nucleus. Receptor surface maps may undergo quite complex transformations, resulting in topographic map deformation. An example is the complex logarithmic transformation between retinal surface and visual cortex surface (e.g., [807, 808]) or superior colliculus map (e.g., [694]). Topographic maps can also result through computations carried out in the brain. A prime example is the map of auditory space in the midbrain [492], which is based on a combination of interaural differences in arrival time, phase differences in continuous sounds, and intensity differences at the two ears as represented centrally through differences in excitation and inhibition [345], but in reference to a topographic map of visual space in the same structure (e.g., [491]). To the experimenter there is no difference between the computational map of auditory space and the topographic map of visual space: Both show a spatially ordered structure. To the animal there is no difference either: The internal order of both is present in the correlation structure. Thus, in the brain all maps are *correlation maps* and there is no real way to distinguish between the two types. Correlation maps represent signal-associated aspects, both sensory and motor, as encoded by individual neurons or small neuronal groups. Topographic maps are only obvious to the investigator, who relates neural activity to the stimuli presented.

Topographic maps are organized on the basis of their sensory receptor or effector surfaces and are therefore likely to be found at the sensory and motor sides of the brain. Functional maps have mainly an organization or ordering through their correlation structure [494, 495]. Neural unit pairs with strong correlations can be considered close in the neural organization; neural units with weak correlations have a larger functional distance. Functional maps may differ for different stimulus conditions [250]. Topographic maps also have a cross-correlation structure, but this is mainly the result of the anatomical ordering and spectrotemporal or spatiotemporal overlap of RFs (Figure 1.11). Here, the principle of obtaining coincident-spike STRFs is illustrated. Typically spikes in two neurons that contribute to their potentially overlapping STRF (i.e., occur within the time–frequency domain of their STRF) are more strongly correlated than those that do not contribute. As can be

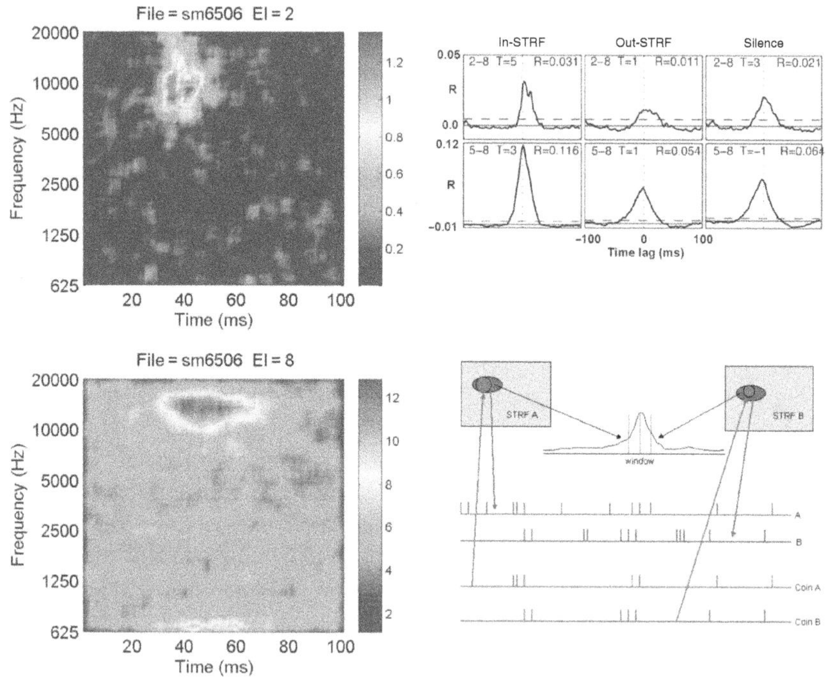

Figure 1.11 Correlation structure of overlapping STRFs in auditory cortex. Left column shows two simultaneously recorded STRFs with partial overlap in the time–frequency domain. The upper STRF shows a responsive area between about 6 and 20 kHz with a CF slightly below 10 kHz and extending in time between 30 and 50 ms after stimulus onset. The lower STRF is generally between 10 and 17 kHz and 30–70 ms and a CF at about 13 kHz. Upper right panels show cross-correlograms between spikes that contributed to the STRFs (In-STRF), between spikes that did not contribute to the STRFs (Out-STRF), and for comparison from the same neurons under spontaneous firing (silence). The right bottom panel shows in cartoon form how coincident-spike STRFs are constructed. Original spike trains in red, only spikes that occur within a given time window from those in the other neuron, and considered as coincident spikes, are shown in green.

seen from the cross-correlograms, spontaneous activity during silence in turn is more strongly correlated than "spontaneous" spikes, that is, those not contributing to the excitatory part of the STRF, during continuous stimulation. Selection of a certain coincidence window allows the construction of STRFs for coincident spikes only.

In Figure 1.12, we show comparisons of STRFs for four sets of recordings comprising activity from four to seven separate electrodes (indicated in the bottom panels) under the conditions of only coincident spikes (within a 10-ms window, top row) and all spikes merged (bottom row). Here, N is the number of recording sites (typically out of 16) that belong to the same cluster and whose spike trains are used in constructing the all-spike and coincident-spike STRFs. It is clear that only in some cases the coincident-spike and all-spike STRFs are similar. For instance, in the third column, where the two STRFs show the highest similarity, the contour of the excitatory part of the coincident-spike STRF overlaps with the all-spike STRF, but differences still exist for the inhibitory parts. The largest differences between coincident-spike and all-spike STRFs are shown in the second column. Here, the coincident-spike STRF forms a small subset of the all-spike STRF but overlaps with its most responsive part. Note the large differences in the inhibitory parts of the two STRFs.

This suggests that different computations can be carried out by the cortex depending on the mode of action of the downstream cells: coincidence detection or temporal integration. Whether neocortical pyramidal cells are acting as coincidence detectors or temporal integrators depends on the degree of synchrony among the synaptic inputs [333, 779]. High-input synchrony leads to the more efficient

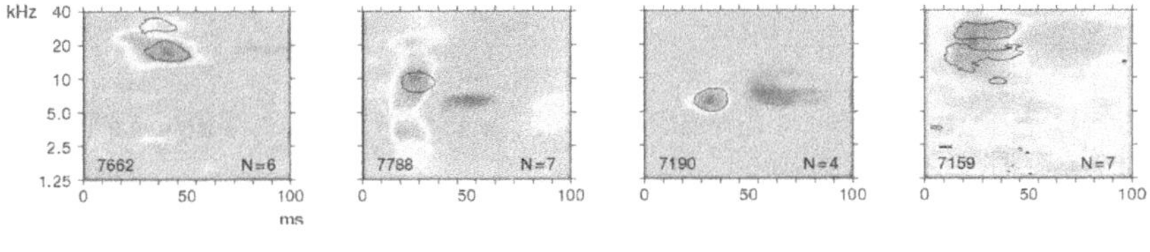

Figure 1.12 Four examples of coincident-spike STRFs (top row) compared, by superimposing its contours, to the all-spike STRFs (bottom row). In the bottom row the number of STRFs that contributed to both types of STRF are indicated (e.g., $N = 6$). Peak STRF values and mean level of activity are indicated at the top of each panel, which give an indication of the different signal-to-noise ratios (SNRs) of the two types of STRF. The percentage of the spikes that contributed to the coincident-spike STRFs is displayed above each set of panels. Warm colors (yellow and red) indicate excitation, cold color (blue) indicates inhibition. Green is typically neutral. (Reprinted from (250), with permission. Copyright © 2006, by the American Physiological Society.)

coincidence detection, whereas low-input synchrony leads to temporal integration. The asynchronous background spikes from other cortical cells could also play an important role in setting the processing mode of the cell [8].

Functional maps are likely to be found in the association areas between sensory and motor areas. However, Vaadia et al. [904] found that about 15% of neurons in primary and secondary auditory cortex of behaving monkeys showed sensorimotor association properties. Besides a strict sensory component in their response, these units showed task-dependent activity as well. Since these units were found widely dispersed throughout the auditory cortex, they suggested that association cortex might overlap with sensory cortex. If this conjecture is correct, it suggests the coexistence of topographic and functional maps in the same cortical structures. Neurons from one type of map may simultaneously take part in another type of map.

Synchronous or correlated activity is the only way in which sensory events propagate through the brain and manifest themselves to the brain; it will also be the way in which such events are remembered. Thus, in a more general sense, simultaneous activity of neurons is also related to memory recall. Activity in representations of different sensory modalities may be correlated on a larger timescale and can be an indication of events in a more general context. The relation between topographic maps of different modalities can be maintained in a functional map. This functional map then constitutes a memory of a situation in which such a multimodal stimulus was present. Later, this map can be activated again and be used to complete a certain pattern of activity or to predict the temporal sequence of a stimulus pattern.

How is such a functional map formed? One may assume that the correlations between the occurrences of events that together form a situation can be stored in terms of connections between the neural units that represent these events. The question can be raised whether there is a sufficient number of connections to store all such items. Since the neocortex has about 10^{11} neurons and about 10^{15} synapses, there seems to be ample space to do so. Functional maps are only characterized by a "correlation structure" among firings from different neurons. They have been attributed as resulting from neural assemblies [316], neural clusters [179], and neural cliques [556]. Functional maps have not yet been visualized by electrophysiological techniques [250].

1.5 CORRELATION IN SENSORY SYSTEMS: CODING, PERCEPTION, AND DEVELOPMENT

In many layered brain structures, spontaneous and highly synchronized neural activity can be found. One of these areas is the hippocampus, where theta wave (4–8-Hz) activity is found that seems to reflect the initiation of purposive movement patterns such as self-generated motion through space [98, 128]. There seems to be less and more localized correlated firing of neurons in the neocortex during active processing compared to the sleep state, where many single units show a high degree of correlation in their firings [213, 663]. Thus, correlated activity does not always accompany active behavior. It could be postulated that during sleep cells in the brain are recruited into large assemblies that generate reverberant activity in the slow-wave or spindle frequency region. Its purpose could be to stabilize memories [856]. We return to these issues in our discussion of correlation in memory systems.

Correlation in Perceptual Coding. Barlow [60] proposed that a major goal of perceptual coding is to produce a minimally redundant neural code, which should facilitate subsequent learning. The information about an underlying signal of interest, such as the visual form or the sound of a predator, may be distributed across many input channels, making it difficult to associate particular stimulus values with distinct responses. A neural code having minimal redundancy should alleviate this problem. If the encoding of the sensory input vector into an N-element feature vector has the property that the N elements are statistically independent, then all that is required to form new associations with some event V (assuming the features are also approximately independent conditioned on V) is knowledge of the conditional probabilities $p(V|y_i)$ for each feature y_i, rather than complete knowledge of the probabilities of events conditional upon each of all combinatorially possible sensory inputs. Atick and Redlich [47] proposed a cost function for Barlow's principle that minimizes the power (redundancy) in the outputs subject to a minimal information loss constraint. Under conditions of high noise (low redundancy), the RFs that emerged were Gaussian-shaped spatial smoothing filters, while at low noise levels (high redundancy) "ON-center OFF-surround" RFs resembling second-order spatial derivative filters emerged. In fact, cells in the mammalian retina and LGN of the thalamus dynamically adjust their filtering characteristics as light levels fluctuate between these two extremes under conditions of low versus high contrast [825, 918]. Moreover, this strategy of adaptive rescaling of neural responses has been shown to be optimal with respect to information transmission [111].

While on the one hand the sensory periphery may strive to remove correlations by lowering redundancy in the neural code, on the other hand there is ample evidence that the brain generates correlated signals for many different purposes, for example, in order to synchronize large-scale networks involved in attention and memory and to provide meaningful commands to the motor system.

Correlation and Its Role in Sensory Development. Genetic information is generally insufficient to provide for more than a crude topographic wiring of the receptor surface onto the cortex in the form of retinotopic maps in visual cortex, tonotopic maps in auditory cortex, and body surface somatotopic maps in the somatosensory cortex. More elaborate topographic orderings, such as those for orientation sensitivity in visual cortex or auditory space, which need to be computed from information at the two ears, in midbrain structures require self-organizing processes based on spontaneous firing [826], lateral inhibition [133], and the ability to change the connection strengths (i.e., that of synapses) between neurons based on correlation of incoming and existing activity [125]. Most rules for modifications in synaptic strength assume that the change is proportional to the change in the correlation or covariance of two neural activity patterns [241]. Evidence for the necessity of temporally correlated activity and the formation of and changes in somatosensory maps [174], tonotopic maps [56, 57, 998], and visual maps [573] is now provided in abundance. In many of these cases the maps and changes therein are guided by experience.

A similar phenomenon may serve to align topographic maps of space for different sensory modalities. Knudsen [491] has shown that auditory and visual maps of young barn owls raised with one ear plugged are in register although the binaural input is distorted. Removing the earplug resulted in a shift of the auditory map of space with respect to the visual one. Apparently, an alignment of the representation of the auditory and visual space occurs in the midbrain (optic tectum) during the early periods of life. Since auditory responses are mainly found in the deep layers of the optic tectum, that is, the visual midbrain, where visuomotor units are also found, it is suggested that localizing sound (auditory) together with the sound sources (visual) results in the alignment of auditory and visual spatial maps. This requires either multimodal neurons [615] or separate sets of auditory and visual neurons with some form of neural interaction between them [778].

The appraisal of the outside world by our senses is done in parallel fashion and results in neural activity patterns organized in topographic maps of the brain. Correlation of nervous activity takes place at many levels. First of all there is the single neuron level. Neurons that have overlapping RFs, defined as the set of sensory receptors that significantly affect its rate of firing, will show a covariance in overall firing rate as well as a coincidence in the occurrence of spikes [250, 886]. Thus, a visual RF is an area on the retina, a somatosensory RF is an area on the body surface, and an auditory RF is a range of sound frequencies.

In the auditory system, overlap of STRFs or a difference in CF, the most sensitive frequency in the RF, is a strong indicator for neural correlation [117, 244, 250] with STRF overlap explaining nearly 40% in the variance of the peak correlation (Figure 1.13). The STRFs represent both the frequency range of sensitivity and the temporal window in which neural activity is elicited. The visual equivalent is the spatiotemporal RF combining the spatial area of sensitivity with the temporal window of evoked neural activity. Usually STRFs are measured using a broadband continuous stimulus such as dynamic ripple noise [486] or multifrequency stimuli consisting of randomly presented gamma tone pips [250]. In the example shown in Figure 1.13, tone pips for each of 81 frequencies over 5 octaves were randomly presented according to a Poisson process, with similar average rate but different realization for each frequency. The topographic mappings will assure that the strength of coincident firing will decrease with distance. Coincident firings are a subset of the firing of the neurons involved and may function to extract relevant information from the neuronal "noise" and improve the SNR [250]. In the visual cortex, neurons in different areas or even different hemispheres show correlated neural activity when they have similar orientation sensitivity or originate from the same eye [260, 502].

Another example, from the field of auditory disorders, where neural synchrony plays an important role is the phenomenon of tinnitus [380], a sensation of a hissing or ringing sound in the ear, in the absence of stimulation. The auditory system has the highest spontaneous activity of all sensory systems, yet we normally do not

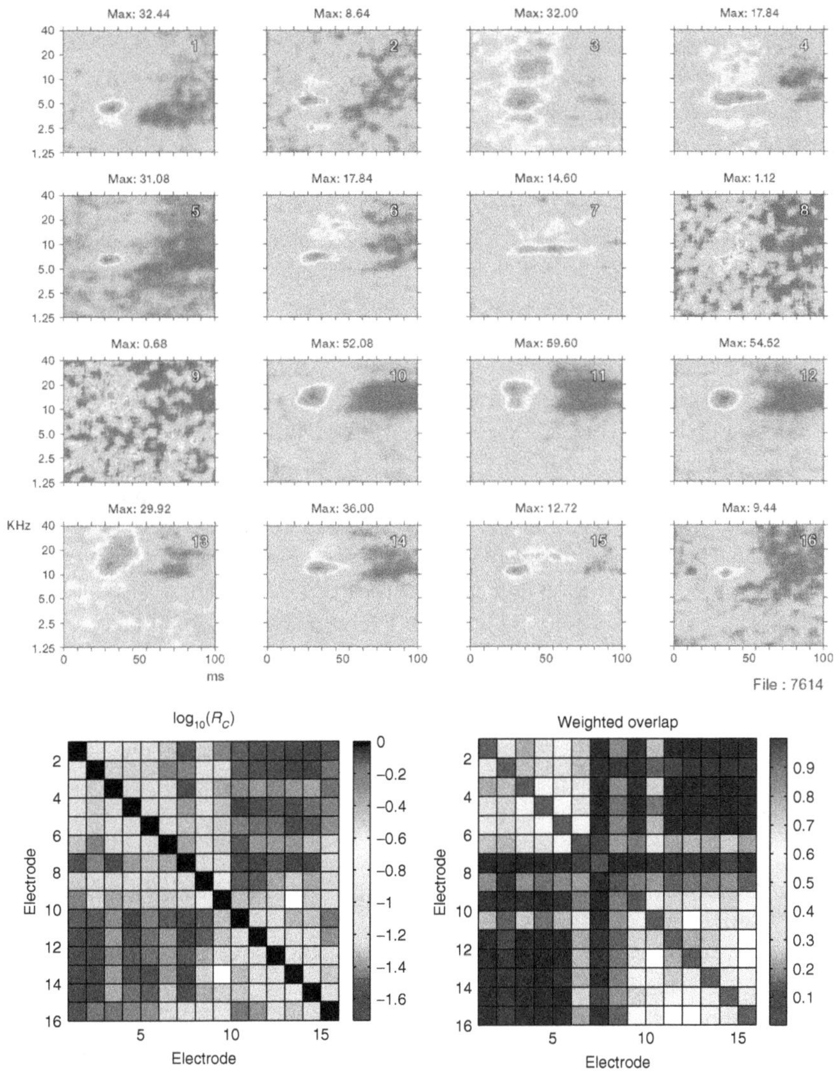

Figure 1.13 The top 16 panels indicate individual STRFs recorded simultaneously with 16 electrodes in the auditory cortex. The left bottom panel shows the $\log_{10}$ of the peak cross-correlation coefficients (R_C) between the spike trains of all electrode pairs. The right bottom panel shows the same for weighted STRF overlap between the STRFs of all electrode pairs.

hear our spontaneous activity, presumably because its firings are not correlated across auditory nerve fibers [445] and very little among nonneighbor auditory cortex neurons [242]. Tinnitus is often present following noise-induced hearing loss, something that is more and more prevalent due to the addiction of a substantial part of the population to overly loud recreational sounds. In contrast to industrial noise exposure, the recreational noise levels are not regulated by any legislation. Animal models of tinnitus usually show a specific reduction in the spontaneous firing rate of auditory nerve fibers tuned to the frequency range of the hearing loss. Tinnitus can be understood on the basis of increased synchrony among neurons [253] in the central auditory system accompanied by increased spontaneous firing rates [463]. Computational models [220, 794] suggest that tinnitus is a byproduct of the action of relatively slow homeostatic mechanisms that tend to stabilize firing rates among neurons [896, 897]. When these homeostatic mechanisms operate at the network level, by upregulating lateral excitation and downregulating lateral inhibition, such models then generate not only increased spontaneous activity but also increased synchrony between neuronal firings [220].

In addition, at the onset of vision, mammals show ocular dominance columns in visual cortex and a segregation of input from the two eyes in different layers in the visual thalamus (i.e., LGN). However, long before onset of vision this layer separation does not exist and cells in the LGN receive input from both eyes. By blocking the formation of action potentials in the retina using tetrodoxin, which blocks sodium channels needed for action potential initiation, the segregation of individual eye input in individual LGN layers is prevented. Similarly, by electrically stimulating both optic nerves synchronously, the segregation is also prevented. Finally, after blocking activity initiation in the retina by tetrodoxin but electrically stimulating the individual optic nerves asynchronously, the segregation of eye input does occur. This all suggests that neural activity that is correlated in individual retinas but asynchronous between retinas is what drives the formation of ocular dominance layers in the LGN, and similarly the formation of ocular dominance columns in visual cortex. This correlated neural activity has to be generated in the retina before the onset of vision (reviewed in [826]).

Correlative Activity in Prevision Retina Shapes Ocular Dominance and Orientation Columns. In a series of studies by Shatz and colleagues, by using a hexagonal sized multielectrode array with 61 electrodes spaced 50 μm apart, recordings were made from up to 100 individual ganglion cells showing bursting periods of about 5 s long and interspersed with 1–2 min of silence. For neighboring electrodes the bursting occurred synchronously, that is, the cross-correlogram of the action potential firings was centered around zero and showed a peak with a half time of about 2.5 s. For more distant electrode pairs the correlogram peak was displaced by a value commensurate with diffusion of an excitable substance at a speed of about 0.2–0.6 mm/s [127, 613]. This results in a wavelike activity pattern that spreads across the retinal surface; the pattern can start anywhere and propagate in any direction. The determining factor is in what direction the local percentage of recruitable cells, that is, those that are not refractory, is large enough to participate in a wavelike activation. This required percentage for wave propagation turned out to be about 30% in modeling studies [127]. The other crucial aspect is a fairly long refractory period (here about 1–2 min). Crucially, these propagating waves disappeared at the onset of vision and the strength of the correlated activity also decreased over the 30-day period that the waves were observed in ferrets by more than a factor 10. In adult ferrets no correlation was detected using these retinal electrode arrays. Therefore, ocular dominance columns are dependent on highly synchronized activity within but not between retinas. This bursting pattern in developing animals may also act to reinforce the topographic accuracy of retinal projections based on the exponentially decreasing peak correlation with distance between ganglion cells. A correlation-based mechanism (i.e., a Hebbian one) in the LGN and visual cortex would favor proximate ganglion cells to connect to the same area in LGN. The initially rough retinotopic projection might thus become gradually more refined with age [973]. This story suggests that correlation of neighboring ganglion cells in the retina ceases to exist once the developmental period is finished. This is not so (reviewed in [612]) because at least three types of correlations can be found between retinal ganglion cells in adult animals. Very narrow correlogram peaks (<1 ms wide) are found between neighboring ganglion cells that share gap junctions (electrical synapses). Intermediate-width correlogram peaks (2–10 ms) are found for those ganglion cells that share gap junction input from the same amacrine cell. This is the most prevalent type. Finally, ganglion cells might share common input from the same bipolar cell through standard chemical synapses and this is reflected in correlogram peaks that are 40–50 ms wide. The shared activity from amacrine cells may encompass ganglion cell clusters over a distance of up to 0.4 mm. In this case amacrine cells provide most of the overlap of the spatiotemporal RFs of the ganglion cells. Each ganglion cell will receive input from many amacrine cells, which typically have small RFs. Coincident firings between two ganglion cells require precise timing, which likely only happens when they get input from the same amacrine cell. For a population of ganglion cells to fire in coincident fashion, they need to have at least one amacrine cell in common. A coincident-spike RF then could reflect that of this particular amacrine cell [796]. Thus, it is likely that it is only the wave pattern of synchronous activity that disappears after eye opening. Orientation columns and directional sensitivity are also formed before eye opening but require subsequent visual activity to maintain these higher order RF properties. That the driving force again is locally synchronized activity in the retina or optic nerve will not be a surprise. Both rearing in darkness and rearing in the absence of visual contrast result in weakening of the individual cortical neuron's selectivity for stimulus orientation and movement direction. Inducing a global synchrony, by electrically stimulating the optic nerve periodically does the same: The individual neuron's orientation and directional responses are greatly reduced. And although orientation columns can still be demonstrated using optical imaging, they lack the normal salience [947].

Formation of Tonotopic and Retinotopic Maps. Higher than normal pair correlations thus appear to be present during map development. When they are local, their effect is beneficial, whereas when they are global, it generally disrupts normal map formation and neuronal properties. Another example where local pair correlations decrease in strength with age is found in area A1 of the cat. Here, the peak strength for neighboring unit pairs recorded on the same electrode remained high up to postnatal day 50 (P50) and then decreased with age [242]. This decrease starts around the same time that frequency-tuning curve bandwidth starts to increase in dorsal and ventral parts of A1 [103] and when the percentage of γ-amino butyric acid (γABA)–ergic neurons in A1 starts to decrease [306]. The tonotopic map in A1 does not appreciably change from P14 until P50 albeit that its gradient (in kilohertz per millimeter) decreases, likely because of cortical growth in that time period [103]. Thus, taking into account that the cat does not hear external sounds of less than 90 dB sound pressure level (SPL) before P9 [245], it appears that there is not much refinement in the tonotopic gradient except for the increases in the bandwidth of frequency tuning in noncentral parts of A1. In subcortical structures such as the inferior colliculus the frequency-tuning curves bandwidth decreases by half over the same period. Thus these changes in frequency-tuning curve bandwidth are occurring in the auditory thalamus and/or the auditory cortex. In visual cortex, a reasonably accurate retinotopy is present at birth in monkey and at the time of eye opening in the kitten, again suggesting that visually evoked activity does not play a major role here [958]. Topographic maps, such as the tonotopic map in A1, can be affected by abnormal neural synchrony during development. Zhang et al. [998] introduced not normally occurring synchronous input to many neurons in the auditory pathway by exposing rat pups to pulsed white noise during the period between P9 and P28. They found a disruption of the individual cortical neuron's frequency selectivity and a disruption of the tonotopicity. A critical period for this effect was suggested by the fact that the same stimulation after P30 did not produce any effect.

Models of Columnar Development Based on Correlated Firings. Columnar organization in visual cortex comes in many shapes. Roughly a column is a three-dimensional structure that cuts perpendicularly through the cortical layers and keeps the same cross section in each layer. The simplest column is cylindrical in shape and is usually called a minicolumn (see Figure 1.5). It is thought to reflect the retinotopic organization in cortex so that a small region in the retina is mapped onto a small region in cortex. The cross section of such a column is also called the RF. Neighboring areas in the retina have neighboring minicolumns in visual cortex giving rise to a retinotopic organization. Spatial RFs like the retinotopic ones typically undergo a characteristic distortion when mapped upon, for example, the superior colliculus and the striate cortex. Schwartz [807–810] has described the mapping on V1 as a logarithmic conformal mapping:

$$w = B \ln(z + A), \tag{1.16}$$

where $w = u + jv$ and $z = x + jy$ represent the cortical and retinal complex coordinates, respectively. The logarithmic conformal mapping maps foveal parts of the RFs such that they are overrepresented at the expense of more peripheral parts of the retina. One of the properties of the logarithmic conformal mapping is that a multiplication of the size of the image plane—the retina—results in a translation in the projection plane—the cortex. This mapping therefore is size invariant, and angles are preserved (Figure 1.14). Thus lines intersecting at right angles on the

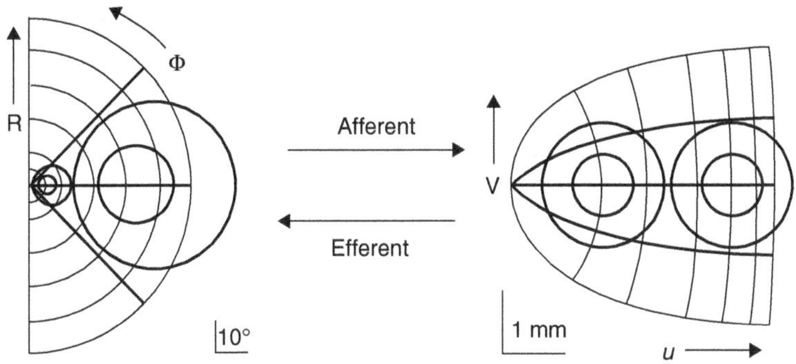

Figure 1.14 The point-to-point mapping between the external world (see half field on the left) and the superior colliculus (on the right). The mapping function used is isotropic ($B_U = 1.5$ mm; $B_V = 1.5$ mm/deg; $A = 3°$; see equations in the text). As a result, the distance from 0 to 80° eccentricity along the horizontal meridian in the colliculus map is about 5 mm. When replotted in the R, Φ coordinates of visual motor space, the corresponding circular visual (movement) fields in the superior colliculus have strikingly different sizes and noticeable skewness along the R dimension. The polar coordinate grid on the left has meridians every 45° and isoeccentricity hemicircles of 5°, 10°, 20°, 30°, 40°, and 50°. (Reprinted from *Vision Research*, Vol. 26, F. P. Ottes et al., Visuomotor fields of the superior colliculus: A quantitative model, pp. 857–873. © 1986, with permission from Elsevier.)

retina will do so in the cortical map as well. Such a mapping has also been shown to apply to the superior colliculus in rhesus monkey [694]. The mapping rule in some detail reads:

$$u = B_u \ln \sqrt{R^2 + A^2 + 2AR\cos\Phi} - B_u \ln A, \tag{1.17a}$$

$$v = B_v \arctan \frac{R\sin\Phi}{R\cos\Phi + A}, \tag{1.17b}$$

where A is the offset in degrees and R is the eccentricity in the visual motor space; B_u and B_v are two scaling constants; parameter A together with B_u/B_v determines the shape of the mapping. For numerical values see the caption of Figure 1.14.

More complex cortical columns are the ocular dominance columns, regions where input from the two eyes alternates (as introduced above). Orientation columns are defined as regions that respond optimally to a narrow range of stimulus orientations and are related to the ocular dominance columns in such a way that regions with a high spatial rate of change of orientation tend to either be aligned along centers of ocular dominance columns or intersect them at right angles.

Nearly all the neural net models for visual cortical map development, and ideally for the retinotopic, ocular dominance and orientation map combined, are based on four common assumptions: (i) Hebbian synapses, (ii) correlated and/or spatially patterned activity in the afferent nerve fibers to the cortex, (iii) fixed synaptic connections between cortical neurons that are excitatory at short distances and inhibitory at longer distances, and (iv) normalization of synaptic strength [869].

The most successful model class is that of the *dimension-reduction models*, such as elastic nets, which make use of competitive Hebbian synapses and are generally based on Kohonen's learning rules for self-organizing maps [497]. Most cortical maps are a combination of a receptor surface map (e.g., retinotopic) with a mapping of other response properties such as orientation selectivity and ocular dominance. Thus, the visual cortex maps several stimulus dimensions onto a two-dimensional surface under the constraint that shape and position of the RFs vary smoothly over the cortical surface. The stimulus dimensions are five: two for retinal position, two for orientation selectivity, and one for ocular dominance. It appears that modeling the formation of retinotopic maps requires quite different learning rules compared to modeling both the orientation and ocular dominance columns [726].

Two types of activity-driven Hebbian learning hypotheses have been used to explain topographic map formation. One is correlation-based learning, which assumes that lateral connections in cortex act linearly such that the activity of cortical neurons is related to their total afferent input. The other is the competitive neural network model where only a small localized region of cortex is active for a given afferent input for a given time and located at the region which receives the strongest total input. This is a "winner-take-all" type of model. Correlation-based learning models cannot predict the emergence of localized RFs but can predict the formation of orientation and ocular dominance columns. The competitive neural network can model all three topographic maps [673].

Specifically, in the correlation-based learning model, the activity of a neuron in layer M is a linear combination of the output of the previous layer L:

$$\mathbf{x}^M = \sum_{i \in L} \mathbf{w}_i x_i^L + \mathbf{a}, \tag{1.18}$$

and the learning rule is

$$\Delta \mathbf{w}_i = \sum_{j \in L} (Q_{ij}^L + b)\mathbf{w}_i + \mathbf{c}, \tag{1.19}$$

where Q_{ij}^L denotes the covariance matrix of points in layer L and $\mathbf{a}, b$, and $\mathbf{c}$ are constants. The competitive neural network model uses feature vectors $\mathbf{x}$ as stimuli, and a point on the cortical surface after learning represents such a feature vector. Other points j on the cortical surface have their connection weights updated according to the following learning rule:

$$\Delta \mathbf{w}_j = \eta h(r)(\mathbf{x} - \mathbf{w}_j), \tag{1.20}$$

where η is a constant and $h(r) = \exp(-r^2/2\sigma^2)$ is a circular symmetric (or elliptic shape) Gaussian function with $r = \|\mathbf{x} - \mathbf{w}_j\|$ denoting the distance between cortical point j and the cortical point closest to the ideal representation of stimulus $\mathbf{x}$. Here the rate of change is proportional to the current level of presynaptic activity as reflected in $(\mathbf{x} - \mathbf{w}_j)$ but conditional on the synapse being close [as reflected in $h(r)$] to the region of cortex that responds most strongly to the stimulus.

It appears that the correlation-based learning model and the competitive neural network model are two extremes of a model with variable intracortical competition. Recapitulating, a Hebbian learning rule can be formulated under the assumption of synaptic competition between cortical neurons. In the limit of weak competition the system becomes linear, no topographic projection emerges, and only the correlations of the input patterns, together with the shape of the interaction function, determine the developing RFs and neural maps. For strong competition a topographic map emerges. In other words, the correlation-based learning model and the competitive neural network model are two extremes of a model with variable intracortical synaptic competition [726].

1.6 CORRELATION IN MEMORY SYSTEMS

Correlation in neural coding and processing is likely to be used in rather different ways by different memory systems. Posterior neocortical areas organize and represent sensory information roughly into hierarchies of topographically organized feature maps, as discussed in the preceding sections. This type of representation may be optimal for extracting and representing the overall statistical structure of the sensory world averaged over a long timescale. The medial temporal lobe (MTL) memory system, on the other hand, exhibits strikingly different properties and plays a unique role in episodic memory. In spite of these differences in timescale, all forms of long-term memory including episodic memory have a common neural mechanism, namely, synaptic plasticity. In contrast, short-term or working memory represents transient information that is stored on a timescale of seconds to minutes and may not involve synaptic changes.

Working memory allows information to be maintained in an active state over a period of seconds to minutes in the form of persistent neural activations. It is thought to result from dynamic interactions between frontal and posterior cortical memory sites. For example, single neurons exhibiting similar, persistent firing during the delay period in spatial working memory tasks have been recorded in both prefrontal and parietal cortices [153]. Functional imaging studies implicate the coactivation of frontoparietal and frontotemporal networks in working memory for auditory [44, 598] and visual [175, 188], information with the specific subregions within these areas dependent on the specific task. Correlated, synchronous firing may play a role in the maintenance of information in working memory. For example, gamma oscillations recorded in electroencephalography (EEG) increase in power in proportion to memory load in a working memory task [407]. Magnetoencephalography (MEG) studies reveal long-range phase synchronization in a network of prefrontal, parietal, and temporal areas implicated in the maintenance of working memory, with the degree of synchronization varying with the attentional demands of the task [341].

Episodic memory refers to memory for specific events situated in particular places and times. In contrast to posterior cortical areas, which recode the sensory signal so as to extract general features and abstract away the specific details of each learning episode, the episodic memory system codes and retrieves the specific details of particular episodes. Thus, the key properties of an associative memory system can be observed: It binds together the elements of a complex event into a single memory trace and subsequently can perform pattern completion, whereby a subset of the elements of the original memory trace can cue the retrieval of the entire episode. The MTL memory system, comprised of the hippocampus and surrounding MTL structures (parahippocampal region, subiculum, entorhinal and perirhinal cortices), is crucial for episodic memory. Hence, people with MTL lesions show severe retrograde and anterograde amnesia for episodic details, while semantic, perceptual and procedural memory, and simple forms of conditioning are spared (e.g., [637]).

Whereas stimulus correlations on a long timescale appear to drive the learning of long-term memory representations in neocortical systems, the hippocampal system must be able to ignore correlations between different events, so as to create unique memory traces for otherwise similar events. For example, the act of driving to work and parking one's car in the same parking lot is extremely similar from one day to the next, and yet finding one's car at the end of the day requires the retrieval of a unique event memory for where the car was parked most recently. Thus the goal of learning in an episodic memory system should be to capture instantaneous correlations between inputs as accurately as possible while decorrelating separate events into nonoverlapping memory traces. Three unique features of the hippocampal memory system account for its ability to carry out this goal: (i) *Anatomy:* Being reciprocally connected to most cortical and subcortical regions, the hippocampal system is well positioned to bind together features from the various cortical maps as well as emotional and other contextual information. It is a massive convergence zone, in terms of both its incoming and outgoing connections. Moreover, the circuitry of the hippocampus, illustrated in Figure 1.15, is strikingly different from that of neocortex. Most notably, the extensive associational fiber pathways, including the CA3 recurrent collaterals and CA3-to-CA1 Shaffer collaterals, may help the hippocampus to perform associative memory and retrieval functions. (ii) *High plasticity:* The ability to encode complex associations ultrarapidly or even with one-shot learning necessitates a very high level of plasticity. Electrophysiological support for this comes from studies of LTP. Long-term potentiation can be induced in the hippocampus within a single session of high-frequency electrical stimulation, whereas neocortical LTP in the awake behaving animal occurs on a much slower timescale, requiring multiple sessions spaced over several days and taking about 15 days to reach asymptotic levels [741]. (iii) *Sparse coding:* Neurons within the hippocampus, under the tight control of inhibitory interneurons, fire at extremely low rates [63, 453]. Using sparse codes, on average, decreases the amount of correlation between neurons and results in highly specific coding at the single-neuron level. For example, hippocampal "place cells" each fire when a rat is in a particular location within a given environment and tend not to respond in other locations in the same environment or in similar locations in different environments. Place-selective neurons have also been recorded in the nonhuman primate [600, 689] and in the human hippocampus [256].

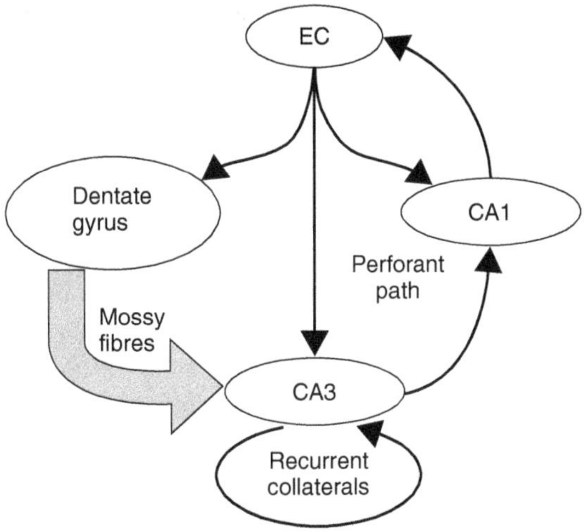

Figure 1.15 The major regions and pathways within the hippocampus. The entorhinal cortex (EC) is reciprocally connected to most cortical areas and in turn projects to all regions within the hippocampus. In addition to receiving direct input from the EC, each region is connected in series in what is known as the trisynaptic circuit, from the EC to the dentate gyrus, CA3, CA1, and back to the EC, thus completing the loop. Some of the most striking and unique anatomical features of the hippocampus are the presence of the mossy fibers, which have the largest and most potent synaptic connections in the mammalian brain, and the massive system of recurrent collateral connections within the CA3 region.

Sparse Coding and Pattern Separation. In 1971 Marr [591] put forward a highly influential computational theory of hippocampal coding. The central ideas in this theory included a rapid, temporary memory store mediated by sparse activations and Hebbian learning, an associative retrieval system mediated by recurrent connections, as well as a gradual consolidation process by which new memories would be transferred into a long-term neocortical store. Many subsequent models have built upon these same basic principles (e.g., [71, 358–360, 460, 550, 604, 605, 610, 693, 769, 892]). Computer simulations have demonstrated that sparse coding serves to remap the input from a space in which many correlated features are present to a new space in which feature correlation is minimized, a process that has come to be known as "pattern separation" [692]. Hence sparse codes are optimal for creating unique event memories with minimal overlap between different memories [591]. When overlap is minimal, interference between different memories is reduced. This allows the hippocampus to employ a very high level of plasticity without suffering unduly from interference [605].

Associative Learning and Pattern Completion. Largely based on predictions from computational models, it is now widely assumed that the associational pathways within the hippocampus, especially recurrent connections within the CA3 region, explain its capacity for pattern completion, that is, the cued retrieval of a complete memory. In support of this notion, place cells maintain their place selectivity even in total darkness [609, 740]. Moreover, knockout mice having selective loss of NMDA receptors in CA3 show normal acquisition and normal place fields in spatial memory tasks but a loss of place-selective firing after partial cue removal [652].

Temporal Sequence Learning and Consolidation. Hippocampal neurons not only encode static "snapshot" memories, they also encode temporal sequences. Strong evidence for this claim comes from cellular recordings suggesting replay of recently experienced event sequences. When a rat runs through a series of locations in an environment, a corresponding series of place cells fire in sequence. Interestingly, during the next few hundred milliseconds, the same set of place cells tend to fire in the same sequence, albeit on a compressed timescale [840]. Similar patterns of very brief temporal sequence replay have also been recorded during slow-wave sleep [535, 648, 712, 964], and very long sequences of up to 2 min have been recorded during rapid-eye-movement (REM) sleep [572], while replay of sequences in reverse order has been reported during periods of inactivity immediately following periods of locomotion in the awake state [287]. One hypothesis as to the functional significance of this sequence replay is that the hippocampus may be involved in long-term memory consolidation, consistent with Marr's theory of hippocampal function as a temporary memory system. The hippocampus, being ideally suited for rapid memory acquisition, could act as a cache for memorizing recently experienced sequences. The repeated replay of these sequences and their corresponding correlated firing patterns in the neocortex via hippocampal–cortical back projections could allow slower, correlation-based learning mechanisms in the neocortex to process the newly learned information. The consolidation hypothesis is controversial, particularly with respect to human episodic memory (e.g., [638]) and animal spatial memory [593], but is supported by evidence from animal lesion studies for at least some types of learning [848].

Oscillatory Firing within the Hippocampus. The hippocampus exhibits at least two distinct modes of firing that appear to have very different behavioral correlates (for a review, see [170]). One mode is highly synchronized to the so-called theta rhythm, a slow, regular rhythm of about 4–8 Hz associated with active wakeful periods and REM sleep. A second predominant mode is characterized by irregular, high-intensity "sharp-wave" events and is seen during quiet wakeful activities as well as in slow-wave sleep. Yet a third mode, the "hippocampal slow oscillation" ($\leq$1 Hz) has also been reported [967]. These different rhythms may help to organize the activities of different subregions of the hippocampus into synchronously firing assemblies to promote learning and memory operations. They may also serve a similar function with regard to coordinating hippocampal–cortical interactions.

During theta mode, hippocampal neurons fire in gamma frequency volleys (40–100 Hz) strongly modulated by the theta rhythm [170]. A number of lines of evidence suggest that the theta oscillation may serve to lock the hippocampus into a memory acquisition/retrieval mode. For example, in many species, theta occurs during alert attentive and exploratory behaviors and REM sleep, but not during quiet waking or consummatory behaviors [965]. Hippocampal theta oscillations have also been observed in human intracranial field recordings during exploration and goal-seeking behavior in a virtual environment [140] as well as during REM sleep [139]. The neuromodulator acetylcholine regulates theta oscillations, and cholinergic agonists enhance both theta oscillations and plasticity [417]. Further, it is the superficial layer neurons of the entorhinal cortex, which convey cortical inputs into the hippocampus, that fire predominantly during theta, whereas the deep-layer neurons responsible for conveying hippocampally generated signals back to the cortex fire more weakly during theta mode [169]. Additionally, electrical stimulation in theta frequency bursts is optimal for LTP induction [526]. Finally, the direction of plasticity is linked to the theta phase: Neurons that fire in phase with theta undergo LTP while neurons firing in antiphase undergo LTD [398, 418, 423, 711]. Thus, it appears that encoding of new information is performed optimally when the incoming information is maximally correlated through the synchronizing effect of the theta oscillations. A related hypothesis is that the hippocampus rapidly switches between encoding and retrieval modes within each theta cycle [357]. In support of this hypothesis, human intracranial recordings have revealed that neurons in numerous brain locations exhibit theta-phase reset during both encoding and retrieval (in response to presentation of study items and test items, respectively), but firing during retrieval is nearly 180° out of phase with that seen during encoding [763]. Whether there is rapid alternation between retrieval and encoding during each theta cycle or switching between these states at a longer timescale remains to be seen.

The theta oscillation may also serve to coordinate assemblies of neurons involved in temporal coding of information. This could explain the phenomenon of theta-phase precession [683]: When a rat first enters the place field of a given place cell, the cell fires at a relatively late phase in the theta cycle, and as the rat moves through the place field, the cell fires at progressively earlier phases. One explanation for phase precession is that strong lateral synaptic connections are formed between neurons coding for nearby places, causing place cells whose fields are about to be entered to receive some lateral activation from place cells whose fields are already occupied [128]. An alternative explanation for phase precession is that it reflects a precise use of spike timing to encode for spatial location, and it is caused by temporal properties of the hippocampal input [681] rather than by lateral interactions. In this view, the theta rhythm would serve as a clock for coordinating the relative timing of neurons, with phase offsets coding for specific place-related information while firing rates correlate with other variables such as running speed [422].

In addition to coordinating circuits within the hippocampus, the theta oscillation may function to coordinate multiple brain regions into functional networks for different purposes. For example, individual neurons in medial prefrontal cortex fire in synchrony with hippocampal theta during foraging and running [424], while hippocampal neurons fire in synchrony with amygdala neurons at theta frequency during retrieval of fearful memories [813]. Increased theta synchrony has been observed in human EEG recordings across prefrontal, MTL, and visual areas during recollection of images compared to object recognition [346]. Thus, correlated firing across multiple brain regions by synchronization to the theta rhythm may be a general principle by which selective regions are recruited into specific tasks.

In the absence of theta activity, during silent wakeful and consummatory behaviors and slow-wave sleep (SWS), hippocampal neurons generate "sharp waves" coinciding with high-frequency "ripples," causing the output region of the hippocampal–cortical interface (deep layers of entorhinal cortex) to fire in synchrony with these sharp wave/ripple events [129, 167, 168]. During sharp waves, the hippocampus appears to act in "top-down" mode, sending information back out to cortex. Sharp waves are generated and propagated via the associational pathways within the hippocampal circuit—the CA3 recurrent collaterals and CA3–CA1 Shaffer collaterals; they arise from highly synchronized population bursts generated in the CA3 triggering aperiodic, high-intensity dendritic field potentials (sharp waves) lasting 40–100 ms in CA1 [167]. Coinciding with the sharp wave events, a 200-Hz oscillatory field potential synchronizes spiking in CA1. The deep layers of the entorhinal cortex, subiculum, and parasubiculum fire strongly during sharp-wave events, propagating activity back out to cortex, while the superficial layers are relatively silent [167]. Thus, during each sharp-wave event, there is a powerful synchronization of the neural circuits connecting the hippocampus to the neocortex [167, 168]. It is widely believed that sharp-wave events during SWS represent hippocampal replay of recently acquired memories for the purpose of memory consolidation. In support of this notion, acetylcholine levels are naturally lowest during SWS, and consolidation of recently learned word pairs is blocked when a cholinergic agonist is injected during SWS but not when the injection occurs during waking periods [305].

As a computational neural network model, the Boltzmann machine, which was first proposed by Hinton and Sejnowski [6, 390], operates in alternating bottom-up and top-down modes reminiscent of the hippocampal–cortical interactions during theta versus sharp waves, suggesting a computational explanation for these two modes. During the "waking" phase, the circuit operates in a data-driven mode and learns to represent the current input pattern via Hebbian learning. During the "sleep" phase, the circuit operates in a generative mode, triggering activity states that may be similar to recently learned memories but that should be "unlearned" via anti-Hebbian learning. Recently, Kali and Dayan [460] developed this idea further in a Boltzmann machine model that demonstrates the feasibility of a rapid memory store within the hippocampus promoting memory consolidation in the neocortex. Moreover, Kali and Dayan proposed that the periodic replay of hippocampally stored memories is required to keep the neocortical and hippocampal memory representations in register with one another.

1.7 CORRELATION IN SENSORIMOTOR LEARNING

Sensory and motor systems are closely connected and form an intrinsic sensorimotor loop/cycle. On the one hand, the motor primitive commands are driven by the sensory events, and on the other hand, the motor events influence the forthcoming sensory inputs. Learning occurs at many levels of the brain from simple conditioning to the learning of complex sensorimotor response mappings. Correlations between sensory inputs and outcomes underlie all types of sensorimotor learning. Specifically, temporal difference learning theory (e.g., see [201]) has proven to be a powerful model for capturing these many levels of learning.

Temporal-Difference (TD) Models of Classical Conditioning and Their Relation to STDP. We remind the reader that a CS is the originally neutral stimulus that, after training, comes to elicit the learned behavior (the CR). This change results from the experimenter-enforced arrangement that the CS temporally precedes and thus predicts reinforcement of the US. Unit firing patterns that develop during conditioning and are specific to the CS (i.e., are absent in response to stimuli that do not predict the US) represent learning-relevant neuronal plasticity. Thus, a CS is a convenient probe for detecting neuronal plasticity, thereby permitting identification of regions and circuits involved in learning processes. In the classical Pavlovian experiment, the CS is the sound of a bell, the US is a plate of food, and the CR is the salivation by the dog involved. Previously, in Section 1.2, we presented an example where the CS was the firings of a presynaptic neuron, the CR was the response of the postsynaptic neuron, and a sound that activated the presynaptic neurons served as the US. The *Rescorla–Wagner rule* [758] for classical conditioning states that "organisms can only learn when events violate their expectations," namely, learning only occurs when the information content of the message is high. In the conditioning experiment, the sound of a bell (although not uncommon to call people to natural or spiritual food) is initially highly unlikely to indicate food for a dog, so it violates expectations. Specifically, the learning rule can be written as

$$\Delta w_{ij}(t) - \eta[r_j(t) - v_j(t)]x_i(t), \qquad (1.21)$$

where $r_j(t)$ is the value of a specialized training signal that induces the desired response of the unit and $v_j(t) = \sum_j w_{ij}(t)x_i(t)$ is the neuron's standard response to an input $x_i(t)$. However, the above model used in animal learning is not a TD model, but all of its predictions can be obtained from TD models.

Specifically, TD models relate to the effect of interstimulus interval between a CS and US and thus cover the class of STDP synaptic plasticity models in principle. In order for an association between the CS and US to occur, they must occur roughly at the same time. This is reflected in an update of association strength, Δw, according to the product of the levels of CS and US processing. Sutton and Barto [867] introduce "reinforcement" as the level of US processing and "eligibility" as the level of CS processing; thus, the learning rule is written as

$$\Delta w = \text{reinforcement} \times \text{eligibility}.$$

Eligibility is always positive, but reinforcement can be either positive or negative. In this context, the Rescorla–Wagner rule can be rewritten as

$$\Delta w_i = \eta(r - V)\beta_i x_i, \tag{1.22}$$

where $r - V$ denotes the reinforcement and $V = \sum_i w_i x_i$ is the predicted signal and $\beta_i x_i$ denotes the eligibility. When $x_i = 0$, then the ith CS, denoted as CS_i, is absent; when $x_i = 1$, CS_i is present. A subsequent improvement over the Rescorla–Wagner model is that all CS in a given trial can produce associative strengths with total sum that is given by $Y = \sum_t r(t) - V(t)$; then the reinforcement is postulated to be equal to dY/dt, which further leads to

$$\Delta w_i = \eta \frac{dY}{dt} \beta_i x_i. \tag{1.23}$$

However, a drawback of this model is that it still does not perform well for long interstimulus interval. As we have seen, classical conditioning can be viewed as a manifestation of the subject's attempt to predict the arrival of the US. In terms of the Rescorla–Wagner model, V is the predicted US level on the trial and r is the actual US level. The difference $r - V$ drives the learning process as the model's reinforcement term. How is this to be extended to a time-dependent form? Here r will change with time within a trial and each successive delayed US within the same trial will have a forgetting effect γr $(0 \leq \gamma < 1)$, and the total prediction becomes nonlinear, $V(t) = r(t+1) + \gamma r(t+2) + \gamma^2 r(t+3) + \ldots$. This results in a modification of the reinforcement term as follows:

$$\Delta w_i = \eta \left[r(t+1) + \gamma V(t+1) - V(t) \right] \beta_i x_i (t+1). \tag{1.24}$$

The reinforcement term can be interpreted as a difference between two predictions, one at time $t+1$ and the other at time t, and can be considered as a discrete-time analog of the prediction dV/dt. This TD model performs well, but its relationship to the STDP synaptic plasticity model is only qualitative. The STDP model works over timescales of the order of 100 ms whereas classical conditioning operates over timescales of one order of magnitude higher. It is believed that in classical conditioning many brain areas are likely involved, with polysynaptic and recurrent circuitry that can effectively extend the timescale of computation without sacrificing the requirement of a very short STDP window at each synapse [88]. Drew and Abbott [229] suggested that cortical up and down states or changes in the spontaneous background activity might generate the required extension of the correlations and the asymmetry in the potentiation and depression part on the behavioral timescale.

Neural Adaptive Information Processing. How does learning alter neuronal representation of sensory events, that is, when animals acquire information about their behavioral significance and meaning? This approach entails an analysis of the responses of neurons to the CS within the appropriate sensory system.

Classical conditioning in the auditory system [194] produces highly specific modification of RFs in auditory cortex. In this case, responses to frequencies near the boundaries but still within the original RF are strengthened at the expense of those to frequencies in the center of the RF. The strengths of the lemniscal input to pyramidal cells are continuously adjusted during learning, whereas the nonlemniscal auditory influence produces a diffuse increase in the excitability of these neurons throughout auditory cortex. Here we note that lemniscal inputs to the cortex synapse generally in layer IV, whereas nonlemniscal inputs synapse with pyramidal cells in layer I. The latter effect is greatly modulated by acetylcholine released diffusely in cortex by electrical stimulation of the basal forebrain. Specificity of RF plasticity (or representational plasticity) is said to result from a modified Hebbian rule, so that increased excitability strengthens some synapses and weakens others within the same neuron.

Effects of Modulatory Neural Systems. Direct neural transmission uses glutamate and GABA as transmitter substances to excite and inhibit, respectively, the receiving neurons. In addition to the direct-acting neurotransmitters, there are numerous modulatory substances including, among others, acetylcholine, dopamine, and serotonin. Modulatory transmitter substances are released in cortex diffusely by systems originating in the forebrain or brainstem. Acetylcholine is released by the nucleus basalis in the basal forebrain, whereas dopamine is released by neurons in the ventral tegmentum area and substantia nigra. Natural release of these modulatory substances during learning or behavior can be mimicked in experimental conditions by electrical stimulation of the releasing structures. The modulatory effects are studied by pairing electrical stimulation of the relevant brain structure with an acoustic stimulus. Pairing nucleus basalis stimulation with narrow-band stimuli for several weeks enhances cortical representation of the sound frequencies [475]. Specifically, pairing nucleus basalis stimulation with periodic tone-pip trains with a rate (15 Hz) above the normal range in cortex enhances responsiveness at this rate; and pairing with periodic tone-pip trains at 5 Hz makes the cortex less responsive at rates it would normally process. This depended on the tone frequency; when it was randomly changed, the effect occurred as described; when the tone frequency was fixed, there was no effect [476]. The effects of the cholinergic and the dopaminergic systems on cortical plasticity are different. Dopaminergic activity may enhance or reduce cortical representation depending on the stimulus contingency [56]. Thus ventral tegmental dopaminergic activity may be essential in contingency-based associative learning. Dopamine has long been thought to play a role in reinforcement learning. Phasic firing of dopamine neurons exhibits a striking correlation with TD error (a reward prediction error signal): A first exposure to a US evokes strong dopamine firing, whereas after repeated CS–US pairings the dopamine response transfers to the CS and is much weaker upon arrival of the US, signaling a near-zero prediction error [805]. On the other hand, when an anticipated reward fails to arrive, dopamine neurons fire below baseline levels, signaling a negative prediction error. Thus the actions of dopamine may constitute the brain's implementation of the TD learning algorithm [64, 865, 866].

Cholinergic effects, unlike dopamine, are largely determined by the spectral and temporal characteristics of the paired sensory stimulus [477, 478]. The cholinergic system could thus be more engaged in stimulus feature-directed perceptual learning [57]. In addition to direct effects on plasticity, acetylcholine suppresses synaptic transmission in cortical feedback pathways [322, 359, 411, 482]. It has been suggested that under conditions of uncertainty, such as when an unexpected stimulus arrives or under high noise levels, acetylcholine upregulates bottom-up, thalamocortical transmission of information, leading to optimal inference and learning [992].

Behavioral Training-Induced STRF Changes. Auditory tasks can influence RF properties of cortical neurons. In a series of such experiments, ferrets were trained on target tone detection and two-tone discrimination tasks and on gap detection and click-rate discrimination. STRF changes were measured online during task performance and facilitative changes occurred within minutes of task onset. During frequency detection or discrimination tasks, there were only spectral changes in the STRF. However, during and following temporal tasks, the STRF showed sharpened temporal aspects [295]. The fact that RF plasticity occurs during very different tasks and learning situations suggested that it represents a general process of information storage and representation [945]. Cortical RF plasticity can be induced within a few trials and the changes paralleled the appearance of the first behavioral signs of learning [237]. Receptive field plasticity has a short-term, fast-learning component which is only demonstrable in a behavioral context [217]. Polley et al. [734] tested whether topographic map plasticity in the adult auditory cortex was controlled by sensory inputs alone and/or by task dependence. Rats trained to attend to intensity cues had an increased proportion of units that were tuned to the target intensity range but showed no change in tonotopic map organization. The degree of topographic map plasticity within the task-relevant stimulus dimension was correlated with the degree of perceptual learning for rats in both tasks.

Learning More Complex Sensorimotor Mappings. According to computational models, learning to associate together a CS and a US can be accomplished by a single neuron. In contrast, learning to perform a complex motor task such as reaching for an object requires the coordination of many muscle groups and numerous brain regions. One way to study how animals accomplish such tasks is to perturb the system. Under such perturbations, the brain must learn to compensate by recalibrating the sensorimotor mapping. For example, it is well known that one can adapt to distorting prisms, even when they completely reverse the visual image, and that the adaptation persists for some period of time after removal of the glasses. Interestingly, wearing laterally displacing prisms during walking results in plasticity, which generalizes to reaching movements [636]. In the auditory domain, when people are fitted with an artificial pinna (outer ear) that perturbs the binaural cues for sound localization, they adapt and eventually regain normal auditory sound localization [396]. However, in contrast to the case of the visual system, upon removal of the artificial pinna the system immediately reverts to its original configuration, demonstrating a capability for maintaining two sets of mappings simultaneously.

It is thought that the difference between the predicted and actual consequences of one's actions is the error signal driving such learning processes. For example, when one puts on distorting prisms and tries to reach for an object under visual guidance, misreaching occurs. The predicted consequence of one's action does not match up with the visual percept. This error is a potent signal for driving plasticity and has been capitalized on in treatment of several neurological disorders. For example, prism adaptation has been used to treat hemispatial neglect [775]. Neglect patients, usually as a result of a right parietal stroke, fail to attend to objects in the left side of space even though they are able to process all of the visual information. Adapting to prisms that shift the visual field in the direction of the "good" side of space results in improvement in neglect symptoms, not only while the prisms are worn but up to 2 h after removal of the glasses. The same principle of altered visual feedback has been used in the treatment of phantom limb pain. The patient is shown a projected, mirrored image of his or her good arm in the place where the paralyzed arm should be and observes the limb movements during an adaptation period, thus providing altered visual feedback that can recalibrate the sensorimotor circuits to some degree; this treatment has been reported to alleviate phantom pain in some patients [746]. Numerous computational models of learning are also based on prediction error. In the case of the TD model [64, 865, 866], the error signal is the difference between received and predicted reinforcement. The TD learning model has been extended to model learning motor actions in more complex domains. The Q-learning algorithm [942] involves learning to predict future reinforcement obtained when choosing among a set of alternative actions. In this case, the agent must learn to predict how much reinforcement can be expected for each combination of actions and states in the environment.

1.8 CORRELATION, FEATURE BINDING, AND ATTENTION

It has been widely believed that very strong correlations in the sensory signals are likely to be learned early in development, effectively becoming hard-wired into neural circuits at early stages of processing. Within spatially localized regions of the visual field, features such as color, orientation, and texture are strongly correlated across space; within the auditory domain, spectral properties are correlated across time. It appears that some neurons are tuned to combinations of such correlated features. For example, people are about 10 times faster at detecting combinations of orientation and color if they are superimposed within the same visuospatial region relative to when they are spatially separated [397]. These findings suggest that correlated features within the same location are coded in combination explicitly at an early stage of processing. However, employing a different neuron to encode every possible combination of stimulus features is clearly infeasible. Thus combinatorial coding cannot be the only solution employed by the brain for detecting stimulus correlations.

How are correlated features detected when there is no neuron already tuned to a given feature combination? The *temporal correlation theory* [922] posits that when features activate a population of neurons coincidentally this temporal coincidence in firing leads to the emergence of a synchronously firing cell assembly. Moreover, von der Malsburg proposed that such transient correlation patterns may become temporarily strengthened by rapid reversible synaptic plasticity. Further, it has been proposed that multiple synchronously oscillating cell assemblies could represent coherent groupings of parts of the same object, offering a solution to the binding problem, a central problem in perception [235, 337] A substantial body of evidence supports a role for synchronization in bottom-up feature integration (for a review, see [871], although there is also evidence to the contrary, as reviewed in Section 1.3). For example, in cat visual cortex, unit recordings revealed periods of oscillatory spike synchronization between neuron pairs in two different visual areas (striate and extrastriate cortices); moreover, the synchrony was greatest for coherent motion of a single stimulus that activated both RFs, weaker for a pair of coherently moving stimuli each within one of the pair's RFs, and weakest for independently moving stimuli [260]. It is unclear how a pattern of synchronous firing could give rise to a categorical object percept, and thus the theory has been much debated (e.g., [820]). Nonetheless, there is broad agreement that the detection of multiple features represented across different neurons is greatly facilitated when those neurons are active more or less synchronously to within a narrow temporal margin.

The temporal correlation theory implies that the synchronous firing of cell assemblies arises via self-organizing, bottom-up processes. An alternative or perhaps complementary view (if both processes take place) is that object perception involves sequential, top-down attention to each feature or object part. Treisman's *feature integration theory* [889, 891] posits that features represented in separate maps must be attended to sequentially in order to be detected conjunctively. An implication of this theory is that individual features may be detected in parallel, whereas a search for a conjunction of features requires a serial search. In support of this notion, Treisman showed that reaction time in visual search for a single feature such as color or orientation is unaffected by the number of distractors in a display, whereas searching for a conjunction of features takes time proportional to the number of distractors. Although the notion of a strict dichotomy between parallel and serial attention has been called into question (e.g., [970]), as has the necessity of attention for object recognition [883], nonetheless, there is substantial evidence for a role for top-down attention in object perception. Perhaps the strongest evidence comes from the study of illusory conjunctions. When people are shown displays of multiple objects, with insufficient time for sequential attention to operate, they frequently make binding errors, for example, reporting having seen a blue "O" in a display that actually contained a red "O" and a blue "H" [889]. Similarly, in the auditory modality, when presented with overlapping streams of high- and low-pitched tones, if subjects were not attending to either stream, they failed to notice deviant tones and did not exhibit EEG mismatch negativity (MMN), whereas attention to the high-pitched tones resulted in MMN for either stream [864]; this suggests that stream segregation, a form of auditory object segmentation, depends upon attention. The top-down application of selective attention to different parts of the same object may serve to synchronize the firing of the corresponding feature detectors and thereby increase the correlations in their outputs. There is ample evidence from EEG recordings that attention during object processing boosts synchronous activity in the gamma-band range (40–130 Hz) (e.g., [204, 768, 872, 884]). Intracranial recordings from awake animals provide further evidence for the synchronizing effects of attention. For example, when macaque V4 neurons are presented simultaneously with stimuli within and outside their RFs, they synchronize much more when the animal's attention is directed to a region within the neurons' RF relative to attention outside the RF [293]. Likewise, in somatosensory cortex, attentional switching between visual and tactile tasks modulates neuronal synchrony, with the degree of synchrony increasing with task difficulty [854]. What might be the role for attentional modulation of neural synchrony? As discussed in Section 1.3, an increase in correlated firing increases the efficiency within which that neuronal representation can be detected at higher levels of processing. Thus, when multiple feature detectors are persistently firing in a correlated manner, they may be more likely to jointly drive a higher level process such as object detection, learning and memory, or a motor response. Another possible role for attention is to increase discriminability. McAdams and Maunsell [603] found that attentional modulation improved the reliability with which V4 neurons could discriminate their preferred orientation. An increase in neuronal reliability could explain the increase in correlated firing resulting from attentional modulation.

Interestingly, recent evidence from human intracranial recordings suggests that attention may affect synchrony in different brain regions in very different ways. For example, attention to a visual stimulus was found to increase stimulus-driven gamma-band oscillations in the fusiform gyrus, whereas attention increased baseline gamma-band oscillations preceding stimulus onset in the lateral occipital sulcus [872]. These authors speculated that the effect of attention during the preparatory phase in the lateral occipital area may have been to put neurons into a state of readiness, thereby allowing their earliest stimulus-driven responses to be synchronized within a narrow temporal window. Many open questions remain regarding the role of correlation and synchrony in feature binding and object perception, and it is currently an active area of research.

1.9 CORRELATION AND CORTICAL MAP CHANGES AFTER PERIPHERAL LESIONS AND BRAIN STIMULATION

As we have seen, activity-dependent reorganization of the cerebral cortex is found during development and learning. In each sensory system, the cortical map can also be induced to undergo large-scale rearrangement after amputation or other surgical or damaging manipulations of the sensory periphery [133, 323, 448]. The best known is the sensation of a phantom limb and in more serious forms that of phantom pain following amputation or deafferentation of a limb in humans. Another less known example is tinnitus that follows partial deafferentation of the output of the cochlea through hearing loss. Although in these cases the effect of cortical reorganization is not beneficial, the neural mechanisms that underlie such deprivation-dependent cortical reorganization may be the same as those responsible for the improvements in perceptual skills that accompany learning. Thus, they warrant a discussion. There are several phases to the reorganization process: an immediate phase of expansion of the representations of parts with intact innervation adjacent to the (partially) deafferented region; a phase lasting weeks or months in which the new representation is consolidated and topographic order at least partly restored; and a late phase of further expansion and use-dependent refinement of internal topography. These common changes can happen in various ways, and we will see that the three major sensory systems, *touch, audition*, and *vision*, differ in how much of the observed topographic map changes in cortex are due to subcortical changes. To appreciate this fully, we have to know the differences in the pathways from receptor to cortex. These are illustrated in Figure 1.16 and discussed below. The same approximate levels in hierarchy are indicated with similar shapes.

Comparative Anatomy of Major Sensory Systems. The sensory system that is confined to the brain is the visual one. The eye with its sensory surface, the retina, is typically considered an extracranial extension of the brain. Briefly,

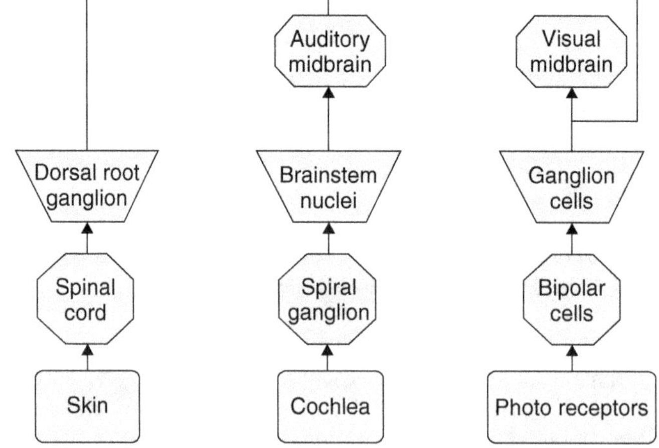

Figure 1.16 Schematic of sensory pathways in somatosensory, auditory, and visual systems (VPN: ventral posterior nucleus; MGB: medial geniculate body; LGN: lateral geniculate nucleus).

the photoreceptors are connected to the bipolar cells which input to the ganglion cells that form the optic nerve. But this is not a one-to-one relay; small groups of neighboring photoreceptors provide common input to individual horizontal cells that mediate lateral inhibition. The bipolar cells that collect input from several neighboring photoreceptors activate several ganglion cells; the amacrine cells connect several ganglion cells together. Thus the retina is a broadly interconnected processing network with the ganglion cells two synapses removed from the receptors. The optic nerve partly activates the superior colliculus in the midbrain but mostly bypasses this to directly innervate the visual thalamic nucleus (i.e., LGN). In this nucleus, input from each eye activates different and mutually exclusive cell layers. The LGN projects mainly to layer IV of V1. The projection of the external world on the retina is topographically mapped onto the thalamus and visual cortex. On the other extreme we find the somatosensory system that has most of its receptors at a long distance from the brain and where a large amount of preprocessing is done outside the brain. Different receptor types reside in the skin of the body surface, information about their activity is transferred to the spinal cord by a large number of nerves that each are responsible for a subset of receptors. Output from the spinal cord ends in the dorsal column nuclei, the cuneate and gracile nuclei, in the brainstem and via the medial lemniscus reaches the ventral posterior nucleus of the thalamus and finally arrives at S1. What sets the somatosensory system apart from the visual one is the extent of divergence of its projections to the dorsal column nuclei and to the thalamus. Thus, a very large number of cells in the dorsal column nuclei and the thalamus come to represent a particular body part, but there is no clear segregation of the outputs of individual nerves representing that body part. In between there is the auditory system with its receptor surfaces clearly outside the brain but close by in the inner ear [248]. Here, there is no map of external space such as in the visual and somatosensory systems, instead sound frequency is mapped "one dimensionally" onto the receptor surface in the cochlea. Thus, the topographic maps of the auditory system are tonotopic or frequency maps and these are in the form of a series of isofrequency sheets, columns that transect the entire cortical area more or less perpendicular to the tonotopic axis. Maps of auditory space are constructed by comparison of sound level and arrival times at the two ears; consequently there are two separate systems, a monaural one and a binaural one, that process information independently. The output of the cochlea is via the auditory nerve, of which each nerve fiber trifurcates at the level of the cochlear nucleus in the brainstem to branch in a tonotopic fashion in each of its three subnuclei. The anterior and posterior ventral cochlear nuclei (VCNs) are purely auditory, but the dorsal cochlear nucleus (DCN) also receives input from the trigeminal nerve among others about the position of the external ear. The output from the DCN terminates in the central nucleus of the inferior colliculus (ICC) of the midbrain. The output from the VCN either goes, via the lateral lemniscus, directly to the ICC (the monaural pathway) or goes to the superior olivary complex in the brainstem where neural activity from the two ears is compared (the binaural pathway) and then to the ICC. The left and right ICC share information and their outputs go to the MGB in the thalamus and then on to A1. Thus, segregation of individual ear output is not as complete as individual eye output in the visual system.

Cortical and Subcortical Topographic Map Reorganization after Peripheral Lesions. For the somatosensory system, Wall et al. [931] reviewed evidence that peripheral injuries cause widespread neurochemical/molecular, functional, and structural alterations in subcortical and cortical substrates of the brain and that cortical changes are but one reflection of global mechanisms that, beginning from the moments after injury, operate at multiple subcortical levels of the somatosensory core. Faggin et al. [270] also suggest both cortical and subcortical reorganization in the somatosensory system. They recorded simultaneously from up to 135 neurons in S1, ventral posterior medial nucleus of the thalamus, and trigeminal brainstem complex of adult rats before and after reversible sensory deactivations by subcutaneous injections of lidocaine. Immediate and simultaneous sensory reorganization was observed at all levels. Thus, peripheral sensory deafferentation triggers a systemwide reorganization. In the visual system of both adult cats and adult macaque monkeys after circumscribed retinal lesions were made, there was no significant sprouting of the retinogeniculate terminals in the lesion projection zone (LPZ). First, the LGN neurons in the LPZ cannot be activated by visual stimuli presented through the lesioned eye and the silent zone closely approximates in size to the normal representation of the lesioned retinal area [268]. Second, the displacements of the RFs observed in the projection zones of retinal lesions in area 17 of cats and macaque monkeys (~6–8 mm) exceeded the expected limits of the lateral spread of geniculocortical afferents (~2 mm). The LPZ in area 17 received their geniculate inputs from the LPZ in the LGN and not from parts of the LGN in which cells responsive to visual stimuli were located. This implicates a cortical mechanism rather than a thalamic mechanism in the topographic reorganization of area 17 [228]. In cat's A1 area, the effect of unilateral localized cochlear lesions in adult cats on the topographic maps of the lesioned cochleas showed large reorganizations [744, 767]. Two to eleven months after a unilateral cochlear lesion affecting the high frequencies, the map of the lesioned cochlea in the contralateral A1 was altered so that the region normally representing the hearing loss frequencies was now occupied by an enlarged representation of lesion-edge frequencies. Along the tonotopic axis the total representation of lesion-edge frequencies could extend up to ~2.6 mm into the hearing loss area, that is, about what one can expect from the thalamocortical divergence. On the basis of threshold sensitivity at the CF, the changes in the map reflect a plastic reorganization rather than simply the residue of prelesion input. In contrast, the map of the unlesioned ipsilateral cochlea, obtained from recordings of the same binaurally sensitive cortical cells, did not differ from those in normal animals. The difference between the ipsilateral and contralateral maps in the region of contralateral map reorganization suggested, in light of the physiology of binaural interactions in the auditory pathway, that the cortical reorganization reflected, at least partly, subcortical changes. To investigate those potential subcortical contributions to cortical reorganization, the frequency organization of the ventral nucleus of the medial geniculate body (MGBv) was investigated 40–186 days following lesioning [464]. In the lesioned animals it was found that, in the region where mid-to-high frequencies are normally represented, there was an "expanded representation" of lesion-edge frequencies. Neuron clusters within these regions of enlarged representation that had "new" characteristic frequencies displayed response properties (latency, bandwidth) very similar to those in normal animals. The tonotopic reorganization observed in MGBv was similar to that seen in A1 and suggested that the auditory thalamus played an important role in cortical plasticity. To additionally examine the contribution of subthalamic changes to the thalamic and cortical map reorganization, the effects of unilateral mechanical cochlear lesions on the frequency organization of the central nucleus of the ICC were examined in adult cats [431]. After recovery periods of 2.5–18 months, the frequency organization of ICC contralateral to the lesioned cochlea was determined separately for the onset and late components of multiunit responses to toneburst stimuli. For the late-response component in all but one penetration through the ICC and for the onset response component in more than half of the penetrations, changes in frequency organization in the lesion projection zone were explicable as the residue of prelesion responses. In half of the penetrations exhibiting nonresidue-type changes in onset response frequency organization, the changes appeared to reflect the unmasking of normally inhibited inputs. In the other half it was unclear whether the changes reflected unmasking or a dynamic process of reorganization. Thus, most of the observed changes were explicable as passive consequences of the lesion, and there was limited evidence for plasticity in the ICC. Immediate unmasking of subthreshold inputs to the ICC was also noted after minute lesions in the spiral ganglion of the auditory nerve [841]. So most of the changes in ICC might never evolve beyond this initial unmasking of excitatory inputs. No evidence was found for reorganization of the topographic maps in the dorsal cochlear nucleus after partial destruction of the cochlea similar to those that cause massive map reorganization in auditory cortex [614, 743], although small regions of the DCN that were deprived of their normal, most sensitive frequency (or CF) input by the cochlear lesion appeared to have acquired new CFs at frequencies at or near the edge of the cochlear lesion. However, because of the elevated thresholds at the new CFs, the changes simply reflected the residue of prelesion input to those sites. The results suggest that the DCN does not exhibit the type of plasticity that has been found in the auditory cortex and even the midbrain. Combined, this suggests that in the somatosensory system topographic map changes may already occur in the spinal cord and definitely in the brainstem nuclei, in the auditory system potential changes are seen in the midbrain and definitely in the thalamus, whereas in the visual system topographic map changes are likely confined to the cortex.

Time Course of Cortical Topographic Map Changes. Immediately after the lesion, neurons in the LPZ area of cortex either fall silent or show dramatically enlarged RFs that extend far outside the prelesioned area. This may represent unmasking of subthreshold excitatory inputs to a suprathreshold driving control by either suppressing inhibitory or potentiating excitatory connections. The second phase, marked by the gradual expansion of the representation of a perilesion receptor surface into the initially silenced region of cortex, takes place over weeks and months. The cortex then develops a new, piecewise-continuous map, continuous up to the border of the lesion and then jumping across. It is during this extended period that axonal growth and synaptogenesis occur [196]. Calford et al. [135] monitored topographic reorganization in the V1 of the cat by recording extracellular activity of cells over the 11 h following the circumscribed outer layer monocular lesion in the retina. In the first hour following the lesion, no neural responses could be elicited within the LPZ by photic stimulation of the lesioned eye. However, in the next 1–3 h after the lesion only 39% of the recording sites within the LPZ remained unresponsive. This percentage further decreased to 31% (3–7 h) and 27% (7–11 h). In these cats, the ectopic RFs recorded from the LPZ within hours after lesioning were up to 10-fold larger than their normal counterparts revealed by stimulating the same cells via the nonlesioned eye. In LPZ regions, 1–2 weeks after bilateral retinal lesions, both spontaneous activity and driven activity were significantly reduced. At the same time, both spontaneous and driven activity significantly increased in cortical regions immediately adjacent to the LPZ (associated with a sharp increase in glutamate immunoreactivity [269]). There are several phases to the reorganization process in somatosensory cortex that are very similar to those in visual cortex: an immediate phase of expansion of the representations of parts with intact innervation adjacent to the deafferented region; a phase lasting weeks or months in which the new representation is consolidated and topographic order restored, and a late phase of further expansion and use-dependent refinement of internal topography. In detail, the time course of changes in somatosensory cortex following median nerve section [616] revealed that large cortical areas were silenced by median nerve transection. Inputs from fragments of dorsal skin were immediately unmasked and had greater than normal RF overlap as a function of distance across the cortical surface. They were transformed over time into very large highly topographic and complete representations of dorsal skin surfaces. Representations of bordering glabrous skin surfaces progressively expanded to occupy larger and larger portions of the former median nerve cortical representation zone. These expanded representations of ulnar nerve–innervated skin surfaces sometimes moved, in entirety, into the former median nerve representational zone. Most of the former median nerve zone was driven by new inputs in a map derived 22 days after nerve section. At 11 days reoccupation was still incomplete. Immediately after amputation of a single exposed digit on the forelimb of the flying fox [134], neurons in the area of cortex receiving inputs from the missing digit were *not* silent but responded to stimulation of adjoining regions of the digit, hand, arm, and wing. In the week following amputation, the enlarged RFs shrank until they covered only the skin around the amputation wound. The immediate response can be interpreted as a removal of inhibition and the subsequent shrinking of the RF might be due to reestablishment of the inhibitory balance in the affected cortex and its inputs. In auditory cortex, immediately after a noise trauma unmasking of excitatory inputs was observed [670]. Initially thresholds were elevated (by about 40 dB) and CFs of units recorded before and immediately after the trauma were shifted to lower values, that is, to the edge of the hearing loss range. Gradually, over a few hours thresholds recovered but CFs did not change, further and essentially all units recorded from acquired new CFs. This confirmed previous findings by Robertson and Irvine [767] where the responses of neuron clusters were examined within hours of making small mechanical cochlear lesions. It was found that shifts in CF toward frequencies spared by the lesions could occur, but thresholds were greatly elevated compared to normal (mean difference was 31.7 dB in five animals). The emergence of driven activity in such regions after prolonged recovery periods in lesioned animals thus suggests that the auditory cortical frequency map undergoes reorganization in cases of partial deafness. Typically, after about 3 weeks reorganization of the cortical tonotopic map is observed [251]. Some features of this reorganization are similar to changes reported in somatosensory cortex after peripheral nerve injury and in visual cortex after retinal lesions, and this form of plasticity may therefore be a feature of all adult sensory systems.

Mechanism of Cortical Topographic Map Changes. In somatosensory cortex, because of the limited extent of the lesion (<2–3 mm), it was proposed that thalamocortical axonal divergence within the cortex represented the neural substrate of reorganization. In auditory cortex, the results by Rajan et al. [744] also pointed to the same extent of reorganization. In visual cortex, the extent of reorganization, ~6–8 mm, was much larger than the lateral spread of thalamocortical afferents. Pyramidal neurons in all cortical areas receive some of their input from afferent connections to the same vertical column. But they also receive inputs from a wide-ranging intracortical network of axons (the horizontal fibers) from more remote pyramidal cells. These horizontal fibers extend about 6-8 mm in cortex, about 2–4 mm radiating outward from each source neuron. The synapse strength of these horizontal fibers can be altered by appropriate patterns of stimulation, that is, by stimulating the horizontal inputs while simultaneously depolarizing the recorded neuron with injected current [394]. The degree of synaptic strengthening depended on the degree of inhibition present: The greater the inhibition, the less effective the synaptic potentiation. The fact that synapses made by horizontal collaterals can be potentiated suggests that synapses that normally play a modulatory role can, under the proper conditions, be strengthened so as to drive their target neurons above the threshold for spiking. After reorganization, the cortex becomes visually responsive again, but via signals conducted horizontally, intracortically, and in an orientation-selective manner from columns of neurons in unaffected regions of cortex outside the original LPZ. For the RFs of cells in the LPZ to shift, the horizontal connections must be strengthened. The way this could be done is by sprouting axon collaterals and by synaptogenesis [323]. The potential synaptic mechanisms that play a role in cortical reorganization are strengthening of existing but weak or subthreshold synaptic inputs and depression or otherwise weakening of existing strong synaptic inputs. Likely bases of map plasticity may lie in the cortical neurons' ability to compensate for changes in excitatory input by regulating turnover of postsynaptic α-amino-3-hydroxy-5-methylisoxazole-4-propionic acid (AMPA) receptors, thereby scaling the size of EPSP amplitudes and thus the overall responses of a neuron to stimulation [896], an effect that is mediated by brain-derived neurotropic factor. There are numerous observations that implicate the inhibitory neurotransmitter GABA in map plasticity. Cortical cells released from inhibition commonly increase the sizes of their RFs, and removal of inhibition is thought to underlie the immediate expansions of RFs of somatosensory cortical neurons after loss of peripheral input by amputation or local anesthesia of a digit, an effect that may depend on loss of tonic control by C-fiber inputs over central inhibitory mechanisms [448].

Correlated Neural Activity and Topographic Map Organizations. The specific patterns of reorganization in cortex were driven by the patterns of correlation in natural sensory stimulation. The experiment reported in [174] fusing digits showed that imposing synchrony in the activation of receptors on both digits drove the reorganization and fusion of cortical RFs. At the cellular level, reorganization was being driven by an activity-dependent process of synaptic strengthening (reminiscent of processes during development). In the visual system RF expansion was accompanied by a selective increase in the strengths of intracortical connections [197]. By monitoring the strength of cross-correlations between pairs of neurons through different stages of RF expansion, it was found that populations of neurons increased the strength of their effective synaptic interconnections in the expanded regions of their RFs. Most of the cross-correlograms, recorded with electrodes separated by 0.1–5 mm, had widths of 5–15 ms and were asymmetric, typical for interactions observed between pairs of V1 neurons in different cortical columns. The increase in effective connection strength was also orientation selective and only occurred between pairs of neurons with similar orientation preferences. Neurons whose orientations differed by more than 30° did not show any increase in their mutual connection strength despite substantial increases and overlap in RF area. All the evidence is consistent with the idea that dynamic RF expansion is mediated through horizontal intracortical connections that link RFs of similar orientation preference. This was corroborated by lesioning studies [136]. Increases in neural synchrony between neurons in the lesion zone were found in auditory cortex immediately after a noise trauma [670] and several weeks to months after the trauma in reorganized cortex [668]. The increased synchrony was specific to reorganization because when reorganization was prevented by targeted acoustic stimulation of the frequency range of the hearing loss, but with hearing loss still present, there was no change in neural synchrony [667, 668]. On the other hand, inducing cortical reorganization in the absence of hearing loss both increased neural synchrony and strengthened horizontal fiber synapses with pyramidal cells in the reorganized area [669].

Behavioral Consequences of Topographic Map Changes Resulting from Receptor Injury. Adult cortex, and potentially also the thalamus, is highly plastic and can change as a result of learning (adaptive plasticity) and as a result of receptor injuries. The map changes that occur may or may not be related to the increased performance after learning [120, 757], and the map changes that occur after peripheral injury may have mostly maladaptive consequences, albeit that some increased performance has been noted. Specifically this was after putative reorganization in auditory cortex where the overrepresentation of the edge frequency was related to improved frequency discrimination [607]. The maladaptive consequences appear to be dominating, as suggested by the correlation of phantom limb sensations and tinnitus with cortical map changes [95, 282, 642]. Assuming that reorganized adult cortex remains plastic, it should be possible to reverse the changes. Because of potential subcortical influences, this should, likely have been done by interference at or close to the receptor level. One successful approach has been to restore the balance of output from the cochlea resulting from high-frequency hearing loss [667]. This was accomplished by continuous stimulation in the frequency region of the noise-induced hearing loss, to compensate for the downregulated spontaneous and driven firing rates. After such a hearing loss, one typically observes reorganization of the cortical tonotopic map, accompanied by increased spontaneous firing rates and increased neural synchrony [666, 668]. The observations after applying the enriched acoustic environment for at least 3 weeks and starting immediately after the trauma included not only absence of cortical reorganization, normal spontaneous firing rates, and unaltered (i.e., normal) neural synchrony but also an improvement of the presumably neurotoxicity-induced hearing loss at frequencies above the hair cell loss frequency range. The sound was composed out of 32 series of tonepips (each for one of the 32 frequencies) with each series an independent realization of a Poisson process with rate of 1.5 Hz, and the co-occurrence of any combination of frequencies was thus completely random. Another take on the driving force of the changes in the cortical activation, besides restoration of the balance between the output from the normal and hearing loss regions in the cochlea, could be that the enriched acoustic environment produced a desynchronization of the inputs to the (to be) reorganized cortex and thus of the activity of that cortical region. There is some doubt about the possibility to use this sound treatment to reverse long-standing reorganizations, that is, those involving axonal sprouting and the formation of new connections. This would suggest that any treatment has to start well before this sprouting process starts.

1.10 DISCUSSION

In this chapter we have reviewed many correlative neural mechanisms and the important roles of correlation in perception, detection, memory, and sensorimotor coordination. We discussed how correlation is detected in sensory inputs in both the single neuron and neural populations and how it is employed in neural systems to encode and transmit information. Specifically, temporally correlated neuronal spike activities emerge from various interactions, such as correlation of firing of neighboring neurons or neural assemblies or synchronous firing of neurons from different subcortical or cortical areas (e.g., retinal ganglion cells, LGN cells, and V1 cells) [16, 833]. We conclude by the observation that correlation is the brain's "basic mechanism to evaluate, to control, and to learn" and "the basis of learning,

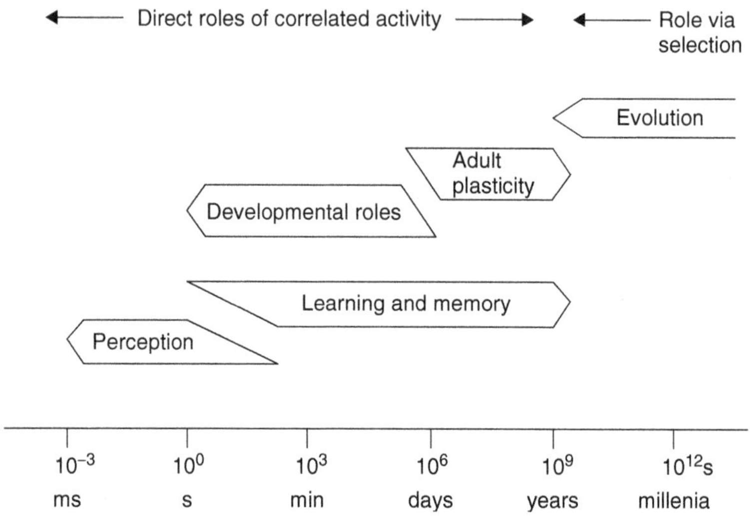

Figure 1.17 An illustrative diagram of timescales of the various roles of correlated activities within the central nervous system of human being. (Reprinted from *Trends in Neuroscience*, Vol. 14, J. E. Cook, Correlated activity in the CNS: A role on every timescale? pp. 397–401. Copyright © 1991, with permission from Elsevier.)

association, pattern recognition, and memory recall in the human nervous system" [241]. In addition, correlation also occurs at both macroscopic and microscopic timescales. According to Cook's categorization [183], correlated activities emerge in the central nervous system on every timescale, ranging from the momentary to the evolutionary (see Figure 1.17).

Learning and synaptic plasticity are the keys to the brain's ability to adapt to the changing environment. Motivated by this neurobiological evidence, we have mentioned several of the most biologically plausible learning algorithms, including Hebbian learning, competitive learning, self-organizing feature maps, STDP, Rescorla–Wagner rule, and TD learning. Despite their different motivational roots, all of these learning rules are based on the principle of correlation. The competitive learning and self-organizing map models share with Hebbian learning the principle that the synaptic strength should change in proportion to presynaptic and postsynaptic correlation. In the case of STDP, the synapse increases its efficacy whenever its input activity correlates with a subsequent change in neuronal output firing. The Rescorla–Wagner and TD learning rules are based on a prediction error signal, specifically, the difference between received and expected reinforcement. When a neural input is correlated with a prediction error, synaptic strengths are adjusted so as to reduce the prediction error. This idea can be generalized to allow for learning based on other prediction error signals. For example, the Kalman filter [461] is a standard method in engineering for training a system to predict a temporally varying signal, and the prediction error drives the parameter adjustment process. Computational neural models based on Kalman filtering are described in greater detail in Chapter 7 of this book. Still more generally, the correlation-based ALOPEX learning algorithm [355, 902] is based on the correlation between an error signal and the parameter changes. Here, the error signal could be any analytically computable cost function. The basic idea is that if a change in a parameter value correlates with an increase in error, then the parameter is more likely to be changed in the opposite direction in the future, whereas a parameter change correlated with a decrease in error is more likely to be repeated in the future. Detailed discussions of the ALOPEX paradigm are presented in Chapter 6 of this book.

Finally, to conclude this chapter, we note that it is the correlative nature of the brain that motivates us to study the notion of correlative learning; moreover, we suggest that the principle of correlation might serve the role of encompassing most (if not all) learning paradigms, many of which will be reviewed in details in Chapter 3.

BIBLIOGRAPHICAL NOTES

For a general background on spiking neurons, textbook treatments can be found in [320, 493]. The classic books [171, 201] serve as excellent sources and references on neuroscience and computational brain models.

Synaptic plasticity is referred to as an adaptive response of the neuron to specific external (stimulus) signals; it is closely related to the notion of "learning." The notion of synaptic plasticity has a long history. The first mention of phenomena that can now be related to modifications in synaptic strengths or number of connections in neural networks can be found in René Descartes' *Traité de l'homme* (*Treatise on Man* (1664) [212], in which he described the concept of a hydraulic nervous system. William James, in his classic work on psychology [436], proposes the law of neural habit, which is also reminiscent of the ideas of Herbert Spencer [844] and Young [990]. In 1949, Donald Hebb published his famous learning postulate in *Organization of Behavior: A Neuropsychological Theory* [377], in which he first hypothesized the correlative mechanism for synaptic plasticity. A general review of Hebb's work can be found in [628, 816]. Hebbian synaptic mechanisms have been reviewed from biological [89, 472], biophysical [122], and physiological [66, 855] perspectives. The discussion of Hebbian synaptic plasticity in hippocampus is found in [472, 584, 817].

The milestone of Hebbian plasticity was further established by the discovery of the phenomenon of LTP [100]. The review of LTP and its relation to Hebbian synaptic plasticity can be found in [121, 586]. The review of spike-timing-dependent Hebbian plasticity can be found in [89].

The notion of the correlative brain was first proposed by [922] and reviewed in more detail in [241]. While it is impossible to include all the bibliographical references regarding the omnipresent importance of correlation and synchronization in population coding, topographic map formation, perception, memory, sensorimotor learning, attention, and feature binding, we will refer the reader to relevant references when discussing specific contents throughout the book.

NOTES

1. The term synapse was first used by Charles S. Sherrington [829] to designate the functional junction between nerve cells.
2. Caianiello [132] also proposed a learning rule known as the *mnemonic equation* for describing the synaptic plasticity. The learning rule, appearing in the form of a differential equation, is given as

$$\frac{d\theta_{ij}}{dt} = \theta_{ij}(t)\left[\alpha x_j(t-k\tau)x_i(t) - \beta\,\mathrm{sgn}(\theta_{ij}(t) - \theta_{ij}(0))\right]\mathrm{sgn}(a_{ij}(t) - \theta_{ij}(0)),$$

   where α and β are constants.
3. The term *self-organization* was first used by Farley and Clark [272] of MIT Lincoln Laboratory in 1954 while they studied the behavior of networks with nonlinear elements. Self-organization was used also by analogy to physics. In physical systems of densely packed units, where the activity of adjacent units is mutually influencing, several self-organizing principles of self-organization are particularly important [929]:

   - Local interactions tend to self-amplify. Synchronous or correlated interactions among coupled units strengthen and spread across ensembles of units, creating coherent activity patterns.
   - Developing activity patterns compete. The strongest (most coherent) patterns vigorously grow at the expense of others. This leads to the formation of activity "domains" of different self-amplifying patterns.
   - Domains of activity tend to cooperate. In spite of the overall competition in the system, domains of correlated activity will tend to coalesce to form larger, coherent activity patterns. If there are no outside influences acting on the system, the activity patterns with the most internal cooperativity and the least competition will win out.

   Von der Malsburg and Singer [929] noted that a fundamental correlate of these three basic principles is that "global order can arise from local interactions . . . ultimately leading to coherent behavior" (p. 71). In a large self-organizing network, a number of competing local domains can coexist, but the tendency of the network is generally toward attaining a globally ordered state. Because of this, even a relatively weak stimulus toward global organization can decisively influence developing local patterns.
4. The notion of RF is a cornerstone in visual physiology. According to H. K. Hartline (1938). The response of single optic nerve fibers of vertebrate eye to illumination of the retina. American Journal of Physiology 121. 400–415. It is "the region of the retina that must be illuminated in order to obtain a response in any given fiber." Nowadays, this notion has been widely extended and generally understood as "the area in which stimulation leads to response of a particular sensory neuron."
5. Sherman and Koch [828] have found that in the cat there are roughly 10^6 fibers from LGN in the thalamus to the visual cortex and about 10^7 fibers in the reverse direction.
6. The term *hippocampus* derives from Greek mythology with the meaning of "horselike sea monster." In anatomy it was named because of its curved "seahorse shape"; the "hippocampus" was first used by the anatomist Giulio Cesare Aranzi (circa 1564) for describing this brain region.

7. William James [436] also emphasized the principle of association that governs activation:

> The amount of activity at any given point in the brain-cortex is the sum of the tendencies of all other points to discharge into it, such tendencies being proportionate (i) to the number of times the excitement of each other point may have accompanied that of the point in question; (ii) to the intensity of such excitements; and (iii) to the absence of any rival point functionally disconnected with the first point, into which the discharges might be diverted.

   These words clearly capture the essence of (i) Hebbian synapse, (ii) presynaptic neural activations, and (iii) the role of inhibition and anti-Hebbian synapse.

8. In Donald Hebb's book, *The Organization of Behavior*, there were three pivotal postulates in the context of synaptic plasticity and learning [377]:

   - The first one is the most celebrated Hebb's postulate formulated as Hebb's rule (p. 62), which provides the basis for adjusting connection weights in biological or artificial neural networks.
   - The second postulate speculates that groups of neurons that tend to fire together form a cell assembly whose activity can persist after the triggering event and serves to represent it (illustrated in Figure 10 of Hebb's book, p. 72).
   - The third postulate states that thinking is the sequential activation of sets of cell assemblies.

9. The phenomenon of LTP was first observed experimentally by Terje Lømo in 1966. Subsequent physiological experiments also validated another phenomenon called long-term depression (LTD), which refers to the weakening of a synapse that lasts from hours to days, as a counterpart of the LTP.

2

CORRELATION IN SIGNAL PROCESSING

Correlation, be it in regards to the *autocorrelation* function of a prescribed signal or the *cross-correlation* function between a pair of somewhat similar signals, is basic to statistical signal processing and time series analysis. Its origin, in the context of random processes, may be traced to a series of papers on Brownian motion and spectral analysis of stochastic processes, which began around 1925 and culminated in the publication of the classic paper "Generalized Harmonic Analysis" by Norbert Wiener in 1930, and with it the discipline of statistical signal processing was established [474, 954, 957]. In addition to popular Fourier analysis, correlation functions were also used in other advanced spectrum analysis methods. In this chapter, we present a systematic review of correlation techniques for *statistical* and *adaptive* signal processing. It is known that a wide class of stationary stochastic processes can be sufficiently characterized by their second-order cumulant statistics, which play a prominent role in statistical and adaptive signal processing. In particular, several key components of signal processing and statistical analysis tools will be covered in this chapter:

- *Spectrum Analysis*: standard and advanced tools for stationary, nonstationary, linear, and nonlinear systems.
- *Filtering*: Wiener filter, least-mean-square filter, recursive least-squares filter, and matched filter.
- *Detection*: correlation-based detection in communication.
- *Estimation*: correlation method for time-delay estimation.

Correlative Learning: A Basis for Brain and Adaptive Systems, by Zhe Chen, Simon Haykin, Jos J. Eggermont, and Suzanna Becker

- *Statistical Analysis*: principal-component analysis, factor analysis, canonical correlation analysis, Fisher linear discriminant analysis, and common spatial pattern analysis.

Some topics are widely known and developed in the literature, and some of them are relatively new. The purpose here is to review these materials in a logical and systematic way, emphasizing the roles of correlation. For the sake of mathematical simplicity, in most cases we restrict our discussion to real-valued random variables in the time domain (although their transforms in the frequency domain are complex); however, the extension to the complex domain is rather straightforward, and we will address it separately in Chapter 5.

2.1 CORRELATION AND SPECTRUM ANALYSIS

The notion of "spectrum" (Latin meaning "appearance") was first introduced by Newton in the study of the distribution of energy emitted by a radiant source arranged in order of wavelength; its meaning was later generalized in Fourier analysis by Hilbert.[1] Spectrum analysis was pioneered by Hermann von Helmholtz for analyzing sound signals; it was further populated in the 1930s with the development of generalized harmonic analysis by Hardy, Wiener, Komolgorov, and others. The theory of generalized harmonic analysis had a major impact on signal analysis and time series analysis, for which one can only observe the correlation functions associated with the signals in practical situations. The spectrum as used here, or more specifically the power spectrum, of a stochastic process describes the distribution of average power of the process as a function of frequency. Although spectrum analysis was initially developed for stationary stochastic processes, many recorded time series signals in the real world are nonstationary or merely locally stationary (see Figure 2.1 for a recurrence plot verification)[2]; therefore, it is imperative to develop advanced spectrum analysis techniques for nonstationary signals. In signal processing, second-order statistics (SOS) and higher order statistics (HOS) are used to characterize the corresponding power spectra or polyspectra using the Fourier transform. In engineering, spectrum analysis has been widely utilized in radar, sonar, geophysics, astronomy, communications, and biomedicine. For discussion simplicity, here we have restricted our discussion to a continuous-time random signal or process unless specified otherwise.

2.1.1 Stationary Process

In signal processing, the most popular signal analysis tool is Fourier spectrum analysis, which is used for a stationary random process.

Definition 2.1 *A stochastic process $x(t)$ is called strict-sense stationary if all of its statistical properties are invariant to a shift of the original; a stochastic process $x(t)$ is called wide-sense stationary if its mean is constant,*

$$\mathbb{E}[x(t)] = \text{const}, \tag{2.1}$$

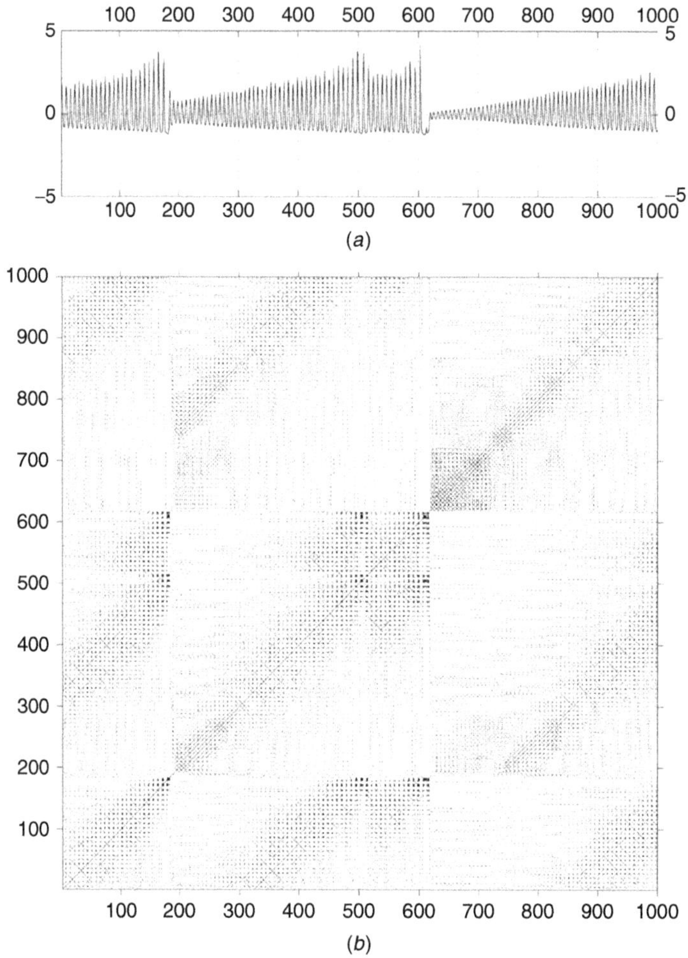

Figure 2.1 (*a*) A nonstationary laser time series. (*b*) Recurrent plot of the laser time series. The density variations near the diagonal of the recurrent plot are a clear indication of nonstationarity.

and its autocorrelation function depends only on the time difference, as shown by

$$C_{xx}(\tau) = \mathbb{E}[x(t_1)x(t_2)] = \mathbb{E}[x(0)x(t_2 - t_1)] \qquad (\tau = t_2 - t_1). \tag{2.2}$$

Unless stated otherwise, we often refer to wide-sense stationary processes as stationary processes.

If the stochastic process is *ergodic*, then it is convenient to define the time-averaged autocorrelation function as

$$\overline{C}_{xx}(\tau) = \lim_{T\to\infty} \frac{1}{T} \int_0^T x(t)x(t+\tau)\, dt. \tag{2.3}$$

It should be noted that the time-averaged autocorrelation function is generally different from the ensemble-averaged (or statistical averaging) autocorrelation function. However, in statistical signal processing, we often assume that the random process is ergodic and thereby use $\overline{C}_{xx}(\tau)$ interchangeably with $C_{xx}(\tau)$, especially when the probability distribution of $x(t)$ is inaccessible. See Appendix A for more mathematical details.

Definition 2.2 *A stochastic process $x(t)$ is called α dependent (where $\alpha > 0$) if all of its values $x(t)$ for $t < t_0$ and for $t > t_0 + \alpha$ are mutually independent; a stochastic process $x(t)$ is called correlation α-dependent if its autocorrelation function satisfies*

$$C_{xx}(t_1, t_2) = 0 \quad for \ \ |t_1 - t_2| > \alpha. \tag{2.4}$$

According to the *Wiener–Khinchin theorem*, the autocorrelation function of a wide-sense stationary stochastic process and its power spectral density (PSD) relate to each other by a pair of Fourier transforms. Mathematically, let $C_{xx}(\tau)$ denote the autocorrelation function of a stationary stochastic process $x(t)$ and let $S_{xx}(\omega)$ denote the PSD of $x(t)$; then we have

$$S_{xx}(\omega) = \int_{-\infty}^{\infty} C_{xx}(\tau) \exp(-j\omega\tau)\, d\tau, \tag{2.5}$$

$$C_{xx}(\tau) = \frac{1}{2\pi} \int_{-\infty}^{\infty} S_{xx}(\omega) \exp(j\omega\tau)\, d\omega, \tag{2.6}$$

where $j = \sqrt{-1}$. The above property is widely used in engineering for spectrum analysis, whose computation is aided by the fast Fourier transform (FFT) algorithm.

Similarly, let $C_{xy}(\tau) = \mathbb{E}[x(t)y(t+\tau)]$ be the cross-correlation function of two stationary stochastic processes $x(t)$ and $y(t)$; correspondingly, we also have

$$S_{xy}(\omega) = \int_{-\infty}^{\infty} C_{xy}(\tau) \exp(-j\omega\tau)\, d\tau, \tag{2.7}$$

$$C_{xy}(\tau) = \frac{1}{2\pi} \int_{-\infty}^{\infty} S_{xy}(\omega) \exp(j\omega\tau)\, d\omega \tag{2.8}$$

When the cross-spectrum $S_{xy}(\omega)$ is normalized by the PSDs $S_{xx}(\omega)$ and $S_{xy}(\omega)$, we obtain the normalized cross-spectrum

$$\rho_{xy}(\omega) = \frac{S_{xy}(\omega)}{\sqrt{S_{xx}(\omega) S_{yy}(\omega)}}, \tag{2.9}$$

which is sometimes also termed *coherency*. The magnitude of $\rho_{xy}(\omega)$ defines the coherence function that indicates the correlation (in the range from 0 and 1) between $x(t)$ and $y(t)$ at any specific frequency ω.

It is also straightforward to generalize the above univariate concepts to a *multivariate stochastic process* or *vector stochastic process*. The vector (m-dimensional) process is defined as a family of m stochastic processes. Let $\mathbf{x}(t) = \{x_i(t)\}_{i=1}^m$ denote a vector process whose components $x_i(t)$ are univariate stochastic processes. The mean function $\boldsymbol{\mu}(t) = \{\mu_i(t)\}$ is also a vector process with elements $\mu_i(t) = \mathbb{E}[x_i(t)]$; and the autocorrelation function of $\mathbf{x}(t)$ is defined as an $m \times m$ matrix function

$$\mathbf{C}_{xx}(t_1, t_2) = \mathbb{E}[\mathbf{x}(t_1)\mathbf{x}^{\mathrm{T}}(t_2)]. \tag{2.10}$$

For stationary processes, the correlation or covariance matrix has a Toeplitz structure in the sense that it has constant entries along the negative-sloping diagonals. An $m \times m$ Toeplitz matrix contains only $2m - 1$ degrees of freedom; such a highly structured Toeplitz matrix is important in linear algebra and statistical signal processing.

■ EXAMPLE 2.1

An autoregressive (AR) process is defined as a process that generates a time series for which representation of the current value of the measured variable involves a weighted sum of past values. The AR processes have been widely used in applications of time series analysis and linear prediction because of the appealing simplicity. In this example we will examine the autocorrelation function property of the linear AR process. In particular, for a narrow-band random signal $x(t)$, let us consider a stationary time-invariant AR model driven by a white Gaussian noise process:

$$a_0 x(t) = -\sum_{i=1}^{p} a_i x(t-i) + \varepsilon(t), \qquad a_0 \neq 0, \tag{2.11}$$

which is referred to an AR(p) model of order p. Alternatively, equation (2.11) can be written as a form of linear prediction (the so-called linear predictive coding):

$$x(t) = -\sum_{i=1}^{p} \tilde{a}_i x(t-i) + \varepsilon(t), \qquad \tilde{a}_i = \frac{a_i}{a_0}. \tag{2.12}$$

Without loss of generality, we assume $\mathbb{E}[\varepsilon(t)] = 0$ and $\mathrm{var}[\varepsilon(t)] = 1$. Multiplying both sides of (2.12) by $x(t-\tau)$ and then taking the statistical expectation, we have

$$-C_{xx}(\tau) = \sum_{i=1}^{p} \tilde{a}_i C_{xx}(i-\tau)(t) \qquad (1 \leq \tau \leq p), \tag{2.13}$$

where we have denoted $C_{xx}(\tau) = \mathbb{E}[x(t)x(t+\tau)]$ and assumed that $\mathbb{E}[x(t-\tau)\varepsilon(t)] = 0$. Equation (2.13) is often known as the *normal equation* or *Yule–Walker equation*; in matrix form, it can be written as

$$\mathbf{r} = \mathbf{C}\mathbf{a}, \tag{2.14}$$

where the autocorrelation matrix $\mathbf{C}$ is a *symmetric, circulant* matrix with elements $C_{ij} = C_{xx}(i-j)$, vector $\mathbf{r}$ is the autocorrelation vector $r_j = C_{xx}(j)$, and vector $\mathbf{a} = [-\tilde{a}_1, \ldots, -\tilde{a}_p]^{\mathrm{T}}$ is the parameter vector.

Without loss of generality, we assume that $a_0 = 1$; then the autocorrelation function of the AR(p) process is defined as

$$\begin{aligned}
C_{xx}(\tau) &= \mathbb{E}[x(t)x(t+\tau)] \\
&= \mathbb{E}\left\{\left[a_1 x(t-1) + a_2 x(t-2) + \cdots + a_p x(t-p)\right]\right. \\
&\quad \times \left.\left[a_1 x(t-1+\tau) + a_2 x(t-2+\tau) + \cdots + a_p x(t-p+\tau)\right]\right\}.
\end{aligned}$$

Expanding the above equation according to the definition of expectation and rearranging the terms, we obtain

$$\begin{aligned}
C_{xx}(\tau) = C_{xx}(\tau)(a_1^2 + a_2^2 + \cdots + a_p^2) \qquad & (p \text{ terms}) \\
+ C_{xx}(\tau-1)(2a_1a_2 + 2a_2a_3 + \cdots + 2a_{p-1}a_p) \qquad & (p-1 \text{ terms}) \\
+ C_{xx}(\tau-2)(2a_1a_3 + 2a_2a_4 + \cdots + 2a_{p-2}a_p] \qquad & (p-2 \text{ terms}) \\
\vdots \qquad & \\
+ C_{xx}(\tau-p+1)(2a_1a_p) \qquad & (1 \text{ term}).
\end{aligned}$$

Hence, the autocorrelation function itself can also be represented as an AR($p-1$) model, with new AR coefficients defined as follows:

$$\begin{aligned}
a_0' &= 1 - (a_1^2 + a_2^2 + \cdots + a_p^2), \\
a_1' &= 2a_1a_2 + 2a_2a_3 + \cdots + 2a_{p-1}a_p, \\
&\vdots \\
a_{p-1}' &= 2a_1a_p.
\end{aligned}$$

For the above AR(p) model (2.11), the transfer function in the z-domain may be formulated as

$$H(z) = \frac{1}{\sum_{k=0}^{p} a_k z^{-k}}, \tag{2.15}$$

where z^{-1} denotes the unit-delay operator. The AR(p) power spectrum of $x(t)$, denoted as $S_{xx}(\omega) \equiv S_{\mathrm{AR}}(\omega)$, is derived by letting $S_{\mathrm{AR}}(\omega) = |H(e^{j\omega})|^2 = |H(z)|^2_{z=e^{j\omega}}$; namely, we have

$$\begin{aligned} S_{\mathrm{AR}}(\omega, \boldsymbol{a}) &= \left| \sum_{k=0}^{p} a_k e^{-j\omega k} \right|^{-2} \\ &= \frac{1}{\left|1 + \sum_{k=1}^{p} a_k e^{-j\omega k}\right|^2} \qquad (-\pi < \omega \leq \pi), \end{aligned} \tag{2.16}$$

where $\boldsymbol{a} = (a_0, a_1, \ldots, a_p)$ specifies the AR parameters and $S_{\mathrm{AR}}^{-1}(\omega, \boldsymbol{a})$ produces a finite $[p + p(p+1)/2]$ sum of orthonormal bases (in terms of $e^{jm\omega}$, $m \in \mathbb{N}$). Furthermore, we can rewrite the parametric power spectrum $S_{\mathrm{AR}}(\omega, \boldsymbol{a})$ as

$$S_{\mathrm{AR}}(\omega, \boldsymbol{a}) \equiv S_{xx}(\omega) = \sum_{t=0}^{\infty} c_t \phi_t(\omega), \tag{2.17}$$

where $\{\phi_t(\omega)\}$ denotes the orthonormal bases and c_t denotes the associated expansion coefficients (note that some coefficients will be zero).

On the other hand, by virtue of the Wiener–Khinchin theorem, the power spectrum of the stationary signal $x(t)$ may be represented by the discrete-time Fourier transform of its autocorrelation function,

$$\begin{aligned} S_{xx}(\omega) &= \sum_{t=-\infty}^{\infty} C_{xx}(t) e^{-j\omega t} \\ &= C_{xx}(0) + \sum_{t=1}^{\infty} 2C_{xx}(t) \cos(\omega t), \end{aligned} \tag{2.18}$$

where the second line follows from the fact that $C_{xx}(t)$ is a symmetric even function. Comparing (2.17) and (2.18), we can derive the corresponding relationship:

$$c_0 = C_{xx}(0), \qquad \phi_0(\omega) = 1, \qquad c_t = 2C_{xx}(t), \qquad \phi_t(\omega) = \cos(\omega t).$$

Let us further consider a special case of the AR(1) model defined in (2.11):

$$x(t) = ax(t-1) + \varepsilon(t) \qquad (|a| < 1).$$

More generally, provided we assume $\mathbb{E}[\varepsilon(t)] = c$ and $\mathrm{var}[\varepsilon(t)] = \sigma^2$ and let $\mu = \mathbb{E}[x(t)]$, then taking the expectation of both sides of the above equation yields

$$\mu = a\mu + c, \tag{2.19}$$

or $\mu = c/(1-a)$. If the white-noise process is zero mean such that $c = 0$, then $\mu = 0$, and the variance of $x(t)$ is given by

$$\operatorname{var}[x(t)] = \mathbb{E}[x^2(t)] - \mu^2 = \frac{\sigma^2}{1-a^2}. \tag{2.20}$$

Moreover, the autocovariance function of the zero-mean stationary signal $x(t)$ is given by

$$\mathbb{E}[x(t)x(t+k)] - \mu^2 = \frac{\sigma^2}{1-a^2} a^{|k|}. \tag{2.21}$$

Hence, the autocovariance function decays with a time constant $-1/\ln|a|$. The PSD function of $x(t)$ is calculated from the discrete-time Fourier transform of the autocovariance function:

$$\begin{aligned} S_{xx}(\omega) &= \frac{1}{\sqrt{2\pi}} \sum_{k=-\infty}^{\infty} \frac{\sigma^2}{1-a^2} a^{|k|} e^{-j\omega k} \\ &= \frac{1}{\sqrt{2\pi}} \left(\frac{\sigma^2}{1+a^2-2a\cos\omega} \right). \end{aligned} \tag{2.22}$$

2.1.2 Nonstationary Process

For nonstationary processes, the statistics of correlation functions depend on time. Specifically, the nonstationary autocorrelation and cross-correlation functions at any pair of fixed times t_1 and t_2 are defined by

$$C_{xx}(t_1, t_2) = \mathbb{E}[x(t_1)x(t_2)],$$
$$C_{xy}(t_1, t_2) = \mathbb{E}[x(t_1)y(t_2)].$$

It can be proved [82] that the following cross-correlation inequality holds:

$$|C_{xy}(t_1, t_2)|^2 \le C_{xx}(t_1, t_1) C_{yy}(t_2, t_2).$$

Provided we let $t_1 = t - \tau/2$ and $t_2 = t + \tau/2$ such that $\tau = t_2 - t_1$ and $t = (t_1 + t_2)/2$, we can define *double-time correlation functions*

$$C_{xx}(t_1, t_2) = \mathbb{E}\left[x\left(t - \frac{1}{2}\tau\right) x\left(t + \frac{1}{2}\tau\right)\right] = \mathbb{E}\left[R_{xx}(t, \tau)\right], \tag{2.23}$$

$$C_{xy}(t_1, t_2) = \mathbb{E}\left[x\left(t - \frac{1}{2}\tau\right) y\left(t + \frac{1}{2}\tau\right)\right] = \mathbb{E}\left[R_{xy}(t, \tau)\right], \tag{2.24}$$

where $R_{xx}(t, \tau) = x(t - \frac{1}{2}\tau)x(t + \frac{1}{2}\tau)$ and $R_{xy}(t, \tau) = x(t - \frac{1}{2}\tau)y(t + \frac{1}{2}\tau)$ define two *local* windowed correlations of the nonstationary signals. In the (t, τ) plane, it is possible to separate nonstationary correlation functions into stationary and nonstationary components. Specifically, one can write

$$\mathbb{E}[R(t,\tau)] = A(t)C(\tau) = A\left(\frac{t_1+t_2}{2}\right)C(t_2-t_1). \tag{2.25}$$

Correspondingly, in spectrum analysis, in order to to characterize a nonstationary time series (random process), we define the *Wigner–Ville distribution* (WVD) as [177]

$$\begin{aligned} W_{xx}(t,\omega) &= \int_{-\infty}^{\infty} x\left(t+\frac{1}{2}\tau\right)x\left(t-\frac{1}{2}\tau\right)\exp(-j\omega\tau)\,d\tau \\ &= \int_{-\infty}^{\infty} R_{xx}(t,\tau)\exp(-j\omega\tau)\,d\tau \end{aligned} \tag{2.26}$$

When the signal $X(t)$ is stationary, namely

$$C_{xx}(t,t+\tau) = \mathbb{E}[x(t)x(t+\tau)] = C_{xx}(0,\tau) \tag{2.27}$$

then the Wigner–Ville spectrum $W_{xx}(t,\omega)$ is equivalent to the PSD $S_{xx}(\omega)$. Figure 2.2 presents an example of applying WVD and short-time Fourier transform to a nonstationary speech signal.

An important property of the Wigner–Ville spectrum is that its marginal distributions in *time* and *frequency* give rise to simple second-order statistics of the random process $x(t)$:

$$\int_{-\infty}^{\infty} W_{xx}(t,\omega)\,dt = S_{xx}(\omega), \tag{2.28}$$

$$\int_{-\infty}^{\infty} W_{xx}(t,\omega)\,d\omega = C_{xx}(t,t) = \mathrm{var}[x(t)]. \tag{2.29}$$

If the signal $x(t)$ is deterministic, then we have

$$\int_{-\infty}^{\infty} W_{xx}(t,\omega)\,dt = |X(\omega)|^2,$$

$$\int_{-\infty}^{\infty} W_{xx}(t,\omega)\,d\omega = |x(t)|^2,$$

where $X(\omega)$ denotes the Fourier transform of $x(t)$ and $W_{xx}(t,\omega)$ is viewed as a *time–frequency distribution* of the signal $x(t)$.

In a manner similar to the stationary process, the eigenanalysis of the autocorrelation function of the nonstationary process can be carried out; see Appendix 2A for details.

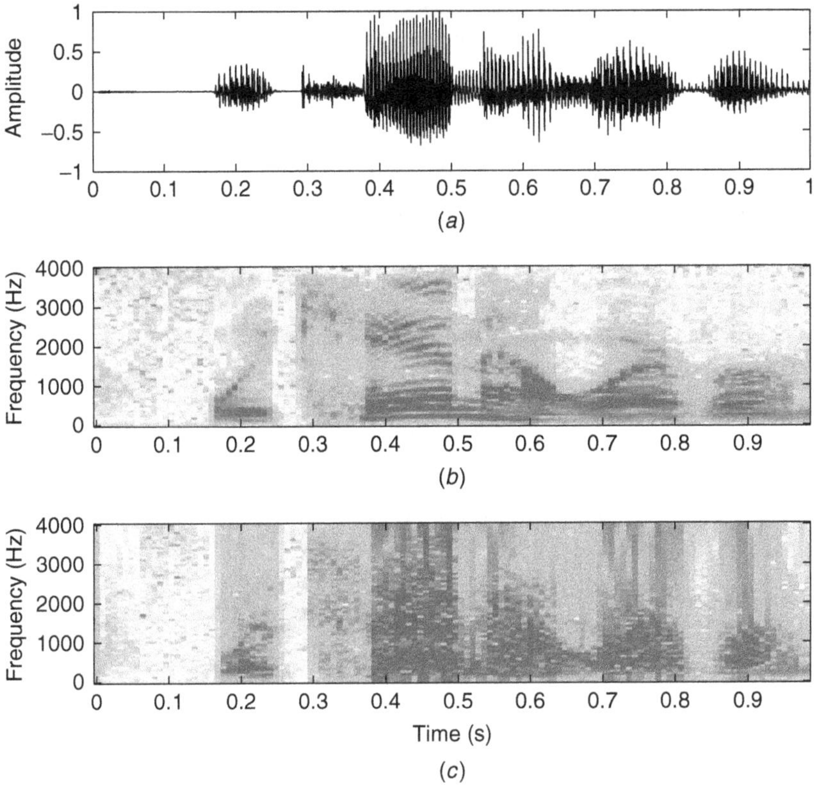

Figure 2.2 Demonstration of the spectrum analysis of a nonstationary speech signal. (*a*) Temporal male speech /we can however/ (with 8 kHz sampling frequency and 1 s duration). (*b*) Speech spectrogram based on short-time Fourier transform with a 128-point FFT and 32-ms Hanning window. (*c*) Wigner–Ville distribution. Note that (*b*) and (*c*) are both properly scaled in the log domain for visualization purpose.

2.1.3 Locally Stationary Process

A locally stationary process is a special class of nonstationary process that might be approximately stationary in a short timescale [587]. Specifically, if stochastic process $x(t)$ is locally stationary within the interval $l(x)$ (namely, $\forall t_0, t \in [t_0 - \frac{1}{2}l(x), t_0 + \frac{1}{2}l(x)]$), the correlation is approximately time invariant,

$$\mathbb{E}[x(t)x(t+\tau)] \approx C_{xx}(t_0; \tau) \quad \text{if } |\tau| \leq \frac{1}{2}l(x). \tag{2.30}$$

Alternatively, let $d(x)$ denote the *decorrelation length* that defines the maximum distance between two correlated points; then

$$\mathbb{E}[x(t)x(t+\tau)] \approx 0 \quad \text{if } |\tau| \geq d(x). \tag{2.31}$$

In addition, a locally stationary process has a decorrelation length that is smaller than half the size $l(x)$ of the stationarity interval:

$$d(x) < \frac{1}{2} l(x). \tag{2.32}$$

Such locally stationary processes are widely used in physics for analyzing real-life data.[3]

In light of the early work by Loève [569], Thomson [881] and Martin and Flandrin [594] introduced the "dynamic spectrum" (or "Loève transform") for a locally stationary process, which is defined as the *expected* WVD of the random process $x(t)$:

$$\begin{aligned}\Gamma(t, \omega) &= \int_{-\infty}^{\infty} C_{xx}(t, \tau) \exp(-j\omega\tau)\, d\tau \\ &= \int_{-\infty}^{\infty} \mathbb{E}\left[x\left(t + \frac{1}{2}\tau\right) x\left(t - \frac{1}{2}\tau\right)\right] \exp(-j\omega\tau)\, d\tau \\ &\equiv \mathbb{E}\left[W_{xx}(t, \omega)\right].\end{aligned} \tag{2.33}$$

Because of the expected value, unlike the standard WVD, $W_{xx}(t, \omega)$, the dynamic spectrum $\Gamma(t, \omega)$ is *nonnegative definite*. Correspondingly, for the time-varying autocorrelation function $C_{xx}(t_1, t_2) = \mathbb{E}[x(t_1)x(t_2)]$, the *generalized spectral density* (or "Loève spectrum") is defined as [375]

$$S_{xx}(\omega_1, \omega_2) = \mathbb{E}[X(\omega_1)X^*(\omega_2)], \tag{2.34}$$

where $X(\omega_1)$ and $X(\omega_2)$ denote the Fourier transforms of $x(t)$ at frequencies ω_1 and ω_2, respectively.

Two important points are noteworthy here:

- Unlike the stationary process, the autocorrelation functions of nonstationary and locally stationary processes cannot be estimated reliably from a few realizations of the random processes. In other words, the traditional periodogram is a poorly *biased* and *inconsistent* estimator of the true time-varying spectrum.
- Recall that the correlation function of the stationary process can be diagonalized with the Fourier integral, in which the cosine and sine basis functions are the eigenvectors (eigenfunctions) of the correlation operator. In contrast, the correlation operator of the nonstationary process is timevarying and impossible to diagonalize. However, it has been known [587] that it is possible to represent such a correlation operator by a sparse matrix; in other words, it is possible to find *local* cosine basis functions to "almost" characterize the eigenvectors.

2.1.4 Cyclostationary Process

Another instance of a stochastic process is the so-called cyclostationary (or periodically stationary) process [308–310]. Mathematically, a stochastic process $x(t)$ is said to be cyclostationary if its mean and correlation are both periodic with a period T, namely,

$$\mathbb{E}[x(t)] = \mathbb{E}[x(t+kT)], \qquad k \in \mathbb{Z}, \tag{2.35}$$

$$\mathbb{E}[x(t_1)x(t_2)] = \mathbb{E}[x(t_1+kT)x(t_2+kT)]. \tag{2.36}$$

For the cyclostationary process $x(t)$, the autocorrelation function is defined as the *average* of a time-dependent correlation function within a period,

$$\overline{C}_{xx}(\tau) = \frac{1}{T}\int_0^T C_{xx}(t;\tau)\,dt, \tag{2.37}$$

and the mean function is also defined as the average mean within a period,

$$\overline{\mu}(t) = \frac{1}{T}\int_0^T \mu(t)\,dt. \tag{2.38}$$

Since $\overline{C}_{xx}(t)$ may be viewed as a random realization of a periodic signal, it can be represented by Fourier series, and the power spectrum of the cyclostationary process is given by

$$S_c(\omega) = \int_0^T \overline{C}_{xx}(\tau)e^{-j\omega\tau}\,d\tau. \tag{2.39}$$

Namely, $S_c(\omega)$ and $\overline{C}_{xx}(\tau)$ consist of a Fourier transform pair.

2.1.5 Hilbert Spectrum Analysis

It is known that the Fourier spectral analysis is limited by two assumptions of the underlying signal: (i) stationarity and (ii) linearity. In the material presented above, we have discussed several spectrum analysis methods to tackle the issue of non-stationarity. In the material presented in this section, we will discuss another spectrum analysis method [414], rooted in the Hilbert transform, for the generic nonlinear, nonstationary signal.

For an arbitrary time series or random signal $x(t)$, its Hilbert transform is defined by (e.g., [348])

$$y(t) = \frac{1}{\pi}\int_{-\infty}^{\infty} \frac{x(\tau)}{t-\tau}\,d\tau, \tag{2.40}$$

where the integral is considered to be the Cauchy principal value[4] because of the possible singularity at $t = \tau$. Given the definition (2.40), $x(t)$ and $y(t)$ form the complex conjugate, and one can define an analytic signal[5] $z(t)$ as

$$z(t) = x(t) + jy(t) = a(t)e^{j\phi(t)} \qquad (j = \sqrt{-1}), \tag{2.41}$$

where

$$a(t) = \sqrt{x^2(t) + y^2(t)}, \qquad \phi(t) = \arctan\left(\frac{y(t)}{x(t)}\right).$$

Essentially, the Hilbert transform can be viewed as the convolution product of $x(t)$ and $1/t$, which emphasizes the local property (thereby tackling nonstationarity) of $x(t)$. Specifically, $a(t)$ defines the envelope and $\phi(t)$ defines the phase, from which one can further define the *instantaneous frequency*:

$$\omega(t) = \frac{d\phi(t)}{dt}. \tag{2.42}$$

Applying the Fourier transform to the analytic signal $z(t)$ yields

$$Z(\omega) = \int_{-\infty}^{\infty} a(t)e^{j\phi(t)}e^{-j\omega t}\,dt = \int_{-\infty}^{\infty} a(t)e^{j(\phi(t)-\omega t)}\,dt, \tag{2.43}$$

where the maximum contribution to $Z(\omega)$ is given by the frequency that satisfies the condition $d\left[\phi(t) - \omega t\right]/dt = 0$, which corresponds to equation (2.42).

Recently, a general spectrum analysis tool called the *Hilbert–Huang transform* (HHT) [413, 414] has been developed for *nonlinear, nonstationary* signals. The basic procedure of the HHT consists of two elements: *empirical mode decomposition* (EMD) and *Hilbert spectral analysis*.[6] The EMD produces adaptive intrinsic mode functions from the observed signal, whereas the Hilbert transform produces a "time–frequency–energy" representation of the signal based on the intrinsic mode functions.

In particular, the Hilbert spectrum is defined on the individual mode functions by computing the Hilbert transform and associated instantaneous frequencies; specifically, the signal $x(t)$ can be represented by [414]

$$x(t) = \sum_{i=1}^{n} a_i(t)\exp\left(j\int \omega_i(t)\,dt\right). \tag{2.44}$$

where n denotes the total number of individual modes [414] from the mode decomposition method. Equation (2.44) defines both the amplitude and the frequency of each component as functions of time; it can be viewed as a form of generalized Fourier expansion that accounts for the non-stationarity. In addition, the expansion in (2.44) separates clearly the amplitude modulation (AM) and frequency modulation (FM), which naturally incorporates the linear and nonlinear properties.

The time–frequency distribution of the amplitude is designated as the Hilbert amplitude spectrum, or Hilbert spectrum, denoted as $H(t, \omega)$; its associated marginal spectrum, $h(\omega)$, is defined as[7]

$$h(\omega) = \int_0^T H(t, \omega)\, dt, \tag{2.45}$$

and the *degree of stationarity*, denoted as $\mathrm{DS}(\omega)$, is defined as [414]

$$\mathrm{DS}(\omega) = \frac{1}{T} \int_0^T \left(1 - \frac{H(t, \omega)}{n(\omega)}\right)^2 dt, \tag{2.46}$$

where $n(\omega)$ denotes the averaged marginal spectrum

$$n(\omega) = \frac{1}{T} h(\omega). \tag{2.47}$$

In addition, given the Hilbert spectrum, the instantaneous energy (IE) density can be defined as

$$\mathrm{IE}(t) = \int_\omega H^2(t, \omega)\, d\omega. \tag{2.48}$$

To illustrate the Hilbert–Huang spectrum analysis, we apply the HHT to a human EEG signal (with sampling rate 250 Hz and duration of 10s) and plot the resulting mode functions and Hilbert spectrum. As shown in Figure 2.3, the Hilbert spectrum can well characterize different "modes" underlying the decomposed mode functions of the real-life noise-contaminated EEG signal.

2.1.6 Higher Order Correlation-Based Bispectra Analysis

Traditional spectrum analysis methods are rooted in second-order moment statistics. However, spectrum analysis may also include higher order moments to define the so-called *bispectra* or *polyspectra* [113, 742]. For instance, one can define the third-order moment of a stochastic process $x(t)$ as

$$C_{xxx}(t_1, t_2, t_3) = \mathbb{E}[x(t_1)x(t_2)x(t_3)]. \tag{2.49}$$

For the time being, let us assume $x(t)$ is a strict-sense stationary process with zero mean; denoting $u = t_1 - t_3$ and $v = t_2 - t_3$ and setting $t_3 = t$, we then have [702]

$$C_{xxx}(t_1, t_2, t_3) = C_{xxx}(u, v) = \mathbb{E}[x(t + u)x(t + v)x(t)]. \tag{2.50}$$

The corresponding bispectrum of the stochastic process $x(t)$, denoted by $S(\mu, \nu)$, is defined as a two-dimensional Fourier transform of its third-order correlation function $C(u, v)$:

$$S_{xxx}(\mu, \nu) = \int_{-\infty}^{\infty} \int_{-\infty}^{\infty} C_{xxx}(u, v) e^{-j(\mu u + \nu v)}\, du\, dv, \tag{2.51}$$

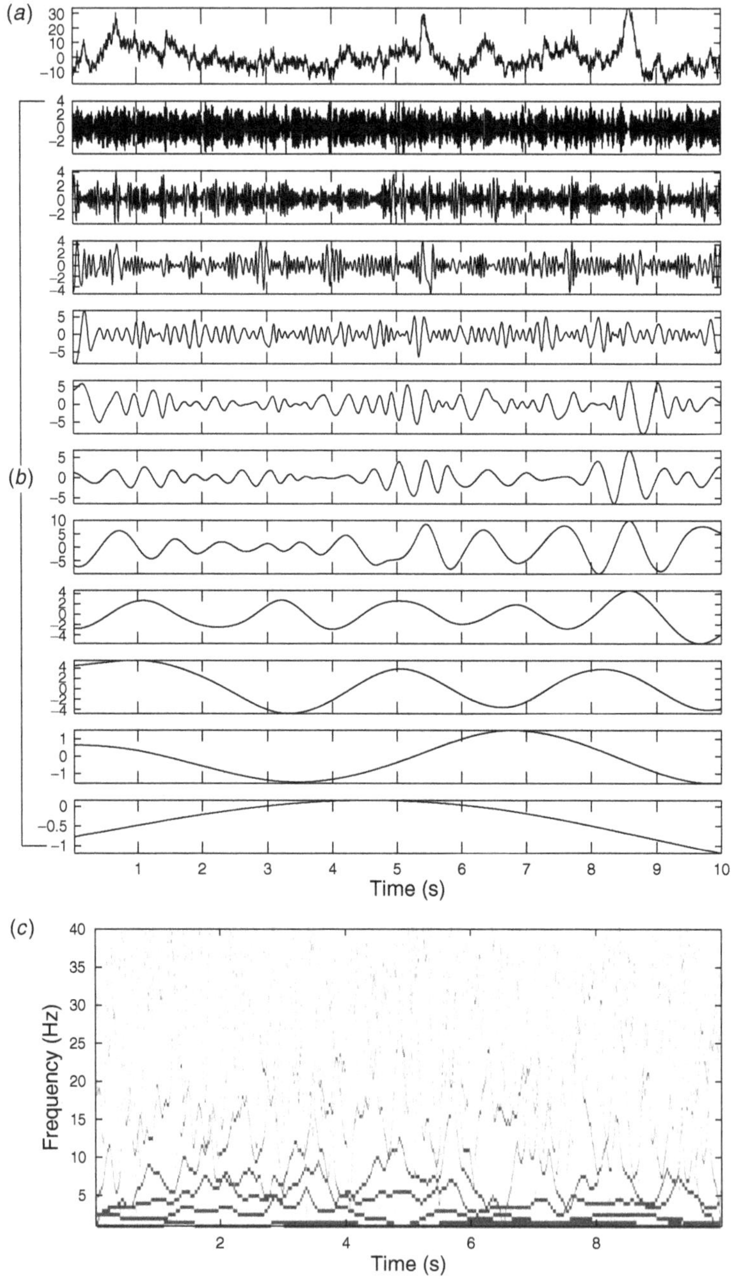

Figure 2.3 An illustration of applying the Hilbert–Huang transform to an EEG signal. (*a*) The raw EEG signal. (*b*) The extracted 11 individual mode functions. (*c*) Skeleton of the Hilbert spectra of the extracted mode functions.

where $S_{xxx}(\mu, \nu)$ is a real function, namely

$$S_{xxx}(-\mu, -\nu) = S^*_{xxx}(\mu, \nu), \tag{2.52}$$

where the asterisk defines the complex conjugate. Let $X(\omega)$ denote the Fourier transform of $x(t)$; then the third-order moment of $X(\omega)$ can be represented in terms of the bispectrum $S_{xxx}(\mu, \nu)$ [702]:

$$\mathbb{E}[X(\mu)X(\nu)X^*(\omega)] = 2\pi S_{xxx}(\mu, \nu)\delta(\mu + \nu - \omega). \tag{2.53}$$

An important property of the third-order correlation function $C_{xxx}(u, v)$ for a stochastic process is that it is invariant to six permutations of the numbers t_1, t_2, and t_3:

$$\begin{aligned} C_{xxx}(u, v) &= C_{xxx}(v, u) = C_{xxx}(-v, u - v) = C_{xxx}(-u, -u + v) \\ &= C_{xxx}(-u + v, -u) = C_{xxx}(u - v, -v). \end{aligned}$$

Correspondingly, the bispectrum also has the following invariance property:

$$\begin{aligned} S_{xxx}(\mu, \nu) &= S_{xxx}(\nu, \mu) = S_{xxx}(-\mu - \nu, \mu) = S_{xxx}(-\mu - \nu, \nu) \\ &= S_{xxx}(\nu, -\mu - \nu) = S_{xxx}(\mu, -\mu - \nu). \end{aligned}$$

In this regard, if any one region in the two-dimensional plane is identified, either (u, v) or (μ, ν), we may also determine the rest of the plane of interest. In practice, the notion of bispectrum can offer some advantages in characterizing the complex (e.g., nonstationary, or non-minimum-phase) system. The application of using bispectra for linear non-minimum-phase system identification may be found in [702].

2.1.7 Higher Order Functions of Time, Frequency, Lag, and Doppler

In signal processing, functions of one or two variables are commonly used: *time* (t), *frequency* (ω), *lag* (τ), and *doppler* (ζ). In time–frequency analysis, quadratic functions of two variables (such as the WVD, spectrogram, or ambiguity function) typically arise.

More generally, it is possible to introduce the one- or two-dimensional *Wigner mappings* to define signal representation functions of three or four variables [688]. Here, we present the corresponding definitions without going into detailed analysis and discussion. With the above notation and letting $x(t)$ and $X(\omega)$ be a Fourier transform pair, four classes of Wigner mapping functions can be defined as follows:

Time–lag–doppler function:

$$Q_x^{\omega}(t,\tau,\zeta)=\int x\left(t+\frac{\eta}{2}+\frac{1}{2}\tau\right)x^*\left(t+\frac{\eta}{2}-\frac{1}{2}\tau\right)\times x^*\left(t-\frac{\eta}{2}+\frac{1}{2}\tau\right)x\left(t-\frac{\eta}{2}-\frac{\tau}{2}\right)e^{-j\eta\zeta}\,d\eta. \tag{2.54}$$

Time–frequency–lag function:

$$Q_x^{\zeta}(t,\omega,\tau)=\int x\left(t+\frac{1}{2}\tau+\frac{\gamma}{4}\right)x^*\left(t-\frac{1}{2}\tau-\frac{\gamma}{4}\right)\times x^*\left(t+\frac{1}{2}\tau-\frac{\gamma}{4}\right)x\left(t-\frac{1}{2}\tau+\frac{\gamma}{4}\right)e^{-j\gamma\omega}\,d\gamma. \tag{2.55}$$

Time–frequency–doppler function:

$$Q_x^{\tau}(t,\omega,\zeta)=\frac{1}{2\pi}\int X\left(\omega+\frac{\zeta}{2}+\frac{\gamma}{4}\right)X^*\left(\omega-\frac{\zeta}{2}-\frac{\gamma}{4}\right)\times X^*\left(\omega+\frac{\zeta}{2}-\frac{\gamma}{4}\right)X\left(\omega-\frac{\zeta}{2}+\frac{\gamma}{4}\right)e^{j\gamma t}\,d\gamma. \tag{2.56}$$

Frequency–lag–doppler function:

$$Q_x^{t}(\omega,\tau,\zeta)=\frac{1}{2\pi}\int X\left(\omega+\frac{\eta}{2}+\frac{\zeta}{2}\right)X^*\left(\omega+\frac{\eta}{2}-\frac{\zeta}{2}\right)\times X^*\left(\omega-\frac{\eta}{2}+\frac{\zeta}{2}\right)X\left(\omega-\frac{\eta}{2}-\frac{\zeta}{2}\right)e^{j\eta\tau t}\,d\eta. \tag{2.57}$$

It was shown that [688]

- $\int Q_x(t,\omega,\tau,\zeta)\,d\omega=Q_x^{\omega}(t,\tau,\zeta)$,
- $\int Q_x(t,\omega,\tau,\zeta)\,d\zeta=Q_x^{\zeta}(t,\omega,\tau)$,
- $\int Q_x(t,\omega,\tau,\zeta)\,d\tau=Q_x^{\tau}(t,\omega,\zeta)$,
- $\int Q_x(t,\omega,\tau,\zeta)\,dt=Q_x^{t}(\omega,\tau,\zeta)$,

where the above-defined four functions are essentially the first-order marginals of the following quartic function of four variables, $Q_x(t,\omega,\tau,\zeta)$, as defined by [688]

$$Q_x(t,\omega,\tau,\zeta)=\int\int x\left(t+\frac{\eta}{2}+\frac{1}{2}\tau+\frac{\gamma}{4}\right)x^*\left(t+\frac{\eta}{2}-\frac{1}{2}\tau-\frac{\gamma}{4}\right)\times x^*\left(t-\frac{\eta}{2}+\frac{\tau}{2}-\frac{\gamma}{4}\right)x\left(t-\frac{\eta}{2}-\frac{\tau}{2}+\frac{\gamma}{4}\right)e^{-j\eta\zeta}e^{-j\gamma\omega}\,d\eta\,d\gamma,$$

or equivalently

$$Q_x(t,\omega,\tau,\zeta) = \left(\frac{1}{2\pi}\right)^2 \int\int X\left(\omega+\frac{\eta}{2}+\frac{\zeta}{2}+\frac{\gamma}{4}\right) X^*\left(\omega+\frac{\eta}{2}-\frac{\zeta}{2}-\frac{\gamma}{4}\right)$$
$$\times X^*\left(\omega-\frac{\eta}{2}+\frac{\zeta}{2}-\frac{\gamma}{4}\right) X\left(\omega-\frac{\eta}{2}-\frac{\zeta}{2}+\frac{\gamma}{4}\right) e^{j\gamma t} e^{j\eta\tau}\, d\eta\, d\gamma.$$

In addition, other functions of one or two variables turn out to be the second- or third-order marginals of the quartic function $Q_x(t,\omega,\tau,\zeta)$. For instance, we have [688]

- $\int\int Q_x(t,\omega,\tau,\zeta)\, d\omega\, d\zeta = |C_{xx}(t,\tau)|^2$,
- $\int\int Q_x(t,\omega,\tau,\zeta)\, d\tau\, d\omega = |W_{xx}(t,\omega)|^2$,
- $\int\int\int Q_x(t,\omega,\tau,\zeta)\, d\omega\, dt\, d\zeta = 2|x(2t)|^2 \otimes |x(2t)|^2$,
- $\int\int\int Q_x(t,\omega,\tau,\zeta)\, d\tau\, dt\, d\zeta = 2|X(2\omega)|^2 \otimes |X(2\omega)|^2$,

where $\otimes$ denotes the convolution operation.

2.1.8 Spectrum Analysis of Random Point Process

The spectrum analysis techniques discussed thus far have been restricted to stochastic processes with numerical (either real- or complex-valued) data, for which a periodogram may be obtained by applying the FFT to the collected (and segmented) data samples. However, this is not applicable to random point processes, which are widely used for describing data that are regarded as a series of events randomly occurring in time. For instance, the neural spike trains discussed in Chapter 1 can be modeled as a random Poisson process. In contrast to the conventional stochastic process that is defined with the Lebesgue measure, the random point process is a probability measure on the space of countable subsets of a probability space [191]. For the time being, we restrict our discussion to the (wide-sense) stationary random point processes.

Let $C(t)$ denote a correlation function of a random point process; the correlation function is defined in terms of the *mean intensity* λ as

$$C(t) = \lambda\delta(t) + \lambda(m(t) - \lambda), \tag{2.58}$$

where the mean intensity parameter λ is defined by [518]

$$\lambda = \lim_{\Delta t\to 0^+} \frac{\Pr\{\text{event in } [t, t+\Delta t]\}}{\Delta t} \tag{2.59}$$

and $m(t)$ denotes the *conditional intensity function* given by

$$m(t) = \lim_{\Delta t\to 0^+} \frac{\Pr\{\text{event in } [u+|t|, u+|t|+\Delta t]\}}{\Delta t}, \qquad |t| > 0, \tag{2.60}$$

which is a symmetric function with zero value at the origin.

In order to conduct spectrum analysis for the random point process, we have to introduce two important concepts in the frequency domain: *spectrum of the intervals* and *spectrum of the counts* [192]:

- The spectrum of the intervals of a point process is the spectrum of the discrete-time series made up from the time intervals between consecutive occurrences.
- The spectrum of the counts of a point process, denoted by $S(\omega)$, is defined as the Fourier transform of the correlation function $C(t)$.

Given the correlation function (2.58), we can define its discretized sequence

$$C_d(k\,\Delta t) = \lambda\,\Delta t^{-1}\,\delta_{0k} + \lambda(m(k\,\Delta t) - \lambda), \qquad (k = 0, \pm 1, \ldots), \tag{2.61}$$

where δ_{ij} denotes the *Kronecker delta*, the discrete spectrum $S_d(\omega)$ is further defined by [518]

$$S_d(\omega) = \Delta t \sum_{k=-\infty}^{\infty} C_d(k\,\Delta t) e^{-jk\omega\Delta t}. \tag{2.62}$$

It follows from (2.58), (2.61), and (2.62) that $S_d(\omega)$ and $S(\omega)$ are related by the equation [518]

$$S_d(\omega) = \lambda + [S(\omega) - \lambda] \otimes \sum_{k=-\infty}^{\infty} \delta\left(\omega + \frac{2\pi k}{\Delta t}\right), \tag{2.63}$$

where $\otimes$ denotes the convolution operation. Hence, by choosing an appropriate time interval Δt, $S_d(\omega)$ will obtain a good approximation of the true spectrum $S(\omega)$ since $|S(\omega - \lambda)|$ decays to zero rapidly with increasing $|\omega|$ for most random processes.

Lago et al. [518] have proposed an AR spectral modeling method for point processes based on estimating the correlation function $C(t)$. Motivated by the *Wold decomposition theorem* and *spectral modeling* [585], they assume that $C_d(k\,\Delta t)$ can be modeled by a p-order AR process such that

$$-C_d(k\,\Delta t) = \sum_{n=1}^{p} a_n C_d(|k-n|\Delta t) \qquad (k = 1, \ldots, p), \tag{2.64}$$

where the order p denotes the number of poles required to fit $S_d(\omega)$ with an all-pole spectrum $S_a(\omega)$:

$$S_a(\omega) = \frac{V}{\left|1 + \sum_{n=1}^{p} a_n e^{-jn\omega\Delta t}\right|^2}, \tag{2.65}$$

where V is a constant that is related to the minimum of the error measure; given $\{C_d(k\,\Delta t)\}$, the AR parameters $\{a_k\}$ can be determined by the *Yule–Walker equations* [369].

Technical details of estimating the conditional intensity and correlation functions of a stationary point process are referred to [114, 191, 192]; see also Appendix 2B for a brief description.

2.2 WIENER FILTER

In addition to spectrum analysis, correlation features just as prominently in filter theory. The term *filter* is commonly used to refer to a system that is designed to extract information about a prescribed quantity of interest from noisy data. In studying harmonic analysis and stochastic processes, Norbert Wiener [957] first proposed the concept of an optimal filter for the processing of a signal that is corrupted by additive noise; such a filter was subsequently referred to as the *Wiener filter* in honor of his pioneering work in statistical signal processing. The Wiener filter has important applications in statistical signal processing, especially for a wide range of wide-sense stationary stochastic processes that invoke only second-order cumulant statistics [459]. The notion of "Wiener filtering" is rather generic and can be defined in either the frequency domain or time domain. Applications of the Wiener filter include, for instance, signal denoising, signal restoration, prediction, and smoothing.

One of the original motivations and applications of the Wiener filter is the problem of prediction. Consider a signal model $x(t) = s(t) + n(t)$, where $s(t)$ denotes a real-valued random process and $n(t)$ denotes additive noise. Now, the goal is to design a filter, defined by the impulse response $h(t)$, to estimate the future value $s(t+\alpha)$ (where $\alpha > 0$) of the random process (note that, when $\alpha = 0$ and $\alpha < 0$, the *prediction* problem changes to the *filtering* and *smoothing* problem, respectively), given the present and past values of the noisy observations $x(t)$:

$$\begin{aligned}\hat{s}(t+\alpha) &= \mathbb{E}[s(t+\alpha)|x(t-\tau);\, \tau \geq 0] \\ &= \int_0^{\infty} h(\beta)x(t-\beta)\,d\beta. \end{aligned} \tag{2.66}$$

In equation (2.66), $\hat{s}(t+\alpha)$ represents the predicted output of a linear time-invariant (LTI) *causal* system[8] [associated with a transfer function $H(z)$] given an input signal $x(t)$. To determine $h(t)$ or $H(z)$, we resort to the *principle of orthogonality*:

1. The estimation error produced by the Wiener filter is orthogonal to the input signal.
2. The error signal is white in the sense that the autocorrelation function of the error signal is an ideal Dirac delta function.

Written in mathematical terms, we have

$$\mathbb{E}\left\{\left[s(t+\alpha) - \int_0^\infty h(\beta)x(t-\beta)\,d\beta\right]x(t-\tau)\right\} = 0 \qquad (\tau \geq 0). \quad (2.67)$$

Rearranging the terms of the above equation, we obtain the continuous-time *Wiener–Hopf equation*[9]

$$C_{sx}(\tau+\alpha) = \int_0^\infty h(\beta)C_{xx}(\tau-\beta)\,d\beta, \qquad (2.68)$$

where $C_{sx}(\tau+\alpha) = \mathbb{E}[s(t+\alpha)x(t-\tau)]$ and $C_{xx}(\tau-\beta) = \mathbb{E}[x(t-\beta)x(t-\tau)]$. The solution of the impulse response $h(t)$ that satisfies (2.68) is known as the *causal Wiener filter*. Let $e(t) = s(t+\alpha) - \hat{s}(t+\alpha)$ denote the prediction error; the Wiener filter is optimal in that it minimizes the mean-square error (MSE):

$$J = \mathbb{E}[e^2(t)] = \mathbb{E}\left[\left|s(t+\alpha) - \mathbb{E}[s(t+\alpha)|x(t-\tau)]\right|^2\right], \qquad (2.69)$$

which obtains the minimum MSE (MMSE) $J_{\min}$.

To obtain the causal Wiener filter $h(t)$, let us first consider the prediction problem in a *noncausal* system, in which the *noncausal Wiener filter* [denoted as $h_0(t)$] satisfies

$$C_{sx}(\tau+\alpha) = \int_{-\infty}^\infty h_0(\beta)C_{xx}(\tau-\beta)\,d\beta. \qquad (2.70)$$

Applying the Fourier transform to both sides, we obtain

$$S_{sx}(\omega)e^{j\omega\alpha} = H_0(\omega)S_{xx}(\omega), \qquad (2.71)$$

and the noncausal Wiener filter in the frequency domain is derived by

$$\begin{aligned} H_0(\omega) &= \frac{S_{sx}(\omega)e^{j\omega\alpha}}{S_{xx}(\omega)} \\ &= \frac{S_{ss}(\omega) + S_{sn}(\omega)}{S_{ss}(\omega) + S_{nn}(\omega) + 2\,\mathrm{Re}[S_{sn}(\omega)]}e^{j\omega\alpha}. \end{aligned} \qquad (2.72)$$

When $s(t)$ and $n(t)$ are uncorrelated, equation (2.72) may be simplified as

$$H_0(\omega) = \frac{S_{ss}(\omega)}{S_{ss}(\omega) + S_{nn}(\omega)}e^{j\omega\alpha}, \qquad (2.73)$$

where the amplitude gain $|S_{ss}(\omega)|/|S_{nn}(\omega)|$ defines the SNR. In this prediction problem, the MMSE of Wiener filtering can be derived as

$$\begin{aligned} J_{\min} &= \frac{1}{2\pi}\int_{-\infty}^{\infty}\left(S_{ss}(\omega) - \frac{|S_{sx}(\omega)|^2}{S_{xx}(\omega)}\right)d\omega \\ &= \frac{1}{2\pi}\int_{-\infty}^{\infty} S_{ss}(\omega)\left(1 - |\rho_{sx}(\omega)|^2\right)d\omega, \end{aligned} \tag{2.74}$$

where

$$\rho_{sx}(\omega) = \frac{S_{sx}(\omega)}{\sqrt{S_{ss}(\omega)S_{xx}(\omega)}} \tag{2.75}$$

is called the *normalized coherence function* whose magnitude $|\rho_{sx}(\omega)|$ is a real function between 0 and 1 that measures the correlation between $s(t)$ and $x(t)$ at each frequency ω. In the special case where $s(t)$ and $n(t)$ are uncorrelated, we also have

$$J_{\min} = \frac{1}{2\pi}\int_{-\infty}^{\infty}\frac{S_{ss}(\omega)S_{nn}(\omega)}{S_{ss}(\omega) + S_{nn}(\omega)}\,d\omega. \tag{2.76}$$

Next, we further pursue the solution for the causal Wiener filter. While the mathematical derivation is somewhat lengthy, the basic idea is that the causal Wiener filter is the causal part of the noncausal Wiener filter if the measurement is white noise. To see this, we assume that $S_{xx}(\omega)$ satisfies the condition

$$\int_{-\infty}^{\infty}\frac{\log|S_{xx}(\omega)|}{1+\omega^2}\,d\omega < \infty, \tag{2.77}$$

which is known as the *Paley–Wiener condition*. It can be shown that $S_{xx}(\omega)$ can be factorized as follows (the so-called *spectral factorization*):

$$S_{xx}(\omega) = S_{xx}^{+}(\omega)S_{xx}^{-}(\omega), \tag{2.78}$$

where $S_{xx}^{+}(\omega)$ and $S_{xx}^{-}(\omega)$ denote the parts of the power spectrum with positive frequency and negative frequency, respectively. Taking the inverse Fourier transform of $S_{xx}^{+}(\omega)$ results in a signal that is zero at negative times (therefore causal), while taking the inverse Fourier transform of $S_{xx}^{-}(\omega)$ results in a signal that is zero at positive times (therefore anticausal).

If $S_{xx}(\omega)$ satisfies the Paley–Wiener condition (2.77), then the signal $x(t)$ is said to have a rational PSD. In the z-domain, alteratively, we can write the following spectral factorization equation [347]:

$$S_{xx}(z) = \sigma_x^2 Q(z)Q\left(\frac{1}{z}\right), \tag{2.79}$$

where σ_x^2 denotes the average power of $x(t)$; $Q(z)$ is a monic, stable, and minimum-phase causal filter (whose poles occur inside the unit circle, i.e., $|z| < 1$). Let $F(z) = 1/[\sigma_x Q(z)]$ be a stable and causal whitening filter; then applying $F(z)$ to $x(t)$ will yield a white noise signal $\varepsilon(t)$, and $C_{\varepsilon\varepsilon}(\tau) = \delta(\tau)$. Substituting $C_{xx}(\tau - \beta)$ with $C_{\varepsilon\varepsilon}(\tau - \beta)$ in the Wiener–Hopf equation (2.70), we obtain

$$h_0^+(\tau) = C_{s\varepsilon}(\tau + \alpha) \qquad (\tau > 0, \alpha > 0), \tag{2.80}$$

where $h_0^+(\tau)$ denotes the impulse response of the white-noise Wiener filter. If we define the causal part of a noncausal filter $h_0^+(t)$ in the z-domain as $H_0^+(z)$, then $H_0^+(z) = [z^\alpha S_{s\varepsilon}(z)]_+$. Given the cross-spectrum between $s(t)$ and $\varepsilon(t)$

$$S_{s\varepsilon}(z) = \frac{S_{sx}(z)}{\sigma_x Q(1/z)}, \tag{2.81}$$

the causal Wiener filter for the prediction problem is derived as [347]

$$H(z) = F(z)H_0^+(z) = \frac{1}{\sigma_x^2 Q(z)} \left[\frac{z^\alpha S_{sx}(z)}{Q(1/z)} \right]_+ . \tag{2.82}$$

That is, we can factorize the causal Wiener filter $H(z)$ as a cascade of whitening filter $F(z)$ and a noncausal Wiener filter $H_0^+(z)$ that is fed with white-noise input. Letting $z = e^{j\omega}$, we obtain the frequency response of the causal Wiener filter.

The notion of Wiener filtering can also be extended for discrete-time random signals. In the discrete-time domain, the Wiener filter corresponds to a linear transversal filter, or a finite-duration impulse response (FIR) filter. Let $\mathbf{x}(t) = [x(t), x(t-1), \ldots, x(t-N+1)]^{\mathrm{T}} \in \mathbb{R}^N$ denote an N-step time-delay input vector and let $\boldsymbol{\theta}(t) = [\theta_0(t), \ldots, \theta_{N-1}(t)]^{\mathrm{T}} \in \mathbb{R}^N$ denote the tap-weight vector; then the desired output $d(t)$ is represented by

$$\begin{aligned} d(t) &= \sum_{k=0}^{N-1} x(t-k)\theta_k(t) + e(t) \\ &= \mathbf{x}^{\mathrm{T}}(t)\boldsymbol{\theta}(t) + e(t), \end{aligned} \tag{2.83}$$

where $e(t)$ denotes the estimation error. Given observation sequences $\{x(t)\}$ and $\{d(t)\}$, the goal of the linear filter is to find an optimal weight vector $\boldsymbol{\theta}$ that achieves the MMSE. According to the (discrete-time) *Wiener–Hopf equation*, the optimal solution is given by the Wiener filter:

$$\begin{aligned} \boldsymbol{\theta}_o &= \left(\mathbb{E}[\mathbf{x}(t)\mathbf{x}^{\mathrm{T}}(t)]\right)^{-1} \left(\mathbb{E}[\mathbf{x}(t)d(t)]\right) \\ &\equiv \mathbf{C}_{xx}^{-1}\mathbf{p}, \end{aligned} \tag{2.84}$$

which equals the product of the inverse of an autocorrelation matrix of the input signal, $\mathbf{C}_{xx}^{-1}$, and the cross-correlation $\mathbf{p}$ between the input and desired output signals.

Specifically, the cost function that the Wiener filter minimizes is a paraboloid function (e.g., [369]):

$$J = \mathbb{E}[d^2(t)] + \boldsymbol{\theta}^{\mathrm{T}}\mathbf{C}_{xx}\boldsymbol{\theta} - \mathbf{p}^{\mathrm{T}}\boldsymbol{\theta} - \boldsymbol{\theta}^{\mathrm{T}}\mathbf{p}. \tag{2.85}$$

Given the Wiener solution (2.84), equation (2.85) achieves the global minimum value (i.e., MMSE):

$$J_{\min} = \mathbb{E}[d^2(t)] + \mathbf{p}^{\mathrm{T}}\mathbf{C}_{xx}^{-1}\mathbf{p}, \tag{2.86}$$

and equation (2.85) may be rewritten as

$$J = J_{\min} + (\boldsymbol{\theta} - \boldsymbol{\theta}_o)^{\mathrm{T}}\mathbf{C}_{xx}^{-1}(\boldsymbol{\theta} - \boldsymbol{\theta}_o). \tag{2.87}$$

Because of its optimality under ideal conditions, the Wiener filter solution often serves as a baseline for performance comparison.

2.3 LEAST-MEAN-SQUARE FILTER

The Wiener filter requires knowledge of the noise and signal statistics (variance or PSD), and the filtering procedure is nonadaptive, both of which may pose some limitations in practice. In order to develop an *adaptive* filter,[10] we design a learning rule that incrementally updates the tapweight to minimize the cost criterion. For this purpose, a simple yet powerful form to approach the solution is the error-correcting least-mean-square (LMS) learning rule [951]:

$$\boldsymbol{\theta}(t+1) = \boldsymbol{\theta}(t) + \eta\mathbf{x}(t)e(t), \tag{2.88}$$

where $e(t)$ denotes the estimation error $e(t) = d(t) - \mathbf{x}^{\mathrm{T}}(t)\boldsymbol{\theta}(t)$, $d(t)$ denotes the desired response, and η is a learning-rate parameter. According to (2.88), the correction term is proportional to the product of the tapinput vector $\mathbf{x}(t)$ and the estimation error $e(t)$. In the limit, as t approaches infinity, the correction term approaches the time-average cross-correlation function $\langle\mathbf{x}(t)e(t)\rangle$, which, in turn, approaches zero in accordance with the *principle of orthogonality*, whereupon the weight vector $\boldsymbol{\theta}(t)$ converges to the Wiener solution given in (2.84). In fact, we may make the following statement ([369], p. 270): "For an ergodic process, the LMS filter asymptotically approaches the Wiener filter, except for an excess mean squared error, as the number of observations approaches infinity." In some sense, the LMS rule may be viewed as a form of Hebbian learning, with the correlation between input and output being replaced by the correlation between tap-delay inputs and estimation error. We will elaborate more on this issue later in Chapter 3.

In the adaptive filter literature [369, 793], there are many variants of the LMS-like error-correcting rule that incorporate *nonlinearity* in terms of either *input* or *error*. In general, the correlative form of an adaptive filter rule is written as follows:

$$\boldsymbol{\theta}(t+1) = \boldsymbol{\theta}(t) + \eta f(\mathbf{x}(t))g(e(t)). \tag{2.89}$$

When f is nonlinear and g is linear, (2.89) takes a form of nonlinearity for the input signal; for instance, $f(\mathbf{x}(t)) = \mathbf{x}(t)/\|\mathbf{x}(t)\|^2$ gives the normalized LMS rule. When f is linear and g is nonlinear, (2.89) takes a form of nonlinearity for the error signal; for instance, the choice of $g(e(t)) = e^3(t)$ defines the least-mean-fourth (LMF) filter. For more variants of the choice of functions f and g, the interested reader is referred to [793] for details.

■ EXAMPLE 2.2

Let us consider an adaptive channel equalization problem [369]. The input signal is a real-valued random Bernoulli sequence $\{u(t)\}$ [namely, $u(t) = \pm 1$] with zero mean and unit variance. The signal is propagated over a time-invariant channel and then corrupted by the additive white noise $v(t)$, where $v(t)$ and $u(t)$ are independent of each other. The adaptive equalizer is aimed at correcting the distortion produced by the Gaussian channel. The block diagram of this experiment is shown in Figure 2.4*a*.

The tap-input of the equalizer at time t is written as

$$x(t) = \sum_{k=1}^{3} h_k x(t-k) + v(t), \tag{2.90}$$

where $v(t)$ is a random Gaussian noise process with variance $\sigma_v^2 = 0.001$ and h_k denotes the impulse response of the channel that is described by the

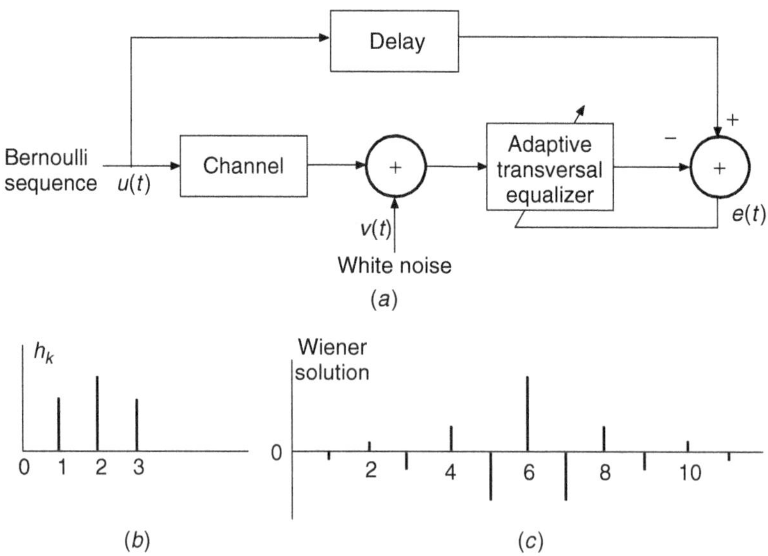

Figure 2.4 (*a*) Block diagram of adaptive equalization experiment. (*b*) The impulse response of the channel. (*c*) The impulse response of optimum transversal equalizer (Wiener solution).

raised cosine function [369]:

$$h_k = \begin{cases} \frac{1}{2}\left[1 + \cos\left(\frac{2\pi}{W}(k-2)\right)\right], & k = 1, 2, 3, \\ 0, & \text{otherwise}, \end{cases} \tag{2.91}$$

where the parameter W controls the amount of amplitude distortion produced by the channel (as well as the eigenvalue spread of the correlation matrix of tap inputs), with the distortion (and also eigenvalue spread) increasing with W. The equalizer has $N = 11$ taps, and the LMS transversal filter is used to model the impulse response that provides an approximate inversion of both minimum-phase and non-minimum-phase components of the channel response. The impulse responses of the channel as well as the optimum transversal equalizer (i.e., Wiener solution) are shown in Figures 2.4*b*,*c*.

In order to calculate the Wiener solution and the theoretical learning curves, we construct the correlation matrix of 11 tap inputs of the equalizer, $\mathbf{x}(t) = [x(t), x(t-1), \ldots, x(t-10)]^T$, that is, a symmetric 11×11 matrix. For the current problem, the input correlation matrix, denoted by $\mathbf{C} = \mathbb{E}[\mathbf{x}(t)\mathbf{x}^T(t)]$, has a *quintdiagonal* structure; namely, the only nonzero elements of $\mathbf{C}$ are on the main diagonal and the four diagonals directly above and below it, two on either side:

$$\mathbf{C} = \begin{bmatrix} r(0) & r(1) & r(2) & 0 & \cdots & 0 \\ r(1) & r(0) & r(1) & r(2) & \cdots & 0 \\ r(2) & r(1) & r(0) & r(1) & \cdots & 0 \\ 0 & r(2) & r(1) & r(0) & \cdots & 0 \\ \vdots & \vdots & \vdots & \vdots & \ddots & \vdots \\ 0 & 0 & 0 & 0 & \cdots & r(0) \end{bmatrix},$$

where

$$\begin{aligned} r(0) &= h_1^2 + h_2^2 + h_3^2 + \sigma_v^2, \\ r(1) &= h_1 h_2 + h_2 h_3, \\ r(2) &= h_1 h_3. \end{aligned}$$

Given the correlation matrix $\mathbf{C}$, the *eigenvalue spread*, defined as the ratio of maximum eigenvalue to the minimum eigenvalue of the correlation matrix, can be calculated as

$$\chi(\mathbf{C}) = \frac{\lambda_{\max}}{\lambda_{\min}}.$$

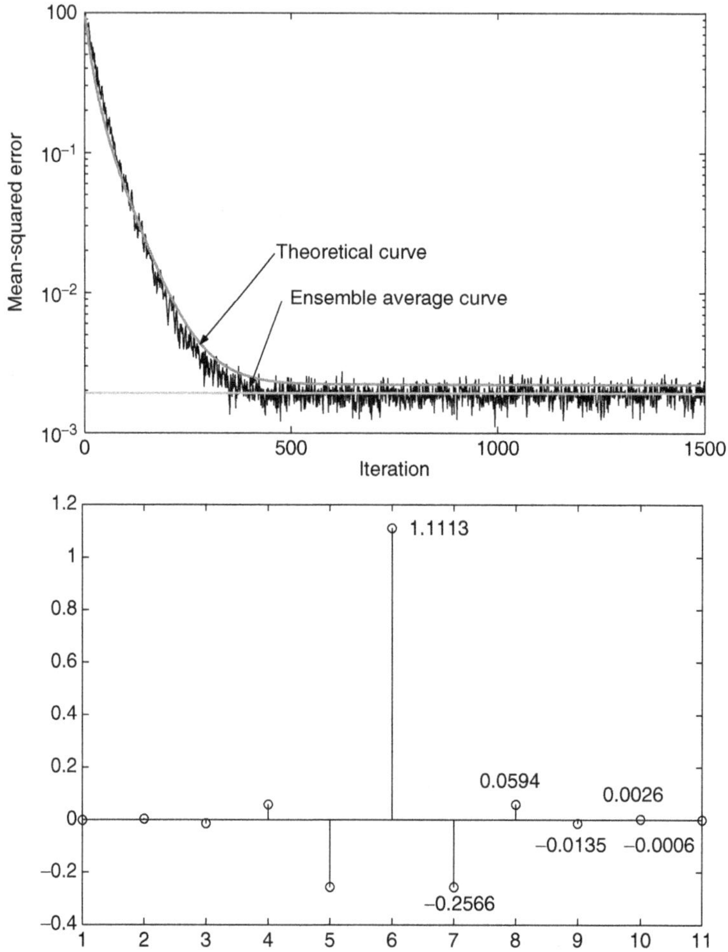

Figure 2.5 An example of asymptotic convergence of the LMS filter to the Wiener filter solution (horizontal straight line). *Top panel:* the ensemble LMS learning curves averaged over 100 independent trials in the adaptive channel equalization example. *Bottom panel:* the estimated impulse response of FIR transversal filter after 1500 iterations.

For a small learning-rate parameter η, the theoretic learning curve of the LMS filter can be derived [369]:

$$\begin{aligned} J(t) &= J_{\min} + \eta J_{\min} \sum_{k=1}^{N} \frac{\lambda_k}{2 - \eta\lambda_k} + \sum_{k=1}^{N} \lambda_k \left(|v_k(0)|^2 - \frac{\eta J_{\min}}{2 - \eta\lambda_k} \right) (1 - \eta\lambda_k)^{2t} \\ &\approx J_{\min} + \frac{\eta J_{\min}}{2} \sum_{k=1}^{N} \lambda_k + \sum_{k=1}^{N} \lambda_k \left(|v_k(0)|^2 - \frac{\eta J_{\min}}{2} \right) (1 - \eta\lambda_k)^{2t}, \end{aligned} \tag{2.92}$$

where λ_k are the eigenvalues calculated from the input correlation matrix $\mathbf{C}$ and $J_{\min}$ is the minimum MSE produced by the Wiener filter as given by (2.86). In (2.92), $v_k(0)$ is the entry of the vector $\mathbf{v}_k(0)$ that is generated by [369]

$$\mathbf{v}(t) = \mathbf{Q}^{\mathrm{T}} \boldsymbol{\varepsilon}_0(t) \tag{2.93}$$

where the orthogonal matrix $\mathbf{Q}$ is obtained by the eigenvalue decomposition (see Appendix C) of the correlation matrix $\mathbf{C}$, written as

$$\mathbf{Q}^{\mathrm{T}} \mathbf{C} \mathbf{Q} = \boldsymbol{\Lambda}, \tag{2.94}$$

where $\boldsymbol{\Lambda}$ is the diagonal matrix containing the eigenvalues in the diagonal and the columns of $\mathbf{Q}$ constitute an orthogonal set of eigenvectors. In (2.93), $\boldsymbol{\varepsilon}_0$ is calculated by

$$\boldsymbol{\varepsilon}_0(0) = \boldsymbol{\theta}_o - \boldsymbol{\theta}(0), \tag{2.95}$$

where $\boldsymbol{\theta}(0)$ denotes the initial weight vector of the filter and $\boldsymbol{\varepsilon}_0(t) = \boldsymbol{\theta}_o - \boldsymbol{\theta}(t)$ and $\boldsymbol{\theta}_o$ denotes the Wienner solution given in (2.84). When time approaches infinity, $t \to \infty$, the learning curve (2.92) will decay to a constant value

$$\begin{aligned} J(\infty) &= J_{\min} + \eta J_{\min} \sum_{k=1}^{N} \frac{\lambda_k}{2 - \eta \lambda_k} \\ &\approx J_{\min} + \frac{\eta J_{\min}}{2} \sum_{k=1}^{N} \lambda_k. \end{aligned} \tag{2.96}$$

In the current experiment, $\chi(\mathbf{C})$ is chosen to be 6.07 (for $W = 2.9$) and a fixed learning-rate parameter $\eta = 0.025$ is used. The experimental learning curve was obtained by ensemble averaging the squared value of the prediction error over 100 independent Monte Carlo trials and for varying t. Given initial parameter vector $\boldsymbol{\theta}(0) = \mathbf{0}$, the results of the learning curve as well as the estimated impulse response are shown in Figure 2.5. As seen in the figure, the theoretical curve fits rather well with the ensemble-average experimental curve.

2.4 RECURSIVE LEAST-SQUARES FILTER

In the adaptive filtering problem, the LMS filter is described by a simple form of correlative learning rule. It can also be extended to a recursive least-squares (RLS) filter by incorporating the computation of the time-varying correlation matrix of the tap-delay input signals into the learning rule [369]. Specifically, let $\mathbf{P}(t) = \mathbf{C}_{xx}^{-1}(t)$, where $\mathbf{C}_{xx}(t)$ denotes the correlation matrix estimate of the input signal. In a recursive estimation fashion, we have

$$\mathbf{C}_{xx}(t) = \lambda \mathbf{C}_{xx}(t-1) + \mathbf{x}(t)\mathbf{x}^{\mathrm{T}}(t), \tag{2.97}$$

where the scalar $0 < \lambda < 1$ is a forgetting factor. In light of the matrix inversion lemma (also called *Woodbury's identity*), we can derive

$$\mathbf{P}(t) = \lambda^{-1}\mathbf{P}(t-1) - \frac{\lambda^{-2}\mathbf{P}(t-1)\mathbf{x}(t)\mathbf{x}^{\mathrm{T}}(t)\mathbf{P}(t-1)}{1 + \lambda^{-1}\mathbf{x}^{\mathrm{T}}(t)\mathbf{P}(t-1)\mathbf{x}(t)}, \tag{2.98}$$

which is known as the *Riccati equation* for the RLS filter. [459]

With the inverse correlation matrix estimate at hand, the RLS filter can be written as

$$\boldsymbol{\theta}(t+1) = \boldsymbol{\theta}(t) + \mathbf{k}(t)e(t), \tag{2.99}$$

where we have defined

$$e(t) = d(t) - \mathbf{x}^{\mathrm{T}}(t)\boldsymbol{\theta}(t-1), \tag{2.100}$$

$$\mathbf{k}(t) = \frac{\mathbf{P}(t-1)\mathbf{x}(t)}{\lambda + \mathbf{x}^{\mathrm{T}}(t)\mathbf{P}(t-1)\mathbf{x}(t)}, \tag{2.101}$$

$$\mathbf{P}(t) = \lambda^{-1}\mathbf{P}(t-1) - \lambda^{-1}\mathbf{k}(t)\mathbf{x}^{\mathrm{T}}(t)\mathbf{P}(t-1). \tag{2.102}$$

The RLS filter can be viewed as a special class of Kalman filter [369]; it can also be understood as an LMS filter with a time-varying learning-rate matrix gain which approximates the inverse of the Hessian matrix (see Appendix 2C for details). The Kalman filter will be discussed in more detail in Chapter 7.

2.5 MATCHED FILTER

A basic problem that often arises in communication systems is that of detecting a pulse transmitted over a channel that is corrupted by additive channel noise. The *matched filter*, designed at the receiver, is aimed at helping to detect and recover the original message signal.

Consider a receiver model that is modeled by a LTI filter with impulse response $h(t)$. The filter input $x(t)$ consists of a pulse (message) signal $s(t)$ corrupted by additive channel noise $w(t)$:

$$x(t) = s(t) + w(t), \qquad 0 \leq t \leq T, \tag{2.103}$$

where T is an arbitrary observation interval. The $w(t)$ is assumed to be the sample function of a white-noise process with zero mean and two-sided PSD $N_0/2$. At the receiver, the filtered output is written as

$$y(t) = s_o(t) + n(t), \tag{2.104}$$

where $s_o(t)$ and $n(t)$ are produced by the signal component $s(t)$ and noise component $w(t)$ of the input $x(t)$, respectively.

Now the goal is to design an optimal filter $h(t)$ that maximizes the *peak pulse SNR*, which is defined as

$$\rho = \frac{|s_o(T)|^2}{\mathbb{E}[n^2(t)]}, \tag{2.105}$$

where $|s_o(T)|^2$ denotes the instantaneous power in the output signal and $\mathbb{E}[n^2(t)]$ denotes the average output noise power. In light of the Fourier transform, we can derive the expression of (2.105) as [366]

$$\rho = \frac{\left|\int_{-\infty}^{\infty} H(\omega)S(\omega)\exp(j2\pi\omega T)\,d\omega\right|^2}{(N_0/2)\int_{-\infty}^{\infty}|H(\omega)|^2\,d\omega}. \tag{2.106}$$

By virtue of *Schwartz's inequality*, it can be shown [366] that the maximum peak pulse SNR is given by

$$\rho_{\max} = \frac{2}{N_0}\int_{-\infty}^{\infty}|S(\omega)|^2\,d\omega, \tag{2.107}$$

in which case the optimal frequency response $H(\omega)$ has the form

$$H_{\text{opt}}(\omega) = cS^*(\omega)\exp(-j2\pi\omega T), \tag{2.108}$$

where $S^*(\omega)$ denotes the complex conjugate of the Fourier transform of the input signal $s(t)$ and c is a scaling factor of appropriate dimension. For a real signal $s(t)$, taking the inverse Fourier transform of (2.108) yields the impulse response of the optimum filter:

$$\begin{aligned} h_{\text{opt}}(t) &= c\int_{-\infty}^{\infty} S^*(\omega)\exp[-j2\pi\omega(T-t)]\,d\omega \\ &= c\int_{-\infty}^{\infty} S(-\omega)\exp[-j2\pi\omega(T-t)]\,d\omega \\ &= c\int_{-\infty}^{\infty} S(\omega)\exp[j2\pi\omega(T-t)]\,d\omega \\ &= cs(T-t). \end{aligned} \tag{2.109}$$

The matched filter is widely used in communications for signal recovery. For example, a well-known example is the design of a correlation receiver for demodulation. Suppose the receiver detector consists of a bank of *correlators* (i.e., product-integrators), each supplied with a corresponding set of coherent reference signals or orthonormal basis functions $\{\phi_j(t)\}$ that are generated locally. The bank of correlators operates on the received signal $x(t)$ within the interval $0 \leq t \leq T$. Using an LTI filter with the impulse response $h_j(t)$, each correlator's filtered output is defined by

$$y_i(t) = \int_{-\infty}^{\infty} x(\tau) h_j(t - \tau)\, d\tau. \tag{2.110}$$

In order to recover the signal, a matched filter is designed to match to a time-reversed and delayed version of the input signal $\phi_j(t)$, namely

$$h_j(t) = \phi_j(T - t). \tag{2.111}$$

Substituting (2.111) into (2.110) yields

$$y_j(t) = \int_{-\infty}^{\infty} x(\tau)\phi_j(T - t + \tau) d\tau. \tag{2.112}$$

Sampling (2.112) at time $t = T$ yields

$$y_j(T) = \int_{-\infty}^{\infty} x(\tau)\phi_j(\tau)\, d\tau = \int_0^T x(\tau)\phi_j(\tau)\, d\tau, \tag{2.113}$$

which produces the output at the jth correlator.

The concept of matched filtering for a one-dimensional signal can also be generalized for a two-dimensional image. The two-dimensional matched filter, being a fixed-size template, is moved around a two-dimensional image to perform a weighted-sum operation between the template values and the image's pixel values. Similar to the one-dimensional case, the two-dimensional matched filter attempts to match the local feature of the image to produce a high degree of correlation [915].

2.6 HIGHER ORDER CORRELATION-BASED FILTERING

As discussed thus far, the canonical correlation notion used in filtering and spectrum analysis is based on second-order statistics. However, it is noteworthy that these concepts are general and by no means limited by second-order correlation statistics. In fact, in order to tackle the nonstationarity of a signal, one may need to include higher order statistics for filtering and spectrum analysis, which aim to enhance the robustness of the conventional methods to outliers.

For instance, the standard Wiener filter is based on second-order correlations and the uncorrelated Gaussian noise assumption. In practice, when the non-Gaussian nature of the signal is invoked, higher order correlation may be robust for signal filtering or denoising. As an example, let us consider a simple noise-corrupted signal model:

$$x(t) = s(t) + n(t), \tag{2.114}$$

where it is assumed here that the white noise $n(t)$ is zero mean and uncorrelated with the zero-mean non-Gaussian signal $s(t)$. Calculating the second- and third-order correlations of the observed signal $x(t)$ respectively yields

$$\begin{aligned} C_{xx}(\tau) &= \sum_{t=0}^{\infty} x(t)x(t+\tau) \\ &= C_{ss}(\tau) + C_{nn}(\tau), \end{aligned} \tag{2.115}$$

$$\begin{aligned} C_{xxx}(\tau) &= \sum_{t=0}^{\infty} x(t)x(t+\tau)x(t+\tau_0) \\ &= \sum_{t=0}^{\infty} s(t)s(t+\tau)s(t+\tau_0) = C_{sss}(\tau), \end{aligned} \tag{2.116}$$

where $\tau > 0$ and τ_0 is a positive constant; the last equality of (2.116) holds because the terms $\langle s^2(t+t_1)n(t+t_2)\rangle$ and $\langle s(t+t_1)n^2(t+t_2)\rangle$ $(\forall t_1, t_2)$ all vanish.

Unlike the matched filter, the desired input signal is usually unknown; therefore the impulse response of the filter needs to be estimated. An ad hoc strategy is to use the correlation statistic to replace the input signal. For instance, using a second-order correlation estimate $\hat{C}_{xx}(\tau)$, the impulse response of the filter can be designed as follows [12]:

$$h(t) = \hat{C}_{xx}(t-T), \qquad t = 0, 1, \ldots, 2T, \tag{2.117}$$

where $2T$ represents the length of the observed signal $x(t)$ for estimating the sample correlation statistic $\hat{C}_{xx}(\tau)$.

In a similar manner, the impulse response of the third-order filter can be designed to be proportional to the estimate of a third-order correlation statistic $\hat{C}_{xxx}(\tau)$ [321]:

$$h(t) = \begin{cases} \hat{C}_{xxx}(T-t), & t = 0, 1, \ldots, T, \\ \hat{C}_{xxx}(t-T), & t = T+1, T+2, \ldots, 2T. \end{cases} \tag{2.118}$$

The institution of such a filter design is justified by the observation that $C_{xxx}(\tau)$ preserves the signal structure and is insensitive to non-Gaussian noise.

Finally, the output of the filter, $y(t)$, is written as

$$y(t) = \gamma \sum_{\tau=0}^{2T} h(\tau)x(t-\tau), \tag{2.119}$$

where γ is a scaling factor that assures the *unity skewness gain* of the filter.

In the previous example, higher order correlation is constructed by naturally including higher-than-two order statistics. In addition, in some applications we can also construct higher order correlation by using certain mathematical tricks (such as "folding" the signal). For instance, given an observed finite-length discrete-time multivariate signal sequences $\{\mathbf{x}(t)\}_{t=1}^{T}$, the conventional second-order correlation matrix $\mathbf{C}_{xx}$ can be estimated as

$$\mathbf{C}_{xx} = \frac{1}{T}\sum_{t=1}^{T}\mathbf{x}(t)\mathbf{x}^{\mathrm{T}}(t). \tag{2.120}$$

Now, we can design a fourth-order correlation matrix $\mathbf{R}$ to replace (2.120) with

$$\mathbf{R} = \frac{2}{T}\sum_{t=1}^{T/2}\mathbf{u}(t)\mathbf{u}^{\mathrm{T}}(t), \tag{2.121}$$

where $\mathbf{u}(t)$ is defined as

$$\mathbf{u}(t) = \mathbf{x}(t) \otimes \mathbf{x}(T - t + 1), \tag{2.122}$$

with $\otimes$ denoting the Hardamard (componentwise) product. By this modification, the correlation matrix $\mathbf{R}$ now consists of fourth-order statistics of the signal. Then it is straightforward to use matrix $\mathbf{R}$ in place of $\mathbf{C}$ in specific signal processing applications.

2.7 CORRELATION DETECTOR

Just like what happens in the brain, correlation detection is also widely used in signal processing and communications. Autocorrelation or cross-correlation methods have been used as feature detectors in numerous applications [525]. According to the nature of the detected signal, a detection scheme can be designed for detecting either a *deterministic* or a *stochastic* signal, which depends on whether the signal is known at the receiver side or not [471].

2.7.1 Coherent Detection

A simple yet popular method of in detecting deterministic signals is so-called coherent detection, which aims at recovering the transmitted or message signals in the presence of noise at the receiver [366]. Suppose the received signal $x(t)$ is corrupted by noise, as shown by

$$x(t) = s_i(t) + w(t), \tag{2.123}$$

where $\{s_i(t)|i = 1, 2, \ldots, M\}$ denotes the set of signals transmitted with equal probability $1/M$ and a specific signal constellation; $w(t)$ denotes the additive white Gaussian noise with zero mean and power spectral density $N_0/2$. To decode the transmitted signals of interest, the received signal is applied to a bank of N correlators, which yields the observation vector $\mathbf{x} = \mathbf{s}_i + \mathbf{w}$. Assuming an additive white Gaussian noise (AWGN) channel model, the received signal points are located inside a "Gaussian-shaped" cloud centered around the message points (denoted by $\{m_i\}$), and the likelihood function can be written as

$$p_{\mathbf{x}}(\mathbf{x}|m_k) = (\pi N_0)^{-N/2} \exp\left[-\frac{1}{N_0}\sum_{j=1}^{N}(x_j - s_{kj})^2\right], \qquad k = 1, \ldots, M, \quad (2.124)$$

such that the estimate $\hat{m} = m_i$ if $p_{\mathbf{x}}(\mathbf{x}|m_k)$ is maximum for $k = i$. Expanding the logarithm of the likelihood function (2.124) yields

$$\begin{aligned} \log p_{\mathbf{x}}(\mathbf{x}|m_k) &= -\frac{1}{N_0}\sum_{j=1}^{N}(x_j - s_{kj})^2 - \frac{N}{2}\log(\pi N_0) \\ &= -\frac{1}{N_0}\left(\sum_{j=1}^{N} x_j^2 - 2\sum_{j=1}^{N} x_j s_{kj} + \sum_{j=1}^{N} s_{kj}^2\right) + C, \end{aligned} \quad (2.125)$$

where C denotes a constant. Since the term $\sum_{j=1}^{N} x_j^2$ is independent of the index k, the decision rule for maximum-likelihood decoding is to search for the maximum value of $2\sum_{j=1}^{N} x_j s_{kj} - \sum_{j=1}^{N} s_{kj}^2$ for all possible k. Notably, the term $\sum_{j=1}^{N} x_j s_{kj}$ represents the inner product (or cross-correlation) between the observation vector $\mathbf{x}$ and signal vector $\mathbf{s}_k$, namely $\langle \mathbf{x}, \mathbf{s}_k \rangle$; for this reason, this type of receiver is called the *correlation receiver* (or correlator-type receiver). A schematic of such a correlation receiver is illustrated in Figure 2.6.

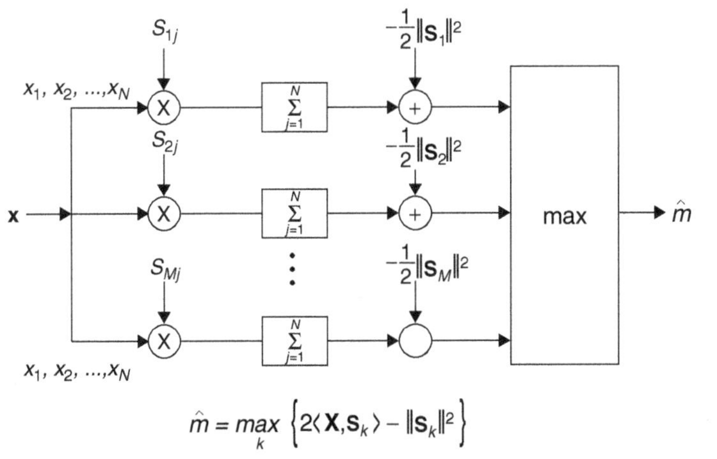

Figure 2.6 A schematic of correlation receiver in coherent detection.

2.7.2 Correlation Filter for Spatial Target Detection

In addition to the application for one-dimensional signals, correlation filters can also be developed for two-dimensional spatial target detection in image analysis—such a correlation filter is known as the *matched spatial filter* (MSF) [909]. The MSF is optimal for the detection of a specific image in the presence of white noise. In the literature, many generalizations of correlation filters have been developed to overcome the limitations of the conventional MSF, such as sensitivity to image distortion and nonrobustness to the colored background noise. In what follows, we describe a robust correlation detection procedure proposed in [510] for a Markov model of background clutter.

Let $\mathbf{x}_i$ denote an $m \times 1$ column vector that is vectorized from an $N \times N$ image (i.e., $m = N^2$) and let $\mathbf{X} = [\mathbf{x}_1, \ldots, \mathbf{x}_\ell]$ be the $m \times \ell$ matrix that contains ℓ observations of $\mathbf{x}$. The goal of a spatial correlation filter is to design a parameter vector $\mathbf{h}$ such that

$$\mathbf{X}^{\mathrm{T}}\mathbf{h} = \mathbf{y}, \tag{2.126}$$

where $\mathbf{y}$ is the specified constraint vector. The solution of the conventional MSF under the additive white-noise assumption is given by

$$\mathbf{h} = \mathbf{X}(\mathbf{X}^{\mathrm{T}}\mathbf{X})^{-1}\mathbf{y}, \tag{2.127}$$

where $\mathbf{X}^{\mathrm{T}}\mathbf{X}$ is essentially the autocorrelation matrix of the input image, except for the scaling factor $1/\ell$. When the background noise is additive and colored (with covariance matrix $\boldsymbol{\Sigma}$), then the optimal filter estimate with a minimum variance is given by [510]

$$\mathbf{h} = \boldsymbol{\Sigma}^{-1}\mathbf{X}(\mathbf{X}^{\mathrm{T}}\boldsymbol{\Sigma}\mathbf{X})^{-1}\mathbf{y}. \tag{2.128}$$

In the presence of cluttered noise in the background, it is simple to model the spatial correlation of images with a one-dimensional exponential model. Specifically, the correlation function for stationary data is modeled as

$$C_{xx}(\tau) = k\exp(-|\tau|\alpha), \tag{2.129}$$

where k is a constant, $1/\alpha$ is the correlation length, and τ is the shift variable. Setting $k = 1$ and $\exp(-\alpha) = \rho$ (where ρ denotes the correlation coefficient) yields

$$C_{xx}(\tau) = \rho^{|\tau|}. \tag{2.130}$$

The correlation matrix $\mathbf{C}(i, j) = C_{xx}(|i - j|)$ is then given by the Toeplitz matrix

$$\mathbf{C} = \begin{bmatrix} 1 & \rho & \rho^2 & \cdots & \rho^{m-1} \\ \rho & 1 & \rho & \cdots & \rho^{m-2} \\ \rho^2 & \rho & 1 & \cdots & \rho^{m-3} \\ \vdots & \vdots & \vdots & \ddots & \vdots \\ \rho^{m-1} & \rho^{m-2} & \rho^{m-3} & \cdots & 1 \end{bmatrix}, \tag{2.131}$$

and its inverse is given by

$$\mathbf{C}^{-1} = \frac{1}{1-\rho^2}\begin{bmatrix} 1 & -\rho & 0 & \cdots & 0 \\ -\rho & 1+\rho^2 & -\rho & \cdots & 0 \\ 0 & -\rho & 1+\rho^2 & \cdots & 0 \\ \vdots & \vdots & \vdots & \ddots & \vdots \\ 0 & \cdots & 0 & -\rho & 1 \end{bmatrix}, \tag{2.132}$$

which is a tridiagonal matrix. In light of the Gaussian–Markov image model, Kumar et al. [510] further generalized the correlation matrix for two-dimensional Markov data; specifically, the $N^2 \times N^2$ correlation matrix is given by

$$\mathbf{C} = \begin{bmatrix} \boldsymbol{\Sigma}_{11} & \rho\boldsymbol{\Sigma}_{11} & \rho^2\boldsymbol{\Sigma}_{11} & \cdots & \rho^{N-1}\boldsymbol{\Sigma}_{11} \\ \rho\boldsymbol{\Sigma}_{22} & \boldsymbol{\Sigma}_{22} & \rho\boldsymbol{\Sigma}_{22} & \cdots & \rho^{N-2}\boldsymbol{\Sigma}_{22} \\ \rho^2\boldsymbol{\Sigma}_{33} & \rho\boldsymbol{\Sigma}_{33} & \boldsymbol{\Sigma}_{33} & \cdots & \rho^{N-3}\boldsymbol{\Sigma}_{33} \\ \vdots & \vdots & \vdots & \ddots & \vdots \\ \rho^{N-1}\boldsymbol{\Sigma}_{NN} & \rho^{N-2}\boldsymbol{\Sigma}_{NN} & \rho^{N-3}\boldsymbol{\Sigma}_{NN} & \cdots & \boldsymbol{\Sigma}_{NN} \end{bmatrix}, \tag{2.133}$$

where $\boldsymbol{\Sigma}_{ij}$ denotes the cross-correlation matrix between the ith and jth rows of the two-dimensional image such that

$$\boldsymbol{\Sigma}_{ij} = \rho^{|i-j|}\boldsymbol{\Sigma}_{ii}, \qquad 1 \le i, j \le N,$$

and $\boldsymbol{\Sigma}_{ii}$ is the same for all i ($1 \le i \le N$) and is identical to (2.131) except for a change in dimensionality. In this case, the correlation matrix $\mathbf{C}$ in (2.133) is block Toeplitz and its inverse can be efficiently calculated. Let $\boldsymbol{\Sigma} = \boldsymbol{\Sigma}_{ii}$ for all i; then the inverse of (2.133) is given by

$$\mathbf{C}^{-1} = \frac{1}{1-\rho^2}\begin{bmatrix} \mathbf{I} & -\rho\boldsymbol{\Sigma}^{-1} & \mathbf{0} & \cdots & \mathbf{0} \\ -\rho\boldsymbol{\Sigma}^{-1} & (1+\rho^2)\Sigma^{-1} & -\rho\boldsymbol{\Sigma}^{-1} & \cdots & \mathbf{0} \\ \mathbf{0} & -\rho\boldsymbol{\Sigma}^{-1} & (1+\rho^2)\boldsymbol{\Sigma}^{-1} & \cdots & \mathbf{0} \\ \vdots & \vdots & \vdots & \ddots & \vdots \\ \mathbf{0} & \cdots & \mathbf{0} & -\rho\boldsymbol{\Sigma}^{-1} & \mathbf{I} \end{bmatrix}. \tag{2.134}$$

Now, given the N^2-dimensional vector $\mathbf{x}$ and the parameter vector $\mathbf{h}$ in (2.128), projecting $\mathbf{C}^{-1}$ onto the data vector space (namely, $\mathbf{z} = \mathbf{C}^{-1}\mathbf{x}$) is equivalent to conducting a two-dimensional spatial convolution operation with the original two-dimensional image. Specifically, Kumar et al. [510] showed that the matrix–vector multiplication operation

$$\mathbf{z}(i) = \frac{1}{1-\rho^2}\left[-\rho\boldsymbol{\Sigma}^{-1}\mathbf{x}(i-1) + (1+\rho^2)\boldsymbol{\Sigma}^{-1}\mathbf{x}(i) - \rho\boldsymbol{\Sigma}^{-1}\mathbf{x}(i+1)\right] \tag{2.135}$$

can be rewritten in terms of the two-dimensional convolution

$$z(i, j) = \omega(i, j) \otimes x(i, j), \tag{2.136}$$

where $x(i, j)$ represents the (i, j)th pixel in the original two-dimensional image (before vectorization) and $\omega(i, j)$ defines the 3×3 mask operator

$$\omega(i, j) = \frac{1}{(1-\rho^2)^2}\begin{bmatrix} \rho^2 & -\rho(1+\rho^2) & \rho^2 \\ -\rho(1+\rho^2) & (1+\rho^2)^2 & -\rho(1+\rho^2) \\ \rho^2 & -\rho(1+\rho^2) & \rho^2 \end{bmatrix}. \tag{2.137}$$

For $\rho = 0$, $\omega(i, j)$ reduces to the Dirac delta operator $\delta(i, j)$, and $z(i, j) = x(i, j)$.

Note that the optimal correlation filter $\mathbf{h}$ [or the mask operator $\omega(i, j)$] is merely dependent on the correlation coefficient ρ. In practice, ρ can be estimated by minimizing the MSE between the *statistical autocorrelation* $C_x(\tau)$ and the two-dimensional *spatial autocorrelation* (see [510]).

2.8 CORRELATION METHOD FOR TIME-DELAY ESTIMATION

In array signal processing, speech processing, or communication, the problem of time-delay estimation (TDE) often arises for localizing the signal source, beamforming, or estimating the signal's direction of arrival (DOA) [152].

Let us consider a classic broadband signal model for the TDE problem as

$$x_1(t) = s(t-k) + w_1(t), \tag{2.138}$$

$$x_2(t) = \alpha s(t-k-\tau) + w_2(t), \tag{2.139}$$

where $x_1(t)$ and $x_2(t)$ denote the output signals of two microphones, α is an attenuation factor due to the propagation effect, τ denotes the time delay between two microphones, k represents the propagation time from the unknown source $s(t)$ to the first microphone, and $w_1(t)$ and $w_2(t)$ denote two zero-mean stationary random noise processes that are mutually uncorrelated and are also both uncorrelated with the broadband signal $s(t)$. The task of the TDE problem is to find an estimate $\hat{\tau}$ of the true delay parameter τ.

The generalized cross-correlation (GCC) method [489] is a classic technique for TDE. Specifically, given two observed signals at two microphones, the following continuous-time GCC function is calculated:

$$C_{\text{GCC}}(\tau) = \int_{-\infty}^{\infty} \Phi(\omega) S_{x_1x_2}(\omega) e^{j2\pi\omega\tau}\, d\omega, \tag{2.140}$$

where $S_{x_1x_2}(\omega) = \mathbb{E}[X_1(\omega)X_2^*(\omega)]$ denotes the cross-spectrum between two signals $x_1(t)$ and $x_2(t)$ and $\Phi(\omega)$ is a weighting function (also called a *prefilter*). The choice of the weighting function is important in determining the TDE performance; it is often chosen according to some criteria. When $\Phi(\omega)$ is an identity function, the standard cross-correlation function is recovered from (2.140); when $\Phi(\omega) = 1/|S_{x_1x_2}(\omega)|$, equation (2.140) reduces to

$$C_{\text{GCC}}(\tau) = \int_{-\infty}^{\infty} \frac{S_{x_1x_2}(\omega)}{|S_{x_1x_2}(\omega)|} e^{j2\pi\omega\tau}\, d\omega, \tag{2.141}$$

which is the so-called *phase transform* (PHAT) algorithm [489]. In the ideal situation of uncorrelated noise and signal, it follows that

$$\frac{S_{x_1x_2}(\omega)}{|S_{x_1x_2}(\omega)|} = e^{-j2\pi\omega\delta_\tau}, \tag{2.142}$$

which provides a delta function at the true delay parameter. The PHAT algorithm requires no statistical characteristics of the signal and noise; it is independent of the source signal $s(t)$, and the weighted cross-spectrum only depends on the channel response—such a property is appealing especially when the characteristics of the signal $s(t)$ vary in time [152].

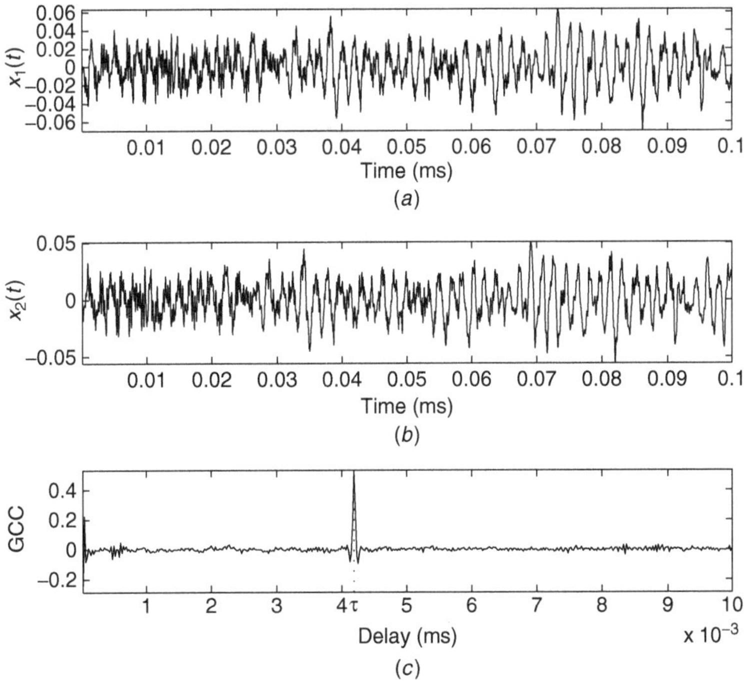

Figure 2.7 An illustration of time-delay estimation with the generalized GCC method. (*a, b*) Two stereo sound signals. (*c*) The position of the maximum peak of GCC is chosen to be the estimated delay parameter.

With the choice of specific weighting function, the optimal time-delay estimate is then derived by

$$\hat{\tau}_{\mathrm{GCC}} = \arg\max_{\tau} C_{\mathrm{GCC}}(\tau). \tag{2.143}$$

In Figure 2.7, we show a simple example where two stereo microphones receive one time-delayed signal in an ideal noise-free, nonreverberant condition, and the GCC method (PHAT algorithm) succeeds in recovering the true delay parameter in the maximum peak of the GCC profile.

2.9 CORRELATION-BASED STATISTICAL ANALYSIS

2.9.1 Principal-Component Analysis

Principal-component analysis (PCA) is a powerful statistical tool for data analysis [447], including feature extraction, dimensionality reduction, and denoising.[11] Stated in words, PCA finds an orthogonal transformation of a number of possibly correlated variables into a smaller number of uncorrelated variables known as principal components.

Given a set of independent and identically distributed (i.i.d.) multivariate data samples $\{\mathbf{x}_i\}_{i=1}^{\ell} \in \mathbb{R}^N$ which are assumed to have zero mean (i.e., the data are centered), the correlation matrix can be estimated from the samples:

$$\mathbf{C} = \mathbb{E}[\mathbf{x}\mathbf{x}^{\mathrm{T}}] \approx \frac{1}{\ell}\sum_{i=1}^{\ell} \mathbf{x}_i\mathbf{x}_i^{\mathrm{T}}. \tag{2.144}$$

Applying the eigenvalue decomposition (EVD) to matrix $\mathbf{C}$ yields[12]

$$\mathbf{C} = \mathbf{U}\boldsymbol{\Lambda}\mathbf{U}^{\mathrm{T}}, \tag{2.145}$$

where $\mathbf{U} = [\mathbf{u}_1, \ldots, \mathbf{u}_N]$ is an orthogonal matrix such that $\mathbf{U}\mathbf{U}^{\mathrm{T}} = \mathbf{I}$ with the columns $\{\mathbf{u}_i\}$ representing the eigenvectors and $\boldsymbol{\Lambda} = \mathrm{diag}\{\lambda_1, \lambda_2, \ldots, \lambda_N\}$ $(\lambda_1 \geq \lambda_2 \geq \cdots \geq \lambda_N)$ is a diagonal matrix containing the associated eigenvalues in the diagonal.

Given the eigenvectors, the data vector $\mathbf{x}$ can be represented by the weighted sum of eigenvectors:

$$\mathbf{x} = \sum_{i=1}^{N} \sqrt{\lambda_i}\mathbf{u}_i \approx \sum_{i=1}^{r} \sqrt{\lambda_i}\mathbf{u}_i \qquad (r < N), \tag{2.146}$$

where the approximation in the second step basically ignores minor component(s) associated with the small eigenvalue(s); by this the dimensionality reduction is achieved.

In light of spectral representation (2.146), taking the autocorrelation of $\mathbf{x}$ yields

$$\begin{aligned}\mathbb{E}[\mathbf{x}\mathbf{x}^{\mathrm{T}}] &= \mathbb{E}\left[\sum_{i=1}^{N}\sqrt{\lambda_i}\mathbf{u}_i\sum_{j=1}^{N}\sqrt{\lambda_j}\mathbf{u}_j^{\mathrm{T}}\right] \\ &= \sum_{i=1}^{N}\lambda_i\mathbf{u}_i\mathbf{u}_i^{\mathrm{T}},\end{aligned} \tag{2.147}$$

which essentially describes the *spectral theorem* (for details of eigenanalysis, see Appendix C).

The PCA technique is widely used for engineering applications, such as image compression, noise reduction, and feature extraction. As an illustration, Figure 2.8 shows the application of PCA in feature extraction for human face images, which results in the so-called eigenfaces [894]. In this example, the eigenfaces were calculated from a subset of 400 faces among the AT&T Olivetti face database. As seen in the figure, the eigenfaces represent the dominant features (in terms of variance) that are located in the subspace of face images.

Although PCA can be solved by the EVD of a correlation matrix, its computational cost is rather high [typically with the order of $\mathcal{O}(N^3)$]; it is therefore desirable to find adaptive procedures to tackle this problem in an efficient fashion, especially when the incoming data arrive sequentially. In Chapter 3, we will revisit the topic of PCA and discuss its various generalizations.

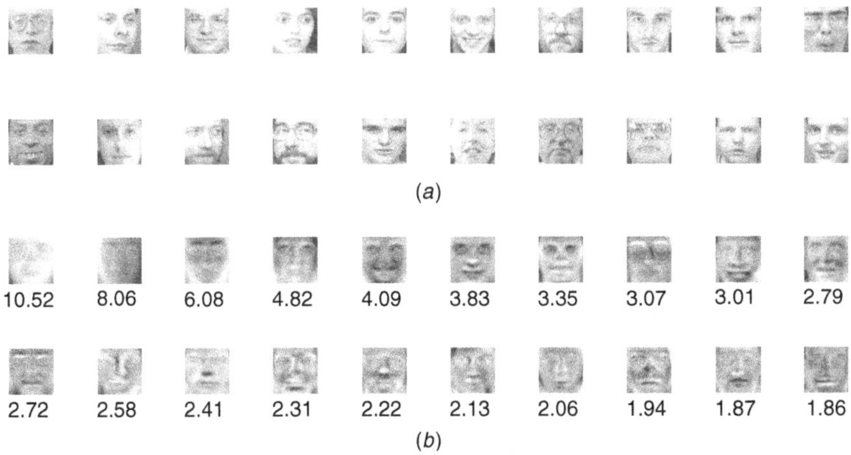

Figure 2.8 A demonstration of PCA for eigenfaces. (*a*) The selected 20 face images from the AT&T Olivetti face database. Each face is represented by a 64 × 64 gray-level (8-bit) image. (*b*) The estimated 20 "eigenfaces" arranged with descending eigenvalues order; the numerical value at the bottom of the eigenface indicates the percentage of the associated eigenvalue among the total eigenspectrum.

2.9.2 Factor Analysis

Factor analysis (FA) is a widely used multivariate analysis technique that aims to represent a set of random variables in terms of a smaller underlying set of factors [353]. In some sense, FA can be viewed as a generalization of PCA in that they are both hidden-variable-directed modeling techniques. However, unlike PCA, which seeks the maximum variance direction among the observations, in the FA model the factors are chosen to account for the correlations between the hidden variables.

Specifically, a simple FA model can be written as

$$x_j = \sum_{k=1}^{r} a_{jk} s_k + n_j \qquad (j = 1, \ldots, m; \;\; r < m) \tag{2.148}$$

or more concisely in the vector form

$$\mathbf{x} = \mathbf{A}\mathbf{s} + \mathbf{n}, \tag{2.149}$$

where $\mathbf{A} = \{a_{jk}\} \in \mathbb{R}^{m \times r}$ denotes the factor loading matrix, $\mathbf{s} \in \mathbb{R}^r$ denotes the latent variable vector (whose elements s_k are known as common factors) with zero mean (this assumption, however, can be relaxed) and covariance $\mathbf{C}$, and $\mathbf{n} \in \mathbb{R}^m$ denotes the additive noise vector with zero mean and covariance $\boldsymbol{\Psi} = \text{diag}\{\psi_1, \ldots, \psi_m\}$. In addition, we assume that the noise is whitened (namely, the matrix $\boldsymbol{\Psi}$ is diagonal), and the noise $\mathbf{n}$ and the latent variables $\mathbf{s}$ are uncorrelated; hence, we may write

$$\mathbb{E}[\mathbf{s}\mathbf{n}^T] = \mathbf{0}, \tag{2.150}$$

$$\mathbb{E}[\mathbf{x}\mathbf{x}^T] = \mathbf{A}\mathbf{C}\mathbf{A}^T + \boldsymbol{\Psi}, \tag{2.151}$$

$$\mathbb{E}[\mathbf{x}\mathbf{s}^T] = \mathbf{A}\mathbf{C}. \tag{2.152}$$

Factor analysis can be viewed as a probabilistic model. For instance, if the noise $\mathbf{n}$ is assumed to be Gaussian, then we obtain the conditional probability $p(\mathbf{x}|\mathbf{s}) = \mathcal{N}(\mathbf{A}\mathbf{s}, \boldsymbol{\Psi})$.

Given a set of i.i.d. observations $\{\mathbf{x}_i\}_{i=1}^{\ell}$, the goal of FA is to estimate the unknown parameters $\mathbf{A}$, $\mathbf{s}$, and $\boldsymbol{\Psi}$. The maximum-likelihood estimate (MLE) solution to FA is available in the literature [450]; an efficient inference procedure for FA is the *expectation–maximization* (EM) algorithm [211, 777] based on MLE (see Appendix E for some background).

Several additional comments are noteworthy:

- The solution of the FA model is not unique. There are many methods for obtaining the factor loadings. To illustrate this point, let us assume that the covariance matrix $\mathbf{C}$ is an identity matrix without loss of generality. Given an orthogonal matrix $\mathbf{U}$, we see that $(\mathbf{AU})(\mathbf{AU})^T = \mathbf{A}\mathbf{A}^T$; namely, the rotation of factors does not change the underlying observations.
- The number of factors, r, is often unknown and a statistical test is needed to find the "correct" model [943].

- Motivated by PCA, principal FA attempts to find the factors that account for the maximum amount of total communality. This is equivalent to finding the eigenvalues and eigenvectors of the reduced correlation matrix

$$\mathbf{R} = (\mathbf{ACA}^{\mathrm{T}} + \boldsymbol{\Psi}) - \boldsymbol{\Psi} = \mathbf{ACA}^{\mathrm{T}}.$$

  In the limit of zero noise, principal FA and PCA are identical. A detailed comparison between FA and PCA, as well as their connection, is discussed in [231, 943].
- In the conventional FA model, the additive noise is assumed to be Gaussian distributed; when the noise is non-Gaussian and the factors are mutually independent, the independent FA model may be derived [49].

2.9.3 Canonical Correlation Analysis

Canonical correlation analysis (CCA) [405] is a statistical method that identifies and quantifies the associations between two sets of variables; it searches for linear combinations of original variables that have *maximal correlation*. The pairs of linear combinations are called *canonical variables* and their correlations are called *canonical correlations*. Hence, canonical correlation measures the strength of linear association between two sets of multivariate variables. In the CCA model, the pairs of maximally correlated linear combinations are chosen such that they are orthogonal to those that have been already identified.

Stated mathematically, suppose we are given two sets of random variables $\mathbf{x}$ and $\mathbf{y}$ stored in two matrices $\mathbf{X} \in \mathbb{R}^{N\times p}$ and $\mathbf{Y} \in \mathbb{R}^{N\times q}$, respectively (where N denotes the dimensionality and $m = p + q$ denotes the total number of observations). Then we can create an augmented variable $\mathbf{z} = [\mathbf{x}^{\mathrm{T}}, \mathbf{y}^{\mathrm{T}}]^{\mathrm{T}}$ and construct a new matrix $\mathbf{Z} \in \mathbb{R}^{N\times m}$, namely $\mathbf{Z} = [\mathbf{X} \mid \mathbf{Y}]$. We now seek to find the linear combinations of the two sets of random variables $\mathbf{x}_i$ and $\mathbf{y}_i$ such that

$$\mathbf{u} = \mathbf{a}^{\mathrm{T}}\mathbf{X} = \sum_{i=1}^{p} a_i \mathbf{x}_i, \tag{2.153}$$

$$\mathbf{v} = \mathbf{b}^{\mathrm{T}}\mathbf{Y} = \sum_{i=1}^{q} b_i \mathbf{y}_i, \tag{2.154}$$

where $\mathbf{u}$ and $\mathbf{v}$ are two basis vectors that are constructed from the two sets of multivariate variables and $\mathbf{a}$ and $\mathbf{b}$ are two associated *regression coefficient* vectors.

With the definition of correlation coefficient

$$\mathrm{corr}(\mathbf{u}, \mathbf{v}) = \frac{\mathrm{cov}(\mathbf{u}, \mathbf{v})}{\sqrt{\mathrm{var}(\mathbf{u})\mathrm{var}(\mathbf{v})}}, \tag{2.155}$$

CCA seeks to find the solution to the following optimization problem:

$$\max_{\mathbf{a},\mathbf{b}} \ \mathrm{corr}(\mathbf{u}, \mathbf{v})$$

subject to

$$\mathrm{var}(\mathbf{u}) = \mathrm{var}(\mathbf{v}) = 1.$$

The dominant canonical correlation, denoted by ρ, is given by

$$\begin{aligned}\rho &= \max_{\mathbf{a},\mathbf{b}} \mathrm{corr}(\mathbf{a}^{\mathrm{T}}\mathbf{X}, \mathbf{b}^{\mathrm{T}}\mathbf{Y}) \\ &= \max_{\mathbf{a},\mathbf{b}} \frac{(\mathbf{a}^{\mathrm{T}}\mathbf{X}, \mathbf{b}^{\mathrm{T}}\mathbf{Y})}{\|\mathbf{a}^{\mathrm{T}}\mathbf{X}\| \cdot \|\mathbf{b}^{\mathrm{T}}\mathbf{Y}\|}. \end{aligned} \tag{2.156}$$

The optimization problem can be solved by a matrix decomposition method. First, the correlation matrix of the augmented variable $\mathbf{z}$ is constructed as

$$\begin{aligned}\mathbf{C}_{\mathbf{zz}} &= \mathbb{E}\left[(\mathbf{x}, \mathbf{y})(\mathbf{x}, \mathbf{y})^{\mathrm{T}}\right] \\ &= \begin{pmatrix} \mathbf{C}_{\mathbf{xx}} & \mathbf{C}_{\mathbf{xy}} \\ \mathbf{C}_{\mathbf{yx}} & \mathbf{C}_{\mathbf{yy}} \end{pmatrix}, \end{aligned} \tag{2.157}$$

where the block correlation matrix $\mathbf{C}_{\mathbf{zz}}$ contains the within-set correlation matrices $\mathbf{C}_{\mathbf{xx}}$ and $\mathbf{C}_{\mathbf{yy}}$ as well as the between-set correlation matrices $\mathbf{C}_{\mathbf{xy}}$ and $\mathbf{C}_{\mathbf{yx}}$ (where $\mathbf{C}_{\mathbf{xy}} = \mathbf{C}_{\mathbf{yx}}^{\mathrm{T}}$). Next, we solve the optimization problem with the Lagrangian method via an eigenvalue equation

$$\mathbf{B}^{-1}\mathbf{A}\mathbf{w} = \rho\mathbf{w}, \tag{2.158}$$

where $\mathbf{w} = [\mathbf{a}^{\mathrm{T}}, \mathbf{b}^{\mathrm{T}}]^{\mathrm{T}}$ denotes the eigenvector and

$$\mathbf{A} = \begin{pmatrix} \mathbf{0} & \mathbf{C}_{\mathbf{xy}} \\ \mathbf{C}_{\mathbf{yx}} & \mathbf{0} \end{pmatrix}, \qquad \mathbf{B} = \begin{pmatrix} \mathbf{C}_{\mathbf{xx}} & \mathbf{0} \\ \mathbf{0} & \mathbf{C}_{\mathbf{yy}} \end{pmatrix}.$$

In light of (2.158) and (2.156), we rewrite the dominant canonical correlation as

$$\rho = \max_{\mathbf{a},\mathbf{b}} \frac{\mathbf{a}^{\mathrm{T}}\mathbf{C}_{\mathbf{xy}}\mathbf{b}}{\sqrt{(\mathbf{a}^{\mathrm{T}}\mathbf{C}_{\mathbf{xx}}\mathbf{a})(\mathbf{b}^{\mathrm{T}}\mathbf{C}_{\mathbf{yy}}\mathbf{b})}}. \tag{2.159}$$

Comments. Several properties of CCA are noteworthy:

- Canonical correlation analysis is a generalization of PCA; indeed, we can show that PCA is a special case of CCA. In PCA, there is only a single set of random variables, hence the augmented variable $\mathbf{z}$ is essentially identical to the original variable $\mathbf{x}$. Specifically, if we let $\mathbf{A} = \mathbf{C}_{\mathbf{xx}}$ and $\mathbf{B} = \mathbf{I}$ (where $\mathbf{I}$ is the identity matrix), then the generalized eigenvalue problem (2.158) reduces to the conventional eigenvalue problem

  $$\mathbf{A}\mathbf{w} = \rho\mathbf{B}\mathbf{w} \longrightarrow \mathbf{C}_{\mathbf{xx}}\mathbf{u} = \lambda\mathbf{u}.$$

  Correspondingly, PCA maximizes the variance of projection of the data (namely, the single set of random variables), whereas CCA maximizes the correlation between a pair (or set) of projections of random variables.

- Canonical correlation analysis is also related to the partial least-squares (PLS) method, which is a linear regression method that seeks a small subset of the input variables whose directions have high variance and high correlation with response. Specifically, similar to CCA, PLS also amounts to solving a generalized eigenvalue problem $\mathbf{Aw} = \rho\mathbf{Bw}$, where

$$\mathbf{A} = \begin{pmatrix} \mathbf{0} & \mathbf{C_{xy}} \\ \mathbf{C_{yx}} & \mathbf{0} \end{pmatrix}, \qquad \mathbf{B} = \begin{pmatrix} \mathbf{I} & \mathbf{0} \\ \mathbf{0} & \mathbf{I} \end{pmatrix}.$$

- From an information-theoretic viewpoint, CCA can be viewed as a method that maximizes the mutual information between two sets of random variables. For simplicity, let us assume that $\mathbf{x} \in \mathbb{R}^p$ and $\mathbf{y} \in \mathbb{R}^q$ are two multivariate Gaussian random variables; then the mutual information between two sets of random variables $\mathbf{x}$ and $\mathbf{y}$ is defined by [509]

$$I(\mathbf{x}; \mathbf{y}) = -\frac{1}{2} \log \left(\frac{\det(\mathbf{C_{zz}})}{\det(\mathbf{C_{xx}}) \det(\mathbf{C_{yy}})} \right), \tag{2.160}$$

where the determinant ratio is known as the "generalized variance" in multivariate analysis. In light of the formulation of CCA, the generalized eigenvalue problem is rewritten as

$$\begin{pmatrix} \mathbf{C_{xx}} & \mathbf{C_{xy}} \\ \mathbf{C_{yx}} & \mathbf{C_{yy}} \end{pmatrix} \begin{pmatrix} \mathbf{a} \\ \mathbf{b} \end{pmatrix} = (1+\rho) \begin{pmatrix} \mathbf{C_{xx}} & \mathbf{0} \\ \mathbf{0} & \mathbf{C_{yy}} \end{pmatrix} \begin{pmatrix} \mathbf{a} \\ \mathbf{b} \end{pmatrix}.$$

The eigenvalues will then be obtained in pairs $\{1 \pm \rho_1, \ldots, 1 \pm \rho_r, 1, \ldots, 1\}$ [where $r = \min\{p, q\}$ and the parameters $(\rho_1, \ldots, \rho_r)$ are the canonical correlation coefficients]. Furthermore, optimization of such a generalized eigenvalue problem is related to the problem of maximizing the mutual information, which is defined by

$$\begin{aligned} I(\mathbf{x}; \mathbf{y}) &= -\frac{1}{2} \log \prod_{i=1}^{r} (1 - \rho_i^2) \\ &= -\frac{1}{2} \sum_{i=1}^{r} \log(1 - \rho_i^2), \end{aligned} \tag{2.161}$$

where $-1 < \rho_i < 1$.

- Let $\rho(\mathbf{x}, \mathbf{y})$ denote the *maximal* canonical correlation between $\mathbf{x} \in \mathbb{R}^p$ and $\mathbf{y} \in \mathbb{R}^q$ and let $I_\rho(\mathbf{x}; \mathbf{y}) = -\frac{1}{2} \log[1 - \rho^2(\mathbf{x}, \mathbf{y})]$; then the following information bounds hold [52]:

$$I_\rho(\mathbf{x}; \mathbf{y}) \leq I(\mathbf{x}; \mathbf{y}) \leq r I_\rho(\mathbf{x}; \mathbf{y}), \tag{2.162}$$

where $r = \min\{p, q\}$.

Generalizations of Linear CCA. It is possible to generalize the CCA model for more than two random variables. Specifically, Kettenring [473] has defined the general CCA as finding the smallest generalized eigenvalue problem for m multivariate random variables $\{\mathbf{x}_1, \ldots, \mathbf{x}_m\}$, and the generalized eigenvalue equation is given as

$$\begin{bmatrix} \mathbf{C}_{11} & \mathbf{C}_{12} & \cdots & \mathbf{C}_{1m} \\ \mathbf{C}_{21} & \mathbf{C}_{22} & \cdots & \mathbf{C}_{2m} \\ \vdots & \vdots & \ddots & \vdots \\ \mathbf{C}_{m1} & \mathbf{C}_{m2} & \cdots & \mathbf{C}_{mm} \end{bmatrix} \begin{bmatrix} \xi_1 \\ \xi_2 \\ \vdots \\ \xi_m \end{bmatrix} = \lambda \begin{bmatrix} \mathbf{C}_{11} & \mathbf{0} & \cdots & \mathbf{0} \\ \mathbf{0} & \mathbf{C}_{22} & \cdots & \mathbf{0} \\ \vdots & \vdots & \ddots & \vdots \\ \mathbf{0} & \mathbf{0} & \cdots & \mathbf{C}_{mm} \end{bmatrix} \begin{bmatrix} \xi_1 \\ \xi_2 \\ \vdots \\ \xi_m \end{bmatrix}$$

or, in short, $\mathbf{C}\boldsymbol{\xi} = \lambda \mathbf{D}\boldsymbol{\xi}$, where we have used the abbreviation $\mathbf{C}_{ij} = \mathbf{C}_{\mathbf{x}_i\mathbf{x}_j}$ in the above equation. Similarly, the mutual information among $\{\mathbf{x}_1, \ldots, \mathbf{x}_m\}$ can be defined as

$$\begin{aligned} I(\mathbf{x}_1, \ldots, \mathbf{x}_m) &= -\frac{1}{2} \log \left(\frac{\det(\mathbf{C})}{\det(\mathbf{C}_{11}) \cdots \det(\mathbf{C}_{mm})} \right) \\ &= -\frac{1}{2} \sum_{i=1}^{r} \log \lambda_i, \end{aligned} \tag{2.163}$$

where the λ_i are the generalized eigenvalues obtained from solving the eigenvalue problem $\mathbf{C}\boldsymbol{\xi} = \lambda \mathbf{D}\boldsymbol{\xi}$ and the ratio $\det(\mathbf{C})/\det(\mathbf{D})$ in (2.163) is often termed a "generalized variance."

Finally, the concept of CCA for a pair of random variables can be generalized to *functional space*. Specifically, instead of considering the correlation function of two random variables, we may analyze the functions of the random variables. Let $\mathcal{F}$ denote the vector space of functions $f : \mathbb{R}^N \mapsto \mathbb{R}$; then we can define the *$\mathcal{F}$-correlation* as the maximal correlation between the random variables $f_1(\mathbf{x})$ and $f_2(\mathbf{y})$ (where $\{\mathbf{x}, \mathbf{y}\} \in \mathbb{R}^N$, $f_1, f_2 \in \mathcal{F}$) as follows [52]:

$$\begin{aligned} \rho &= \max_{f_1, f_2 \in \mathcal{F}} \operatorname{corr}(f_1(\mathbf{x}), f_2(\mathbf{y})) \\ &= \max_{f_1, f_2 \in \mathcal{F}} \frac{\operatorname{cov}(f_1(\mathbf{x}), f_2(\mathbf{y}))}{\operatorname{var}(f_1(\mathbf{x}))^{1/2} \operatorname{var}(f_2(\mathbf{y}))^{1/2}}. \end{aligned} \tag{2.164}$$

If the random vectors $\mathbf{x}$ and $\mathbf{y}$ are mutually independent, then the $\mathcal{F}$-correlation will be zero. We will revisit this topic and discuss the kernelized version of the CCA model later in Chapter 4.

■ EXAMPLE 2.3

In this example, we revisit the AR model (Example 2.1) and show its connection with the CCA model [210]. Let us consider a first-order multivariate (vector) AR model

$$\mathbf{x}(t) = \mathbf{A}\mathbf{x}(t-1) + \mathbf{e}(t), \tag{2.165}$$

where $\mathbf{A} \in \mathbb{R}^{N \times N}$ is a constant coefficient matrix and $\mathbf{e}(t) \in \mathbb{R}^N$ represents a zero-mean multivariate Gaussian white-noise process. Let us assume $\mathbf{x}(t)$ is stationary and uncorrelated with $\mathbf{e}(t)$. In light of the *Yuler–Walker equation*, we can estimate the matrix **A** as

$$\mathbf{A} = \mathbf{C}_1 \mathbf{C}_0^{-1}, \tag{2.166}$$

where $\mathbf{C}_1 = \mathbb{E}[\mathbf{x}(t)\mathbf{x}^{\mathrm{T}}(t+1)] = \mathbb{E}[\mathbf{x}(t)\mathbf{x}^{\mathrm{T}}(t-1)]$ and $\mathbf{C}_0 = \mathbb{E}[\mathbf{x}(t)\mathbf{x}^{\mathrm{T}}(t)]$.

Now, let $\mathbf{x}(t)$ and $\mathbf{x}(t-1)$ be the two sets of random vectors treated in the CCA model. We want to find the optimal projections of these two random variables in order to maximize their correlation coefficient. Specifically, let $u(t) = \mathbf{a}^{\mathrm{T}}\mathbf{x}(t)$ and $v(t) = \mathbf{b}^{\mathrm{T}}\mathbf{x}(t-1)$; we will find the optimal **a** and **b** such that the correlation coefficient $\rho = \mathrm{corr}(u(t), v(t))$ is maximized. As known in the earlier discussion, the solution is obtained by solving a generalized eigenvalue problem as follows:

$$\begin{pmatrix} \mathbf{C}_0 & \mathbf{0} \\ \mathbf{0} & \mathbf{C}_0 \end{pmatrix}^{-1} \begin{pmatrix} \mathbf{0} & \mathbf{C}_1 \\ \mathbf{C}_1 & \mathbf{0} \end{pmatrix} \mathbf{w} = \rho \mathbf{w},$$

where $\mathbf{w} = [\mathbf{a}^{\mathrm{T}}, \mathbf{b}^{\mathrm{T}}]^{\mathrm{T}}$. More specifically, we can write $\mathbf{C}_0^{-1}\mathbf{C}_1\mathbf{a} = \rho\mathbf{a}$ and $\mathbf{C}_0^{-1}\mathbf{C}_1\mathbf{b} = \rho\mathbf{b}$. If we arrange the eigenvectors in a nondecreasing order and put them into two singular matrices $\mathbf{W}_1 = [\mathbf{a}_1, \ldots, \mathbf{a}_N]$ and $\mathbf{W}_2 = [\mathbf{b}_1, \ldots, \mathbf{b}_N]$, we further apply the orthogonal matrices to the random vector $\mathbf{x}(t)$:

$$\mathbf{u}(t) = \mathbf{W}_1^{\mathrm{T}}\mathbf{x}(t), \qquad \mathbf{v}(t) = \mathbf{W}_2^{\mathrm{T}}\mathbf{x}(t-1). \tag{2.167}$$

Given $\mathbf{v}(t)$, we can derive a regressor model between $\mathbf{u}(t)$ and $\mathbf{v}(t)$:

$$\mathbf{u}(t) = \left(\langle \mathbf{u}\mathbf{v}^{\mathrm{T}} \rangle \langle \mathbf{v}\mathbf{v}^{\mathrm{T}} \rangle^{-1} \right) \mathbf{v}(t). \tag{2.168}$$

If we define $\langle \mathbf{v}\mathbf{v}^{\mathrm{T}} \rangle = \mathbf{W}_2\mathbf{C}_0\mathbf{W}_2^{\mathrm{T}} = \mathbf{I}$ and $\langle \mathbf{u}\mathbf{v}^{\mathrm{T}} \rangle = \mathbf{W}_1\mathbf{C}_1\mathbf{W}_2^{\mathrm{T}} = \mathbf{R}$, then we further have

$$\mathbf{u}(t) = \mathbf{R}\mathbf{v}(t), \tag{2.169}$$

which bears the same mathematical form as the predictive model in equation (2.165) except that **R** is now a diagonal matrix. Hence, the canonical patterns of CCA are related to a transformation that diagonalizes the vector AR(1) model.

Moreover, their relationship can also be revealed if we apply the singular value decomposition (SVD) (see Appendix C) to matrix **A** of the multivariate AR model, namely

$$\mathbf{A} = \mathbf{U}\mathbf{S}\mathbf{V}^{\mathrm{T}}, \tag{2.170}$$

where $\mathbf{U} = [\mathbf{u}_1, \ldots, \mathbf{u}_N]$ and $\mathbf{V} = [\mathbf{v}_1, \ldots, \mathbf{v}_N]$ and the column vectors of $\mathbf{U}$ and $\mathbf{V}$ correspond to the left and right singular vectors, respectively; $\mathbf{S}$ is a diagonal matrix containing the singular values $\{s_1, \ldots, s_N\}$ in the diagonal. Specifically, the left and right singular vectors satisfy the following eigenequations:

$$\mathbf{A}\mathbf{A}^{\mathrm{T}}\mathbf{u}_i = s_i^2\mathbf{u}_i, \qquad \mathbf{A}^{\mathrm{T}}\mathbf{A}\mathbf{v}_i = s_i^2\mathbf{v}_i \qquad (i = 1, \ldots, N). \tag{2.171}$$

Suppose $\mathbf{x}(t)$ is whitened such that its linear transformation $\tilde{\mathbf{x}}(t) = \mathbf{D}\mathbf{x}(t)$ satisfies $\tilde{\mathbf{C}}_0 = \mathbb{E}[\tilde{\mathbf{x}}(t)\tilde{\mathbf{x}}^{\mathrm{T}}(t)] = \mathbf{I}$ and $\tilde{\mathbf{C}}_1 = \mathbb{E}[\tilde{\mathbf{x}}(t)\tilde{\mathbf{x}}^{\mathrm{T}}(t-1)]$; then the new eigenequations of the CCA model reduce to $\tilde{\mathbf{C}}_1\mathbf{a} = \rho\mathbf{a}$ (or equivalently $\tilde{\mathbf{C}}_1\tilde{\mathbf{C}}_1^{\mathrm{T}}\mathbf{a} = \rho^2\mathbf{a}$) and $\tilde{\mathbf{C}}_1\mathbf{b} = \rho\mathbf{b}$ (or equivalently $\tilde{\mathbf{C}}_1^{\mathrm{T}}\tilde{\mathbf{C}}_1\mathbf{b} = \rho^2\mathbf{b}$). This again shows the mathematical equivalence between the canonical patterns $\mathbf{a}$ and $\mathbf{b}$ and the singular vectors of the multivariate AR(1) model [210].

2.9.4 Fisher Linear Discriminant Analysis

Fisher linear discriminant analysis (LDA) is a widely used statistical method for pattern classification (e.g., [231]). Given a data set with two classes, the goal of the LDA is to find the best feature (or feature set) that optimally discriminates the two categories.

Specifically, let $\mathbf{x} \in \mathbb{R}^N$ and $\mathbf{y} \in \mathbb{R}^N$ be two random variables; the issue of interest is to seek a linear discriminant characterized by $\mathbf{w} \in \mathbb{R}^N$ in order to maximize the so-called *Fisher discriminant ratio* or *Rayleigh quotient*[13]:

$$\begin{aligned}\rho &= \frac{\mathbf{w}^{\mathrm{T}}\mathbf{C}_b\mathbf{w}}{\mathbf{w}^{\mathrm{T}}\mathbf{C}_w\mathbf{w}} \\ &= \frac{\mathbf{w}^{\mathrm{T}}(\boldsymbol{\mu}_x - \boldsymbol{\mu}_y)(\boldsymbol{\mu}_x - \boldsymbol{\mu}_y)^{\mathrm{T}}\mathbf{w}}{\mathbf{w}^{\mathrm{T}}(\boldsymbol{\Sigma}_x + \boldsymbol{\Sigma}_y)\mathbf{w}} = \frac{[\mathbf{w}^{\mathrm{T}}(\boldsymbol{\mu}_x - \boldsymbol{\mu}_y)]^2}{\mathbf{w}^{\mathrm{T}}(\boldsymbol{\Sigma}_x + \boldsymbol{\Sigma}_y)\mathbf{w}},\end{aligned} \tag{2.172}$$

where $\mathbf{C}_b = (\boldsymbol{\mu}_x - \boldsymbol{\mu}_y)(\boldsymbol{\mu}_x - \boldsymbol{\mu}_y)^{\mathrm{T}}$ and $\mathbf{C}_w = \boldsymbol{\Sigma}_x + \boldsymbol{\Sigma}_y$ denote, respectively, the between-class and within-class covariance matrices; $\boldsymbol{\mu}_x$ and $\boldsymbol{\Sigma}_x$ (or $\boldsymbol{\mu}_y$ and $\boldsymbol{\Sigma}_y$) denote the mean and covariance of the random vector $\mathbf{x}$ (or $\mathbf{y}$), respectively. In specific terms, the linear discriminant attempts to find a direction that maximizes the projected class mean—the numerator of (2.172)—while minimizing the class variance in this direction—the denominator of (2.172).

Notably, since $\mathbf{w}^{\mathrm{T}}\mathbf{C}_w\mathbf{w} \geq 0$, maximization of the Fisher discriminant ratio can be converted into the following constrained optimization problem:

$$\begin{aligned}&\min_{\mathbf{w}} L(\mathbf{w}, \lambda) \\ &L(\mathbf{w}, \lambda) = -\mathbf{w}^{\mathrm{T}}\mathbf{C}_b\mathbf{w} + \lambda(\mathbf{w}^{\mathrm{T}}\mathbf{C}_w\mathbf{w} - 1),\end{aligned} \tag{2.173}$$

where $\lambda > 0$ denotes the Lagrangian multiplier that imposes the constraint $\mathbf{w}^{\mathrm{T}}\mathbf{C}_w\mathbf{w} = 1$ while maximizing $\mathbf{w}^{\mathrm{T}}\mathbf{C}_b\mathbf{w}$. Minimizing (2.173) is then equivalent to setting $\partial L(\mathbf{w}, \lambda)/\partial \mathbf{w} = 0$, which in turn is equivalent to solving the generalized eigenvalue problem

$$\mathbf{C}_b \mathbf{w} = \lambda \mathbf{C}_w \mathbf{w}, \tag{2.174}$$

where λ may be viewed as the corresponding maximum eigenvalue of the above eigenvalue equation.

As a consequence, the optimal discriminant that maximizes the Fisher discriminant ratio ρ is given by [231]

$$\mathbf{w}_{\text{opt}} = (\boldsymbol{\Sigma}_x + \boldsymbol{\Sigma}_y)^{-1}(\boldsymbol{\mu}_x - \boldsymbol{\mu}_y) \tag{2.175}$$

and the corresponding maximum discriminant ratio is

$$\rho_{\max} = (\boldsymbol{\mu}_x - \boldsymbol{\mu}_y)^{\mathrm{T}}(\boldsymbol{\Sigma}_x + \boldsymbol{\Sigma}_y)^{-1}(\boldsymbol{\mu}_x - \boldsymbol{\mu}_y), \tag{2.176}$$

which is quadratic with respect to $\boldsymbol{\mu}_x - \boldsymbol{\mu}_y$ and linearly proportional to the inverse of the within-class covariance matrix $\mathbf{C}_w$.

Finally, the linear classifier obtained from LDA defines a hyperplane, which is described by the equation

$$h = \text{sign}(\mathbf{w}^{\mathrm{T}}\mathbf{z} + b), \tag{2.177}$$

where $\mathbf{w}$ denotes the normal vector obtained from (2.175), $b \in \mathbb{R}$ denotes an offset parameter, $\mathbf{z} \in \mathbb{R}^N$ denotes the test input vector, and $h = \pm 1$ specifies the decision direction along the hyperplane.

Fisher LDA can be generalized to multiple-class or nonlinear discriminant analysis [231]; we will discuss these extensions in Chapter 4.

2.9.5 Common Spatial Pattern Analysis

Common spatial pattern (CSP) analysis is a statistical algorithm that designs an optimal spatial filter that discriminates two different classes of patterns under two different conditions. It was originally developed for discriminating two populations of multichannel EEG signals recorded in imaginary hand movement [748].

Let $\mathbf{X}_l$ ($l \in \{1, 2\}$) denote an $N \times T$ matrix under a specific condition l, where N denotes the number of channels (or electrodes) and T denotes the total number of samples per channel. The *normalized* spatial correlation/covariance matrix of the data is defined by

$$\mathbf{C}_l = \frac{\mathbb{E}[\mathbf{X}_l \mathbf{X}_l^{\mathrm{T}}]}{\text{tr}\left(\mathbb{E}[\mathbf{X}_l \mathbf{X}_l^{\mathrm{T}}]\right)}, \tag{2.178}$$

where $\text{tr}(\cdot)$ denotes the trace operator and the average is taken over multiple independent trials under respective conditions.

Let $\mathbf{C} = \mathbf{C}_1 + \mathbf{C}_2$; applying EVD to matrix $\mathbf{C}$ yields

$$\mathbf{C} = \mathbf{P}\boldsymbol{\Sigma}\mathbf{P}^{\mathrm{T}}, \tag{2.179}$$

where $\mathbf{P}$ denotes the orthogonal matrix that contains eigenvectors in column vectors and $\boldsymbol{\Sigma}$ denotes the diagonal matrix containing eigenvalues. Let $\mathbf{S}_1 = \mathbf{P}\mathbf{C}_1\mathbf{P}^{\mathrm{T}}$ and $\mathbf{S}_2 = \mathbf{P}\mathbf{C}_2\mathbf{P}^{\mathrm{T}}$; it can be proved that [940]:

- Matrices $\mathbf{S}_1$ and $\mathbf{S}_2$ share the common eigenvectors, namely

$$\mathbf{S}_1 = \mathbf{U}\boldsymbol{\Sigma}_1\mathbf{U}^{\mathrm{T}}, \qquad \mathbf{S}_2 = \mathbf{U}\boldsymbol{\Sigma}_2\mathbf{U}^{\mathrm{T}}.$$

- The sum of the eigenvalue matrices induced from $\mathbf{S}_1$ and $\mathbf{S}_2$ yields an identity matrix:

$$\boldsymbol{\Sigma}_1 + \boldsymbol{\Sigma}_2 = \mathbf{I}.$$

Notably, the eigenvalue matrices $\boldsymbol{\Sigma}_1$ and $\boldsymbol{\Sigma}_2$ are complementary: The eigenvector with the largest eigenvalue for $\mathbf{S}_1$ has the least eigenvalue for $\mathbf{S}_2$, and vice versa. Hence, the projection of the whitened multichannel data onto the first and last eigenvectors in the matrix $\mathbf{U}$ yields the optimal discriminating features that characterize specific patterns between two conditions (i.e., for $l \in \{1, 2\}$).

For feature extraction, the projection matrix $\mathbf{W} = (\mathbf{U}^{\mathrm{T}}\mathbf{P})^{\mathrm{T}}$ is defined and further mapped to a new trial (denoted by $\mathbf{X}^k$) of EEG recordings by linear projection (i.e., $\mathbf{Z}^k = \mathbf{W}\mathbf{X}^k$). The interpretation of $\mathbf{W}$ is twofold [748]: On the one hand, the rows of matrix $\mathbf{W}$ can be viewed as the stationary spatial filter; on the other hand, the columns of $\mathbf{W}^{-1}$ can be viewed as the common spatial patterns or the time-invariant EEG source distribution patterns.

The CSP algorithm and its extensions have been widely used in the brain–computer interface (BCI) [223, 544, 718] for feature extraction. As a spatial filter, the CSP algorithm can find the optimal (in the MMSE sense) spatial direction that maximizes the variance for one class and that at the same time minimizes the variance for the other class. As a demonstration, in Figure 2.9, we present a simple example for applying the CSP algorithm for feature extraction in discriminating (left-hand vs. right-hand) imagery movement using real-life 32-channel EEG recordings from one human subject. Specifically, the features are the log power (variance), $f_i^k = \log\left[\mathrm{var}(Z_i^k)\right]$ (where the projections Z_i^k are given by the rows of $\mathbf{W}$ that are associated with the largest and the smallest eigenvalues); the features are then fed into a linear classifier (e.g., Fisher LDA) for binary classification. As seen in the figure, the classification results are almost perfect for both training and testing samples.

Essentially, the CSP algorithm seeks to find a spatial direction (i.e., a spatial filter) that maximizes the variance for one class and minimizes the variance for the other class. Mathematically this is formulated as a constrained optimization problem:

$$\max_{\mathbf{w}} \mathbf{w}^{\mathrm{T}}\mathbf{C}_1\mathbf{w} \quad \text{subject to} \quad \mathbf{w}^{\mathrm{T}}(\mathbf{C}_1 + \mathbf{C}_2)\mathbf{w} = 1. \tag{2.180}$$

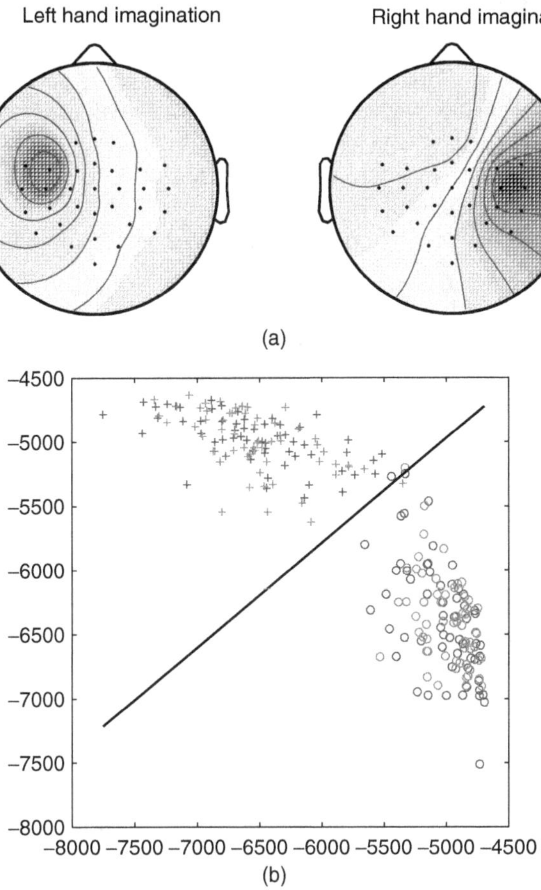

Figure 2.9 An application of the CSP/LDA algorithms for left-hand/right-hand motor imagery two-category classification in a BCI application. (*a*) Spatial patterns of imagined left and right hand movements; the black dots indicate the 32 channel locations. (*b*) The decision boundary (straight line) obtained from the Fisher LDA algorithm for two classes (represented by circles and crosses). In this example, 99.17 and 98.33% correct classification rates for 120 training samples (blue) and 120 testing samples (red) were obtained, respectively.

This, in turn, can be formulated as solving the generalized eigenvalue problem

$$\lambda(\mathbf{C}_1 + \mathbf{C}_2)\mathbf{w} = \mathbf{C}_1\mathbf{w},$$

or equivalently

$$\lambda\mathbf{w} = (\mathbf{C}_1 + \mathbf{C}_2)^{-1}\mathbf{C}_1\mathbf{w},$$

where the eigenvector **w** associated with the maximum eigenvalue λ is the desired direction that maximizes the variance for class 1.

2.10 DISCUSSION

In this chapter, we have discussed a number of correlation-based methods in signal processing and engineering applications: *spectrum analysis, filtering, prediction, detection, time-delay estimation*, and *statistical data analysis (dimensionality reduction, classification)*. As seen throughout the chapter, correlation has played the pivotal role in these methods.

We conclude the chapter by highlighting several important observations:

- Classical statistical signal processing methods are commonly based on the assumptions of *stationarity, Gaussianity*, and *linearity*. However, some of these assumptions are not always fully justifiable in real-life engineering applications. Therefore, developing robust signal processing techniques for tackling nonstationarity, non-Gaussianity, and nonlinearity is a general trend in these areas. As we have partially shown, a few spectrum analysis methods have been developed for handling the nonstationary and nonlinear nature of the signal. To overcome the non-Gaussianity and enhance the robustness of the signal processing techniques, it is a common practice to include higher order statistics in signal filtering, spectrum analysis, or signal detection.
- In order to improve the robustness to nonstationarity, it is more desirable to rely on *adaptive* signal processing techniques, which continuously and recursively estimate the statistics of incoming signal or data in a dynamic environment—this involves the fundamental concept of "learning." We have given a simple example, the LMS filter, in this chapter; more discussions of correlation-based learning will be presented in the next chapter. On the other hand, in order to tackle the nonlinearity underlying the signal or data, it is common to replace the standard linear model with nonlinear models such as artificial neural networks or kernel machines, which we will also discuss in Chapters 3 and 4.
- The essence of many correlation-based statistical analysis methods, such as PCA, FA, CCA, LDA, and CSP, is to solve an eigenvalue or generalized eigenvalue problem; hence the optimal solution to such a problem has a closed form. Nonlinear generalizations of these concepts will be discussed in Chapter 4 in the context of kernel learning.

APPENDIX 2A: EIGENANALYSIS OF AUTOCORRELATION FUNCTION OF NONSTATIONARY PROCESS

Let $x(t)$ be a zero-mean nonstationary univariate random process which has a two-dimensional autocorrelation function $C_{xx}(t_1, t_2)$ and a two-dimensional power spectrum $S_{xx}(\omega_1, \omega_2)$.

It is known that the autocorrelation function is positive semidefinite in the sense that, for any sequence $t_1, \ldots, t_n$ and complex constants $\alpha_1, \ldots, \alpha_n$, the following inequality always holds:

$$\sum_{i=1}^{n}\sum_{j=1}^{n}\alpha_i^*\alpha_j C_{xx}(t_i, t_j) \geq 0, \tag{2.A.1}$$

where α_i^* denotes the complex conjugate of α_i. If $C_{xx}(t_i, t_j)$ is continuous, then we also have

$$\int_{-\infty}^{\infty}\int_{-\infty}^{\infty} x^*(t_1)C_{xx}(t_1, t_2)x(t_2)\,dt_1\,dt_2 \geq 0. \tag{2.A.2}$$

In light of *Mercer's theorem*, eigenanalysis states that the autocorrelation function can be represented by its eigenfunctions and eigenvalues:

$$C_{xx}(t_1, t_2) = \sum_{i=1}^{\infty}\lambda_i\phi_i(t_1)\phi_i^*(t_2). \tag{2.A.3}$$

Likewise, applying the Fourier transform to $C_{xx}(t_1, t_2)$ yields the eigenfunction in the frequency domain, which defines the *generalized spectral density* in (2.34):

$$S_{xx}(\omega_1, \omega_2) = \sum_{i=1}^{\infty}\lambda_i\phi_i(\omega_1)\phi_i^*(\omega_2). \tag{2.A.4}$$

By virtue of *Parseval's theorem*, it follows that $S_{xx}(\omega_1, \omega_2)$ is also positive semidefinite.

In light of the *Schwartz inequality*, it further follows that

$$|S_{xx}(\omega_1, \omega_2)|^2 \leq S_{xx}(\omega_1, \omega_1)S_{xx}(\omega_2, \omega_2), \tag{2.A.5}$$

where $S_{xx}(\omega, \omega) \equiv S_{xx}(\omega)$ is indeed the marginal distribution of the WVD in the frequency domain:

$$S_{xx}(\omega) = \int_{-\infty}^{\infty} W_{xx}(t, \omega)\,dt. \tag{2.A.6}$$

APPENDIX 2B: ESTIMATION OF INTENSITY AND CORRELATION FUNCTIONS OF STATIONARY RANDOM POINT PROCESS

Estimation of the intensity and correlation (or covariance) functions of a stationary point process has been discussed at length in [114, 191, 192, 518].*

*Due to space limitation, we can only briefly describe some basic results here; the material in this section is excerpted and modified from the cited references.

Suppose that there are N events occurring in the time interval $[0, T]$ at times t_k $(k = 1, 2, \ldots, N)$; let $\rho(t)$ define a train of delta functions

$$\rho(t) = \sum_{k=1}^{N} \delta(t - t_k); \tag{2.B.1}$$

then the autocorrelation function of $\rho(t)$ is defined as

$$C_\rho(t) = \frac{1}{N} \int_{0^-}^{T-|t|} \rho(\tau)\rho(\tau + |t|)\, d\tau, \tag{2.B.2}$$

which is known to be positive semidefinite.

A histogram-based estimation approach for the conditional intensity function for $0 < |t| \leq T$ is based on the sum of contiguous times between events assembled in the statistic

$$\tilde{m}(t) = \frac{1}{N} \sum_{n=1}^{N-1} \sum_{k=1}^{N-n} \delta(t_{n+k} - t_n - |t|), \qquad 0 < |t| \leq T. \tag{2.B.3}$$

In practice, it is common to apply a weight function to smooth the $\tilde{m}(t)$. For instance, it was suggested in [192] that, given a small bin Δt, a smoothed estimate $\hat{m}(t)$ may be obtained from the integral average

$$\hat{m}\left(n\,\Delta t + \frac{\Delta t}{2}\right) = \frac{1}{\Delta t} \int_{n\Delta t}^{(n+1)\Delta t} \tilde{m}(\tau)\, d\tau. \tag{2.B.4}$$

In order to keep the estimate unbiased, the length of the interval, T, has to be sufficiently long such that $T \geq |n\,\Delta t|$.

In order to derive the conditional density and correlation functions at integer multiples of Δt, it is convenient to introduce the *complete conditional intensity function* $\kappa(t)$, which is defined by

$$\kappa(t) = \delta(t) + m(t) \tag{2.B.5}$$

with $m(t)$ being defined in (2.60). The estimate of the complete conditional intensity function $\kappa(t)$ based on the histogram can be regarded as an approximation of $C_\rho(t)$.

Similarly, we may define

$$\tilde{\kappa}(t) = \delta(t) + \tilde{m}(t) \tag{2.B.6}$$

with $\tilde{m}(t)$ being defined in (2.B.3). A smoothed estimate of $\tilde{\kappa}(t)$, denoted as $\hat{\kappa}(t)$, can be obtained by convoluting a weight function $w(t)$ with $\tilde{m}(t)$, which then yields

$$\hat{\kappa}(t) = \delta(t) + \tilde{m}(t) \otimes w(t), \tag{2.B.7}$$

where $w(t)$ is subject to a unit-area constraint such that $\int w(t)\,dt = 1$. Equation (2.B.4) is indeed a special case of *Daniell's weight function* [192].

From (2.B.4), the smoothed estimate for $\kappa(t)$ at integer multiples of Δt may be written as

$$\hat{\kappa}_d(k\,\Delta t) = \frac{1}{\Delta t}\left(\delta_{ok} + \int_{k\Delta t-\Delta t/2}^{k\Delta t+\Delta t/2} \tilde{m}(\tau)\,d\tau\right). \tag{2.B.8}$$

Correspondingly, a smoothed estimate for the correlation function at integer multiples of Δt may be obtained as

$$\hat{C}_d(k\,\Delta t) = \hat{\lambda}(\hat{\kappa}_d(k\,\Delta t) - \hat{\lambda}), \tag{2.B.9}$$

where $\hat{\lambda} = N/T$ denotes the estimated mean intensity of the point process.

In a similar manner, the cross-intensity and cross-correlation functions for a pair of random point processes can also be derived. An example of using such statistics for analyzing sensory encoding using pairs of neural spike trains is referred to [849].

APPENDIX 2C: DERIVATION OF LEARNING RULES WITH QUASI-NEWTON METHOD

Let us assume that the goal of a filter is to minimize a quadratic cost function $J(t)$ whose second-order Taylor series expansion yields

$$J(\boldsymbol{\theta}(t) + \Delta\,\boldsymbol{\theta}(t)) \approx J(\boldsymbol{\theta}(t)) + \mathbf{g}^{\mathrm{T}}(t)\,\Delta\boldsymbol{\theta}(t) + \frac{1}{2}\Delta\boldsymbol{\theta}^{\mathrm{T}}(t)\mathbf{H}(t)\,\Delta\boldsymbol{\theta}(t), \tag{2.C.1}$$

where $\mathbf{g}(t) = \partial J(t)/\partial\boldsymbol{\theta}$ denotes the gradient vector and $\mathbf{H}(t) = \partial^2 J(t)/(\partial\boldsymbol{\theta}\,\partial\boldsymbol{\theta}^{\mathrm{T}})$ denotes the Hessian matrix.

Given the above quadratic approximation, the Newton method specifies the optimal learning rule for a parameter vector $\boldsymbol{\theta}$, as shown by

$$\boldsymbol{\theta}(t) = \boldsymbol{\theta}(t-1) - \mathbf{H}^{-1}(t)\mathbf{g}(t), \tag{2.C.2}$$

which requires the knowledge of the inverse of the Hessian matrix. For a linear filter (or linear neuron) $y(t) = \mathbf{x}^{\mathrm{T}}(t)\boldsymbol{\theta}(t)$, the Hessian matrix reduces to the *correlation matrix* of the input data, namely, $\mathbf{H}(t) \equiv \mathbf{C}_{xx}(t)$.

The quasi-Newton method tries to sidestep the difficulty of direct estimation of the Hessian matrix; instead, it updates the Hessian matrix sequentially:

$$\mathbf{H}(t) = \mathbf{H}(t-1) + \mathbf{x}(t)\mathbf{x}^{\mathrm{T}}(t). \tag{2.C.3}$$

By letting the learning rate be (or proportional to) the inverse of the Hesssian, namely, $\eta(t) = \mathbf{H}^{-1}(t)$, in light of *the matrix inverse lemma* (also called

Woodbury's equality), we can derive the following optimal learning rule for a linear filter [161]:

$$\boldsymbol{\eta}(t) = \boldsymbol{\eta}(t-1) - \frac{\boldsymbol{\eta}(t-1)\mathbf{x}(t)\mathbf{x}^{\mathrm{T}}(t)\boldsymbol{\eta}(t-1)}{1 + \mathbf{x}^{\mathrm{T}}(t)\boldsymbol{\eta}(t-1)\mathbf{x}(t)}, \tag{2.C.4}$$

$$\boldsymbol{\theta}(t) = \boldsymbol{\theta}(t-1) + \boldsymbol{\eta}(t)\mathbf{x}(t)e(t), \tag{2.C.5}$$

where $e(t) = d(t) - y(t)$ denotes the error signal, $\boldsymbol{\eta}(t)$ is a learning-rate matrix, and the term $\boldsymbol{\eta}(t)\mathbf{x}(t)$ plays a role similar to that of the Kalman gain. Observe that the learning rule (2.C.5) also bears resemblance to the RLS filter and Kalman filter [161].

For a nonlinear filter (or nonlinear neuron model), say $y(t) = f(\mathbf{x}^{\mathrm{T}}(t)\boldsymbol{\theta}(t))$, we can approximate the online Hessian matrix by

$$\mathbf{H}(t) \approx \mathbf{H}(t-1) + \mathbf{g}(t)\mathbf{g}^{\mathrm{T}}(t), \tag{2.C.6}$$

from which the following update rule is derived [161]:

$$\boldsymbol{\eta}(t) = \boldsymbol{\eta}(t-1) - \frac{\boldsymbol{\eta}(t-1)\mathbf{g}(t)\mathbf{g}^{\mathrm{T}}(t)\boldsymbol{\eta}(t-1)}{1 + \mathbf{g}^{\mathrm{T}}(t)\boldsymbol{\eta}(t-1)\mathbf{g}(t)}, \tag{2.C.7}$$

$$\boldsymbol{\theta}(t) = \boldsymbol{\theta}(t-1) + \boldsymbol{\eta}(t)\,\nabla_{\boldsymbol{\theta}} f(\mathbf{x}^{\mathrm{T}}(t)\boldsymbol{\theta}(t))e(t), \tag{2.C.8}$$

where $e(t) = d(t) - y(t) = d(t) - f(\mathbf{x}^{\mathrm{T}}(t)\boldsymbol{\theta}(t))$.

BIBLIOGRAPHICAL NOTES

Classical signal processing was pioneered and developed independently by Wiener, Kolmogorov, and Khinchin [474, 954], among many others. In particular, generalized harmonic analysis and Fourier transform have laid the foundation of statistical signal processing, for which the autocorrelation and cross-correlation functions have played important roles in spectrum analysis, filter design, signal detection, time-delay estimation, and so on. In studying cybernetics, Wiener [955] also proposed to use the autocorrelation function for analyzing the spectrum of brain waves.

Spectrum analysis of stationary processes was established independently by Wiener and Kolmogorov; general second-order theory of nonstationary signals was established in the 1940s by Loève [569]. In particular, correlation functions have been widely used for spectrum analysis for various stochastic processes (e.g., [307, 702, 707]). Collected volumes on spectrum analysis can be found in [361, 362, 371]. A review of the generalized time–frequency distribution may be found in [176, 177]. The Hilbert–Huang transform was first developed in [414] to tackle the nonlinear and nonstationary problem of spectrum analysis.

The concept of prediction dates back to the 1940s, when Norbert Wiener developed a mathematical theory for calculating the best filters and predictors for detecting signals hidden in noise [957]. Following Wiener and Shannon's pioneering work, Elias [258] proposed the notion of "predictive coding" in the context of signal coding.

The original work of the LMS filter is credited to Widrow and Hoff [951]. The formulation of the Kalman filter appeared in the same year [461]. Both of these filters have survived the test of time. Textbook treatments of adaptive filters, including the LMS filter, the Kalman filter, and their numerous variants, can be found in [369, 793, 953].

Correlation-based detection methods are widely used in signal processing and communications [455, 525]. The matched filter is an optimal filter that helps to detect and recover noise-corrupted known message signals in communication systems [366]. Excellent resources of detection theory are referred to [458, 471]. The GCC method was first proposed by Knapp and Carter [489] for time-delay estimation; an overview of time-delay estimation techniques can be found in [152].

Correlation matrices have deep roots in statistical analysis. Principal-component analysis [447], FA [353], and CCA [405] are three representative examples. The Fisher LDA algorithm is a correlation-based pattern classification technique that optimally discriminates two pattern categories. The CSP algorithm was originally developed for discriminating multichannel EEG patterns of imagined hand movements [748]; it has been widely used in BCI applications [224, 718].

NOTES

1. Reportedly the word *spectrum* was adopted by the mathematician David Hilbert from the 1897 article by Wilhelm Wirtinger in the study of operator theory.
2. A recurrence plot [236] is a useful statistical tool for time series analysis in physics. The main role of the recurrence plot is to reveal the nonstationarity of a time series as well as to indicate the degree of aperiodicity. Generally, for a stationary time series the recurrent plot is homogeneous along the main diagonal. The correlation integral is the "density" of points on a recurrence plot, that is, the number of points divided by the total number of points in the plot.
3. In physics, it is common to distinguish two classes of correlation functions: (i) the off-critical correlation function $\langle x(t)x(t+\tau)\rangle \sim e^{-|\tau|/\tau_d}$ [where $|\tau| \leq \frac{1}{2}l(x)$], and (ii) the on-critical powerlaw correlation function $\langle x(t)x(t+\tau)\rangle \sim |\tau|^{-\tau_d}$ [where $|\tau| \geq d(x)$].
4. The Cauchy principal-value is also known as the principal-value integral, which is often defined around the singular points in a limiting case. For instance, the Cauchy principal value of a finite integral of a function f about a point c, with $a \leq c \leq b$, is defined as

$$\mathcal{P}\int_a^b f(x)\,dx \equiv \lim_{\varepsilon\to 0^+}\left[\int_a^{c-\epsilon} f(x)\,dx + \int_{c+\varepsilon}^b f(x)\,dx\right].$$

5. A signal that has no negative-frequency components is called an analytic signal. The Hilbert transform essentially filters out all negative frequencies of the input signal.
6. The HHT is an empirically based data analysis method. Given a real-valued signal $x(t)$, the HHT starts with the EMD and generates a set of adaptive bases called intrinsic mode functions $\{x_i(t)\}_{i=1}^N$, which are often physically meaningful. An intrinsic mode function is defined as any function that has the same number of zero crossings and extrema and also has symmetric envelopes defined by the local maxima and minima. The intrinsic mode functions are sequentially extracted by applying a "sifting" procedure; the process

continues until a certain stopping criterion is met or an intrinsic mode function with no more than two extrema is found.

7. Note that the frequency in $h(\omega)$ has a different meaning from Fourier spectral analysis. In the Fourier representation, the existence of energy at a frequency ω represents a component of a sine or cosine wave that persists through the time span of the signal (or time series). In the Hilbert marginal spectrum representation, the existence of energy at the frequency ω means that in the whole time span of the data there is a higher likelihood for such a wave to appear locally. Above all, the Fourier spectrum is somewhat meaningless physically when the signal is highly nonstationary, although the short-time Fourier transform (STFT) may be used for characterizing the spectrum (i.e., spectrogram) of locally stationary signals within the window length.
8. A causal system is a system with output and internal states that depend only on the current and previous input values. This property is referred to as *causality*. A system that has some dependence on input values from the future (in addition to possible dependence on past or current input values) is termed a noncausal system, and a system that depends solely on future input values is an anticausal system.
9. An integration of this form is called a *Fredholm equation*. The theory on the existence of a solution to the Fredholm equation is well established in the literature; numerical methods are often used to solve such an equation.
10. An adaptive filter is defined as a self-designing system that relies for its operation on a recursive algorithm, which makes it possible for the filter to perform satisfactorily in an environment where knowledge of the relevant statistics is not available. According to its operation requirement, adaptive filters are often classified into:

    - *Supervised adaptive filters*, which require the availability of a training sequence that provides different realizations of a desired response for specific input signals.
    - *Unsupervised adaptive filters*, which perform adjustments of free parameters without the need for a desired response. The filter design requires specific self-organizing principles to guide the parameter adjustment.

    Depending on the filtering operation, adaptive filters can be either *linear* or *nonlinear*; see [369, 793] for more discussions.
11. In the context of statistical signal processing, PCA is also known as the *Karhunen–Loéve transform*.
12. Eigenvalues and diagonalization of a symmetric square matrix (or more generally, operator) were discovered in 1926 by the mathematician Augustin Louis Cauchy in the process of finding normal forms for quadratic functions. Later, John von Neumann established a more general spectrum theorem stating that every real, symmetric matrix is diagonalizable.
13. Note that optimization of the standard Fisher discriminant ratio relies on the estimation of parameters $\boldsymbol{\mu}_x, \boldsymbol{\mu}_y, \boldsymbol{\Sigma}_x, \boldsymbol{\Sigma}_y$, which are often estimated from the finite samples of labeled data. To enhance the robustness of the discriminant or to reduce the sensitivity to outliers, one may alternatively seek to optimize a minimax criterion [523]:

    $$\rho = \frac{|\mathbf{w}^{\mathrm{T}}(\boldsymbol{\mu}_x - \boldsymbol{\mu}_y)|}{\sqrt{\mathbf{w}^{\mathrm{T}}\boldsymbol{\Sigma}_x\mathbf{w}} + \sqrt{\mathbf{w}^{\mathrm{T}}\boldsymbol{\Sigma}_y\mathbf{w}}},$$

    which is similar to the standard Fisher discriminant ratio (2.172) and can be tackled via convex optimization methods.

3

CORRELATION-BASED NEURAL LEARNING AND MACHINE LEARNING

The development of computational neural models and learning algorithms is a central task in computational neuroscience. Following the preceding discussions of correlation in Chapters 1 and 2, this chapter presents a comprehensive overview of correlation-based computational neural models in the literature as well as numerous correlation-based neural learning and machine learning paradigms, as to be seen, many of which have gone far beyond the original Hebbian postulate of learning.

As will be shown in Sections 3.1 and 3.2, the correlation-based synaptic plasticity and learning rules reviewed here include all three of the major machine learning paradigms—*unsupervised, supervised*, and *reinforcement* learning—used widely in computational models of brain function and in artificial and adaptive systems that imitate adaptive functions or modules of the brain. Specifically, several classes of learning rules are covered:

- Unsupervised Hebb-type learning: competitive learning, Bienenstock–Cooper–Munro (BCM) learning, PCA learning and its generalizations, wake–sleep learning and Boltzmann learning;
- Supervised error-driven learning: the perceptron learning rule and LMS rule
- Temporal Hebbian learning, TD learning, and models that integrate reinforcement-driven and Hebbian learning

Correlative Learning: A Basis for Brain and Adaptive Systems, by Zhe Chen, Simon Haykin, Jos J. Eggermont, and Suzanna Becker

- Unsupervised information-theoretic learning: Linsker's rule, Imax rule, local decorrelating learning rule, blind source separation (BSS), independent-component analysis (ICA), and slow feature analysis (SFA).

Although the above-mentioned learning rules are rooted in different backgrounds, it is our intention to underscore their inherent connections and to illustrate the utility of these algorithms by drawing on examples from state-of-the-art research. In particular, we emphasize their links to Hebbian plasticity and common roots in correlation-based learning principles as well as their underlying biological motivation.

In Section 3.3, many correlation-based computational neural models are reviewed that span a wide range of sensory, motor, perceptual, and cognitive brain functions, including associative memory, coincidence detection, sound localization and segregation in the auditory system, topographic map formation in the visual system, feature binding for sensory perception, as well as sensorimotor control in the cerebellum.

3.1 CORRELATION AS A MATHEMATICAL BASIS FOR LEARNING

In Chapter 1, we presented a brief overview of the roots of correlative learning, including Hebb's original postulate of learning. In this section, we will give a comprehensive overview of the various synaptic learning rules in support of our claim that correlation can serve as a mathematical basis for many learning algorithms for modeling either biological or artificial intelligent systems.

From a mathematical viewpoint, learning can be regarded as an optimization problem. When viewed in this light, there are three major goals in developing models that learn: (i) to develop an appropriate objective function, (ii) to find an optimization procedure to minimize (or maximize) the objective function, and (iii) to use this optimization procedure to find good parameter values that constitute the minima (or maxima) of the objective function. In neurobiological systems, learning is a synaptic adaptation process, so the parameters to be optimized are the synaptic connection strengths. Hence, in both artificial and biological neural networks, learning can be viewed as a mathematical procedure that results in optimal synaptic rules and, in applying these rules, results in optimal synaptic strengths. On the other hand, biological systems may be more likely to find the optimal synaptic learning rules through a combination of evolution and learning.

3.1.1 Hebbian and Anti-Hebbian Rules (Revisited)

In Chapter 1, we discussed Hebbian synaptic plasticity and several of its variants. For convenience, let us rewrite Hebb's postulate of learning [equation 1.12] here:

$$\Delta\theta_{ij}(t) = \eta\langle x_i(t)y_j(t)\rangle, \tag{3.1}$$

which states that the modification of the synaptic strength, $\Delta\theta_{ij}$, is in *positive* proportion to the correlation between the presynaptic activity x_i and postsynaptic activity y_j. The Hebbian rule given by equation (3.1) is one of the simplest possible formulations of synaptic plasticity; its mathematical analysis is given in Appendix 3A. This rule is in keeping with Hebb's original postulate that the efficacy of the connection between neurons A and B should increase in proportion to the degree to which neuron A repeatedly takes part in the firing of neuron B. However, this rule only allows for synaptic strengthening. In order to satisfy biological constraints, there must also be some mechanism for synaptic weakening.

A necessary counterpart of Hebbian plasticity, known as anti-Hebbian learning, may be expressed as

$$\Delta\theta_{ij}(t) = -\eta\langle x_i(t)y_j(t)\rangle, \tag{3.2}$$

which states that the modification of the synaptic strength is in *negative* proportion to the correlation between the presynaptic and postsynaptic activities. From a biological viewpoint, anti-Hebbian learning is necessary to limit and stabilize synaptic growth. From a mathematical perspective, we can show that, without the addition of an anti-Hebbian term, pure Hebbian learning will lead to an instability of the synapses and, by combining Hebbian and anti-Hebbian terms, the learning process becomes stable (see Appendix 3B for details).

3.1.2 Covariance Rule

As mentioned above, the original Hebbian postulate only allows for an increase in synaptic weight between synchronously firing neurons. To prevent unlimited growth, it is necessary to extend the Hebb's rule to allow for weight decreases when neurons fire asynchronously. To take care of this matter, Sejnowski [814, 815] proposed a covariance-based learning rule:

$$\Delta\theta_{ij} = \eta(x_i - \langle x_i\rangle)(y_j - \langle y_j\rangle), \tag{3.3}$$

where θ_{ij} denotes the strength of the synapse between neurons i and j and $\langle x_i\rangle$ and $\langle y_j\rangle$ represent the mean pre- and postsynaptic activities, respectively. Taking a time average of the change in synaptic weight of (3.3) yields

$$\langle\Delta\theta_{ij}\rangle = \eta(\langle x_i y_j\rangle - \langle x_i\rangle\langle y_j\rangle), \tag{3.4}$$

where the first term on the right-hand side denotes the Hebbian synapse and the second term may be viewed as an activity-dependent "threshold" that varies with the product of time-averaged pre- and postsynaptic activity levels. If, on average, the presynaptic activity x_i is independent on the postsynaptic activity y_j, namely $\langle x_i y_j\rangle = \langle x_i\rangle\langle y_j\rangle$, then no change in synaptic strength should occur.

The Hebbian covariance rule dictates that, when neurons fire synchronously in a correlated manner, their connection strengths should increase, whereas if their firing patterns are anticorrelated, then the weights should decrease. This is indeed consistent with the LTD phenomenon evidenced in the hippocampus [850]. Willshaw and Dayan [962] showed the optimality of the covariance rule (3.3) for storing patterns in correlation matrix memories. As we will show later, many synaptic learning rules, including the error-correcting LMS rule, Oja's local PCA rule, and the BCM rule, are all special cases of the covariance rule.

3.1.3 Grossberg's Gated Steepest Descent

In the context of synaptic plasticity, Grossberg [344] has reviewed many neural learning laws and unified them with a so-called *gated steepest descent* (GSD) rule. Specifically, the GSD rule is described by the differential equation

$$\frac{d\theta}{dt} = f(x)[-c\theta + g(y)], \tag{3.5}$$

where θ denotes the unknown synaptic parameter, x denotes the activity of a presynaptic (or postsynaptic) cell, y denotes the activity of a postsynaptic (or presynaptic) cell, $f(\cdot)$ and $g(\cdot)$ represent the linear or nonlinear functions that regulate the presynaptic or postsynaptic neurons, and c is a constant coefficient.

Discretizing (3.5) yields the following difference equation:

$$\begin{aligned}\theta(t+1) &= \theta(t) + \eta f(x)[g(y) - c\theta(t)] \\ &= \theta(t) + \eta f(x)g(y) - \eta c f(x)\theta(t),\end{aligned} \tag{3.6}$$

where η is a small positive learning-rate parameter, the second term on the right-hand side of (3.6) is a Hebb-like correlation term, and the third term, $-f(x)\theta(t)$, is a weight-decay term that imposes synaptic stability to satisfy physiological constraints. This learning rule (3.6) has been proposed by many authors in various forms (e.g., see reviews in [122, 548]). Figure 3.1 presents a graphical illustration of the GSD rule as well as the classic Hebb's rule.

Whereas Hebbian learning only allows the synaptic strength to increase, while anti-Hebbian learning only allows the synaptic weights to decrease, the GSD law integrates both Hebbian and anti-Hebbian properties and thereby avoids the problems of weight explosion/implosion inherent in either of these rules individually. It has a general correlational form which can be either linear or nonlinear. As we will see later, many neural synaptic rules, including competitive learning and instar/outstar learning, are special cases of (3.6).

For instance, Grossberg's "instar" learning rule can be expressed by the equation

$$\theta_i(t+1) = \theta_i(t) + \eta\Theta(x_i(t))[x_i(t)y(t) - \theta_i(t)], \tag{3.7}$$

where $\Theta(\xi)$ is a Heaviside (or unit step) function that is 1 only when $\xi > 0$ and 0 otherwise; its rule is to force $\Delta\theta_i(t+1) = 0$ when $x_i(t) \leq 0$. As the learning process goes on and after a period of time, the weight will converge to the time average of the product $x_i y$ (i.e., the correlation), as shown by [378]: $\theta_i(\infty) \to \langle x_i y \rangle$.

3.1.4 Competitive Learning Rule

Competitive learning, as an important ingredient of self-organizing systems, is a form of correlative learning. In the computational neuroscience literature, many computational models of competitive learning have been proposed (e.g., [302, 343, 498, 782, 921, 950]; for a review, see [67, 296]). Although different computational models may have different learning rules, the common goal of competitive learning algorithms is to learn a certain number of parameter vectors (synaptic strengths) in a possibly high-dimensional space. The distribution of these vectors should reflect to some degree the probability distribution of the input data [296]. Representative goals of competitive learning include

- Formation of topographic maps [499, 597, 675]
- Vector quantization [499]
- Feature extraction [782, 802]
- Clustering [302, 611] and density estimation

Competitive learning methods can be categorized according to the type of activation function they use, which can be either "hard competition" or "soft competition." The hard-competitive learning, or *winner-take-all* (WTA) version of competitive learning, comprises methods where each input datum only determines the adaptation of one winning neuron. In contrast, soft-competitive learning allows parallel adaptation of multiple neurons when presenting the input data. The term "soft-competitive

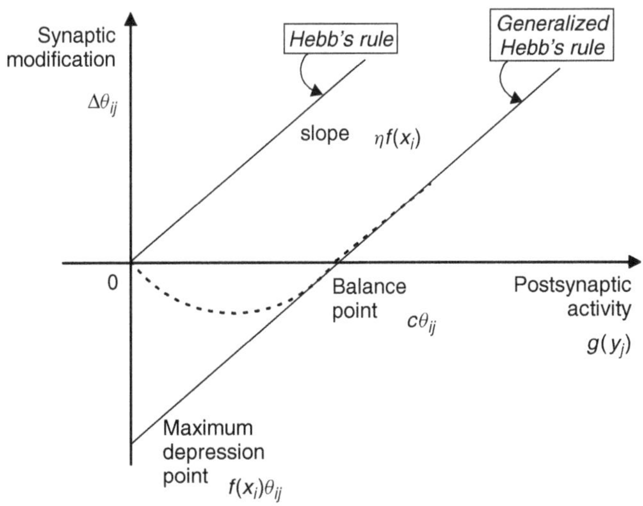

Figure 3.1 Graphical illustration of synaptic modification rules. The ordinate represents the change of synaptic weight, $\Delta\theta_{ij}$, while the abscissa represents the postsynaptic activity $g(y_j)$. The intersections of the GSD rule with the abscissa and ordinate define the *balance point* $c\theta_{ij}$ and the *maximum depression point* $f(x_i)\theta_{ij}$, respectively. The dashed curve can be viewed as the BCM rule.

learning" was first proposed by Nowlan [672], whose algorithm employed a normalized activation function whereby each neuron's activation represents the probability of that neuron accounting for the data. Each neuron adapts its weights in proportion to this probability. It turns out that this algorithm is equivalent to fitting a Gaussian mixture density model to the data. Other representative soft-competitive learning algorithms include the *neural gas* algorithm [596] and competitive Hebbian learning algorithm [595] (as well as their generalizations, e.g., [296, 597]).

One of the most popular forms of competitive learning is the so-called *self-organizing map* (SOM). Specifically, its learning rule may be formulated as [499]

$$\boldsymbol{\theta}_j(t+1) = \boldsymbol{\theta}_j(t) + \eta h_{j,i(\mathbf{x})}[\mathbf{x}(t) - \boldsymbol{\theta}_j(t)], \tag{3.8}$$

where $h_{j,i(\mathbf{x})}$ is a neighborhood (such as an isotropic Gaussian) function that defines the WTA region around the winning neuron i; the neurons (with indices j) within the region (including the winning neuron i) are allowed to update (i.e., they are excitatory), whereas the others are inhibitory. As an example, Figure 3.2 presents an illustration of applying the SOM rule (3.8) for learning the topology of two-dimensional ring-shaped input patterns. Given randomly initialized weights, the learning process converged within 200 iterations with a learning-rate parameter $\eta = 0.001$. As seen from the figure, the learned 10×10 mesh grid has approximated quite well both the *topology* and the *density* of the data.

Depending on the specific versions of the algorithm, competitive learning may have either a supervised or unsupervised form, respectively, given by

$$\boldsymbol{\theta}_w(t+1) = \boldsymbol{\theta}_w(t) + \eta y(t)[\mathbf{x}(t) - \boldsymbol{\theta}_w(t)], \tag{3.9a}$$

$$\boldsymbol{\theta}_w(t+1) = \boldsymbol{\theta}_w(t) + \eta[\mathbf{x}(t) - \boldsymbol{\theta}_w(t)], \tag{3.9b}$$

where $\mathbf{x}(t)$ represents the input vector that is often normalized, namely, $\|\mathbf{x}(t)\| = 1$; $y(t)$ denotes the scalar output; and $\boldsymbol{\theta}_w$ denotes the synaptic strength associated with the "winning" neurons. Observe that equations (3.9a) and (3.9b) are special forms

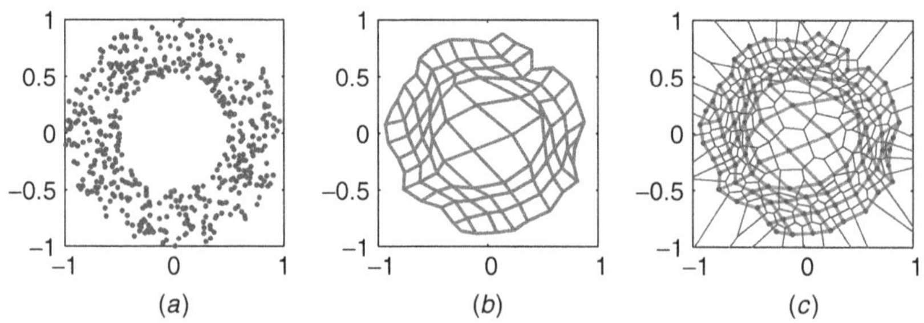

Figure 3.2 (*a*) The 500 uniformly distributed two-dimensional data points in a ring-shaped region. (*b*) The learned 10 × 10 mesh grid that resembles the topology (and density) of the data. (*c*) The weight space separated by Voronoi regions.

of (3.6) when functions f and g are both linear in (3.9a) and f is a constant in (3.9b). Specifically, equation (3.9a) is essentially another form of the instar rule, which can be decomposed into two terms: the first term $\eta \mathbf{x}(t)y(t)$ is a Hebbian term, and the second term $-\eta \boldsymbol{\theta}_w(t)y(t)$ is a weight-decay term. When the output $y(t)$ in equation (3.9a) consists of an externally provided supervisory signal or class label [namely, $y(t) = \pm 1$], the competitive learning rule can be used for supervised learning, such as learning vector quantization (LVQ) (e.g., [499]). When $y(t)$ is the soft activation output $[0 \leq y(t) \leq 1]$ of a neuron, equation (3.9a) is used in soft-competitive learning, where each neuron learns in proportion to how active it is, and the activation function implements a soft competition. When $y(t) \equiv 1$ [i.e. $y(t)$ is the activation of the single winning neuron, assuming a hard WTA competition], equation (3.9a) reduces to the unsupervised form (3.9b). The unsupervised competitive learning rule (3.9b) may be used for learning the data topology as in the natural gas algorithm [596] or the mean vector of clustered data as in K-means clustering [611].

Therefore, competitive self-organization can be seen as a form of Hebbian learning in a network with competitive interactions (e.g. defined by lateral inhibition or a WTA activation function), with a decay term that guarantees normalization; this property can be interpreted as a conservation of metabolic resources. Due to its simplicity and biological relevance, competitive Hebbian learning may be closely related to spike-timing-dependent synaptic plasticity [843].

3.1.5 BCM Learning Rule

In studying visual cortical plasticity, Bienenstock, Cooper, and Munro [93] proposed a synaptic modification hypothesis, in which there is a trade-off between Hebbian and anti-Hebbian properties mediated by the addition of a sliding modification threshold. Also, based on correlative learning, the BCM learning rule is defined as follows:

$$\Delta\theta_i = \eta\phi(y, \xi)x_i, \tag{3.10}$$

where x_i denotes the presynaptic activity and $\phi(\cdot)$ is a nonlinear function of the postsynaptic activity y that has two zero crossings, one at $y = 0$ and the other at $y = \xi$; the variable ξ represents the dynamic threshold, which is a superlinear function of the recent history of cell activity. Thus if one neuron has been active recently, its threshold will be raised, allowing other neurons to win the competition, while other neurons whose activities have been low in recent history will have their thresholds lowered.

For instance, one specific example of (3.10) may be described by the differential equation

$$\tau\frac{d\theta_i}{dt} = x_i(y - \xi)y \quad \text{or} \quad \tau\frac{d\boldsymbol{\theta}}{dt} = \mathbf{x}(y - \xi)y, \tag{3.11}$$

in which $\phi(y, \xi) = (y - \xi)y$ and $\xi = \langle|\mathbf{x}^{\mathrm{T}}\boldsymbol{\theta}|^2\rangle$. To generalize (3.11) to multiple postsynaptic neurons, we can introduce an inhibitory mechanism

$$\overline{y}_j = y_j - \eta \sum_{k \neq j} y_k; \tag{3.12}$$

then (3.11) can be rewritten as

$$\tau \frac{d\theta_{ij}}{dt} = x_i(\overline{y}_j - \xi)\overline{y}_j. \tag{3.13}$$

Note that (3.11) allows for both Hebbian and anti-Hebbian learning, according to whether the postsynaptic activity is greater than or less than a movable threshold that depends on the neuron's recent firing history. This allows for neurons that have not been firing in a long time (and hence have a very low threshold of firing) to gain a competitive advantage, thereby overcoming a problem with standard competitive learning that some units may capture all the activation while other units remain dormant. The BCM theory is claimed to be biologically plausible and has been used for formation of visual receptive fields [532] as well as feature extraction [97, 186, 430, 830].

The same mathematical structure as in BCM learning is realized by correlative learning with inhibitory neurons. This was proposed by Amari and Takeuchi [32] earlier than the BCM theory. Recently, the role of inhibition in neural self-organization has been given much attention. Specifically, in Amari–Takeuchi's model, the neuron has an inhibitory input $x_o(t)$ with an associated synaptic weight $\theta_{oj}(t)$, so that the neuron's output is written as

$$y_j = f\left(\sum_i x_i\theta_{ij} + x_o\theta_{oj}\right). \tag{3.14}$$

For this neuron, the learning rule is Hebbian for the excitatory synapses and anti-Hebbian for the inhibitory synapses, namely,

$$\Delta\theta_{ij} = \eta x_i y_j, \tag{3.15a}$$

$$\Delta\theta_{oj} = -\eta x_o y_j. \tag{3.15b}$$

Then, the neuron self-organizes to be responsive to characteristic features of the ensemble of signals.

3.1.6 Local PCA Learning Rule

As discussed in Chapter 2, PCA requires an operation of matrix decomposition of the correlation or covariance matrix. When the dimensionality of data, m, is large, the memory storage of the correlation matrix can be prohibitive, and the computation is also costly [with complexity $\mathcal{O}(m^3)$ given the correlation matrix];

besides, the matrix decomposition method is offline and therefore it is less appealing for sequential data.

In order to extract the dominant principal component, Oja [676] proposed an online self-organizing learning rule for the first principal component, which is also referred to as *maximum eigenfiltering*.[1] Oja's PCA rule is local and computationally efficient, while keeping the Euclidean norm of a neuron's incoming synaptic weight vector at unity. Specifically, the output neuron signal $y(t)$ is expressed by

$$y(t) = \sum_{i=1}^{m} \theta_i(t)x_i(t) = \mathbf{x}^{\mathrm{T}}(t)\boldsymbol{\theta}(t), \tag{3.16}$$

where $x_i(t)$ denotes the ith presynaptic neuron input. Motivated by the eigenvalue decomposition, we wish to have

$$\mathbf{C}_{xx}\boldsymbol{\theta} = \lambda\boldsymbol{\theta} \quad \text{subject to} \quad \|\boldsymbol{\theta}\| = 1,$$

where $\mathbf{C}_{xx} = \mathbb{E}[\mathbf{x}(t)\mathbf{x}^{\mathrm{T}}(t)]$ is the autocorrelation matrix of the input $\mathbf{x}$, and λ is the associated eigenvalue subject to. If we use the instantaneous value to replace the expectation, namely, $\mathbf{x}(t)\mathbf{x}^{\mathrm{T}}(t)\boldsymbol{\theta} = \lambda\boldsymbol{\theta}$, then we recover the correlation relationship: $\boldsymbol{\theta} = (1/\lambda)\mathbf{x}(t)y(t)$. In order to reveal the Hebbian nature of this weight update rule, consider the online version of Oja's weight update equation:

$$\theta_i(t+1) = \frac{\theta_i(t) + \eta y(t)x_i(t)}{\left\{\sum_{k=1}^{m}\left[\theta_k(t) + \eta y(t)x_k(t)\right]^2\right\}^{1/2}}, \tag{3.17}$$

which, for a sufficiently small learning-rate parameter η, can be approximated by a simpler Hebbian rule with a decay term:

$$\theta_i(t+1) = \theta_i(t) + \eta y(t)[x_i(t) - y(t)\theta_i(t)]. \tag{3.18}$$

Oja's learning rule (3.18) has the important property that the weight vector converges to the predominant eigenvector (i.e., that corresponding to the maximum eigenvalue) of the covariance matrix of the input vector, in other words, the first principal component of the input distribution [679]; this can be analyzed within the framework of stochastic approximation (see Appendix B for background). It is noteworthy that it is possible to extract the rest of the principal components by projecting the data onto a subspace that is perpendicular to the largest eigenvector and then employing the same Oja's rule for the second largest component, and so on; this "deflation" learning process is similar to the *Gram–Schmidt orthogonalization* procedure.

Oja's learning rule has been further extended to multiple output neurons with orthogonal weight vectors [677, 680, 789]. In particular, Sanger [789] proposed a *generalized Hebbian algorithm* (GHA) to extract multiple principal components.

Let $\theta_{ji}(t)$ denote the synaptic weight that connects the ith input $x_i(t)$ and the jth output $y_j(t)$ ($j = 1, 2, \ldots, n$, where n is the desired number of principal components) and is define $y_j(t)$ as a linear sum of the m input signals:

$$y_j(t) = \sum_{i=1}^{m} \theta_{ji}(t) x_i(t). \tag{3.19}$$

Then the GHA rule is given as

$$\Delta\theta_{ji}(t) = \eta \left(y_j(t) x_i(t) - y_j(t) \sum_{k=1}^{j} \theta_{ki}(t) y_k(t) \right). \tag{3.20}$$

Written in matrix form, let $\mathbf{W}(t) = [\boldsymbol{\theta}_1(t), \boldsymbol{\theta}_2(t), \ldots, \boldsymbol{\theta}_n(t)]^{\mathrm{T}}$ denote an $n \times m$ synaptic matrix whose row vector is defined as $\boldsymbol{\theta}_k = [\theta_{k1}, \theta_{k2}, \ldots, \theta_{km}]$ ($k = 1, 2, \ldots, n$); then the GHA rule can be rewritten as

$$\Delta\mathbf{W}(t) = \eta \left(\mathbf{y}(t)\mathbf{x}^{\mathrm{T}}(t) - \mathrm{LT}[\mathbf{y}(t)\mathbf{y}^{\mathrm{T}}(t)]\mathbf{W}(t) \right), \tag{3.21}$$

where $\mathbf{x}(t) = [x_1(t), \ldots, x_m(t)]^{\mathrm{T}}$ and $\mathbf{y}(t) = [y_1(t), \ldots, y_n(t)]^{\mathrm{T}}$ are the $m \times 1$ and $n \times 1$ column vectors, respectively; and the operator $\mathrm{LT}[\cdot]$ sets all the elements above the diagonal of its matrix argument to zero (i.e., making the matrix *lower triangular*). Sanger's learning algorithm can also be viewed as an online version of the Hebbian learning (similar to Oja's rule) combined with the Gram–Schmidt orthogonalization procedure.

■ EXAMPLE 3.1

In this example, we use the GHA to illustrate the application of PCA for image compression [222, 364]. The *Mona Lisa* image used here was digitalized to form a 256×256 image with 256 gray levels (Figure 3.3*a*); the intensity of the pixels is normalized to lie within the range [0, 1]. The image is coded using a linear feedforward neural network with a single layer of eight neurons, each with 64 inputs. To train the neural network, 8×8 nonoverlapping blocks of the image were used. The learning rule (3.21) is conducted with 1000 epochs (i.e., 1000 scans of the image) with learning-rate parameter $\eta = 10^{-4}$.

Upon completing the learning process, Figure 3.3*b* shows the 8×8 masks representing the synaptic weights learned by the network. Each of the eight masks displays the set of synaptic weights associated with a particular neuron of the network. Specifically, excitatory (positive) synaptic weights are shown white, whereas inhibitory (negative) synaptic weights are shown black; gray indicates zero weights.

Given the compressed image (Figure 3.3*c*), we can further quantize the image for the sake of efficient storage and transmission. For instance, if we

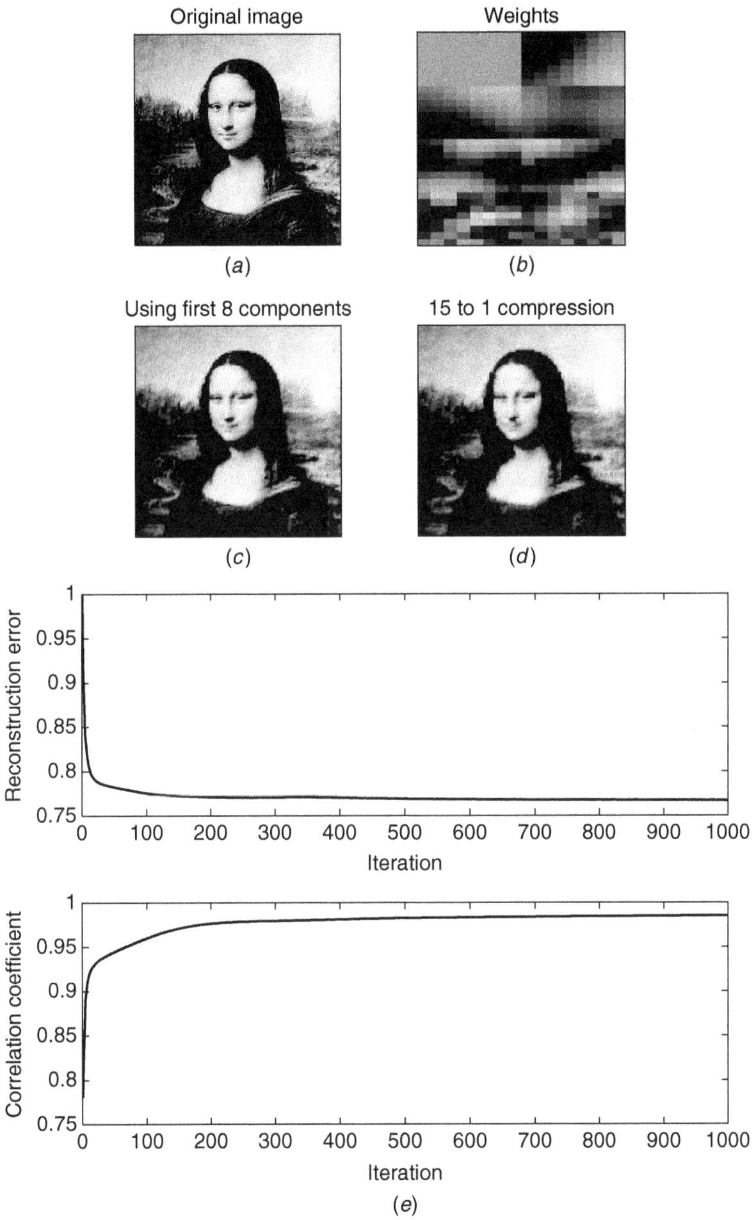

Figure 3.3 A demonstration of PCA learning for image compression. (*a*) A gray-scale image of *Mona Lisa*. (*b*) The 8 × 8 masks representing the synaptic weights (the columns of the weight matrix $\mathbf{W}^T$) learned by the GHA. (*c*) Reconstructed image using the learned eight principal components without quantization. (*d*) Reconstructed image with 15 to 1 compression ratio using quantization, resulting in a data rate of 0.53 bits per pixel. (*e*) The panels illustrate the learning curves of reconstruction error and correlation coefficient between the original and reconstructed images.

assign the bits to the mask by [7 7 6 4 3 3 2 2], based on this representation, a total of 34 bits were needed to code each 8×8 blocks of pixels, resulting in a data rate of 0.53 bits per pixel and a sum square reconstruction error of 12.97. The resultant quantized image is shown in Figure 3.3*d*.

During the learning process, it is observed that the normalized mean-squared reconstruction error curve gradually decreases; correspondingly, the correlation coefficient between the original and reconstructed images gradually increases.

3.1.7 Generalizations of PCA Learning

It is noted that the standard PCA learning rules (e.g., Oja's rule and Sanger's GHA) rely upon the exclusive use of feedforward connections, as shown in Figure 3.4*a* (although the weight orthogonalization procedure of the GHA implicitly requires lateral propagation of information between the output neurons); moreover, they assume linear neurons and static dynamics. We can extend PCA learning by relaxing one of these assumptions, which will lead to several noteworthy generalizations:

- Using lateral connections
- Imposing a triangular structure constraint
- Introducing temporal dynamics
- Using quadratic neurons

In what follows, we will elaborate each of the above generalizations.

Lateral Connections. In contrast to Oja's rule and the GHA, the *adaptive principal-components extraction* (APEX) algorithm [216, 511] uses both feefor-ward and lateral connections (see Figure 3.4*b*) and iteratively computes the jth principal component given the first $j-1$ principal components. Specifically, the jth neuron's output consists of both feedforward and lateral inputs

$$y_j(t) = \mathbf{x}^{\mathrm{T}}(t)\boldsymbol{\theta}_j(t) + \mathbf{y}_{j-1}(t)\mathbf{a}_j(t), \tag{3.22}$$

where $\mathbf{a}_j$ denotes the feedback lateral connections and $\mathbf{y}_{j-1}(t) = [y_1(t), \ldots, y_{j-1}(t)]^{\mathrm{T}}$ denotes the augmented vector that consists of the previous $j-1$ neurons' outputs. Then the update equations for parameter vectors $\boldsymbol{\theta}$ and $\mathbf{a}$ are defined, respectively, as

$$\Delta\boldsymbol{\theta}_j(t) = \eta\left[y_j(t)\mathbf{x}(t) - y_j^2(t)\boldsymbol{\theta}_j(t)\right], \tag{3.23}$$

$$\Delta\mathbf{a}_j(t) = -\eta\left[y_j(t)\mathbf{y}_{j-1}(t) + y_j^2(t)\mathbf{a}_j(t)\right], \tag{3.24}$$

which are both special cases of correlative learning rules. Equation (3.23) is essentially a generalization of Oja's rule, with $y_j(t)\mathbf{x}(t)$ being a Hebbian term. Equation (3.24) is mainly anti-Hebbian because of the term $-y_j(t)\mathbf{y}_{j-1}(t)$; the other term $y_j^2(t)\mathbf{a}_j(t)$ is included for the sake of stability.

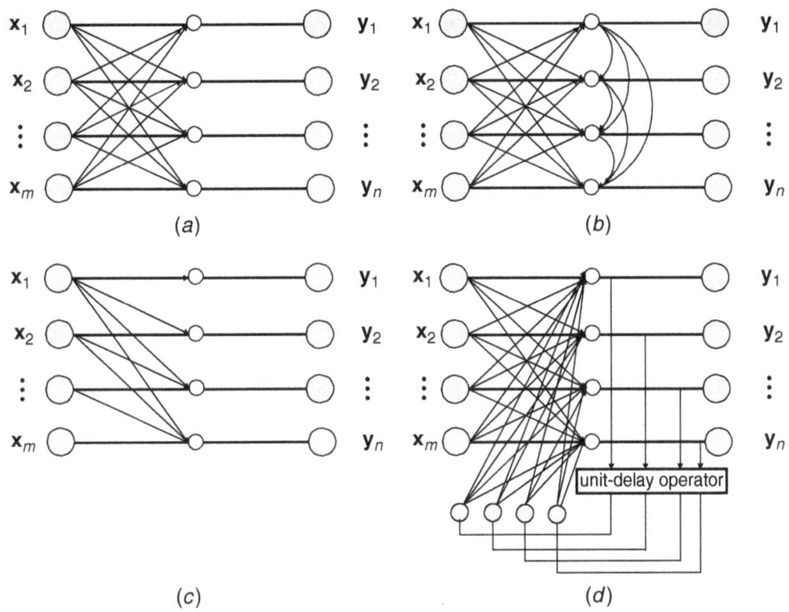

Figure 3.4 The network architectures for PCA: (*a*) feedforward; (*b*) feedforward with lateral connections; (*c*) triangular; (*d*) recurrent.

Triangular Network. It is also possible to perform decorrelation in certain topologically constrained linear networks [698]. The essential idea is that, instead of performing an eigenvalue decomposition of the correlation matrix, one may employ other matrix factorization methods, such as the *Cholesky decomposition*. Specifically, let $\mathbf{C}$ denote a symmetric positive-definite matrix equal to the covariance matrix of the input data $\mathbf{x}$; then matrix $\mathbf{C}$ may be factorized as

$$\mathbf{C} = \mathbf{L}\mathbf{L}^{\mathrm{T}}, \tag{3.25}$$

where $\mathbf{L}$ denotes a lower triangular matrix (namely, the elements above the diagonal are zeros). The goal is then to learn a transformation matrix $\mathbf{S}$ (which is defined as the inverse of the Cholesky factor $\mathbf{L}^{-1}$) such that the output is represented by $\mathbf{y} = \mathbf{S}\mathbf{x}$ and $\mathbb{E}[\mathbf{y}\mathbf{y}^{\mathrm{T}}] = \mathbf{I}$. By imposing topological constraints (see Figure 3.4*c*), anti-Hebbian learning rules emerge from the triangular network [698].

Recurrent Network. In addition to feedforward and lateral structures, PCA can also be implemented in a recurrent network (see Figure 3.4*d*), which is referred to as "recursive PCA" [919]. Specifically, let $\mathbf{x} \in \mathbb{R}^m$ and $\mathbf{y} \in \mathbb{R}^n$ denote, respectively, the input and output of the recurrent network; the simple recurrent network can be described by the following linear dynamic equation:

$$\mathbf{y}(t) = \mathbf{W}\mathbf{x}(t) + \sqrt{\alpha}\mathbf{V}\mathbf{y}(t-1), \tag{3.26}$$

where $\mathbf{W} \in \mathbb{R}^{n \times m}$ and $\mathbf{V} \in \mathbb{R}^{n \times n}$ are two matrices of synaptic connections and $\alpha \in [0, 1)$ is a positive gain constant. The matrix $\mathbf{V}$ is further assumed to have eigenvalues not greater than 1; therefore, the output $\mathbf{y}$ is bounded, stable, and asymptotically independent of its initial condition. Maximizing the variance of the output $\mathbf{y}(t)$ intuitively can potentially uncover a rich internal dynamics within the input signals.

Let $\mathbf{z}(t) = [\mathbf{x}^{\mathrm{T}}(t), \sqrt{\alpha}\mathbf{y}^{\mathrm{T}}(t-1)]^{\mathrm{T}}$; then equation (3.26) can be rewritten as

$$\mathbf{y}(t) = \tilde{\mathbf{W}}\mathbf{z}(t), \tag{3.27}$$

where $\tilde{\mathbf{W}} = [\mathbf{W}, \mathbf{V}]$. Applying Oja's rule to the vector $\mathbf{z}(t)$ yields the following update rules for recursive PCA [919]:

$$\Delta w_{ij}(t) = \eta y_i(t) \left(x_j(t) - \sum_{k=1}^{n} w_{kj}(t) y_k(t) \right), \tag{3.28}$$

$$\Delta v_{ij}(t) = \eta y_i(t) \left(\sqrt{\alpha} y_j(t-1) - \sum_{k=1}^{n} v_{kj}(t) y_k(t) \right). \tag{3.29}$$

It is found that upon convergence of the recursive PCA the rows of $\tilde{\mathbf{W}}$ will span the principal subspace of $\mathbf{z}$, and $\tilde{\mathbf{W}}^{\mathrm{T}}\tilde{\mathbf{W}}$ is a projection onto the n-subspace spanning the dimensions of highest variance of $\mathbf{z}$ [919].

The recursive version of PCA has some unique properties that are not shared with the standard PCA, such as the phenomena of local minimum and bifurcation, learning dynamics, and the greater representational power obtained by the algorithm's capability for capturing temporal context [919].

Quadratic PCA. In contrast to Oja's linear PCA model, a nonlinear extension of PCA with quadratic neurons (hence referred to as quadratic PCA) has also been developed [326, 877]. Specifically, the single neuron output for the quadratic PCA is described by

$$\begin{aligned} y &= \sum_{i=1}^{m} \theta_i x_i + \sum_{i=1}^{m} \sum_{j=1}^{m} w_{ij} x_i x_j \\ &= \boldsymbol{\theta}^{\mathrm{T}}\mathbf{x} + \mathbf{x}^{\mathrm{T}}\mathbf{W}\mathbf{x}, \end{aligned} \tag{3.30}$$

where the first term on the right-hand side is the same as the corresponding term in the learning rule for the linear neuron of the standard PCA model and the second term of the right-hand side accounts for the quadratic contributions of two neurons that are coupled by synaptic connection w_{ij}. The learning rules for adapting the unknown parameters can be derived as

$$\Delta \theta_i = \eta(y x_i - y^2 \theta_i), \tag{3.31a}$$

$$\Delta w_{ij} = \eta(y x_i x_j - y^2 w_{ij}). \tag{3.31b}$$

Let $\tilde{\boldsymbol{\theta}} = (\{\theta_i\}, \{w_{ij}\})$ denote a reorganized $(m + m^2)$-dimensional vector and let $\tilde{\mathbf{C}}$ denote an $(m + m^2) \times (m + m^2)$ augmented matrix

$$\tilde{\mathbf{C}} = \begin{pmatrix} \mathbf{C}_2 & \mathbf{C}_3 \\ \dot{\mathbf{C}}_3 & \mathbf{C}_4 \end{pmatrix}, \tag{3.32}$$

where $(\mathbf{C}_2)_{ij} = \mathbb{E}[x_i x_j]$, $(\mathbf{C}_3)_{ijk} = \mathbb{E}[x_i x_j x_k]$, and $(\mathbf{C}_4)_{ijkl} = \mathbb{E}[x_i x_j x_k x_l]$. With these new notations, equations (3.31a) and (3.31b) can then be unified into one update equation as follows [877]:

$$\Delta\tilde{\boldsymbol{\theta}} = \eta(\tilde{\mathbf{C}}\tilde{\boldsymbol{\theta}} - \lambda\tilde{\boldsymbol{\theta}}) \tag{3.33}$$

with $\lambda = \tilde{\boldsymbol{\theta}}^{\mathrm{T}}\tilde{\mathbf{C}}\tilde{\boldsymbol{\theta}}$. Upon convergence, $\Delta\tilde{\boldsymbol{\theta}} = \mathbf{0}$, and this is essentially solving an eigenvalue problem $\tilde{\mathbf{C}}\tilde{\boldsymbol{\theta}} = \lambda\tilde{\boldsymbol{\theta}}$, which, however, unlike the linear PCA, invokes up to fourth-order moment statistics of the inputs.

Minor-Component Analysis. An opposite but related problem to PCA is to find the *least important* component associated with the smallest eigenvalue. This problem, often referred to as *minor-component analysis* (MCA) [678], may be viewed as the opposite of PCA; therefore it is expected that the algebraic sign of the associated update equation will be reversed. Specifically, an anti-Hebbian MCA learning rule was proposed as follows [982]:

$$\Delta\theta_i(t+1) = -\eta[y(t)x_i(t) - y^2(t)\theta_i(t)]. \tag{3.34}$$

It can be shown that if the smallest eigenvalue of the correlation matrix $\mathbf{C} = \mathbb{E}[\mathbf{x}(t)\mathbf{x}^{\mathrm{T}}(t)]$ is $\lambda_{\min}$ with multiplicity 1, then

$$\lim_{t\to\infty} \boldsymbol{\theta}(t) = \eta\mathbf{u}_{\min}, \tag{3.35}$$

where $\mathbf{u}_{\min}$ is the eigenvector of $\mathbf{C}$ associated with the minimum eigenvalue $\lambda_{\min}$.

In a similar context, Luo et al. [578] also proposed a learning rule for MCA. By minimizing the *Rayleigh quotient* of the weight vector, an alternative MCA learning rule may be derived,

$$\Delta\theta_i(t+1) = -\eta\left[\sum_{j=1}^{m}\theta_j^2(t)y(t)x_i(t) - y^2(t)\theta_i(t)\right]. \tag{3.36}$$

Finally, other extensions of PCA-type learning rules include the cross-coupled Hebbian learning rule for SVD [215], a higher order correlation-based version of Oja's learning rule [876], and the kernelized Hebbian PCA learning rule, the latter of which will be discussed in Chapter 4.

3.1.8 CCA Learning Rule

As discussed earlier in Chapter 2, CCA can be viewed as a generalization of PCA, and PCA is a degenerate case of CCA. Naturally, CCA can also be implemented with adaptive learning rules in a similar manner to PCA. For simplicity, we only discuss CCA in the context of two sets of random variables, which are denoted by $\mathbf{x}_1 = [x_{11}, \ldots, x_{1n}]^{\mathrm{T}}$ and $\mathbf{x}_2 = [x_{21}, \ldots, x_{2m}]^{\mathrm{T}}$.

Consider the multiple input–single output (MISO) case; let y_1 and y_2 denote, respectively, the linear combination of the variables from $\mathbf{x}_1$ and $\mathbf{x}_2$:

$$y_1 = \sum_{j=1}^{n} \theta_{1j} x_{1j} = \boldsymbol{\theta}_1^{\mathrm{T}} \mathbf{x}_1, \qquad y_2 = \sum_{j=1}^{m} \theta_{2j} x_{2j} = \boldsymbol{\theta}_2^{\mathrm{T}} \mathbf{x}_2.$$

The goal of CCA is to find $\boldsymbol{\theta}_1 = [\theta_{11}, \ldots, \theta_{1n}]^{\mathrm{T}}$ and $\boldsymbol{\theta}_2 = [\theta_{21}, \ldots, \theta_{2m}]^{\mathrm{T}}$ such that the maximum correlation between y_1 and y_2 is achieved.

Typically, a unit-variance constraint on y_1 and y_2 is imposed to avoid degenerate solutions. Using the method of Lagrange multipliers, an objective function (to be maximized) is defined as

$$J = \langle y_1 y_2 \rangle + \frac{1}{2}\lambda_1 (y_1^2 - 1) + \frac{1}{2}\lambda_2 (y_2^2 - 1)$$

where λ_1 and λ_2 are two Lagrangian multiplier coefficients and $\langle \cdot \rangle$ denotes the statistical average over the observed data. Alternating optimization of $\boldsymbol{\theta}_1, \lambda_1$ and $\boldsymbol{\theta}_2, \lambda_2$ will yield a monotonic increase of the objective function,

$$\Delta \boldsymbol{\theta}_1 = \eta \frac{\partial J}{\partial \boldsymbol{\theta}_1} = \eta \mathbf{x}_1 (y_2 - \lambda_1 y_1), \tag{3.37}$$

$$\Delta \lambda_1 = \gamma \frac{\partial J}{\partial \lambda_1} = \gamma (y_1^2 - 1), \tag{3.38}$$

$$\Delta \boldsymbol{\theta}_2 = \eta \frac{\partial J}{\partial \boldsymbol{\theta}_2} = \eta \mathbf{x}_2 (y_1 - \lambda_2 y_2), \tag{3.39}$$

$$\Delta \lambda_2 = \gamma \frac{\partial J}{\partial \lambda_2} = \gamma (y_2^2 - 1), \tag{3.40}$$

where η and γ are small step-size parameters; in order to assure convergence, it is often required that $\gamma \gg \eta$.

The extensions of CCA to the general cases of multiple input–multiple output (MIMO) neurons, nonlinear neural networks, and nonlinear canonical correlation were discussed in [412, 519, 520, 716]. The Imax algorithm, discussed later in Section 3.2.4, can also be shown to be a MIMO nonlinear generalization of CCA [69].

3.1.9 Wake- Sleep Learning Rule for Factor Analysis

The "wake–sleep" learning rule [387, 657] is a simple learning algorithm for unsupervised neural networks (such as the Helmholtz machine) or stochastic models with hidden variables that employ an unsupervised version of the delta rule. In [657], Neal and Dayan proposed a delta rule wake–sleep learning procedure for fitting a factor analysis model.

The factor analysis model can be viewed as a simple Helmholtz machine with two layers of linear units which consists of a generalized model that is defined by a hidden factor and further corrupted by Gaussian noise. Let $\mathbf{x} \in \mathbb{R}^n$ denote the real-valued visible input vector and y be a real-valued hidden scalar variable[2] that is normally distributed (i.e., $y \sim \mathcal{N}(0, 1)$); and assumed $\mathbf{x}$ to be generated by the following linear *generative model*:

$$\mathbf{x} = \boldsymbol{\mu} + y\mathbf{g} + \boldsymbol{\varepsilon}, \tag{3.41}$$

where the vector $\mathbf{g} \in \mathbb{R}^n$ denotes the "factor loading," $\boldsymbol{\mu} \in \mathbb{R}^n$ denotes the mean vector and is often assumed to be zero without loss of generality, and $\boldsymbol{\varepsilon} \sim \mathcal{N}(\mathbf{0}, \boldsymbol{\Sigma})$ is a noise vector with zero mean and diagonal covariance matrix $\boldsymbol{\Sigma} = \text{diag}\{\sigma_1, \ldots, \sigma_n^2\}$. In addition, the hidden variable y is defined by a *recognition model*, written as

$$y = \mathbf{r}^{\mathrm{T}}\mathbf{x} + \nu, \tag{3.42}$$

where $\mathbf{r} \in \mathbb{R}^n$ denotes the "top-down" weight vector and $\nu \sim \mathcal{N}(0, \sigma^2)$ denotes additive Gaussian noise with zero mean and variance σ^2.

Specifically, Neal and Dayan [657] derived a simple wake–sleep learning rule for alternatingly updating the loading factor $\mathbf{g}$ and the noise covariance matrix $\boldsymbol{\Sigma}$ in the "wake phase" and updating $\mathbf{r}$ and σ^2 in the "sleep phase":

$$\mathbf{g}(t+1) = \mathbf{g}(t) + \eta\left[\mathbf{x}^{(c)}(t) - \mathbf{g}(t)y^{(c)}(t)\right]y^{(c)}(t), \tag{3.43}$$

$$\sigma_i^2(t+1) = \alpha\sigma_i^2(t) + (1-\alpha)\left[\mathbf{x}^{(c)}(t) - g_i(t)y^{(c)}(t)\right]^2, \tag{3.44}$$

$$\mathbf{r}(t+1) = \mathbf{r}(t) + \eta\left[y^{(f)}(t) - \mathbf{r}^{\mathrm{T}}(t)\mathbf{x}^{(f)}(t)\right]\mathbf{x}^{(f)}(t), \tag{3.45}$$

$$\sigma^2(t+1) = \alpha\sigma^2(t) + (1-\alpha)\left[y^{(f)}(t) - \mathbf{r}^{\mathrm{T}}(t)\mathbf{x}^{(f)}(t)\right]^2, \tag{3.46}$$

where the scalars η and α denote two step-size parameters; the superscripts (c) and (f) (on both $\mathbf{x}$ and y) are used to discriminate the data being completely observed in the external world from the fantasy data being produced by the estimated generative model.

Under regular conditions and with appropriate step-size parameters, the above-described wake–sleep learning rule leads to the convergence to an MLE in a similar fashion as the EM algorithm [429, 657]. The combined "bottom-up" (recognition model) and "top-down" (generative model) learning paradigm, both having roughly Hebbian forms, may be applicable to learning to model sensory data in hierarchically structured cortical circuits [150, 387, 541].

3.1.10 Boltzmann Learning Rule

The Boltzmann machine can be seen as the stochastic, generative counterpart of the discrete Hopfield network, with an important difference being that it allows for hidden units and hence is capable of learning internal representations of the data. Motivated by the energy function employed in the Hopfield network, Hinton and Sejnowski [390] derived a simple and local learning rule for inferring the "hidden states" that can produce observed samples from a learned Boltzmann (or Gibbs) distribution

$$p(\mathbf{x}) = \frac{1}{Z}\exp\left(-\frac{1}{T}\mathbf{x}^{\mathrm{T}}\mathbf{W}\mathbf{x}\right), \tag{3.47}$$

where $\mathbf{x} = \{x_i\}_{i=1}^n$ denotes the discrete state, Z denotes the partition function, T denotes the temperature parameter, and $\mathbf{W} = \{w_{ij}\}_{i,j=1}^n$ denotes the symmetric positive-definite weight matrix that defines the potential energy

$$J(\mathbf{x}) = -\frac{1}{2}\mathbf{x}^{\mathrm{T}}\mathbf{W}\mathbf{x} = -\frac{1}{2}\sum_i \sum_{j,j\neq i} w_{ij}x_i x_j,$$

in which $w_{ii} = 0$ for all i. In general, the Boltzmann machine contains "visible" units and "hidden" units, where the visible units are those that receive information from the observed data from the environment and the hidden units are supposed to capture the internal structure of the data. Each unit in the Boltzmann machine computes an energy gap that results from flipping a single unit (say unit i) from 0 (or -1) to $+1$, denoted as ΔJ_i, as given by

$$\Delta J_i = \sum_j w_{ij}x_j, \tag{3.48}$$

and the "turning on" probability for the ith unit is given by

$$p(x_i) = \frac{1}{1+\exp(-\Delta J_i/T)}.$$

With an information-theoretic measure criterion, the Boltzmann machine employs a *contrastive* Hebbian learning rule, combining a Hebbian learning term in the "positive" phase (or the *clamped* state) of learning and an anti-Hebbian term in the "negative" phase (or the *free* state) [6, 391]:

$$\Delta w_{ij} = \eta(\langle x_i\, x_j\rangle_+ - \langle x_i\, x_j\rangle_-), \tag{3.49}$$

where $\langle\cdot\rangle$ denotes sample expectation averaged over the observed noisy samples. In the positive ("learning") phase, the activities of the visible units are fully constrained by the training patterns, whereas in the negative ("unlearning") phase, depending on the model, some or all of the visible units' states are generated by a Monte Carlo sampling procedure. In both phases, the states of the unclamped neurons in the model must be repeatedly stochastically updated until the network settles into an equilibrium state. To avoid the settling procedure becoming trapped in local minima, a simulated annealing procedure is used whereby the settling process starts at a very high temperature where state updates are very random and the temperature is gradually lowered until equilibrium is reached. The learning converges extremely slowly due to the simulated annealing required at every learning iteration as well as the very large space of states that must be sampled in the unclamped (negative) phase. An important, more recent innovation that overcomes these difficulties is the restricted Boltzmann machine with "brief Gibbs sampling" [385, 388]. When a balance between the clamped and free states is achieved, the Boltzmann machine learning procedure approaches the equilibrium state and the learning process terminates, namely $\Delta w_{ij} = 0$ for all i and j. As can be seen by examining (3.49), the learning rule is local and conceptually simple; however, the convergence process may be very slow in practice. The conventional Boltzmann learning procedure uses pairwise correlation, but it is possible to generalize to a higher order Boltzmann machine by using higher order correlations in the context of mean-field theory [465]. Finally, although the hidden states in the standard Boltzmann machine are commonly binary and discrete, extensions to the continuous-valued restricted Boltzmann machine have also been developed [384, 386, 878].

3.1.11 Perceptron Rule and Error-Correcting Learning Rule

In the late 1950s, Frank Rosenblatt proposed a model called the *perceptron* [773], which was initially applied to the simulation of early visual processing, feature extraction, and classification. The perceptron model uses layers of thresholded linear neurons. Specifically, let x_j and y_i denote the jth input neuron and ith output neuron, respectively, and let d_i denote the binary-valued (0,1) target output associated with the ith neuron whose output activation is y_i; then the synaptic weight, θ_{ij}, is updated by the rule

$$\Delta\theta_{ij}(t) = \eta x_j(t)[d_i(t) - y_i(t)], \tag{3.50}$$

where $y_i = 1$ if $y_i \geq b$ (where b denotes a threshold value) and $y_i = 0$ otherwise. The perceptron rule is the first supervised learning rule for neural networks published in the literature; under certain conditions, including most importantly the restriction that the pattern classes are linearly separable, its convergence is guaranteed [773]. Despite its restriction to solving linearly separable problems and its lack of convergence for nonlinear classification problems [630], the perceptron learning rule has provided the foundation for many more advanced learning rules in subsequent years.

As we have discussed in Chapter 2, the LMS filter can be seen as an adaptive linear neuron that approximates the solution of the Wiener filter. Here, we provide more analysis for this simple yet efficient error-correcting learning rule. For the reader's convenience, the MIMO version of the LMS rule (2.88) is rewritten here as

$$\begin{aligned}\Delta\theta_{ij}(t) &= \eta x_j(t)[d_i(t) - y_i(t)] \\ &= \eta x_j(t)d_i(t) - \eta x_j(t)y_i(t),\end{aligned} \tag{3.51}$$

where $x_j(t)$ and $y_i(t)$ denote the network's jth input and ith output signals, respectively, and $d_i(t)$ denotes the desired output signal associated with $y_i(t)$. Suppose that $x_j(t)$ and $d_i(t)$ are random signals; then taking the time average of both sides of (3.51) yields

$$\langle\Delta\theta_{ij}(t)\rangle = \eta\langle x_j(t)d_i(t)\rangle - \eta\langle x_j(t)y_i(t)\rangle, \tag{3.52}$$

where the first correlation term on the right-hand side of (3.52) is a *forced* Hebbian rule, whereas the second correlation term represents an anti-Hebbian rule.

Note that when the random signal $d_i(t)$ is zero mean and orthogonal (i.e., uncorrelated) to $x_j(t)$, the Hebbian term in (3.52) will become zero; accordingly, (3.52) reduces to

$$\langle\Delta\theta_{ij}(t)\rangle = -\eta\langle x_j(t)y_i(t)\rangle, \tag{3.53}$$

which is a pure anti-Hebbian rule. From another perspective, we can view the unsupervised anti-Hebbian learning as a *stochastic* version of (3.52) in which the desired output signal $d_i(t)$ is assumed to be a zero-mean random noise process. A mathematical proof of such a statement is given in Appendix 3C.

In the literature, the error-correcting LMS rule (3.51) is also known as the *stochastic gradient descent* or *delta rule*. Moreover, although the LMS learning rule was initially developed for operating with a linear neuron, a "generalized delta rule" has been extended to nonlinear neurons and multilayer networks using backpropagation [780, 948].

Several additional points are noteworthy:

- The error-correcting LMS rule can be combined with a conventional Hebbian rule in specific scenarios; this is often advantageous since the incorporation of the supervised mode generally accelerates the convergence of unsupervised Hebbian learning. For instance, it is possible to integrate the LMS rule to learn the correlation memory matrix [35]:

  $$\mathbf{M}(t+1) = \mathbf{M}(t) + \eta[\mathbf{y}_k - \mathbf{M}(t)\mathbf{x}_k]\mathbf{x}_k^{\mathrm{T}}, \tag{3.54}$$

  where $\mathbf{M}$ denotes the connection weight matrix between the input pattern $\mathbf{x}$ and the output pattern $\mathbf{y}$. It can be shown that in the case of autoassociation where $\mathbf{y}_k = \mathbf{x}_k$, as time $t \to \infty$, we have

  $$\mathbf{M}(\infty)\mathbf{x}_k = \mathbf{x}_k. \tag{3.55}$$

  In words, upon convergence of learning, $\mathbf{x}_k$ will correspond to the eigenvector of the matrix $\mathbf{M}(\infty)$, with an associative eigenvalue of unity.

- Under special circumstances, the error-correcting learning rule (i.e., LMS or backpropagation) and contrastive Hebbian learning [6, 715] are equivalent for training neural networks. Movellan [639] first showed that contrastive Hebbian learning is equivalent to the generalized delta rule for networks with a single layer. Recently, Xie and Seung [978] also showed that when the multilayer perceptron (MLP) has linear output units and weak feedback connections the change in network states caused by clamping the output neurons is the same as the error signal produced by backpropagation to within a scalar factor.

3.1.12 Differential Hebbian Rule and Temporal Hebbian Learning

It has been known for a long time that the original Hebbian postulate does not explicitly address the feedback mechanism of synaptic plasticity; namely, the previous presynaptic and/or postsynaptic terms generally do not influence the current or future synaptic modification. To overcome this pitfall, Klopf [488] and Kosko [505] proposed a generalized version of Hebbian synaptic plasticity. In particular, the synaptic modulation is made proportional to the *temporal rates* of the presynaptic and postsynaptic activities, which emphasizes the succession order in time. In the literature, this modification is called the *differential Hebbian rule* because the changes in synaptic weight are driven by a conjunction of the short-term temporal changes in the presynaptic inputs and postsynaptic output, as shown by

$$\Delta\theta_{ij} = \eta\, \Delta x_i\, \Delta y_j, \tag{3.56}$$

where Δx_i and Δy_j denote the temporal changes of presynaptic and postsynaptic activities, respectively. Since a time interval is incorporated into (3.56) by correlating earlier changes in the presynaptic activity with later changes in the postsynaptic activity, the differential Hebbian rule allows us to learn a *causal* relationship based upon the temporal events; namely, the present (or future) event can be associated with the history of past events, which is analogous to the concept of *Pavlovian classical conditioning* [201]. Alternatively, the differential Hebbian learning rule may be written in another form [378]:

$$\Delta\theta_{ij}(t) = -a\theta_{ij}(t) + \left[-b\theta_{ij}(t) + cx_i(t)y_j(t)\right]\Theta(\Delta y_j(t))\Theta(-\Delta x_i(t)), \tag{3.57}$$

where $\Theta(\cdot)$ denotes the Heaviside or unit step function with $\Theta(\xi) = 1$ for $\xi > 0$ and $\Theta(\xi) = 0$ otherwise; a, b, and c are positive constants and $a \ll b$. The purpose of a is to force the weights that are never or rarely increased to eventually approach zero; the term $-b\theta_{ij}(t) + cx_i(t)y_j(t)$ plays the usual role of Hebbian plasticity, and it is gated by the product of two unit step functions, which equals 1 only when $\Delta y_j > 0$ and $\Delta x_i < 0$. Such an asymmetric window ensures the formation of a temporal relationship between x_i and y_j in spatiotemporal learning.

As a special case, the differential rule (3.56) has the form

$$\Delta\theta_{ij} = \eta x_i\, \Delta y_j, \tag{3.58}$$

which states that the synaptic change is proportional to the presynaptic activity and the temporal rate of the postsynaptic activity. An example of such a learning rule was used in the classical conditioning model by Sutton and Barto [866]; specifically, the learning rule has the subsequent form

$$\Delta\theta(t) = \eta\overline{x}(t)[y(t) - y(t-1)], \tag{3.59}$$

where the first term on the right-hand side, $\overline{x}(t)$, denotes the *eligibility trace* of the input,

$$\overline{x}(t) = \lambda\overline{x}(t-1) + x(t) \qquad (0 < \lambda < 1),$$

and the second term on the right-hand side, $y(t) - y(t-1)$, denotes the temporal change of the output. This kind of correlative learning was later generalized by Sutton [865] in temporal-difference learning (which we will discuss later in this chapter).

Several variants of differential Hebbian learning are noteworthy:

- Mitchison [632] developed an anti-Hebbian form of a differential Hebbian rule,

$$\Delta\theta_i = -\eta\,\Delta x_i\,\Delta y + \alpha x_i y, \tag{3.60}$$

  subject to the weight normalization constraint $\|\theta_i\|^2 = 1$. Such an anti-Hebbian differential learning rule was used for generating center-surround receptive fields that remove linear spatiotemporal variations of input patterns.
- Földiak [284, 285] developed a hybrid differential Hebbian rule for competitive learning,

$$\Delta\theta_{ij}(t) = \eta\overline{y}_j(t)[x_i(t) - \theta_{ij}(t)], \tag{3.61}$$

  where the postsynaptic output is defined by the eligibility trace $\overline{y}_j(t) = \alpha\overline{y}_j(t-1) + y_j(t)$ and $\overline{y}_j(t)x_i(t)$ specifies a temporal Hebbian term. The winning postsynaptic neurons y_j in (3.61) can be selected out of hard or soft competition. In [285], spatial invariance was converted into a temporal feature by presenting transformation sequences within the invariance classes, and the temporal smoothness constraint was incorporated into (3.61). Such a "trace tracking" rule forces the output neuron to develop invariant responses to the input patterns that tend to occur close together in time; for this reason, equation (3.61) may also be used for learning object invariance and feature decorrelation [771, 772, 861].
- Roberts [765] proposed a differential spike-timing-dependent Hebbian learning rule with temporally asymmetric characteristics; the key feature of the learning rule is that the synaptic efficacy is approximately proportional to the

rate of the postsynaptic spike probability. Specifically, the synaptic rule is described as follows:

$$\langle \Delta\theta(t) \rangle \propto \sum_k c_k \frac{\partial^k}{\partial y^k} g(y,t) \approx \eta \frac{\partial g(y,t)}{\partial y}, \tag{3.62}$$

where y denotes a spike and $g(y,t)$ denotes the probability of a spike firing at time t in the postsynaptic neuron. In [765], $g(y,t)$ was defined by the probability of the membrane potential exceeding the threshold $V_0 = V_{\text{th}}$:

$$g(t) = g(V_0(t)) = \int_0^\infty p(V - V_0(t), \sigma)\, dV. \tag{3.63}$$

Provided the probability density $p(V - V_0(t))$ is Gaussian,

$$p(V - V_0, \sigma) = \frac{1}{\sqrt{2\pi}\sigma} \exp\left[-\frac{(V - V_0)^2}{2\sigma^2} \right], \tag{3.64}$$

then substituting (3.64) into (3.63) yields a "sigmoid-shaped" *complementary error function* with a threshold value of $\frac{1}{2}$. Equation (3.62) represents macroscopic results (on the timescale of several conditioning cycles) that follow from the microscopic temporal rules (on the timescale of the interspike interval); it also describes the probabilistic nature of the synaptic modification quantity within the spike time window [766].

Notably, differential Hebbian learning is in spirit close to the STDP learning rule [752, 765] in that the synaptic modification is proportional to the time derivatives of presynaptic and postsynaptic firing rates rather than the firing rates per se.

Hebbian learning can be used not only for stationary or static data but also for temporal sequence data in a dynamic environment. In particular, learning the timescale is crucial to tackle dynamical systems. Time-dependent Hebbian plasticity is important to model self-organizing cortical functions and has been widely used in learning recurrent neural networks. In general, we can write the generalized Hebbian rule in the form of a differential equation associated with a time constant τ:

$$\tau \frac{d\boldsymbol{\theta}}{dt} = \mathbf{x}(t) f(y(t - \Delta t)), \tag{3.65}$$

where $\mathbf{x}(t)$ and $y(t)$ represent the presynaptic and postsynaptic activities, respectively; f is a generic function of the postsynaptic output; and Δt denotes a positive time-delay constant. Varying the time constant τ of (3.65) will result in a different "speed scale" of the learning process; however, τ is often much greater than the time constant of the system dynamics. Notably, the temporal Hebbian rule is capable of *trace learning*, where the term "trace" refers to the history of synaptic activity.

If we integrate (3.65) from $t = 0$ to $t = T$, then the resulting synaptic plasticity is defined by

$$\boldsymbol{\theta} = \frac{1}{T}\int_0^T dt\, \mathbf{x}(t) \int_{-\infty}^{\infty} d\tau\, f(y(t-\tau)). \tag{3.66}$$

A modified form of (3.65) is to impose a time-varying threshold on the input $\mathbf{x}(t)$ within a sliding time window in order to assure stability of the learning rule:

$$\tau \frac{d\boldsymbol{\theta}}{dt} = [\mathbf{x}(t) - \overline{\mathbf{x}}(t)]\, f(y(t - \Delta t)), \tag{3.67}$$

where $\overline{\mathbf{x}}(t) = \langle \mathbf{x}(t) \rangle$ denotes the time average of the presynaptic input $\mathbf{x}(t)$.

The temporal Hebbian learning rule (3.65) can be viewed as a generalization of the differential Hebbian rule; in many cases, the boundary between them is fuzzy. In addition, temporal Hebbian learning has close ties with temporal difference and classic reinforcement learning, which we discuss next.[3]

3.1.13 Temporal Difference and Reinforcement Learning

TD Learning. Reinforcement learning [868] can also be viewed as a correlative learning process. As partly discussed in Chapter 1, the early theory of reinforcement learning was motivated from the classical conditioning model. The overall goal of reinforcement learning is to learn a good approximation of a *value function*, which indicates the expected sum of future rewards, where a reward received at τ steps into the future is often discounted by an exponential factor γ^{τ}. Unlike the conventional dynamic programming method, reinforcement learning usually assumes no knowledge of the state transition probability and the environment; thus learning is achieved via an online trial-and-error procedure. According to the preceding definition of the value function, for consistency, the value predicted at the current time step should equal the reward received at the next time step plus the discounted value predicted at the next time step. This consistency condition motivates the update equation for learning the value function in the TD learning algorithm (a special form of reinforcement learning). At each time step the target for the value $V(t)$ is taken to be $\gamma V(t+1) + r(t+1)$ [where $r(t+1)$ denotes the reward at time $t+1$]. The difference between this target value and the current value $V(t)$ is known as the TD error signal. The development of TD learning was partially motivated by error-correcting supervised learning [865]; hence, it takes the same functional form as the LMS rule (2.88) in that the weights are updated in proportion to the correlation between the input stimuli and the error in predicting a reinforcement signal [865]:

$$\begin{aligned}\boldsymbol{\theta}(t+1) &= \boldsymbol{\theta}(t) + \eta\left[r(t+1) + \gamma V(s(t+1)) - V(s(t))\right] \nabla_{\boldsymbol{\theta}} V(s(t)) \\ &= \boldsymbol{\theta}(t) + \eta \mathbf{x}(t)\varepsilon(t),\end{aligned} \tag{3.68}$$

where $V(s(t)) = \mathbf{x}^{\mathrm{T}}(t)\boldsymbol{\theta}(t)$ represents the *linear* value function and $\nabla_{\boldsymbol{\theta}} V(s(t)) = \mathbf{x}(t)$ denotes the its gradient vector. The vector s denotes the observable state, and $\mathbf{x}(t) = [x(t), x(t-1), \ldots, x(t-\tau)]^{\mathrm{T}}$ denotes the tap-delayed feature vector associated with the state. [4] The term $\varepsilon(t)$ represents the TD error, as defined by

$$\begin{aligned}\varepsilon(t) &= r(t+1) + \gamma V(s(t+1)) - V(s(t)) \\ &\triangleq r(t+1) + \gamma V(t+1) - V(t)\end{aligned} \tag{3.69}$$

where $0 < \gamma \leq 1$ is the discount factor. The learning rule (3.68) has the effect that when more reward is received than expected [i.e., $\varepsilon(t) > 0$], $\boldsymbol{\theta}$ is incremented in proportion to the correlation between the unexpected reward (i.e., positive TD error) and input state; on the other hand, if less reward is received than expected, $\boldsymbol{\theta}$ is incremented in proportion to the correlation between the penalty (i.e., negative TD error) and input state. The TD error in (3.69) essentially approximates the difference between the actual and predicted total future reward, which is based on a "bootstrapping" method for the value function. Note that

$$\begin{aligned}V(t) &\triangleq \sum_{\tau>0}^{T} r(t+\tau) = r(t+1) + \sum_{\tau>1}^{T} r(t+\tau) \\ &= r(t+1) + V(t+1),\end{aligned} \tag{3.70}$$

where the value function $V(t)$ is interpreted as a prediction of the total future reward expected from time t onward to the end of the trial. The neurobiological plausibility of TD learning as an example of classical conditioning is discussed in [201, 868]. Specifically, it was suggested that the value function $V(t)$ provides a plausible mechanism by which animals may use prediction to optimize the behavior when rewards are delayed (i.e., solving the so-called *temporal credit assignment problem*) and explains a wide range of psychological and neurobiological data. According to [201], the TD error might be represented by the activity of dopaminergic neurons in the ventral tegmental area in the midbrain; neurophysiological evidence indicates that the dopamine signal acts to gate and regulate the neural plasticity.[5] Recently, Rao and Sejnowski [752, 753] have proposed a spike-timing-dependent version of the TD learning rule. Basically, if the presynaptic spike precedes (within a time window of 10 ms) the postsynaptic spike, the synaptic weight increases; if the order is reversed, the synaptic weight decreases. The temporally asymmetric timing window enables the spike-timing-dependent Hebb-like rule to learn temporal sequences and predict future events, which may be a fundamental function of the human brain.

Local Reinforcement Learning. One of the criticisms of error-correcting learning algorithms such as backpropagation is its biological implausibility in terms of a backpropagated *quantitative* error signal [349, 691]. To overcome this drawback, Mazzoni et al. [602] proposed a local and biologically plausible learning rule that uses a *qualitative* reward signal to modify the synaptic connections:

$$\Delta\theta_{ij} = \rho r(y_i - p_i)x_j + \lambda\rho(1-r)(1-y_i-p_i)x_j, \tag{3.71}$$

where ρ and λ are two scalar constants, θ_{ij} denotes the strength of the synaptic weight between the input signal x_j and output signal y_i, p_i denotes the probability of the ith neuron firing, and r represents a scalar reinforcement signal between 0 and 1. The first term on the right-hand side of (3.71) computes the reward portion of the learning rule, whereas the second term is the penalty portion. Ignoring the constant terms and the stochastic component, equation (3.71) changes the synaptic weight by correlating three terms, namely, *reinforcement signal, presynaptic activity*, and *postsynaptic activity*. Such a *tri-Hebbian* term is believed to be important in modeling synaptic plasticity at both the microscopic and macroscopic levels. A correct response (large value of r) will strengthen θ_{ij}, whereas an incorrect response (small value of r) will weaken θ_{ij}; the reward value can be calculated from the averaged output error [602].

Reinforcement Hebbian Learning. It should be clarified that although Hebbian synaptic plasticity is only dependent on local information, the synaptic modulation process may be subject to modulation by global signals, whose roles are to enable the induction or consolidation of changes at synapses that have met the criteria for Hebbian modification [122] under the constraint of other factors such as attention or global reinforcement. For instance, a global reinforcement signal (e.g., "correct" or "incorrect") that is transmitted broadly and diffusely through chemical substances (i.e., neurotransmitters) can modulate synaptic modification within a large population of activated synapses. Because of this reason, some recent efforts have been devoted to combining Hebbian learning and reinforcement learning. A simple and direct approach is to revise the original Hebbian postulate by adding an additional multiplicative factor r (where $r \in \{0, 1\}$) to the generalized Hebbian rule (3.6) as follows:

$$\Delta\theta(t) = \eta r f(x)[g(y) - \theta(t)]. \tag{3.72}$$

Equation (3.71) is an instance of (3.72) that integrates a tri-Hebbian term. As another example, Alspector et al. [17] suggested the use of "excess reinforcement" to introduce global reinforcement control; the synaptic modification rule consists of correlative Hebbian and anti-Hebbian terms:

$$\Delta\theta_{ij}(t) = \eta\left[\langle r x_i y_j\rangle - \langle r\rangle\langle x_i y_j\rangle\right], \tag{3.73}$$

where $r \in \{-1, +1\}$ denotes the reward signal that is set to $+1$ if the output is correct and -1 otherwise.

In a related context, Bosman et al. [105] suggested a biologically inspired neural network model that integrates Hebb's rule with reinforcement learning to overcome the "path interference" problem. Specifically, the generic learning rule was given by

$$\Delta\theta_{ij} = \eta\left[a_{ij}(r) + b_{ij}(r)x_i + c_{ij}(r)y_j + d_{ij}(r)x_i y_j\right], \tag{3.74}$$

where $a_{ij}(r), b_{ij}(r)$, and $d_{ij}(r)$ denote the coefficients that relate to the local presynaptic or postsynaptic signals as well as the reinforcement signal r (where $r \in \{0, 1\}$).

In general, reward-based learning uses evaluative (i.e., behavioral) feedback and relates closely to the animal conditioning literature as well as to the neuroscientific literature regarding the dopaminergic reward system of the brain, whereas conventional correlation-based learning is oriented to nonevaluative feedback. Recently, Wörgötter and Porr [975] have reviewed the reward-based and correlation-based learning methods and attempted to unify them within the same framework. Notably, two important observations were pointed out in [975]:

- The "eligibility trace" estimation that was widely used in reinforcement learning is a correlation-based process. The concept of such traces is related to the neural differential Hebbian learning context. Hence, TD-based learning [e.g., TD(λ) or Q(λ) learning] is formally related to Hebbian learning.
- Rewards may be correlated with the sensory events; therefore, correlation-based processes can be introduced into the self-evaluative "actor–critic" model for the sake of closed-loop control in reinforcement learning.

■ EXAMPLE 3.2

In this example, we illustrate how to utilize TD learning for solving a simple temporal credit assignment problem. The example taken from [865] consists of a Markov chain with five "normal" states (B, C, D, E, F) and two additional "absorbing" states (start state A and end state G), as shown in Figure 3.5. The rewards for states A and G are 0 and 1, respectively. The transition probabilities of moving from each state to its adjacent (either left or right) state (except for the absorbing states) are all equal to $\frac{1}{2}$. In addition, the transition probabilities of ending in state G from each state are

$$P_T = \left[0,\ \frac{1}{6},\ \frac{2}{6},\ \frac{3}{6},\ \frac{4}{6},\ \frac{5}{6},\ 0\right].$$

The goal of this problem is to learn the transition probabilities P_T with a series of random walks (while each session from the start to the end is called one trial).

The problem of temporal credit assignment arises since the reward is only given at the end of each random walk such that the final reward associated with moving into each state must be estimated long after that state has been left. In the context of TD learning, the value function here is modeled as a linear combination of state vector and the weighted coefficients

$$V(\mathbf{x}) = \mathbf{x}^{\mathrm{T}}\boldsymbol{\theta},$$

where $\mathbf{x}$ is a five-dimensional binary state vector that encodes the state identification number (e.g., $[1, 0, 0, 0, 0]$ represents state B and $[0, 0, 0, 0, 1]$

represents state F), while $\boldsymbol{\theta}$ denotes the parameter vector that represents the nonzero transition probabilities in P_T (for states B through F). In our experiment, a total number of 500 trials were run. The learning-rate parameter used in (3.68) was initially set as 0.05 and gradually annealed by 1% after each trial. The initial parameter vector is set as $[0.2, 0.2, 0.2, 0.2, 0.2]$; the desired (optimal) estimate should be $[0.1667, 0.3333, 0.5, 0.6667, 0.8333]$. Upon completion of 500 trials of learning, we obtain the final estimate $\boldsymbol{\theta} = [0.1531, 0.3262, 0.4844, 0.6348, 0.7802]$; the learning error curve and the estimate comparison are illustrated in Figure 3.5.

3.1.14 General Correlative Learning and Potential Function

Thus far in this section we have discussed a number of correlative learning rules. It would be interesting to see if all of them can be unified within the same mathematical framework. Indeed, a general correlative learning rule was formulated by Amari [20] as follows. Given a presynaptic input signal $x_i(t)$ and a general learning (or reinforcement) signal $r_j(t)$ presented to the postsynaptic neuron j, the correlative learning rule is written in the form [20, 32]

$$\Delta\theta_{ij} = \eta\langle x_i(t) r_j(t)\rangle, \tag{3.75}$$

or in the online form by taking the decay factor into account

$$\Delta\theta_{ij}(t+1) = (1-\varepsilon)\theta_{ij}(t) + \eta x_i(t) r_j(t), \tag{3.76}$$

where $0 < \varepsilon < 1$ is a forgetting (decay) factor. In either case, the synaptic weights converge roughly to the correlation of x_i and r_j, namely,

$$\theta_{ij} = \text{const} \times \langle x_i(t) r_j(t)\rangle. \tag{3.77}$$

Here, the learning signal $r_j(t)$ may depend on the input signal $x_i(t)$, the postsynaptic output signal $y_j(t)$, the extra supervised signal $z_j(t)$ coming from the outside teacher, or the temporal error signal $\varepsilon(t)$ estimated from temporal difference. Many learning algorithms that we have discussed and will discuss later can be unified in the above general correlative learning form. We illustrate this with several examples here:

- Hebbian or anti-Hebbian learning: $r_j(t) = y_j(t)$ or $r_j(t) = -y_j(t)$.
- Perceptron learning: $r_j(t) = y_j(t) - z_j(t)$, where $y_j(t)$ is the output of the binary neuron and $z_j(t)$ is the desired binary output (either 0 or 1).
- LMS error-correcting learning: $r_j(t) = \sum_i \theta_{ij} x_i(t) - z_j(t)$, where $z_j(t)$ is the teacher signal; the synaptic weight matrix converges to the minimizer of the squared error $\langle (z_j - \sum_i \theta_{ij} x_i)^2 \rangle$.
- PCA learning: $r(t) = \sum \theta_i(t) x_i(t)$; the principal component of the covariance matrix $\mathbf{C} = \langle \mathbf{x}(t)\mathbf{x}^{\mathrm{T}}(t)\rangle$ is obtained under the normalization condition.

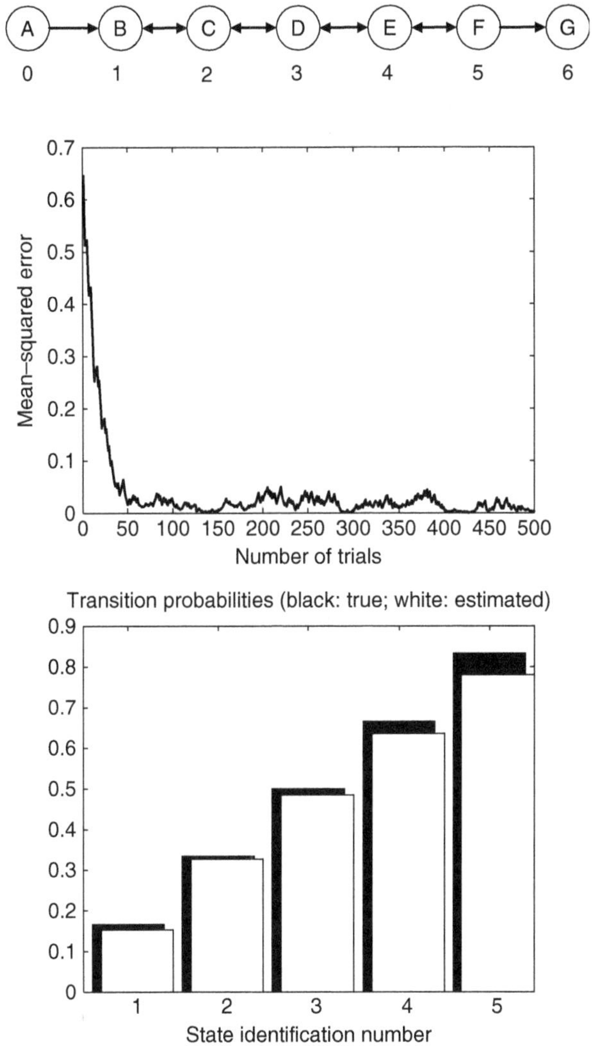

Figure 3.5 *Upper panel:* a seven-state Markov chain, where the number underneath each state indicates the state identification number. *Middle panel:* the learning error curve. *Bottom panel:* the true and estimated transition probabilities.

- TD learning: $r(t) = \varepsilon(t)$, where $\varepsilon(t)$ is the TD error that is estimated by the TD method.
- Associative memory learning: $r_j(t) = x_j(t)$ such that the synaptic weights learn the autocorrelation matrix $\theta_{ij} = (1/T)\sum_{t=1}^{T} x_i(t)x_j(t)$; this serves as the basis of correlation memory matrix (to be discussed in Section 3.3).
- Temporal association learning: $r_j(t) = x_j(t+1)$, in this case, the synaptic weights will be equal to $\theta_{ij} = [1/(T-1)]\sum_{t=1}^{T-1} x_i(t)x_j(t+1)$, which can

learn the auto- and cross-covariance of temporal patterns (to be discussed in Section 3.3).

It should be noted that there are many alternative choices of learning signal r_j for the development of different correlative learning algorithms [20].

The above correlative learning scheme can be analyzed by using the learning potential function. In the case of supervised learning, when the learning signal is a function of $u = \sum_j \theta_i x_i$ and z such that $r = r\left(\boldsymbol{\theta}^{\mathrm{T}}\mathbf{x}, z\right)$, the potential is defined by [20]

$$R(\boldsymbol{\theta}, \mathbf{x}, z) = \int^{\boldsymbol{\theta}^T \mathbf{x}} r(u, z)\, du, \tag{3.78}$$

such that $\partial R/\partial \boldsymbol{\theta} = r(u, z)\mathbf{x}$ is a Hebbian term. Then, the learning algorithm is written as the gradient descent form

$$\boldsymbol{\theta}(t+1) = (1-\varepsilon)\boldsymbol{\theta}(t) - \eta \frac{\partial R}{\partial \boldsymbol{\theta}}. \tag{3.79}$$

Equation (3.79) is a stochastic difference equation, and $\boldsymbol{\theta}$ will converge to a local minimum of the expected potential function $L(\boldsymbol{\theta}) = \mathbb{E}[R(\boldsymbol{\theta}, \mathbf{x}, z)]$, where the expectation is taken with respect to $\mathbf{x}$ and z. This equation also serves as the basis for analyzing the convergence of the learning process for the synaptic weights; taking the expectation of (3.79), we will have $\langle \Delta \boldsymbol{\theta} \rangle = -\eta[\partial L(\boldsymbol{\theta})/\partial \boldsymbol{\theta}]$.

3.2 INFORMATION-THEORETIC LEARNING

The synaptic modification or parameter update rules discussed in the preceding section cover a wide range of adaptive learning mechanisms, starting from self-organizing Hebbian learning and going on to supervised error-correction learning, unsupervised competitive learning, and reinforcement learning. Although these learning mechanisms originate from different principles, they all share the common *correlative* property in one form or another. In a related context, we may identify another class of learning rules that operate by virtue of the *decorrelative* property, rooted in information-theoretic measures, for which we have caste them under the umbrella of information-theoretic learning, numerous examples of which are found in the BSS and ICA literature. In this section, we will review some representative information-theoretic learning algorithms that are well fitted for modeling the functional roles of sensory systems. In the course of formulating the learning rules, correlation has again played a critical role in characterizing the emergent self-organizing behavior.

Before describing specific correlative or decorrelative learning rules, the common features of information-theoretic learning rules are summarized here:

- They are unsupervised.
- They are Hebb-like or correlation based.

- They directly use or implicitly define information-theoretic criteria as objective functions for optimization.
- They mostly involve the estimation of second- or higher order statistics or estimation of the probability density function (see Appendix D for a discussion).
- They are widely used in modeling self-organizing neural perceptual systems.

3.2.1 Mutual Information versus Correlation

In the context of information theory, mutual information is the most popular quantitative measure that characterizes the mutual dependence between two random variables.[6] Mutual information is often regarded as a *generalized measure* of correlation. It is known that the conventional correlation coefficient is based on second-order statistics; for two random variables x_i and x_j, their correlation coefficient is defined as

$$\rho(x_i, x_j) = \frac{\text{cov}(x_i, x_j)}{\sqrt{\text{var}(x_i)\,\text{var}(x_j)}}. \tag{3.80}$$

When x_i and x_j are *jointly Gaussian* distributed, the correlation coefficient is related to the mutual information by the equation

$$\begin{aligned} I(x_i, x_j) &\equiv \int p(x_i, x_j) \log\left(\frac{p(x_i, x_j)}{p(x_i)p(x_j)}\right) dx_i\, dx_j \\ &= -\frac{1}{2}\log\big(1 - \rho^2(x_i, x_j)\big). \end{aligned} \tag{3.81}$$

As seen, a correlation coefficient between x_i and x_j that is small in absolute value implies that there is little mutual information between them, and the maximum mutual information is achieved when $\rho(x_i, x_j) = \pm 1$.

3.2.2 Barlow's Postulate

According to Barlow [60], a goal of sensory coding is to find an efficient (factorial or minimally redundant) code for data compression, dimensionality reduction, and feature extraction. Mathematically, given an N-dimensional sensory input signal $\mathbf{x}$, sensory coding uses an unsupervised learning rule to find a (linear or nonlinear) mapping with a transformation function $\mathbf{F}$,

$$\mathbf{y} = \mathbf{F}(\mathbf{x}), \tag{3.82}$$

such that the components of the d-dimensional ($d \leq N$) output are statistically mutually uncorrelated, namely

$$\mathbb{E}[y_m y_n] = \mathbb{E}[y_m]\mathbb{E}[y_n] \qquad (m \neq n), \tag{3.83}$$

or statistically independent, namely

$$p(\mathbf{y}) = \prod_{k=1}^{d} p(y_k), \tag{3.84}$$

and the information is transmitted forward with minimum information loss.

From an information-theoretic viewpoint, minimizing the information loss can be understood as maximizing the mutual information between the input and output signals, which is defined as

$$\begin{aligned} I(\mathbf{x};\mathbf{y}) &= H(\mathbf{x}) - H(\mathbf{x}|\mathbf{y}) \\ &= H(\mathbf{y}) - H(\mathbf{y}|\mathbf{x}), \end{aligned} \tag{3.85}$$

where $H(\cdot)$ denotes the Shannon entropy [823]

$$H(\mathbf{x}) = -\int_{-\infty}^{\infty} p(\mathbf{x}) \log p(\mathbf{x})\, d\mathbf{x}$$

and $H(\mathbf{x}|\mathbf{y})$ and $H(\mathbf{y}|\mathbf{x})$ denote the conditional entropies.

Given the transformation function $\mathbf{F}$ of (3.82), it can be shown that

$$p(\mathbf{y}) \leq \frac{p(\mathbf{x})}{\sqrt{\det\left[\mathbf{J}^{\mathrm{T}}\mathbf{J}\right]}}, \tag{3.86}$$

$$H(\mathbf{y}) \leq H(\mathbf{x}) + \frac{1}{2}\int d\mathbf{x}\, p(\mathbf{x}) \log\!\left(\det\left[\mathbf{J}^{\mathrm{T}}\mathbf{J}\right]\right), \tag{3.87}$$

where $\mathbf{J} = \partial\mathbf{F}/\partial\mathbf{x}$ and $\det[\cdot]$ denotes the determinant (whose argument is a square matrix). The equality in (3.87) holds if and only if the function mapping $\mathbf{F}$ is bijective and reversible—the simplest example of such a function is a linear mapping function, namely, $\mathbf{y} = \mathbf{F}(\mathbf{x}) = \mathbf{W}\mathbf{x}$.

In what follows, we will present examples of information-theoretic learning algorithms and elaborate their links to Barlow's postulate.

3.2.3 Hebbian Learning and Maximum Entropy

In a series of seminal papers, Linsker [557–560] has suggested using Hebb-like (covariance) learning rules for simulating the development of visual receptive fields. Specifically, according to Linsker's postulate, the change of the synaptic weights (subject to certain bounding constraints) is given by

$$\Delta\boldsymbol{\theta} = \eta(\mathbf{x}y + a\mathbf{x} + \mathbf{b}y + \mathbf{d}), \tag{3.88}$$

where $\mathbf{x} \in \mathbb{R}^N$ denotes the presynaptic input, $y = \boldsymbol{\theta}^{\mathrm{T}}\mathbf{x}$ denote the postsynaptic output, and a, $\mathbf{b}$, and $\mathbf{d}$ denote the constant terms with appropriate dimensions.

Taking the average of both sides of (3.88) yields

$$\langle \Delta\boldsymbol{\theta} \rangle = \eta \left[\mathbf{C}_{xx}\boldsymbol{\theta} + a\boldsymbol{\mu} + \mathbf{b}\boldsymbol{\theta}^{\mathrm{T}}\boldsymbol{\mu} + \mathbf{d} \right], \tag{3.89}$$

where $\boldsymbol{\mu} = \langle \mathbf{x} \rangle$ denotes the mean vector of the random presynaptic input. Provided that $a > 0$, $\mathbf{b} = a\mathbf{1}$ (where $\mathbf{1}$ is an all-1 column vector), and $\mathbf{d} = \mathbf{0}$, (3.89) reduces to

$$\langle \Delta\boldsymbol{\theta} \rangle = \eta \left[\mathbf{C}_{xx}\boldsymbol{\theta} + a(\boldsymbol{\mu} - \boldsymbol{\theta}^{\mathrm{T}}\boldsymbol{\mu}\mathbf{1}) \right]. \tag{3.90}$$

If we assume that the components of $\mathbf{x}$ are random and share the same mean, namely $\mu_k = \mu$ $(k = 1, 2, \ldots, N)$, then (3.90) can be further simplified to

$$\langle \Delta\boldsymbol{\theta} \rangle = \eta \left[\mathbf{C}_{xx}\boldsymbol{\theta} + a\mu \left(1 - \sum_{k=1}^{N} \theta_k \right) \mathbf{1} \right]. \tag{3.91}$$

It can be shown that equation (3.91) is obtained by minimizing the following objective function:

$$J = -\frac{1}{2}\boldsymbol{\theta}^{\mathrm{T}}\mathbf{C}_{xx}\boldsymbol{\theta} + \frac{a\mu}{2}\left(1 - \sum_{k=1}^{N} \theta_k \right)^2.$$

The synaptic weights developed under Linsker's rule are closely related to the correlation matrix of the input neuron activities. The weight dynamics analysis of such a learning rule in terms of the eigenvectors of the autocorrelation matrix $\mathbf{C}_{xx}$ was detailed in [581].

Linsker [561, 562] also studied the sensory coding achieved by a single linear neuron whose postsynaptic output is represented as a linear sum of presynaptic inputs plus white Gaussian noise, namely,

$$y = \boldsymbol{\theta}^{\mathrm{T}}\mathbf{x} + \upsilon, \tag{3.92}$$

where $\upsilon \sim \mathcal{N}(0, \sigma_\upsilon^2)$. Provided that the multivariate input signal $\mathbf{x}$ is Gaussian with zero mean and covariance $\mathbf{C}_{xx}$, namely $\mathbf{x} \sim \mathcal{N}(\mathbf{0}, \mathbf{C}_{xx})$, it follows that the output $\mathbf{y}$ is also Gaussian with zero mean and variance $\mathrm{var}[y] = \boldsymbol{\theta}^{\mathrm{T}}\mathbf{C}_{xx}\boldsymbol{\theta} + \sigma_\upsilon^2$. Then, the entropy of output signal y, denoted as $H(y)$, can be calculated as

$$H(y) = \frac{1}{2}\left[1 + \log(2\pi\boldsymbol{\theta}^{\mathrm{T}}\mathbf{C}_{xx}\boldsymbol{\theta} + 2\pi\sigma_\upsilon^2) \right]. \tag{3.93}$$

In addition, the mutual information between the output y and input $\mathbf{x}$, denoted by $I(y; \mathbf{x})$, is defined as

$$I(y; \mathbf{x}) = H(y) - H(y|\mathbf{x}), \tag{3.94}$$

where the conditional entropy $H(y|\mathbf{x})$ is equal to the entropy of the noise (or residual error)

$$H(y|\mathbf{x}) = H(v) = \frac{1}{2}\left[1 + \log(2\pi\sigma_v^2)\right]. \tag{3.95}$$

Substituting (3.93) and (3.95) into (3.94) yields

$$I(y;\mathbf{x}) = \frac{1}{2}\log\frac{\boldsymbol{\theta}^{\mathrm{T}}\mathbf{C}_{xx}\boldsymbol{\theta} + \sigma_v^2}{\sigma_v^2} = \frac{1}{2}\log\left(1 + \frac{\boldsymbol{\theta}^{\mathrm{T}}\mathbf{C}_{xx}\boldsymbol{\theta}}{\sigma_v^2}\right), \tag{3.96}$$

where the variance ratio $\boldsymbol{\theta}^{\mathrm{T}}\mathbf{C}_{xx}\boldsymbol{\theta}/\sigma_v^2$ may be viewed as a measure of the SNR. Therefore, applying a simple form of Hebbian learning to the linear neuron, as described by the Hebbian rule

$$\boldsymbol{\theta}(t+1) = \boldsymbol{\theta}(t) + \eta y(t)\mathbf{x}(t), \tag{3.97}$$

implicitly minimizes the cost function $J = -\frac{1}{2}\boldsymbol{\theta}^{\mathrm{T}}\mathbf{C}_{xx}\boldsymbol{\theta}$; this is tantamount to maximizing the variance of the output, which, in turn, is equivalent to maximizing the entropy of the output as well as maximizing the mutual information (3.96). When the Gaussian assumption of the neuron's input is invalid, then maximizing the output's variance is not sufficient to maximize the output entropy, which implies that we have to rely on other methods to approximate the entropy $H(y)$.

Several additional comments are noteworthy:

- In a similar vein to Linsker's work, Yuille et al. [995] proposed a local Hebbian rule for developing orientation-selective cortical cells. Their learning rule, different from Linsker's, includes a weight constraint term that prevents divergence in the learning. Specifically, the learning rule has the form

$$\boldsymbol{\theta}(t+1) = \boldsymbol{\theta}(t) + \eta\left[y(t)\mathbf{x}(t) - \|\boldsymbol{\theta}(t)\|^2\boldsymbol{\theta}(t)\right], \tag{3.98}$$

  which, in turn, implicitly minimizes the objective function

$$J = -\frac{1}{2}\boldsymbol{\theta}^{\mathrm{T}}\mathbf{C}_{xx}\boldsymbol{\theta} + \frac{1}{4}\|\boldsymbol{\theta}\|^4.$$

- Applying the maximum-entropy (MaxEnt) principle to the noiseless linear neuron output is equivalent to performing PCA for the Gaussian input data.
- The MaxEnt or *Infomax* principle provided the motivation for Bell and Sejnowski [78] to develop an ICA algorithm discussed later in this chapter.

3.2.4 Imax Algorithm

Sensory inputs are often coherent over space (or time). How to extract the underlying higher order features or regularities in the sensory input is the major goal of perceptual learning. Motivated by the early work of Linsker [560] and Pearlmutter and Hinton [713], Becker and Hinton [74] proposed the Imax principle for unsupervised learning, which dictates that signals of interest should have high mutual information across different sensory channels. For instance, if two networks each produce a single output from two separate but neighboring modules (e.g., small patches of retina), say y_1 and y_2, then the goal is to maximize the mutual information $I(y_1, y_2)$ between these two neighboring outputs to produce a coherent outcome.

In general, the input may be high dimensional and may require a nonlinear transformation in order to extract the features of interest, and the network might also have a hierarchical architecture (see Figure 3.6*a*). For binary inputs, the mutual information can be estimated by calculating the marginal and the joint entropy of y_1 and y_2 according to

$$I(y_1, y_2) = H(y_1) + H(y_2) - H(y_1, y_2).$$

For real-valued inputs, estimating pdf $p(y_1, y_2)$ and its marginal distributions is not easy. However, if we assume two outputs are Gaussian distributed and the output noise obeys an i.i.d. additive Gaussian distribution, then the expression $I(y_1, y_2)$ may be analytically calculated (or approximated)

$$I(y_1; y_2) \approx \frac{1}{2} \log \left(\frac{\text{var}(y_1 + y_2)}{\text{var}(y_1 - y_2)} \right), \tag{3.99}$$

where $\text{var}(y_1 + y_2)$ denotes the variance of the sum of two outputs and $\text{var}(y_1 - y_2)$ denotes the variance of the difference between two outputs. If we assume one output is a noisy version of the other, then (3.99) can be viewed as a measure of

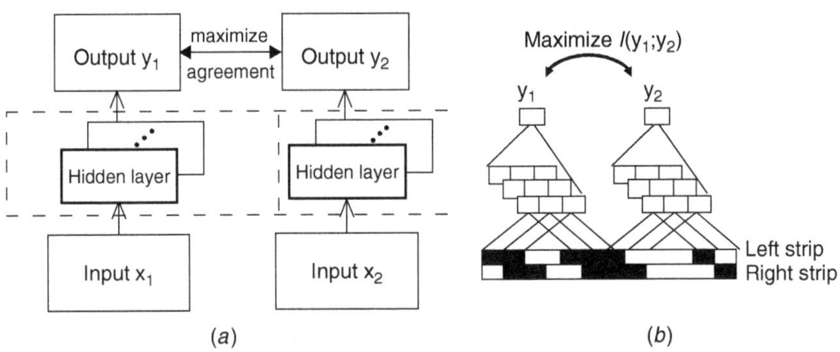

Figure 3.6 (*a*) The Imax learning principle maximizes the mutual information between features y_1 and y_2 extracted from different input channels. (*b*) The Imax architecture used to learn stereo features from binary images.

the SNR. In [74], the Imax principle was applied successfully to extract stereo disparity, which is one of the many cues important for depth perception, from random-dot stereograms (with either binary- or continuous-valued shifts); Becker and Hinton showed that using a network architecture with multiple stages of processing in a nonlinear neural circuit (with hidden layer and hidden units) and applying gradient ascent with respect to the parameters could extract binocular disparity (see Figure 3.6*b*). Notably, in this example, the neurons can learn from their mutual neighbors, which allows the network to discover an interpolation for visual scenes.

Subsequently, Zemel and Hinton [996] extended this idea to allow for more than one output per module. For instance, they used four outputs per module to identify four degrees of freedom in two-dimensional objects: *size, orientation, horizontal and vertical positions*; in their case, the mutual information measure was defined as

$$I(\mathbf{y}_1, \mathbf{y}_2) = \frac{1}{2}\log\left(\frac{\det(\mathbf{C}_{\mathbf{y}_1+\mathbf{y}_2})}{\det(\mathbf{C}_{\mathbf{y}_1-\mathbf{y}_2})}\right), \tag{3.100}$$

where $\mathbf{C}_{\mathbf{y}_1+\mathbf{y}_2}$ and $\mathbf{C}_{\mathbf{y}_1-\mathbf{y}_2}$ denote the covariance matrices of the sum and the difference, respectively, between two output vectors $\mathbf{y}_1$ and $\mathbf{y}_2$.

In applications, the Imax algorithm has been used to learn temporally coherent features [69, 70, 857]; similar algorithms for binary units were also developed by Kay and colleagues [469, 470, 722]. For further discussion of the Imax principle in unsupervised learning, see [69, 72, 75].

3.2.5 Local Decorrelative Learning

Decorrelation is widely believed to serve as a basic self-organizing principle for preprocessing sensory input in the cortex [61]; specifically, decorrelation can be attained by lateral inhibition and anti-Hebbian learning. The PCA learning discussed earlier indeed belongs to a specific class of local decorrelative learning algorithms. In this context, Becker and Plumbley [75] (see also [216, 877]) have reviewed a number of unsupervised learning procedures, and we briefly describe a few examples here.

For instance, Barlow and Földiák [61] proposed using a local anti-Hebbian learning rule for a recurrent network with lateral inhibitory connections (Figure 3.7*a*). For an N-dimensional input $\mathbf{x}$, the network attempts to produce an N-dimensional decorrelated output $\mathbf{y}$:

$$\mathbf{y}(t) = \mathbf{x}(t) - \mathbf{V}(t)\mathbf{y}(t), \tag{3.101}$$

where $\mathbf{V} \in \mathbb{R}^{N\times N}$ denotes a lateral synaptic weight matrix. At the equilibrium point, the output should satisfy the equation

$$\mathbf{y} = (\mathbf{I} + \mathbf{V})^{-1}\mathbf{x}, \tag{3.102}$$

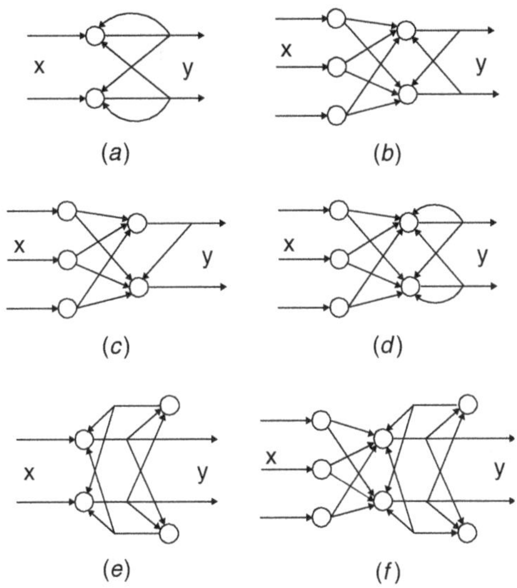

Figure 3.7 Linear recurrent decorrelating network architectures.

in which the matrix $\mathbf{I}+\mathbf{V}$ is assumed to be positive definite. If we assume the feedback connection matrix $\mathbf{V}$ is symmetric and has all zeros in the diagonal, then the synaptic weights v_{jk} can be updated via a local decorrelative learning rule [61],

$$\Delta v_{jk} = \eta y_j y_k \qquad (j \neq k), \tag{3.103}$$

or in matrix form

$$\Delta \mathbf{V} = \eta \times \text{off-diag}(\mathbf{y}\mathbf{y}^{\mathrm{T}}), \tag{3.104}$$

which will force $\mathbf{V}$ to have a symmetric structure; this algorithm converges when $\mathbb{E}[y_j y_k] = 0$ for all $j \neq k$ such that the output units are mutually uncorrelated.

Földiák [283] further generalized this architecture with an additional layer (see Figure 3.7*b*) to extract uncorrelated nonredundant outputs. Specifically, the output equation is

$$\mathbf{y}(t) = \mathbf{W}(t)\mathbf{x}(t) + \mathbf{V}(t)\mathbf{y}(t). \tag{3.105}$$

At the equilibrium point, it follows that

$$\mathbf{y} = (\mathbf{I} - \mathbf{V})^{-1}\mathbf{W}\mathbf{x}. \tag{3.106}$$

The learning rule for the weight matrices $\mathbf{W}$ and $\mathbf{V}$ is similar to Oja's rule (3.18) and (3.103); the learning dynamics was also detailed in [543]. Notably, Földiák's network architecture has symmetric lateral connections that are different from the asymmetric recurrent connections in the APEX network described earlier (see Figure 3.7*c*). Also note that equation (3.105) is different from equation (3.26) in recursive PCA, the latter of which has the term $\mathbf{V}(t)\mathbf{y}(t-1)$ on the right-hand side.

Plumbley [728] suggested a similar architecture to Földiák's network, but with additional *self-inhibitory* connections (namely, v_{jj} are not zeros); see Figure 3.7*d*. In this case, the learning rules are described as [75]

$$\Delta w_{ij}(t+1) = \eta_w \left[y_j(t) x_i(t) - \alpha w_{ij}(t) \right], \tag{3.107a}$$

$$\Delta v_{jk}(t+1) = \eta_v \left[y_j(t) y_k(t) - \beta \delta_{jk} \right], \tag{3.107b}$$

where δ_{jk} denotes the Kronecker delta that is 1 when $j = k$ and zero otherwise; η_w and η_v are two learning-rate parameters and $\eta_v \gg \eta_w$. In this case, the outputs will converge to an uncorrelated, equal variance set that spans the principal subspace.

In [877] (also in [48]), a slightly different recurrent network architecture (Figure 3.7*e*) was suggested for decorrelating the outputs. In such a case, the recurrent network's dynamics can be described by the equations

$$\mathbf{y} = \mathbf{x} - \mathbf{V}\mathbf{z}, \tag{3.108}$$

$$\mathbf{z} = \mathbf{V}^{\mathrm{T}}\mathbf{y}, \tag{3.109}$$

which further leads to

$$\mathbf{y} = (\mathbf{I} + \mathbf{V}\mathbf{V}^{\mathrm{T}})^{-1}\mathbf{x}. \tag{3.110}$$

With self-inhibitory connections, the synaptic weights can be adapted with a local Hebbian rule

$$\Delta v_{jk} = \eta(y_j z_k - v_{jk}). \tag{3.111}$$

Let $\mathbf{C}_{yy} = \langle \mathbf{y}\mathbf{y}^{\mathrm{T}} \rangle$; then the above equation can be rewritten in the matrix form

$$\Delta \mathbf{V} = \eta(\mathbf{C}_{yy} - \mathbf{I})\mathbf{V}. \tag{3.112}$$

The algorithm will converge when $\mathbf{C}_{yy} = \mathbf{I}$; namely, the outputs are mutually uncorrelated and have equal variances. Plumbley [728] further generalized such a network structure by adding an additional layer (see Figure 3.7*f*). The new network equation is given by

$$\mathbf{y} = (\mathbf{I} + \mathbf{V}\mathbf{V}^{\mathrm{T}})^{-1}\mathbf{W}\mathbf{x}, \tag{3.113}$$

and the learning rules for $\mathbf{W}$ and $\mathbf{V}$ are given, respectively, by (3.107a) and the equation

$$\Delta v_{jk}(t+1) = \eta_v \left[y_j(t) z_k(t) - \beta v_{jk}(t) \right] \tag{3.114}$$

in place of (3.107b).

3.2.6 Blind Source Separation

Mathematically, the BSS problem can be described as an instantaneous linear mixing process followed by an unmixing process. Specifically, the linear mixing is described as

$$\mathbf{x} = \mathbf{A}\mathbf{s} + \mathbf{n}, \tag{3.115}$$

where $\mathbf{s}$ is an $m \times 1$ source vector, $\mathbf{A} \in \mathbb{R}^{n\times m}$ $(n \geq m)$ is an unknown mixing matrix, $\mathbf{x}$ is an $n \times 1$ mixed-signal vector, and $\mathbf{n}$ is an $n \times 1$ independent noise vector (in the simplest case, $\mathbf{n}$ vanishes to zero). The goal of BSS is to find a demixing matrix $\mathbf{W}$ to recover the original signals in $\mathbf{s}$ (up to certain scaling and permutation ambiguities),[7] which is represented by the equation

$$\mathbf{y} = \mathbf{W}\mathbf{x} = \mathbf{W}(\mathbf{A}\mathbf{s}). \tag{3.116}$$

To design *temporal* BSS algorithms, one can use several different criteria, such as the second- or higher order statistics, independence, or nonstationarity. Note that "uncorrelated" implies that the autocorrelation matrix $\mathbb{E}[\mathbf{s}(t)\mathbf{s}^{\mathrm{T}}(t+\tau)]$ is diagonal. Because statistical independence implies a lack of correlation but not vice versa, ICA algorithms can be applied to the BSS problem in a straightforward way but BSS algorithms might not be sufficient for the ICA problem.

Given a noiseless linear mixing model (i.e., $\mathbf{n} \to \mathbf{0}$), let us assume the mixing sources have zero mean; then the correlation (or covariance) matrix of the mixed signals can be represented as

$$\mathbf{C}_{xx}(\tau) = \mathbf{A}\mathbf{C}_{ss}(\tau)\mathbf{A}^{\mathrm{T}}, \tag{3.117}$$

where $\mathbf{C}_{ss}$ is a diagonal matrix (since $\mathbf{s}$ only contains mutually uncorrelated or independent sources). The mixing matrix $\mathbf{A}$ can be found by performing a unitary eigenvalue decomposition of $\mathbf{C}_{xx}$, and the corresponding eigenvectors will be the columns of the mixing matrix. In theory, any covariance matrix at nonzero lag is sufficient to estimate the mixing matrix [946, 989]; this fact motivated Molgedey and Schuster [633] and Belouchrani et al. [81] to use a set of covariance matrices for joint diagonalization in the context of BSS.

In particular, taking the expectation of both sides of (3.116), we obtain the *time-delayed correlation matrix* [633]

$$\mathbf{C}_{yy}(\tau) = \mathbf{W}\mathbf{C}_{xx}(\tau)\mathbf{W}^{\mathrm{T}} = \mathbf{W}\mathbf{A}\mathbf{C}_{ss}(\tau)\mathbf{A}^{\mathrm{T}}\mathbf{W}^{\mathrm{T}}. \tag{3.118}$$

If the separated sources are uncorrelated, then $\mathbf{C}_{yy}$ will be close to a diagonal matrix, denoted by $\mathbf{D}(\tau)$. Hence, minimizing the departure of $\mathbf{C}_{yy}$ from being diagonal may yield a possible solution. Let us assume the following cost function to be minimized:

$$J(t) = \sum_{\tau=0}^{d} \left\| \mathbf{W}\mathbf{C}_{xx}(\tau)\mathbf{W}^{\mathrm{T}} - \mathbf{D}(\tau) \right\|_F^2, \tag{3.119}$$

where $\|\cdot\|_F$ denotes the *Frobenius norm* of the matrix and d denotes the total number of delayed covariance matrices (with different values of τ). Applying the gradient descent rule to (3.119) yields the following learning rule:

$$\Delta\mathbf{W} = -\eta \sum_{\tau=0}^{d} [\mathbf{W}\mathbf{C}_{xx}(t;\tau) + \mathbf{W}^{\mathrm{T}} - \mathbf{D}(t;\tau)]\mathbf{W}[\mathbf{C}_{xx}(t;\tau) + \mathbf{C}_{xx}^{\mathrm{T}}(t;\tau)] \quad (3.120)$$

Under the special circumstance where $\tau = d = 0$ and $\mathbf{D}(\tau) = \mathbf{I}$, we can derive

$$\Delta\mathbf{W}(t+1) = -\eta\langle \mathbf{y}(t)\mathbf{y}^{\mathrm{T}}(t) - \mathbf{I}\rangle \mathbf{W}(t)\mathbf{C}_{xx}(t) \quad (3.121)$$

which is the adaptation rule used for blind decorrelation. If we further constrain $\mathbf{C}_{xx}(t) = \mathbf{I}$, then (3.121) reduces to the learning rule of Silva and Almeida [834]:

$$\Delta\mathbf{W}(t+1) = -\eta\langle \mathbf{y}(t)\mathbf{y}^{\mathrm{T}}(t) - \mathbf{I}\rangle \mathbf{W}(t). \quad (3.122)$$

Alternatively, Cichocki et al. [173] also proposed a. *locally adaptive* algorithm in place of the globally adaptive rule (3.122) for data whitening; specifically, their learning rule is described by

$$\Delta\mathbf{W}(t+1) = -\eta\langle \mathbf{y}(t)\mathbf{y}^{\mathrm{T}}(t) - \mathbf{I}\rangle, \quad (3.123)$$

which is Hebb-like and can be easily implemented in hardware due to its local memory requirement. Theoretical analysis and comparison between (3.122) and (3.123) are given in [227].

Another popular batch-type blind separation learning algorithm based on second-order statistics is the so-called AMUSE (Algorithm for Multiple Unknown Signals Extraction) [887]. The learning procedure in the AMUSE consists of two steps. In the first step, a whitening procedure is applied to the input signal $\mathbf{x}(t)$, which is described by

$$\mathbf{z}(t) = \mathbf{Q}\mathbf{x}(t) = \mathbf{S}^{-1/2}\mathbf{G}^{\mathrm{T}}\mathbf{x}(t), \quad (3.124)$$

where $\mathbf{Q} = \mathbf{S}^{-1/2}\mathbf{G}^{\mathrm{T}}$ and $\mathbf{G}$ is obtained from the EVD: $\mathbf{C}_{xx} = \mathbb{E}[\mathbf{x}(t)\mathbf{x}^{\mathrm{T}}(t)] = \mathbf{G}\mathbf{S}\mathbf{G}^{\mathrm{T}}$. In the second step, SVD is applied to the (p-lag) *time-delayed* correlation matrix (here we assume $p = 1$)

$$\mathbf{C}_{zz}(p) = \mathbb{E}[\mathbf{z}(t)\mathbf{z}^{\mathrm{T}}(t-1)] = \mathbf{U}\boldsymbol{\Sigma}\mathbf{V}^{\mathrm{T}}, \quad (3.125)$$

where $\boldsymbol{\Sigma}$ is the diagonal matrix that contains singular values and $\mathbf{U}$ and $\mathbf{V}$ are two new orthogonal matrices. Finally, the separation matrix $\mathbf{W}$ is calculated analytically in accordance with

$$\mathbf{W} = \mathbf{U}^{\mathrm{T}}\mathbf{Q}. \quad (3.126)$$

The AMUSE algorithm is a batch (i.e., noniterative) type of BSS algorithm and is well suited for separating temporally correlated signals.

From the algebraic viewpoint, Parra and Sajda [704] presented a unified view of the BSS problem and showed that it can be formulated as a generalized EVD problem with different assumptions of (non-Gaussian, nonstationary, or nonwhite mutually independent) sources; specifically, the solution for the demixing matrix is given by the generalized eigenvectors that simultaneously diagonalize the covariance matrix of the observations and an additional symmetric matrix whose form depends upon the particular assumptions being made.

3.2.7 Independent-Component Analysis

Similar to BSS, ICA also assumes an instantaneous linear mixing model, either in *time* or in *space*. The goal of ICA is similar to that of BSS except that its criterion is slightly different in that ICA exploits the higher order statistics.[8] The typical assumptions made in ICA are that the sources are *statistically independent* and *non-Gaussian* (or at most one Gaussian source). The following properties can be inferred and often used to characterize the statistical independence between the sources $\{s_1, \ldots, s_m\}$:

$$p(s_1, \ldots, s_m) = \prod_{i=1}^{m} p(s_i), \tag{3.127}$$

$$\mathbb{E}_{s_i s_j}[s_i^p(t) s_j^q(t+\tau)] = \mathbb{E}_{s_i}[s_i^p(t)] \mathbb{E}_{s_j}[s_j^q(t+\tau)] \qquad (p, q \in \mathbb{N}), \tag{3.128}$$

$$\mathbb{E}_{s_i s_j}[f(s_i(t)) g(s_j(t))] = \mathbb{E}_{s_i}[f(s_i(t))] \mathbb{E}_{s_j}[g(s_j(t))] \qquad (\forall\ f, g). \tag{3.129}$$

It is noteworthy that when the Gaussian assumption is valid the statistical independence assumption reduces to a lack of correlation and ICA degenerates to PCA as a special case. Essentially, equations (3.128) and (3.129) describe the *nonlinear decorrelation*, which is typically weaker than independence; however, as shown below, nonlinearity is a natural way to extend the Hebbian learning from "being decorrelative" to "being independent."

Nonlinear PCA Hebbian Learning. As we discussed earlier, Oja's PCA rule is limited to linear neurons. It is possible to generalize the idea of Hebbian learning to nonlinear neurons. Suppose $J(t) = J(\boldsymbol{\theta}(t))$ is an objective function to be maximized (e.g., the normalized kurtosis function),[9] and let $\partial J(t)/\partial \boldsymbol{\theta} = \psi(y(t))\mathbf{x}(t)$, where $\psi(\cdot)$ denotes the derivative of the objective function $J(t)$; we may similarly derive the nonlinear PCA Hebbian learning rule [680], given in vector form as

$$\boldsymbol{\theta}(t+1) = \boldsymbol{\theta}(t) + \eta \frac{\partial J(t)}{\partial \boldsymbol{\theta}} = \boldsymbol{\theta}(t) + \eta \psi(y(t))\mathbf{x}(t) \tag{3.130}$$

followed by a normalization step $\boldsymbol{\theta}(t+1) \leftarrow \boldsymbol{\theta}(t+1)/\|\boldsymbol{\theta}(t+1)\|$. For a small learning rate η, equation (3.130) can be approximated by

$$\Delta\boldsymbol{\theta}(t+1) = \eta[\mathbf{I} - \boldsymbol{\theta}(t)\boldsymbol{\theta}^{\mathrm{T}}(t)]\mathbf{x}(t)\psi(y(t)), \tag{3.131}$$

which can be viewed as the nonlinear analog of Oja's linear PCA rule.

To extend (3.131) to multiple neurons, Oja et al. [680] and Karhunen and Jourtensalo [466] proposed the following learning rule in the context of source separation:

$$\Delta \mathbf{W} = \eta[\mathbf{I} - \mathbf{W}\mathbf{W}^{\mathrm{T}}]\mathbf{x}\psi(\mathbf{W}\mathbf{x}), \tag{3.132}$$

where we have used vector notation $\mathbf{y} = \mathbf{W}\mathbf{x}$ in place of $y = \boldsymbol{\theta}^{\mathrm{T}}\mathbf{x}$ in (3.131). Roughly speaking, imposing strong nonlinear decorrelation (i.e., nonlinear PCA) with an appropriate nonlinearity would yield an approximate independence between random variables. This can be viewed as an ad hoc version of ICA.

Infomax. Based on the maximum entropy (MaxEnt) principle and motivated by Linsker's work, Bell and Sejnowski [78] proposed the so-called Infomax ICA algorithm for maximizing the output entropy. Assuming an instantaneous, noiseless, linear mapping $\mathbf{x} = \mathbf{A}\mathbf{s}$ (where $\mathbf{A} \in \mathbb{R}^{m \times m}$ is a square mixing matrix), in light of the deterministic linear equation (3.116), the entropy of the demixing output, $\mathbf{y} = \mathbf{W}\mathbf{x}$, is calculated as

$$H(\mathbf{y}) = H(\mathbf{x}) + \log|\det(\mathbf{W})|, \tag{3.133}$$

where $H(\mathbf{x})$ represents the entropy of the input signal that is independent of $\mathbf{W}$ (hence it can be dropped off in the learning procedure). To derive the learning rule for $\mathbf{W}$, Bell and Sejnowski [78] used a nonlinear vector-valued function $\psi(\cdot)$ (the so-called "activation function" or "negative score function") to approximate the cumulative distribution function of $\mathbf{y}$ in order to maximize its resultant entropy.[10] Specifically, by virtue of the independence measure (3.127), it is natural to minimize the Kullback–Leibler (KL) divergence between $p(\mathbf{y}; \mathbf{W})$ and $\prod_{i=1}^{m} p_{y_i}(y_i; \mathbf{W})$:

$$\begin{aligned} D(\mathbf{W}; \mathbf{y}) &= \int p(\mathbf{y}) \log\left(\frac{p(\mathbf{y})}{\prod_{i=1}^{m} p_{y_i}(y_i)}\right) d\mathbf{y} \\ &= -H(\mathbf{y}) + \sum_{i=1}^{m} H(y_i). \end{aligned} \tag{3.134}$$

In light of (3.133) and (3.134), the following learning rule can be derived:

$$\begin{aligned} \Delta \mathbf{W}(t+1) &= \eta\left[\mathbf{W}^{-\mathrm{T}}(t) - \psi(\mathbf{y}(t))\mathbf{x}^{\mathrm{T}}(t)\right] \\ &= \eta\left[\mathbf{I} - \psi(\mathbf{y}(t))\mathbf{y}^{\mathrm{T}}(t)\right]\mathbf{W}^{-\mathrm{T}}(t), \end{aligned} \tag{3.135}$$

where $\mathbf{W}^{-\mathrm{T}}$ denotes the inverse of the transposed matrix $\mathbf{W}^{\mathrm{T}}$. It is noteworthy that the learning rule (3.135) is not *fully local* in that it involves the matrix $\mathbf{W}$ on the right-hand side. Additionally, it requires the operation of a matrix inverse that is seemingly biologically implausible.

To overcome this problem, Linsker [563] proposed a fully local Hebbian learning rule that enables information maximization for *arbitrary* input distributions. In so doing, Linsker introduced an auxiliary vector $\mathbf{v} \in \mathbb{R}^m$ and an extra set of synaptic weights, $\mathbf{F} \in \mathbb{R}^{m \times m}$ as feedback connections to sidestep the direct calculation of the matrix inverse $\mathbf{W}^{-\mathrm{T}}$. Specifically, the auxiliary vector $\mathbf{v}$ is represented by

$$\mathbf{v}(t) = \mathbf{y}(t) + \mathbf{F}(t)\mathbf{v}(t-1) = \mathbf{W}\mathbf{x}(t) + \mathbf{F}(t)\mathbf{v}(t-1), \tag{3.136}$$

and the feedback weights are updated as

$$\Delta\mathbf{F}(t+1) = \eta\left[-\alpha\mathbf{y}(t)\mathbf{y}^{\mathrm{T}}(t) + \mathbf{I} - \mathbf{F}(t)\right], \tag{3.137}$$

where α is a constant parameter that ensures the convergence of (3.137). Iterating (3.136) and (3.137) alternatingly will cause the activations to gradually approach the equilibrium points upon convergence, as indicated here,[11]

$$\lim_{t\to\infty} \mathbf{F}(t) \to \mathbf{I} - \alpha\langle\mathbf{x}\mathbf{x}^{\mathrm{T}}\rangle,$$

$$\lim_{t\to\infty} \mathbf{v}(t) \to (\mathbf{I} - \mathbf{F})^{-1}\mathbf{W}\mathbf{x},$$

by means of which we further obtain

$$\lim_{t\to\infty} \alpha\langle\mathbf{v}(t)\mathbf{x}^{\mathrm{T}}\rangle \to \alpha(\mathbf{I} - \mathbf{F})^{-1}\mathbf{W}\langle\mathbf{x}\mathbf{x}^{\mathrm{T}}\rangle = \mathbf{W}^{-\mathrm{T}},$$

which then provides an approximate solution to the estimation of the matrix inverse $\mathbf{W}^{-\mathrm{T}}$.

Natural Gradient Learning. Natural gradient learning is a generalization of the stochastic gradient descent rule in that it exploits the concept of *Riemannian geometry* [25, 31]. In the ICA context, the online natural gradient rule for updating the square demixing matrix $\mathbf{W}$ is described by the following rule [29]:

$$\Delta\mathbf{W}(t+1) = \eta[\mathbf{I} - \psi(\mathbf{y}(t))\mathbf{y}^{\mathrm{T}}(t)]\mathbf{W}(t), \tag{3.138}$$

which is essentially a variant of (3.135) in light of the *equivariant property* from information geometry [29, 147]. Note that (3.135) and (3.138) are both decorrelative anti-Hebbian rules. Specifically, as learning goes on, the outer product $\psi(\mathbf{y}(t))\mathbf{y}^{\mathrm{T}}(t)$ gradually approximates the cross-correlation matrix between the output signals $\mathbf{y}(t)$ and its nonlinearly transformed version $\psi(\mathbf{y}(t))$. After a sufficiently large number of time steps, the correlation matrix approximates the identity matrix, whereupon the incremental change in the demixing matrix $\Delta\mathbf{W}(t+1)$ is reduced to zero and the algorithm converges. With appropriately chosen learning rate η and activation function ψ, the learning rule (3.138) is stable and is guaranteed to reach a feasible solution [27].

Two additional points are noteworthy:

- In the original formulation of the natural gradient algorithm, the demixing matrix $\mathbf{W}$ is assumed to be a square matrix; however, this assumption can be relaxed. Variants of the natural gradient algorithm for over- and undercomplete cases were also developed in [26, 28, 148].
- The conventional ICA methods assume a feedforward linear network architecture. Extensions to recurrent networks with lateral inhibition were also discussed in [324, 832].

Projection Pursuit. The goal of sensory coding is to exploit the intrinsic (e.g., sparse or factorial) structures underlying the high-dimensional sensory data. Hence, feature extraction becomes a fundamental role in sensory processing. According to the theory of *exploratory projection pursuit* (EPP) [291], the search for the interesting structure in data space can be achieved by seeking deviation from the Gaussian distribution in the projected space. Based on this theory, projection pursuit optimizes an objective function that measures the deviation from the Gaussian distribution. Examples of such a metric often involves higher order cumulant statistics, such as kurtosis and skewness. Hence, projection pursuit can be used for blind source extraction and separation [860]. For instance, Girolami and Fyfe [324] used *negentropy* and *kurtosis* as the projection pursuit indices and proposed the following learning rule:

$$\Delta\mathbf{W}(t+1) = \eta[\mathbf{I} \pm \tanh(\mathbf{y})\mathbf{y}^{\mathrm{T}} - \alpha\mathbf{y}\mathbf{y}^{\mathrm{T}}]\mathbf{W}(t), \tag{3.139}$$

which can be viewed as a generalization of the natural gradient algorithm [542]. The choice of the algebraic sign, $\pm$, depends on the the kurtosis of the sources, either positive (for super-Gaussian or leptokurtic sources) or negative (for sub-Gaussian or platykurtic sources).

Complexity Pursuit. As an extension of projection pursuit, complexity pursuit [425, 859, 860] introduces the notion of temporal complexity (or predictability, or coding complexity) for temporally structured signals. Complexity pursuit can be used for extracting a signal or multiple signals given the linear mixture of source signals. Specifically, let $\overline{\mathbf{C}} \in \mathbb{R}^{m\times m}$ and $\tilde{\mathbf{C}} \in \mathbb{R}^{m\times m}$ denote, respectively, the long-term and short-term covariance matrices between the m signal mixtures, denoted by an m-dimensional vector $\mathbf{x}$. Let $y_i = \boldsymbol{\theta}_i^{\mathrm{T}}\mathbf{x}$ be the one extracted signal at the output. Then one can maximize the following objective function in order to extract the *most predictable* source signal:

$$J = \log\frac{\boldsymbol{\theta}_i^{\mathrm{T}}\overline{\mathbf{C}}\boldsymbol{\theta}_i}{\boldsymbol{\theta}_i^{\mathrm{T}}\tilde{\mathbf{C}}\boldsymbol{\theta}_i} \triangleq \log\frac{V_i}{U_i}, \tag{3.140}$$

where $V_i = \boldsymbol{\theta}_i^{\mathrm{T}} \overline{\mathbf{C}} \boldsymbol{\theta}_i$ and $U_i = \boldsymbol{\theta}_i^{\mathrm{T}} \tilde{\mathbf{C}} \boldsymbol{\theta}_i$. Applying stochastic gradient ascent to (3.140) yields the following learning rule:

$$\Delta \boldsymbol{\theta}_i = \eta \frac{\partial J}{\partial \boldsymbol{\theta}_i} = \eta \left(\overline{\mathbf{C}} \frac{2\boldsymbol{\theta}_i}{V_i} - \tilde{\mathbf{C}} \frac{2\boldsymbol{\theta}_i}{U_i} \right). \tag{3.141}$$

For simultaneous extraction of m signals, it was shown in [859, 860] that this is essentially solving a generalized eigenvalue problem. Specifically, setting the gradient $\partial J / \partial \boldsymbol{\theta}_i$ to zero yields

$$\overline{\mathbf{C}} \boldsymbol{\theta}_i = \tilde{\mathbf{C}} \frac{V_i}{U_i} \boldsymbol{\theta}_i, \tag{3.142}$$

for which the solution $\{\boldsymbol{\theta}_i\}$ defines the eigenvectors of the matrix $\tilde{\mathbf{C}}^{-1} \overline{\mathbf{C}}$ with the corresponding eigenvalues $\lambda_i = V_i / U_i$.

Higher Order ICA. Although ICA assumes the hidden components are mutually independent, this is often not the case in practice. Consequently, the assumption of independence can be relaxed down to higher order decorrelation.[12] For instance, using the measure of higher order decorrelation (3.128), the second-order BSS algorithms, such as AMUSE [887] or SOBI (second-order blind identification) [81], can be extended to ICA for separating independent, non-Gaussian source signals. Specifically, upon the first-stage EVD, we obtain the uncorrelated signals of $\mathbf{z}(t)$ from (3.124); then, instead of using the time-delayed correlation matrix in (3.125), we construct a contracted *quadricovariance matrix*

$$\begin{aligned} \mathbf{C}_{\mathbf{z}}(\mathbf{T}) = \mathbb{E}[\mathbf{z}^{\mathrm{T}}(t) \mathbf{T} \mathbf{z}(t) \mathbf{z}(t) \mathbf{z}^{\mathrm{T}}(t)] - \mathbf{C}_{zz}(0) \mathbf{T} \mathbf{C}_{zz}(0) \\ + \operatorname{tr}(\mathbf{T} \mathbf{C}_{zz}(0)) \, \mathbf{C}_{zz}(0) - \mathbf{C}_{zz}(0) \mathbf{T}^{\mathrm{T}} \mathbf{C}_{zz}(0), \end{aligned} \tag{3.143}$$

where tr(·) denotes the matrix trace operator, $\mathbf{C}_{zz}(0) = \mathbb{E}[\mathbf{z}(t)\mathbf{z}^{\mathrm{T}}(t)]$, and $\mathbf{T}$ represents a freely chosen symmetric positive-definite matrix (typically, $\mathbf{T}$ is an identity matrix $\mathbf{I}$, or $\mathbf{T} = \mathbf{e}\mathbf{e}^{\mathrm{T}}$, where $\mathbf{e}$ denotes the vector of a unitary matrix). If we let $\mathbf{T} = \mathbf{I}$, then (3.143) is rewritten as

$$\begin{aligned} \mathbf{C}_{\mathbf{z}}(\mathbf{I}) = \mathbb{E}[\mathbf{z}^{\mathrm{T}}(t) \mathbf{z}(t) \mathbf{z}(t) \mathbf{z}^{\mathrm{T}}(t)] - 2\mathbf{C}_{zz}(0) \mathbf{C}_{zz}(0) \\ + \operatorname{tr}(\mathbf{C}_{zz}(0)) \, \mathbf{C}_{zz}(0). \end{aligned} \tag{3.144}$$

Applying the EVD to the quadricovariance matrix of (3.144) yields

$$\mathbf{C}_{\mathbf{z}}(\mathbf{I}) = \mathbf{U} \boldsymbol{\Lambda} \mathbf{U}^{\mathrm{T}}, \tag{3.145}$$

where $\mathbf{U}$ denotes the orthogonal eigenvector matrix that contains the eigenvectors $\mathbf{u}_i$ ($i = 1, \ldots, n$) as column vectors and $\boldsymbol{\Lambda} = \operatorname{diag}\{\lambda_1 \|\mathbf{u}_1\|^2, \ldots, \lambda_n \|\mathbf{u}_n\|^2\}$ is the associated diagonal matrix, with $\lambda_i = \kappa_4(s_i) = \mathbb{E}[s_i^4] - 3(\mathbb{E}[s_i^2])^2$ as the fourth-order kurtosis statistic of the ith zero-mean source signal s_i. When the original source signals are non-Gaussian and have distinct kurtosis statistics, then the EVD of matrix $\mathbf{C}_\mathbf{z}(\mathbf{I})$ is unique in the sense that all the eigenvalues inside the diagonal matrix $\mathbf{\Lambda}$ are distinct, and we may estimate the mixing matrix analytically by $\hat{\mathbf{A}} = \mathbf{U}$. The above procedure described here is known as the FOBI (fourth-order blind identification) algorithm [142, 172, 654].

3.2.8 Slow Feature Analysis

Slow feature analysis (SFA) is an unsupervised learning method that was proposed for learning invariance in the visual cortex [966]. Slow feature analysis is appealing for signal analysis and object recognition since it was conjectured that slowly varying features can be an approximation of invariant features for temporally structured signals. The idea behind SFA is to subject the input signal to a nonlinear transformation and then apply PCA to the transformed signal as well as its time derivative. Slow feature analysis is guaranteed to find the optimal solution within a functional family and can learn to extract a large number of *decorrelated* features, which are ordered by their degrees of invariance.

Specifically, let $\mathbf{y}(t) \in \mathbb{R}^n$ denote the transformed vectorial signal from a vectorial input signal $\mathbf{x}(t) \in \mathbb{R}^m$:

$$\mathbf{y}(t) = \mathbf{g}(\mathbf{x}(t)), \tag{3.146}$$

where $\mathbf{g}(\cdot)$ is a vector-valued (componentwise) function that consists of a weighted sum of N (usually $N > \max\{m, n\}$) nonlinear functions

$$g_i(\mathbf{x}) = \sum_{j=1}^{N} \theta_{ij} h_j(\mathbf{x}). \tag{3.147}$$

The jth output component may be represented as

$$\begin{aligned} y_i(t) &= g_i(\mathbf{x}(t)) = \boldsymbol{\theta}_i^{\mathrm{T}} \mathbf{h}(\mathbf{x}(t)) \\ &= \boldsymbol{\theta}_i^{\mathrm{T}} \mathbf{z}(t), \end{aligned} \tag{3.148}$$

where $\mathbf{z}(t) = \mathbf{h}(\mathbf{x}(t))$. The goal of SFA is then to minimize the variance of the time derivative of y_i^2, denoted by $\langle \dot{y}_i^2(t) \rangle$, subject to three constraints on the output signal $y_i(t)$:

$$\langle y_i(t) \rangle = \boldsymbol{\theta}_i^{\mathrm{T}} \langle \mathbf{z} \rangle = 0 \quad \text{(zero mean)}, \tag{3.149}$$

$$\langle y_i^2(t) \rangle = \boldsymbol{\theta}_i^{\mathrm{T}} \langle \mathbf{z}\mathbf{z}^{\mathrm{T}} \rangle \boldsymbol{\theta}_i = 1 \quad \text{(unit variance)}, \tag{3.150}$$

$$\forall j < i: \qquad \langle y_j(t) y_i(t) \rangle = \boldsymbol{\theta}_j^{\mathrm{T}} \langle \mathbf{z}\mathbf{z}^{\mathrm{T}} \rangle \boldsymbol{\theta}_i = 0 \quad \text{(decorrelation)}. \tag{3.151}$$

The decorrelation property can be fulfilled if the weight vectors $\{\boldsymbol{\theta}_i\}$ are mutually orthogonal. Therefore, SFA reduces to a typical eigenvalue computation problem: finding the *least important* component (i.e., with the smallest eigenvalue) of the autocorrelation matrix of the time derivative of $\mathbf{z}(t)$, namely $\langle \dot{\mathbf{z}}\dot{\mathbf{z}}^{\mathrm{T}} \rangle$, whereas the weight vectors correspond to the associated eigenvectors. In the literature, this problem is known as MCA, which has been discussed in the preceding section of this chapter. Interestingly, it has been shown recently [99] that linear SFA is functionally equivalent to the time-delayed second-order BSS algorithm (e.g., [633]).

In this section thus far, we have presented a brief overview of information-theoretic learning algorithms within the unsupervised learning framework, all of which share the *decorrelative principle* underlying Barlow's postulate in perceptual learning. For the reader's convenience, a short list of information-theoretic learning rules and their associated cost functions is given in Table 3.1.

■ EXAMPLE 3.3

In this example, we use an information-theoretic ICA learning rule to mimic visual receptive fields which single cells use for encoding natural images. Following Bell and Sejnowski [79], we strive to discover the "independent components" that are used as edge filters in image coding [79].[13]

Specifically, given selected image patches randomly drawn from some gray-scale natural images, with intensity scale in [0, 255] and normalized to the range [0, 1] (see Figure 3.8 for two selected examples), we assume that the image formation process is subject to a linear superposition of some basis vectors. Mathematically, it can be represented as a linear mapping $\mathbf{X} = \mathbf{AS}$, where $\mathbf{X}$ represents the "mixed source" matrix that contains the

Table 3.1 Summary of Information-Theoretic Learning Rules and Associative Cost Functions (All Assumed Minimized)

Learning Rule	Comment	Cost Function
Oja's PCA rule (3.18)	Hebbian	$-\boldsymbol{\theta}^{\mathrm{T}}\mathbf{C}_{xx}\boldsymbol{\theta}$ s.t. $\|\boldsymbol{\theta}\| = 1$
Luo et al.'s MCA rule (3.36)	Anti-Hebbian	$\boldsymbol{\theta}^{\mathrm{T}}\mathbf{C}_{xx}\boldsymbol{\theta}/\|\boldsymbol{\theta}\|^2$
Linsker's rule (3.91)	Hebbian	$-\frac{1}{2}\boldsymbol{\theta}^{\mathrm{T}}\mathbf{C}_{xx}\boldsymbol{\theta} + \frac{1}{2}a\mu(1 - \sum_{k=1}^{N}\theta_k)^2$
Yuille et al.'s rule (3.98)	Hebbian	$-\frac{1}{2}\boldsymbol{\theta}^{\mathrm{T}}\mathbf{C}_{xx}\boldsymbol{\theta} + \frac{1}{4}\|\boldsymbol{\theta}\|^4$
Linear Hebb's rule (3.97)	MaxEnt	$-\frac{1}{2}\log(1 + \boldsymbol{\theta}^{\mathrm{T}}\mathbf{C}_{xx}\boldsymbol{\theta}/\sigma_v^2)$
Nonlinear Hebb's rule (3.130)	Nonlinear PCA	Kurtosis contrast function
Blind decorrelation (3.120)	BSS	$\sum_{\tau=0}^{d}\|\mathbf{W}\mathbf{C}_{xx}(\tau)\mathbf{W}^{\mathrm{T}} - \mathbf{D}(\tau)\|_F^2$
Infomax (3.135)	MaxEnt ICA	$-\log\lvert\det(\mathbf{W})\rvert + \sum_{i=1}^{m}H(y_i)$
Projection pursuit (3.139)	BSS/ICA	Negentropy, normalized kurtosis
Complexity pursuit (3.141)	BSS	$-\log[\boldsymbol{\theta}_i^{\mathrm{T}}\overline{\mathbf{C}}\boldsymbol{\theta}_i/\boldsymbol{\theta}_i^{\mathrm{T}}\tilde{\mathbf{C}}\boldsymbol{\theta}_i]$

Figure 3.8 Two selected natural images (from ftp://ftp.cnl.salk.edu/pub/tony/VRimages). The small rectangle illustrates the size of the image patch.

observed image patches in the rows with each column representing a vectorized 16×16 image patch, $\mathbf{S}$ represents the "original image code," and $\mathbf{A}$ is a square mixing matrix the columns of which can be viewed as basis vectors that encode the visual stimuli. Linear superposition of these basis vectors (with independent weighting coefficients) reconstructs the image formation process. Now, we wish to find an inverse mapping $\mathbf{Y} = \mathbf{WX}$ such that $\mathbf{Y}$ is equal to $\mathbf{S}$ (subject to scaling and permutation ambiguities). In our experimental setup, $\mathbf{X}$ is a $256 \times 20,000$ matrix and $\mathbf{A}$ and $\mathbf{W}$ are both 256×256 matrices.

Upon prewhitening the data and applying the ICA learning rule (3.138) for 10,000 iterations (with hyperbolic tangent score function and an initial learning rate $\eta = 0.005$), we obtained the demixing matrix $\mathbf{W}$. The product $\mathbf{WA}$ ideally will be a diagonal matrix (after permutation arrangement). Then we invert $\mathbf{W}$ to obtain the mixing matrix $\mathbf{A} = \mathbf{W}^{-1}$. The learned basis vectors of $\mathbf{A}$, arranged as a set of 16×16 images, consist of *oriented* and *localized* Gabor-like filters (see Figure 3.9). It is argued that these orientation-selective filters appear similar to the receptive fields of simple cells in the primary visual cortex (V1) [79, 685]. This example can be viewed as an application of the spatial ICA technique [860].

3.2.9 Energy-Efficient Hebbian Learning

In the preceding presentations of information-theoretic learning algorithms, we have witnessed that a simple form of Hebbian learning (e.g., [560, 647]) can develop an information-efficient neuronal code. This may be a good model of what takes place in the developing perceptual system, in which the neurons seek to maximize

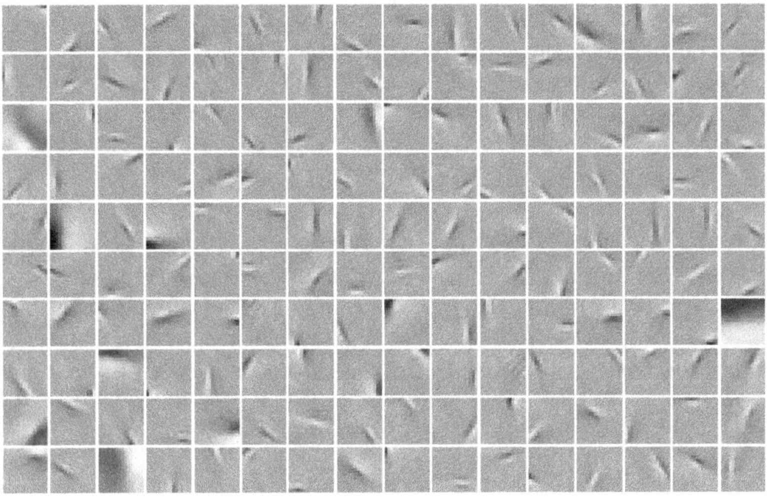

Figure 3.9 The independent basis functions of natural images. Each patch corresponds to one column of the estimated mixing matrix **A**. It appears that some basis images preserve Gabor filter-like receptive fields, which are local and may be viewed as edge or bar feature detectors.

the SNR or information transfer or minimize the mutual information between the presynaptic and postsynaptic outputs. A further constraint in biological systems is the fact that the metabolism of biological neurons and synapses demands various degrees of energy consumption dependent on the wiring length, rate of firing, and biochemical processes. In particular, information is metabolically expensive to process and transmit, suggesting that energy-efficient neural codes should be favored [528–530, 549].

For the first time, Heerema and van Leeuwen [379] derived the Hebbian learning rule from an energy-saving viewpoint. From a physics perspective, they used a binary neuron model (i.e., with only "0" and "1" states) and derived a Hebbian rule by starting with the following two assumptions:

- Biological assumption: It is least probable to modify a synapse if the presynaptic neuron is inactive.
- Physical assumption: The change of a synapse, whether it be a strengthening or a weakening one, can only be achieved by adding energy to the system; stated in mathematical terms, we may say

$$\Delta J[\boldsymbol{\theta}(t+1)] = \sum_{i} \sum_{j \in N_i} c_i \left[\theta_{ij}(t+1) - \theta_{ij}(t)\right]^2,$$

where ΔJ denotes the energy change, the positive constants c_i are characteristic of the ith neuron, and N_i denotes the local neighborhood of neuron i.

Based on these two assumptions, Heerema and van Leeuwen expressed the physical assumption by the following equation for deriving the energy-saving learning rules that can be either nonlocal or local (the local one is based on an approximation of the nonlocal version). Specifically, the local learning rule may be described as

$$\Delta\theta_{ij}(t) = \eta\,\{\kappa - [h_i(t) - b_i]\,(2x_i - 1)\}\,(2x_i - 1)y_j, \tag{3.152}$$

where x_i and y_j denote the pre- and postsynaptic neurons' activities, respectively; the function h describes the potential difference between the interior and the exterior of a neuron at its axon hillock; η is a learning-rate parameter; b_i is a threshold potential constant; and κ is also a constant. Although the derivation of (3.152) is physically oriented and the biological grounds are not fully justified, it is still invaluable in that it offers us a new way of thinking, in terms of *energy economy* and *energy efficiency*, from the perspective of a biological system with metabolic constraints.

3.2.10 Discussion

Categorization. To summarize Sections 3.1 and 3.2, we have reviewed and derived a variety of correlation-based learning algorithms which cover numerous learning paradigms that include unsupervised, supervised, and reinforcement learning. At this point, it is worth making some comparisons and comments. Specifically, we may categorize these learning rules under three different criteria:

- *Local versus Nonlocal Rules*: By locality, we mean local in *time* or local in *space*. Hebb's postulate and many variants of Hebbian learning that have been proposed are meant to be local in both time and space; namely, synaptic modification only relies upon the local information available in the presynaptic and postsynaptic neurons. Such a property makes Hebb's rule (including its various extensions) simple yet biologically plausible. However, locality is a double-edged sword. The constraint of being fully local severely limits the power of the learning rule. In the design of adaptive learning systems for practical applications, we may wish to remove this restriction. In fact, most learning rules reviewed in this chapter are only *generalized Hebbian*; they are correlative or associative in spirit, but some of them require global information from other synapses or require a global feedback error (or reward) signal. In general, a learning rule derived from a global cost function cannot always be converted into a local rule, except for a few special cases (e.g., [563]).
- *Hebbian versus Error-Driven Rules*: Unlike Hebbian learning, error-driven learning usually invokes an error signal that is derived from either a supervised or an unsupervised objective function. A major criticism of error-driven learning is its lack of biological justification: where does the error come from and how can synapses know about it? For a discussion of related issues, see [691]. It is also our belief that these two learning frameworks should be integrated together to complement each other; indeed, we have seen an example

in this chapter [the error-driven correlation memory learning rule in (3.54)]. In addition, we will show another example of such a learning paradigm, known as ALOPEX, in Chapter 6.

- *Correlation-Based versus Reward-Based Rules:* Correlation-based learning rules are mostly of the purely Hebbian form, while reward-based learning algorithms augment the Hebbian term with a (multiplicative) reward factor or a reward prediction error term, as in TD learning. As reviewed in [975], these two learning paradigms may be unified and integrated, as discussed earlier in Section 3.1.13. In addition, the reinforcement signal might not appear directly in the learning rule, but it can be used for modulating or driving the input representation, which further influences the Hebbian synaptic plasticity (e.g., [649]).

For the reader's convenience, we summarize and compare representative learning rules and their attributes in Table 3.2.

Synaptic Inhibition. Within the neocortex, there are a large number of inhibitory contacts at the soma and dendrites of cortical pyramidal cells. Lateral inhibition among the excitatory cells plays a crucial role in the development of the receptive fields and synaptic plasticity. Lateral inhibition allows cells to compete with each other and function in an energy-efficient fashion (i.e., with low activity levels, thereby satisfying metabolic constraints). Competition implies that if the efficacy of a synapse increases, then that of other synapses must decrease. Such a desirable "normalization" property can be attained in a number of possible ways, including via inhibition.

In addition to its neurobiological roots, from a computational viewpoint, inhibition is important for stablizing the learning process, which may prevent the weights

Table 3.2 Comparison of Representative Learning Rules and Their Attributes

Learning Rule	Local	Supervisory	Reward Driven	Biologically Inspired
Instar rule	Yes	No	No	Yes
SOM rule	Yes	No	No	Yes
BCM rule	Yes	No	No	Yes
Oja's rule	Yes	No	No	Yes
Wake–sleep rule	Yes	No	No	Yes
Boltzmann rule	Yes	No	No	Yes
LMS rule	No	Yes	No	No
Temporal Hebbian rule	Yes	No	No	Yes
TD rule	Yes	No	Yes	Yes
Linsker's rule	Yes	No	No	Yes
Imax	Yes	No	No	Yes
Infomax	No	No	No	Yes

growing infinitely without bound. Generally, synaptic inhibition may be categorized in two ways:

- *Postsynaptic versus Presynaptic Inhibition:* Inhibition can be either postsynaptic or presynaptic. Postsynaptic refers to *post-integration inhibition*, in which the competition is achieved at the level of the cell body or soma by using lateral connections among a population of neurons. Postsynaptic inhibition is the most common way to implement WTA competition in neural network models. On the other hand, presynaptic inhibition implies *preintegration inhibition*, in which the inhibitory interneurons attempt to block their own preferred inputs from activating other neurons before the integration takes place in the soma. Such a presynaptic inhibition allows a neural network to respond simultaneously to multiple stimuli, to distinguish overlapping stimuli, and to deal correctly with ambiguous stimuli [846, 847]. The WTA competition mechanism can also be implemented by presynaptic inhibition [994].
- *Divisive versus Subtractive Inhibition:* Inhibition can have either a divisive or a subtractive form. Inhibition can be implemented through the activation function or through the weight update equation. For instance, the "soft-competitive" (e.g., softmax) and "hard-competitive" (e.g., WTA) activation functions are two ways to obtain divisive inhibition. Another such example is through divisive normalization (e.g., [811]), which we will also use in a case study later in Chapter 7. Weight normalization can be viewed as an alternative to inhibition that prevents unlimited synaptic growth. In addition to the standard divisive form (e.g., [627, 921]), examples of the subtractive form of weight normalization include Sejnowski's covariance Hebbian learning [814, 815] and the weight-decay term of Oja's rule [676]. For further discussion of the issue of weight normalization, the reader is referred to [201].

Sparse Coding. Sparse coding is a term often used to refer to the representation of sensory stimuli with a sparse pattern of activation in neurons. Although there is a large amount of neurons involved at the early stage of sensory processing and stimulus coding, not all neurons fire together. If the ratio of the number of firing neurons against the number of inactive neurons is small, then the neuronal code is said to be sparse [201].

According to Barlow [60, 61], the major goal of stimulus coding at early stages of perceptual processing (e.g., in the retina, LGN, and V1, and similarly for other sensory modalities) is to reduce the high level of the redundancy in the sensory information. To do so, neurons should have an economical and efficient coding scheme for sensory representations. Barlow's early hypothesis of sparse coding is that the activity of a small number of neurons selected from a very large population forms a distributed representation of the sensory input [59]. He further suggested that the coding economy is brought about by reducing the frequency of impulses in neurons carrying the representation rather than by reducing the number of neurons involved [58]. According to Shannon's information theory, sparse or factorial codes have minimum entropy [62, 277]. On the other hand, the cost of coding is intuitively related to channel capacity, which is defined in terms of the channel bandwidth and SNR. Because of the stochastic nature of neuronal codes, it is inevitable that coding errors occur in accounting for different attributes of sensory stimuli. However, maintaining an accurate (i.e., with a high-SNR) representation of information using neurons with noisy firing functions might not be energy efficient [530]. Therefore, in order to balance the trade-off between coding accuracy and coding efficiency, it is necessary to use some degree of redundancy in the form of distributed representations.

It is now widely accepted that sparse coding is information efficient and plays an important role in encoding sensory information in the visual [286, 917] and auditory [552] cortices. Because of its importance, computational neuroscientists have devoted considerable effort to discovering the computational mechanisms behind sparse coding, at early stages of sensory processing [685–687] as well as at higher levels, such as higher visual processing in the inferotemporal (IT) cortex [991].

Different learning algorithms for generating sparse codes can be distinguished in four ways:

- *Local Hebbian and Anti-Hebbian Learning*: As first demonstrated in [284], a simple anti-Hebbian learning rule is capable of generating sparse coding. Specifically, Földiák [284] proposed a linear feedforward network with lateral connections which used a Hebbian rule to adapt the feedforward connections and an anti-Hebbian rule to adapt the lateral connections. More recently, Falconbridge et al. [271] used a slightly modified yet biologically plausible correlative rule for learning the same network architecture and reported similar observations of sparse codes; the receptive fields learned by the network exhibit Gabor-like filters that resemble the receptive fields seen in V1.
- *Regularized Hebbian Learning*: One of the earliest algorithms for generating sparse codes was proposed by Olshausen and colleagues [553, 684–687], who showed that sparse coding of natural images produces localized and oriented basis filters that resemble the receptive fields of simple cells in V1. Using a linear feedforward network, their simple Hebbian learning rule was combined with a regularization term that incorporates the sparsity constraint. The sparseness of the neuronal code is controlled by a heavy-tailed prior distribution imposed on the coefficients. Recently, this idea was extended to a bilinear network model to further enhance the invariance representation [340].
- *Nonnegative Sparse Coding*: Motivated by the fact that the neuron's firing rate is purely nonnegative, Lee and Seung [538, 539] suggested a coding scheme that combines both *sparsity* and *nonnegativity* constraints for a single-layer linear network. Their proposed learning rule is local and multiplicative, which may be viewed as a special form of correlative learning (in the logarithm domain). It has been shown in [538] that the receptive fields induced by nonnegative coding exhibit a localized (nonholographic) and sparse distributed representation. Recently, Li et al. [554] and Hoyer [408, 409] further

proposed several nonnegative sparse coding algorithms that explicitly impose the sparsity constraint and allow control over the degree of sparsity.

- *ICA and Energy-Based Model Learning*: Bell and Sejnowski [79] applied their Infomax-based ICA algorithm to image coding and reported that the independent components of the natural scenes resemble edge filters (see Figure 3.9); such Gabor-like filters are believed to be a good model of the spatiotemporal receptive fields of simples cells in V1 [907, 908]. Hyv̈arinen and Hoyer [426] extended sparse coding to complex cells, for which they used a two-layer neural network to simulate the responses of complex cells with contour coding. Specifically, the complex cells' responses are calculated in a feed-forward manner using an energy model and nonnegative sparse coding, and these responses are subsequently analyzed by a higher order sparse coding layer in the network. Such a hierarchical coding structure therefore offers the capability of modeling nonlinear and higher order neuronal functions (such as contour integration) in higher levels of the visual system.

3.3 CORRELATION-BASED COMPUTATIONAL NEURAL MODELS

3.3.1 Correlation Matrix Memory

As discussed earlier in Chapter 1, associative memory is an important function of the correlative brain. An associative memory system has the ability to encode patterns by associating together the pattern elements. Another important property of associative memory is the ability to recall a stored pattern given a subset of the pattern elements, which is referred to as cued recall or pattern completion. One of the proposals for model of associative memory was put forward by Gabor [304], based on the holographic principle, which suggested that a complete object can be reconstructed by a fragment or parts of the object itself. Later, Steinbuch [851, 852] proposed the *learning matrix* for memory storage. The potential value of learning matrices was further discussed in [853]. Since then a great many associative memory models have been proposed in the computational neural network literature [20, 33, 34, 185, 496, 498, 570, 651]. The common feature of these associative memory models is their basis in correlation for either *pattern association* or *pattern completion*. Starting with this section, we will present a brief overview of different computational models in the following order:

- Correlation matrix memory
- Hopfield network
- Brain-state-in-a-box (BSB) model
- Autoencoder network

In the late 1960s and early 1970s, Anderson [33, 34] proposed a linear associative memory model for pattern recognition, and the correlative learning rule forms the building block for a class of associative memory models called *correlation matrix memory*. The correlation matrix memory, in general, establishes an associative mapping: $\mathbf{x}_k \mapsto \mathbf{y}_k$; in other words, it associates together pairs of input vectors $\mathbf{x}_k \in \mathbb{R}^m$ and output vectors $\mathbf{y}_k \in \mathbb{R}^n$ $(k = 1, \ldots, \ell)$ into an $m \times n$ memory matrix $\mathbf{M}$ by

$$\mathbf{M} = \sum_{k=1}^{\ell} \mathbf{y}_k \mathbf{x}_k^{\mathrm{T}}, \tag{3.153}$$

which is also referred to as the *outer product rule* that is precisely equivalent to Hebb's rule in vector/matrix notation. When $\mathbf{y} = \mathbf{x}$, we have an *autoassociative* memory; otherwise (3.153) is referred to as a *heteroassociative* memory. In sequential form, (3.153) can be constructed recursively:

$$\mathbf{M}_k = \mathbf{M}_{k-1} + \mathbf{y}_k \mathbf{x}_k^{\mathrm{T}}, \qquad k = 1, 2, \ldots, \ell. \tag{3.154}$$

To achieve memory recall of a pattern, say $\mathbf{x}_j$, we multiply it with the memory matrix, which results in

$$\mathbf{y} = \mathbf{M}\mathbf{x}_j. \tag{3.155}$$

Substituting (3.153) into (3.155) yields

$$\begin{aligned} \mathbf{y} &= \sum_{k=1}^{\ell} \mathbf{y}_k \mathbf{x}_k^{\mathrm{T}} \mathbf{x}_j = \sum_{k=1}^{\ell} (\mathbf{x}_k^{\mathrm{T}} \mathbf{x}_j) \mathbf{y}_k \\ &= (\mathbf{x}_j^{\mathrm{T}} \mathbf{x}_j) \mathbf{y}_j + \sum_{k=1; k \neq j}^{\ell} (\mathbf{x}_k^{\mathrm{T}} \mathbf{x}_j) \mathbf{y}_k, \end{aligned} \tag{3.156}$$

where $\mathbf{x}_k^{\mathrm{T}} \mathbf{x}_j$ is the *inner product* between the past input observations and the recalled input pattern. The second line in equation (3.156) reveals that the resultant output can be broken down into two terms, the first due to the desired output associated with the given input $\mathbf{x}_j$ and the second due to cross-talk terms between the given input $\mathbf{x}_j$ and the other stored patterns. When the cross-talk terms predominate in the memory matrix $\mathbf{M}$, there will be recall errors. The output of the recall process will be exactly correct when (i) all the input vectors are of length 1 and (ii) the cross-talk term is zero. The latter is achieved when all the input vectors are mutually orthogonal. Hence, the maximum number of patterns that can be stored and exactly retrieved is equal to m, the dimensionality of the input.

To test the model for associative retrieval, we apply a new (unseen, or noise-corrupted) pattern, say $\mathbf{x}'$, through the memory matrix

$$\mathbf{y} = f(\mathbf{M}\mathbf{x}'), \tag{3.157}$$

where the function $f(\cdot)$ may be linear or nonlinear (e.g., the signum function). To the extent that the memory matrix has captured the correlative structure of

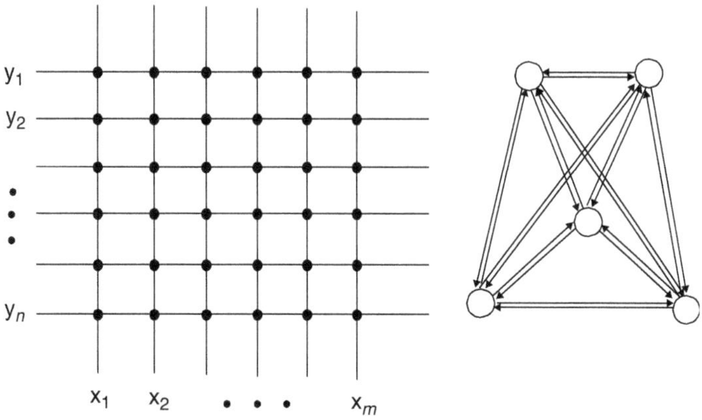

(*a*) (*b*)

Figure 3.10 (*a*) Associative memory. (*b*) Hopfield network.

the training patterns, it should be able to perform pattern completion. When f is nonlinear, the nonlinear associative memory model has an improved capability of error correction [20, 24].

In general, the memory matrix $\mathbf{M}$ can be decomposed into a sum of two matrices:

$$\mathbf{M} = \mathbf{R} + \mathbf{A}, \tag{3.158}$$

where the *diagonal* matrix $\mathbf{R}$ serves the purpose of *recognition* and the *off-diagonal* matrix $\mathbf{A}$ serves the purpose of *association*; these two matrices are complementary and jointly establish the relationship between patterns $\mathbf{x}$ and $\mathbf{y}$.

3.3.2 Hopfield Network

A *nonlinear* associative memory model having the same correlation matrix memory structure given by equation (3.153) was proposed by Amari [19], where the output takes on values of 1 or -1. This is a *recurrently connected* autoassociative memory model. Hopfield [399] developed a related model by analogy with the spin model of physics, which later has become known as the *discrete Hopfield network.* With the tool of statistical mechanics, Hopfield was able to show that this model forms memories by creating fixed-point attractors around the stored patterns and performs memory retrieval by settling into attractor states via a process of energy minimization.

The discrete Hopfield model is a *content-addressable memory* (CAM), meaning that the content of the memory itself (i.e., a partial or noise-corrupted version of a stored memory pattern) may be used to retrieve the full stored memory. Given a recurrent Hopfield network that consists of N neurons, in the *storage phase*, the symmetric weight matrix **W** of the CAM is calculated as

$$\mathbf{W} = \frac{1}{N}\sum_{i=1}^{\ell} \mathbf{x}_i \mathbf{x}_i^{\mathrm{T}}, \tag{3.159}$$

where $\mathbf{x} = [x_1, x_2, \ldots, x_N]^{\mathrm{T}}$ ($x_i \in \{-1, +1\}$) denotes the N-dimensional *bipolar* fundamental memory vector. In order to encourage sparse activation of states, equation (3.159) was also later modified to

$$\mathbf{W} = \frac{1}{N}\sum_{i=1}^{\ell} \mathbf{x}_i \mathbf{x}_i^{\mathrm{T}} - \ell\mathbf{I}, \tag{3.160}$$

where **I** denotes the identity matrix. In the *retrieval phase*, an asynchronous updating procedure for the state vector, denoted by $\mathbf{y} \in \{\pm 1\}^N$, is applied as follows:

$$\mathbf{y} = \mathrm{sgn}(\mathbf{W}\mathbf{y} + \mathbf{b}), \tag{3.161}$$

where **b** is a bias vector, and sgn(·) is the signum function. When the neurons in a Hopfield network are repeatedly updated in random order according to (3.161), the state update process will minimize a Lyapunov energy function and eventually converge to a fixed-point attractor. Specifically, with the symmetric weight constraints ($w_{ij} = w_{ji}$ and $w_{ii} = 0$), the Lyapunov function is defined as

$$\begin{aligned} J &= -\frac{1}{2}\sum_{i=1}^{N}\sum_{j=1}^{N} w_{ij} y_i y_j \\ &= -\frac{1}{2}\mathbf{y}^{\mathrm{T}}\mathbf{W}\mathbf{y}. \end{aligned} \tag{3.162}$$

A major limitation of the discrete Hopfield network is its low capacity. Theoretically, the maximum number of patterns that can be stored and exactly retrieved is $N/(2 \ln N)$ (e.g., see [364, 381]). Another limitation of the original Hopfield network is its restriction to binary neurons. In a subsequent model, Hopfield generalized this model to allow for neurons with continuous-valued, graded responses [402]. A final limitation of the Hopfield network is the fact that it is autoassociative. To overcome this limitation, Kosko [506] extended the Hopfield network by introducing an additional layer to perform recurrent autoassociation and heteroassociation. Such a two-layer heteroassociative recurrent network architecture is called the *bidirectional associative memory* (BAM) and also uses a local correlative learning rule to train the connection weight matrix to associate pattern pairs. Specifically, the $N \times m$ memory matrix $\mathbf{M} : \mathbb{R}^N \to \mathbb{R}^m$ is learned by the outer product rule

$$\mathbf{M} = \sum_{i=1}^{\ell} \mathbf{a}_i \mathbf{b}_i^{\mathrm{T}}, \tag{3.163}$$

where $\mathbf{a}_i$ and $\mathbf{b}_i$ are the bipolar modes of $\mathbf{x}_i \in \mathbb{R}^N$ and $\mathbf{y}_i \in \mathbb{R}^m$, respectively. As in the case of the associative model described at the beginning of this section, if the inputs $\mathbf{a}_1, \mathbf{a}_2, \ldots, \mathbf{a}_\ell$ are mutually orthogonal, namely

$$\mathbf{a}_i^{\mathrm{T}}\mathbf{a}_j = \begin{cases} 1, & i = j, \\ 0, & i \neq j, \end{cases} \tag{3.164}$$

then it follows that

$$\begin{aligned} \mathbf{a}_i^{\mathrm{T}}\mathbf{M} &= \mathbf{a}_i^{\mathrm{T}}\left(\sum_{j=1}^{\ell} \mathbf{a}_j \mathbf{b}_j^{\mathrm{T}}\right) \\ &= \mathbf{a}_i^{\mathrm{T}}\mathbf{a}_i\mathbf{b}_i^{\mathrm{T}} + \sum_{j=1, j\neq i}^{\ell} \mathbf{a}_i^{\mathrm{T}}\mathbf{a}_j\mathbf{b}_j^{\mathrm{T}} \\ &= \mathbf{b}_i^{\mathrm{T}}. \end{aligned} \tag{3.165}$$

In the BAM, the network output is also fed back to the input nodes. The procedure is repeated until an equilibrium point is reached at which $\mathbf{M}$ is said to be *bidirectionally stable*. The associated Lyapunov energy function in the BAM model is defined as $J = -\mathbf{a}^{\mathrm{T}}\mathbf{M}\mathbf{b}$.

The associative memory models (correlation matrix memory, discrete Hopfield network, and BAM) described thus far all attempt to remember a set of given signal patterns and recall any of them. In other words, they are all static pattern memory models, which do not deal with temporally dynamic inputs. A dynamic pattern recollection model was first proposed by Amari [19]. Suppose that a sequence of temporal patterns $\mathbf{x}(1), \mathbf{x}(2), \ldots, \mathbf{x}(T)$ is given. The temporal cross-covariance memory matrix is constructed as [19]

$$\mathbf{M} = \frac{1}{T}\sum_{t=1}^{T} \mathbf{x}(t)\mathbf{x}^{\mathrm{T}}(t+1). \tag{3.166}$$

This can be approximated online by a temporal correlative learning rule

$$\Delta\theta_{ij} = \eta x_i(t) x_j(t+1). \tag{3.167}$$

Such a dynamic model memorizes the temporal pattern sequence in a specific order, and it recalls the entire sequence one by one as the dynamics proceeds. The dynamics is similar as (3.161) but with a new asymmetric matrix $\mathbf{M}$:

$$\mathbf{z}(t+1) = \operatorname{sgn}(\mathbf{M}\mathbf{z}(t)), \tag{3.168}$$

where $\mathbf{z}(1)$ is the key pattern that is a noisy version of $\mathbf{x}(1)$.

The detailed dynamical process of recalling as well as the capacity of such a dynamic associative memory model were reported in [30].

3.3.3 Brain-State-in-a-Box Model

The BSB model, first described by Anderson [37, 40], is an auto associative recurrent network. Let $\mathbf{W} = \{w_{ij}\}$ denote an $N \times N$ symmetric synaptic weight matrix; then the BSB is described by the following pair of equations:

$$\mathbf{y}(t) = \mathbf{x}(t) + \beta \mathbf{W}\mathbf{x}(t), \tag{3.169}$$

$$\mathbf{x}(t+1) = \psi(\mathbf{y}(t)), \tag{3.170}$$

where β is a small positive constant called the *feedback factor*, $\mathbf{x}(t)$ represents an N-dimensional state vector of the BSB at time t, and $\psi(\cdot)$ is a piecewise linear activiation function. The BSB model is a *dynamic* associative memory model similar to the Hopfield network in the sense that it settles into an attractor state, thereby minimizing a Lyapunov energy function

$$J = -\frac{\beta}{2}\sum_{i=1}^{N}\sum_{j=1}^{N} w_{ij}x_i x_j = -\frac{\beta}{2}\mathbf{x}^{\mathrm{T}}\mathbf{W}\mathbf{x}. \tag{3.171}$$

Since the BSB may be viewed as an attractor network, in which the stable corners of the unit hypercube act as point attractors, it can be used as an unsupervised learning algorithm for pattern association [38]. For instance, let $\{\mathbf{x}_i\}_{i=1}^{\ell}$ denote a set of training patterns; then during the learning process the synaptic weights are adapted by the error-correcting learning rule

$$\Delta\mathbf{W} \leftarrow \eta(\mathbf{x}_i - \mathbf{W}\mathbf{x}_i)\mathbf{x}_i. \tag{3.172}$$

When the learning task is accomplished (i.e., $\Delta\mathbf{W} = \mathbf{0}$), linear association is established, meaning that

$$\mathbf{W}\mathbf{x}_i = \mathbf{x}_i \qquad (i = 1, \ldots, \ell). \tag{3.173}$$

In light of (3.173), it appears that the goal of the learning process is to force the linear associator to develop a particular set of eigenvectors (defined by training patterns $\{\mathbf{x}_i\}_{i=1}^{\ell}$) with eigenvalues equal to unity.

3.3.4 Autoencoder Network

Correlation memory can be categorized into autoassociation and heteroassociation. In contrast to the heteroassociation in the correlation memory matrix, the autoencoder network focuses on the autoassociation task. Basically, the autoencoder network attempts to link an input pattern with itself or reconstruct the input in the output. This autoassociation is particularly useful for pattern completion, noise reduction, and data compression.

In light of the associative memory, suppose we construct the memory matrix

$$\mathbf{M} = \sum_{k=1}^{\ell} \mathbf{x}_k \mathbf{x}_k^{\mathrm{T}} \tag{3.174}$$

such that $\mathbf{M}/\ell$ is an approximation of the autocorrelation matrix of the input data. Then, applying $\mathbf{M}$ to a new pattern $\mathbf{x}$ yields

$$\begin{aligned} \mathbf{M}\mathbf{x} &= \sum_{k=1}^{\ell} \mathbf{x}_k \mathbf{x}_k^{\mathrm{T}} \mathbf{x} \\ &\approx \ell \mathbf{C}_{xx} \mathbf{x}. \end{aligned} \tag{3.175}$$

Achieving autoassociation means that

$$\mathbf{C}_{xx}\mathbf{x} = \frac{1}{\ell}\mathbf{x}, \tag{3.176}$$

which essentially describes an eigenvalue equation. Hence, solving the autoassociation problem is an eigenvalue decomposition problem.

Now, the goal is to learn a weight matrix, denoted by $\mathbf{W}$, which attempts to approximate the transpose of the memory matrix $\mathbf{M}$, namely $\mathbf{W} \approx \mathbf{M}^{\mathrm{T}}$, such that

$$\begin{aligned} \hat{\mathbf{x}} &= \mathbf{W}^{\mathrm{T}}\mathbf{z} = \mathbf{W}^{\mathrm{T}}(\mathbf{W}\mathbf{x}) \\ &\approx \mathbf{M}\mathbf{M}^{\mathrm{T}}\mathbf{x} \approx \mathbf{x}. \end{aligned} \tag{3.177}$$

where $\mathbf{Z} = W_x$ denotes the linear network output. The optimality of the solution is measured by the reconstruction error

$$\begin{aligned} J &= \mathbb{E}[\|\hat{\mathbf{x}} - \mathbf{x}\|^2] \\ &= \mathbb{E}[\mathrm{tr}\left((\hat{\mathbf{x}} - \mathbf{x})(\hat{\mathbf{x}} - \mathbf{x})^{\mathrm{T}}\right)] \\ &= \mathrm{tr}\left(\mathbb{E}[\mathbf{x}\mathbf{x}^{\mathrm{T}} + \hat{\mathbf{x}}\hat{\mathbf{x}}^{\mathrm{T}} - \mathbf{x}\hat{\mathbf{x}}^{\mathrm{T}} - \hat{\mathbf{x}}\mathbf{x}^{\mathrm{T}}]\right) \\ &= \mathrm{tr}\left(\mathbf{C}_{xx}\right) + \mathrm{tr}\left(\mathbf{W}^{\mathrm{T}}\mathbf{W}\mathbf{C}_{xx}\mathbf{W}^{\mathrm{T}}\mathbf{W}\right) - 2\,\mathrm{tr}\left(\mathbf{W}\mathbf{C}_{xx}\mathbf{W}^{\mathrm{T}}\right), \end{aligned} \tag{3.178}$$

where the last line of (3.178) follows from the fact that

$$\mathrm{tr}\left(\mathbf{C}_{xx}\mathbf{W}^{\mathrm{T}}\mathbf{W}\right) = \mathrm{tr}\left(\mathbf{W}^{\mathrm{T}}\mathbf{W}\mathbf{C}_{xx}\right) = \mathrm{tr}\left(\mathbf{W}\mathbf{C}_{xx}\mathbf{W}^{\mathrm{T}}\right).$$

Therefore, the *unsupervised* Hebbian learning for the autoencoder can be interpreted as a special form of *supervised* learning that minimizes the reconstruction error between the original input and its reconstructed version $\hat{\mathbf{x}} = \mathbf{W}^{\mathrm{T}}\mathbf{z}$.

The autoencoder can be viewed as a multilayer linear network (see Figure 3.11), which is also referred to as a PCA network [54], because the weights discovered by

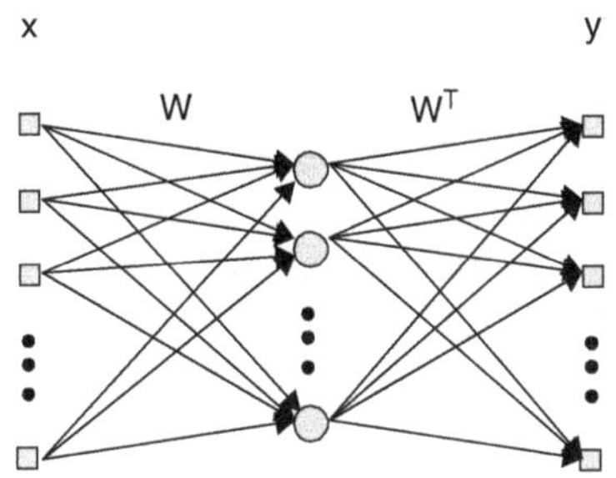

Figure 3.11 Illustration of the autoencoder network for PCA.

the n hidden-layer units span the same subspace as the first n eigenvectors (subject to rotation of the subspace). For instance, let us consider a two-layer linear network. Let $\mathbf{z} = \mathbf{W}\mathbf{x}$ denote the output of the hidden neurons, and let $\mathbf{W}$ and $\mathbf{W}^{\mathrm{T}}$ denote the input-to-hidden and hidden-to-output weight matrices, respectively. Then the learning rule for updating the connection weights is given by

$$\Delta\mathbf{W}^{\mathrm{T}}(t) = \eta\left[\mathbf{x}(t) - \mathbf{W}^{\mathrm{T}}(t)\mathbf{z}(t)\right]\mathbf{z}(t), \tag{3.179}$$

or in scalar form, we obtain

$$\Delta w_{ij}(t) = \eta\left(x_i(t) - \sum_j w_{ij}(t)z_j(t)\right)z_j(t). \tag{3.180}$$

As shown in [54], the error surface of the autoencoder is nonconvex and has saddle points but no local minima; the error landscape has a unique minimum corresponding to the projection onto the subspace spanned by the first principal eigenvectors of the covariance matrix associated with the training data, while other saddle points correspond to projections onto subspaces generated by higher order eigenvectors.

All of the associative memory models discussed so far share a common limitation, that is, a very low memory capacity. Similarly, the PCA-type models (including the autoassociator) can find a maximum of n principal components, where n is the dimensionality of the input. The way to overcome this limitation is to add two features: (i) nonlinearity and (ii) hidden layers. The Boltzmann machine, discussed earlier in Section 3.1.10, is a generalization of the Hopfield network to a multilayered network and can be trained as an autoassociator by having one layer of visible units and one layer of hidden units. The autoencoder network discussed above can be trained with multiple hidden layers and nonlinearity using the backpropagation learning algorithm, though the result will no longer be equivalent to PCA.

3.3.5 Novelty Filter

As discussed several times earlier, decorrelation may play an important role in serving as a basic self-organizing principle in the neocortex. To model this functionality, Kohonen and Oja [500] proposed a correlation-based novelty filter that uses a local, unsupervised Hebbian learning rule for decorrelating the features of input patterns. The novelty filter is essentially a recurrent linear network (see Figure 3.12) with lateral connections between the output neurons. Specifically, the activation of the ith output unit is represented as

$$y_i = x_i + \sum_{j=1}^{m} w_{ij} y_j, \tag{3.181}$$

where $\{w_{ij}\}$ denote the output-to-output lateral connection weights, with initial values all set to zero. The network is then trained by repeatedly presenting patterns from the training set to the input neurons subject to the unit-variance constraint of the output, namely $\langle y_i^2 \rangle = 1$. The synaptic strengths are then modified according to the following symmetric, anti-Hebbian learning rule:

$$\Delta w_{ij}(t) = \begin{cases} -\eta y_i(t) y_j(t) & \text{if } i \neq j, \\ 0 & \text{otherwise.} \end{cases} \tag{3.182}$$

If the learning-rate parameter η is sufficiently small, then as $t \to \infty$ the synaptic weights between two output neurons will change in (negative) proportion to the correlation between the activities of the output neurons, averaged over the training patterns. Therefore, if two output units are initially positively correlated, the inhibition between them will gradually increase, thereby reducing the correlation. Eventually, the network may settle into a stable state, in which case it satisfies that $\langle w_{ij} \rangle = 0$ and $\langle y_i y_j \rangle = 0$ for all $i \neq j$; in addition, under the unit-variance constraint $\langle y_i^2 \rangle = 1$, the correlation matrix of the network output $\mathbf{y}$ will approximate an identity matrix. This property therefore can be used for data whitening and feature extraction.

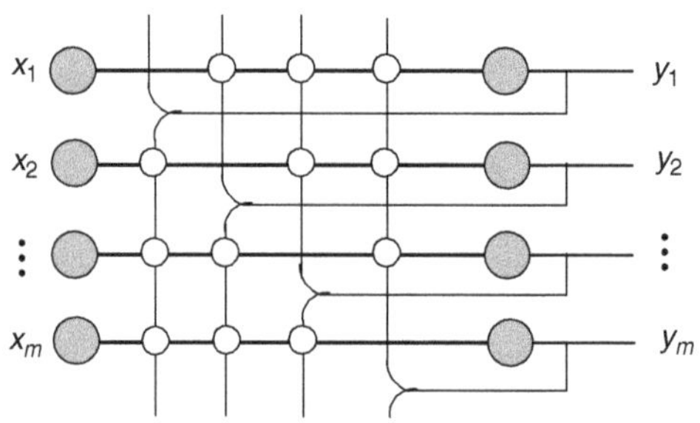

Figure 3.12 An illustrative diagram of the novelty filter. The output neurons are connected by lateral, inhibitory connections that are used to decorrelate the output units.

In the steady state, we can rewrite the network equation in matrix form,

$$\mathbf{y} = \mathbf{x} + \mathbf{W}\mathbf{y}. \tag{3.183}$$

That is, upon passing the transient state, the network output is calculated by

$$\mathbf{y} = (\mathbf{I} - \mathbf{W})^{-1}\mathbf{x} = \mathbf{T}\mathbf{x}, \tag{3.184}$$

where $\mathbf{T} = (\mathbf{I} - \mathbf{W})^{-1}$ represents a transformation matrix. In order to satisfy the stability of the linear system (3.184), the matrix $\mathbf{I} - \mathbf{W}$ has to be nonsingular [namely, $\det(\mathbf{I} - \mathbf{W}) \neq 0$] or the matrix $\mathbf{T}$ must have bounded eigenvalues.

3.3.6 Neuronal Synchrony and Binding

As discussed earlier in Chapter 1, synchronized firing is omnipresent within populations of neurons [334],[14] and it is widely believed that neuronal synchrony plays an important role in information processing within the cortex (e.g., [335, 337, 501, 503]). One hypothesis regarding the role of input synchrony is that neurons can be viewed as coincidence detectors (instead of temporal integrators) when performing perceptual tasks.

According to von der Malsburg [926], binding is a very general problem that applies to all types of knowledge representations, from the most basic perceptual representation to the most complex cognitive representation. Binding may be either in a *static* form or a *dynamic* form. One hypothesis is that dynamic binding is under the control of an attention mechanism which is used to control the synchronized activities of different assemblies of neurons and how the finite binding resource is allocated among the assemblies [836, 837]. One of the most popular dynamic binding theories is based on *temporal synchrony*, hence the reference to it as "*temporal binding*." The hypothesis of temporal binding states that different attributes (e.g., different features of a visual object) are bound together by means of synchronized firing of neurons that encode those different features. As the firing patterns of each neuronal assembly are independent from each other (e.g., by firing in another phase), they can form multiple distributed representations of feature conjunctions at the same time. See Figure 3.13 for an illustrative example. The binding by synchronization can also work across large separations between different cortical areas and, by this, establish a bridge between modules that encode different attributes.

Temporal binding theory was originally proposed by von der Malsburg [922] in his illuminating technical report "The Correlation Theory of Brain Function," in which he also suggested that the binding mechanism could be accomplished by a temporary strengthening of synapses between correlated neurons via a Hebbian mechanism. Such a synapse was referred to as the "*Malsburg synapse*" by Francis Crick [193] to distinguish it from the conventional "Hebbian synapse." Moreover, the synchronized mechanism allows the neurons to be linked in multiple active groups simultaneously and form a topological network. Specifically, von der Malsburg proposed a *dynamic link architecture* (see, e.g., [922, 925]) to

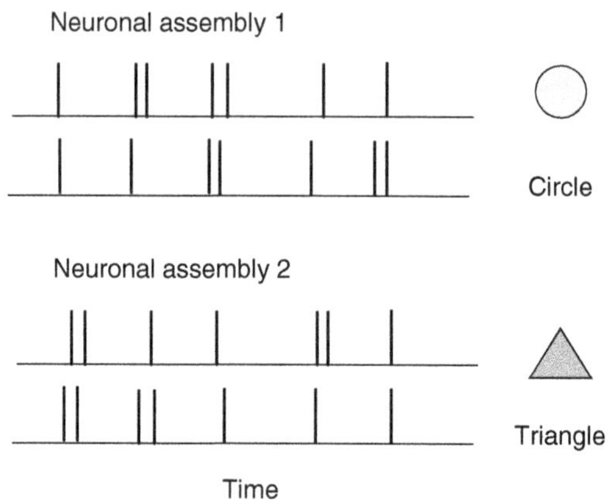

Figure 3.13 Feature (shape and color) encoding in neuronal assemblies via temporal binding. The first neuronal assembly encodes the "circle" shape and "light" color, whereas the second neuronal assembly encodes the "triangle" shape and "dark" color; the association of shape and color is represented and bound by synchronized activities of the neurons within each of two neuronal assemblies such that separate objects (i.e., the light-color circle and dark-color triangle) can be encoded simultaneously.

solve the temporal binding problem by letting neural signals fluctuate in time and by synchronizing those sets of neurons that are to be bound together into a higher level concept. With the same idea, von der Malsburg and Schneider [927] proposed a solution to the *cocktail party problem*.[15] In particular, they developed a neural cocktail party processor that uses synchronization (such as the sound onset synchrony) and desynchronization to segment the sensory inputs. It is noteworthy that von der Malsburg's correlation theory is equally applicable to the feature-binding problem in visual, auditory, or sensorimotor systems [259, 501, 890, 924, 926].

Related to von der Malsburg's theory of feature binding through synchronous oscillations is the notion of *synfire chains*. A synfire chain, as first proposed by Abeles in 1982 [3, 4], consists of neurons in a group firing synchronously, passing on their activations to another group of neurons, which then fire synchronously, and so on. Moreover, Bienenstock [92] suggested that the dynamics of cortex on the 1-ms timescale may be described as the activation of circuits of the synfirechain type. According to this theory, a pattern is characterized by the propagation of volleys of nearly synchronous spikes along a synfire chain. The microstructure of cortical connectivity, shaped by Hebbian synaptic plasticity, is a superposition of synfire chains, while a neuron participates in many distinct chains. At any given time, a large number of synfire chains are simultaneously active, and synchronization is made possible by weak synaptic coupling between chains. The fundamental computational unit in the cortex may be a wavelike spatiotemporal pattern of synfire-type activation, and the binding mechanism underlying compositionality in cognition may be the accurate synchronization of synfire waves that propagate simultaneously on distinct, weakly coupled, synfire chains in cortical connectivity.

Similar to von der Malsburg's theory, the synfire chain has been proposed as a neural mechanism for dynamic grouping [5].

3.3.7 Oscillatory Correlation

In computational neuroscience, the development of temporal correlation theory has been motivated by electrophysiological evidence of synchronized oscillations in auditory, visual, and olfactory cortices. Motivated by the early work of von der Malsburg [922, 927], the theory was further extended to different sensory domains whereby phases of neural oscillators are used to encode the binding of sensory components [935, 938]. For example, Brown and Wang [119, 937] developed a two-layer oscillator network (see Figure 3.14) that performs stream segregation based on *oscillatory correlation* as a possible basis for performing computational auditory scene analysis. In their oscillatory correlation-based model, a stream is represented by a population of synchronized relaxation oscillators, each of which corresponds to an auditory feature, and different streams are represented by desynchronized oscillator populations. Lateral connections between oscillators encode the harmonicity and proximity in time and frequency.

The theory of oscillatory correlation is an active research topic for addressing the binding problem. Recently, Wang [936] presented a survey of oscillatory correlation theory and computational neural models that are capable of performing figure–ground segregation. In the oscillatory correlation theory, *time* plays an important role in binding, as different segments of a signal or pattern unfold in time [936]. Exploring the time dimension for sensory processing and scene analysis remains a future research challenge for computational neuroscience and neural computation.

3.3.8 Modeling Auditory Functions

Correlation plays an important role in the auditory system. Specifically, the roles of autocorrelation and cross-correlation are omnipresent in various stages of spatial hearing, binaural processing, pitch estimation, and coincidence detection (see [372, 373] for an overview).

For instance, a central task of spatial hearing is sound localization. A classic model for sound localization was developed by Jeffress [441] using the binaural cue *interaural time difference* (ITD). In Jeffress's model (see Figure 3.15), the use

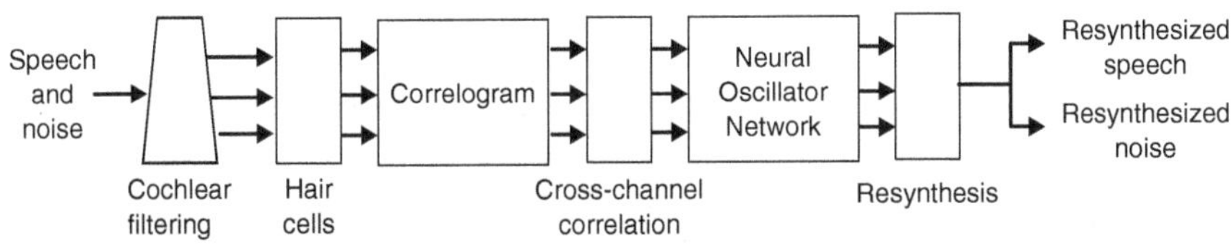

Figure 3.14 A schematic of neural correlated oscillator. (Adapted, with permission, from [937], *IEEE Transactions on Neural Networks*. Copyright © 1999 by IEEE.)

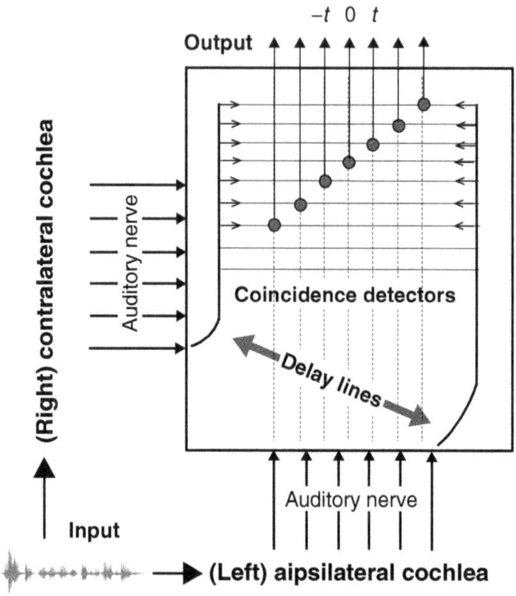

Figure 3.15 An illustration of Jeffress's model for coincidence detection. The sound waves propagate and arrive at the two ears with slight delays, and neuronal signals travel along transmission lines to an array of coincidence detectors. The coincidence detectors respond if signals from both sides arrive simultaneously. Due to transmission delays, the position of the activated coincidence detector depends upon the location of the sound source.

of cross-correlation is proposed to calculate the ITD in the auditory system and explain how it represents the ITD that is calculated from the signals received at the two ears. The sound processing and representation in Jeffress's model are simple and neurobiologically plausible [151, 451]. Gerstner et al. [318] used a spike-time-dependent Hebbian learning rule with a 20–100-ms timescale to demonstrate its role in delay tuning and temporal coding for auditory systems. The temporal window employed in their Hebbian rule enabled it to learn the spike-timing correlation such that the model is capable of forming and selecting delay lines. Specifically, the spike-timing-dependent Hebbian rule is described as follows [320]:

$$\frac{d}{dt}\theta_{ij}(t) = S_j(t)\left[a_1^{\text{pre}} + \int_0^{\infty} W(\tau)S_i(t-\tau)\,d\tau\right] + S_i(t)\int_0^{\infty} W(-\tau)S_j(t-\tau)\,d\tau, \tag{3.185}$$

where $S_j(t) = \sum_k \delta(t - t_j^{(k)})$ denotes the presynaptic spike train and $S_i(t) = \sum_k \delta(t - t_i^{(k)})$ denotes the postsynaptic spike train. The term a_1^{pre} in (3.185) is a small positive value. The learning rule (3.185) is only adapted when θ_{ij} is located within the region $(0, \theta_{\max}]$. The temporal window $W(\tau)$ is asymmetric and has a negative integral, namely $\int W(\tau)\,d\tau < 0$. The combination of a learning window with negative integral and a positive non-Hebbian term a_1^{pre} leads to a stabilization of the postsynaptic firing rate [320].

There is empirical evidence that the auditory system uses both temporal and spatial coincidence detection for various auditory functions, including periodicity pitch perception and sound localization [451, 822]. In general, spatiotemporal coincidence can be modeled by the cross-correlation function between two signals. Specifically, for a nonstationary sound (e.g., speech) signal, the *normalized* interaural cross-correlation function is defined as

$$C_{lr}(i, j, \tau) = \frac{\sum_{k=0}^{T-1} l_i(j-k) r_i(j-k-\tau)}{\sqrt{\sum_{k=0}^{T-1} l_i^2(j-k) r_i^2(j-k-\tau)}},$$

where $C_{lr}(i, j, \tau)$ denotes the cross-correlation coefficient at lag τ for the ith frequency channel and the jth time instance; l and r denote the auditory peripheral signals at the left and right ears, respectively; and T denotes the window length.

In light of the *Wiener–Khinchin theorem*, the normalized interaural cross-correlation function can be efficiently computed by using the FFT algorithm, which will result in a two-dimensional time–frequency map known as the cross-correlogram. The cross-correlogram visually depicts the interaural time difference between the two ears. The human brain is known to be extremely efficient in taking advantage of such a binaural cue for sound localization. Figure 3.16 presents an illustration of binaural auditory processing for real-life recorded stereo audio signals using interaural cross-correlation, which shows that the correlation varies according to frequency and internal delay.

In addition to cross-correlation, autocorrelation may also play a role in auditory functions such as pitch extraction [221]. First, the sound is decomposed into independent frequency channels via a bank of gamma tone filters.[16] Then the output of each channel is correlated with a delayed version of the same signal. This can be illustrated via an autocorrelogram with the horizontal axis representing time lag and the vertical axis representing frequency. For a periodic signal, the peak will appear at the integer multiples of the period. Specifically, the short-term *normalized* autocorrelation function can be defined as

$$C(i, j, \tau) = \frac{\sum_{k=0}^{T-1} x_i(j-k) x_i(j-k-\tau)}{\sum_{k=0}^{T-1} x_i^2(j-k)},$$

where $x_i(j)$ represents the jth sample of the signal at the ith frequency channel, τ denotes the time lag, and T denotes the (rectangular) window length.[17] The dynamic range of $C(i, j, \tau)$ is restricted within $[-1, 1]$ by normalizing the instantaneous energy of the ith channel. Using the FFT, the normalized autocorrelation function can be further represented by a two-dimensional time–frequency map known as the autocorrelogram, denoted by $\text{ACG}(\tau, f)$ (where τ denotes the time lag and $f = \{f_i\}$ denote different frequency bands).

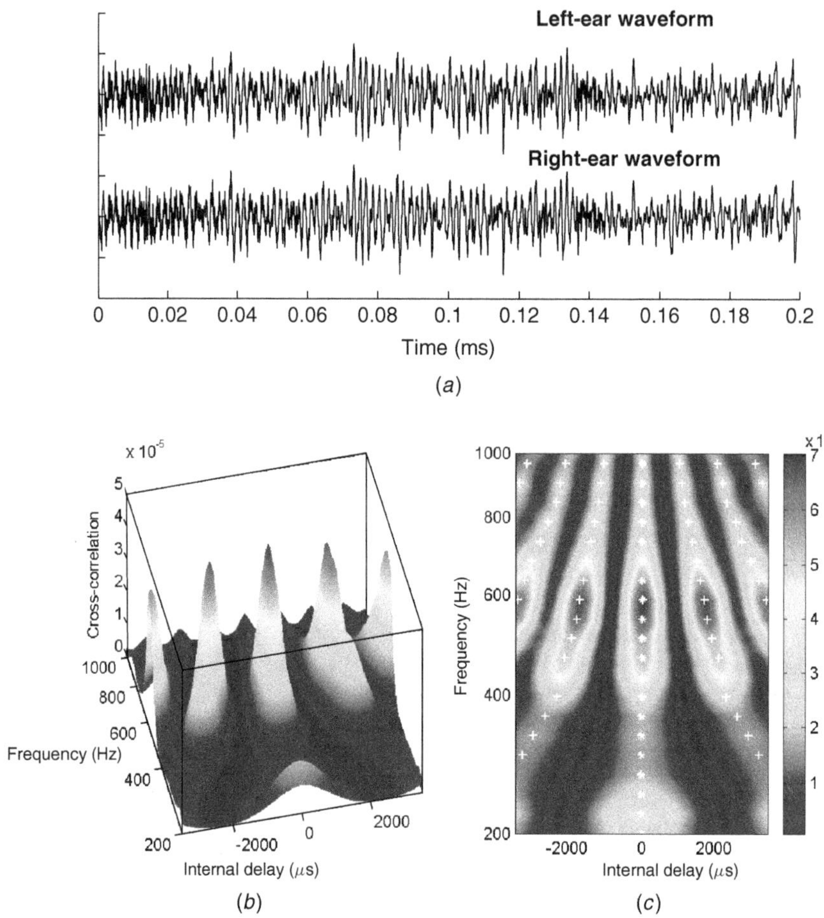

Figure 3.16 An illustration of binaural auditory processing using cross-correlation. (*a*) The waveforms recorded at the front end of two ears with sampling frequency 48 kHz. (*b*) The three-dimensional correlogram. (*c*) The two-dimensional correlogram with marked local maxima.

The summary autocorrelation index is then introduced to sum over the values across all the frequency bands in the two-dimensional autocorrelogram, as shown by

$$C(\tau) = \sum_i \mathrm{ACG}(\tau, f_i),$$

which will produce a one-dimensional plot with respect to the time lag τ. As an example, Figure 3.17 presents a simple illustration of autocorrelation analysis for two synthetic vowel signals /a/ and /u/, with their fundamental frequencies (i.e., pitches) centered at 200 and 300 Hz, respectively.

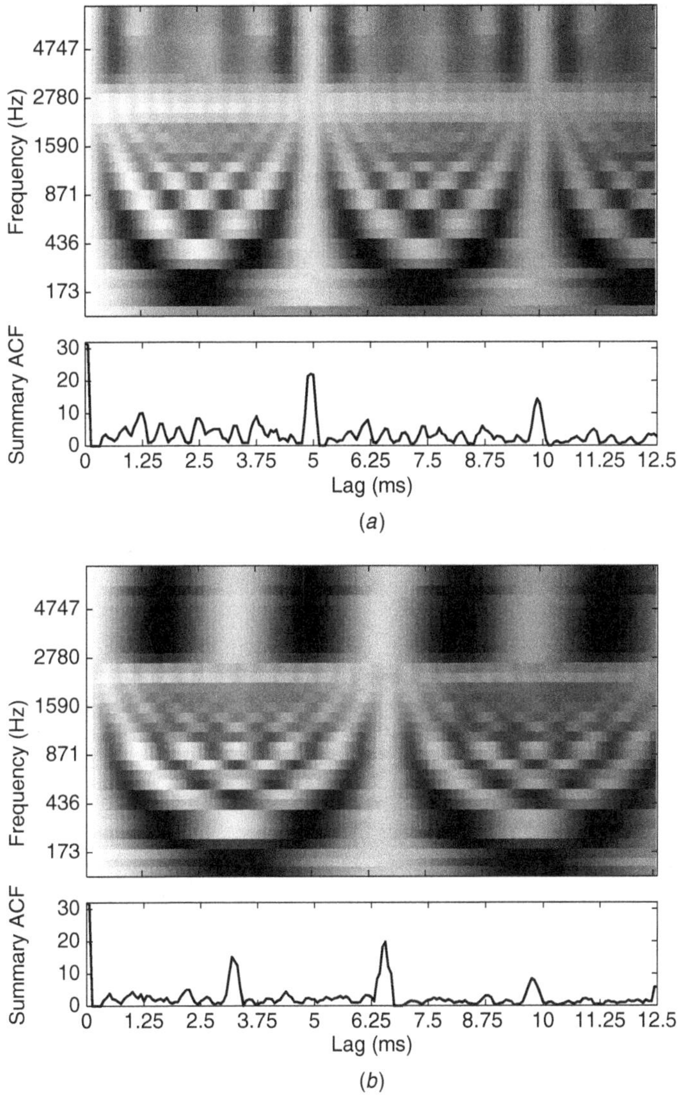

Figure 3.17 (*a*) Autocorrelogram and summary autocorrelation function (ACF) for the vowel /a/ with 200 Hz central frequency. The short-term autocorrelation is estimated with a window length of 20 ms and sampling frequency 16 kHz. It can be seen at time lag 5 ms that there is a common periodicity across most frequency bands, which indicates the fundamental frequency 200 Hz. Peaks also appear at lags of fundamental period multipliers such as 10 ms, 15 ms, and so on. (*b*) The same analysis applied to vowel /u/ with 300 Hz central frequency. (Courtesy of Rong Dong. Taken From (221) with permission.)

3.3.9 Correlations in the Olfactory System

In addition to the auditory system, correlation theory is equally important to the olfactory system. In the mammalian brain, the olfactory system consists of the olfactory bulb and olfactory cortex. In modeling the olfactory bulb, Freeman et al. [290] proposed an input correlation learning rule for generating and classifying patterns in olfactory systems. Specifically, the olfactory bulb was modeled by an array of coupled nonlinear oscillators (the so-called KII[18] set) that are driven by a set of differential equations [289]. An input correlation learning rule, being a modified Hebbian rule, was used to modify the interconnection strengths inside the model. To describe it mathematically, let θ_{ij} denote the *excitatory* coupling parameter from the ith neuron to the jth neuron; then the "input correlation rule" is described by

$$\theta_{ij} = \begin{cases} C_{\text{high}} & \text{if } x_i x_j \neq 0, \\ C_{\text{low}} & \text{otherwise}, \end{cases}$$

where x_i and x_j, respectively, denote the binary input patterns (i.e., $x_i, x_j \in \{0, 1\}$) and C_{high} and C_{low} denote the respective predefined high and low constants. When multiple input channels are nonzero, the network of strongly coupled oscillators forms a binary template for the input pattern (i.e., the odorant) and the template consists of the set of strongly interconnected neurons. It is claimed in [290] that this simple correlation rule enables the neural network to exhibit the desired properties of pattern generation and recognition in the olfactory bulb.

In studying the olfactory system, Hopfield and colleagues [116, 400, 401] have suggested using relative timing of action potentials to encode stimuli in *concentration-invariant* olfactory recognition tasks, and the synaptic learning in the olfactory bulb follows a STDP computation. In [401], Hopfield and Brody showed that the STDP is also capable of self-repairing for odor recognition. Let x_k denote the inputs that encode the pairings between presynaptic and postsynaptic spikes, let $\theta_k = W(k\delta_t)$ denote the synaptic weights that characterize the function connecting pre- and postsynaptic neurons to produce spikes in the time bin of length δ_t indexed by k within specific spiking time intervals, and let y denote the output neuron that represents the predicted probability of a presynaptic neuron belonging to the class appropriate for the connection. Then the neuron's input and output may be established by the equation

$$y = f\left(\sum_k \theta_k x_k\right),$$

where f is a logistic sigmoid function. Minimizing the Kullback–Leibler (KL) divergence between the network-defined probability y and the actual probability distribution would yield a simple Hebbian rule for synaptic adaptation

$$\delta\theta_k \propto (d - y)x_k,$$

where d is a binary value (0 or 1) that depends on the actual firing condition. It was shown in [401] that such a derived learning rule yields a synaptic choice function $W(\delta_t)$ with qualitative similarity to that of STDP [87].

3.3.10 Correlations in the Visual System

Topographic Map Formation. Of all the neocortical areas, the visual cortex is best understood in terms of neuronal response properties, in large part due to the Nobel Prize–winning work of David Hubel and Torsten Wiesel [416]. Following on their influential work, a great deal of effort has been dedicated to understanding the formation of visual feature maps which encode features such as ocular dominance, orientation, direction of movement, spatial frequency, and binocular disparity. As discussed in Chapter 1, it is known that the neurons in V1 lying along a column orthogonal to the cortical surface exhibit similar responses to similar visual stimuli, and the responses vary in the tangential direction parallel to the surface. The ordered three-dimensional structure is referred to as the neural map, and the cortical column corresponds to the vertical arrangement of neurons that have similar response properties. Most models of cortical map formation assume that visual experience drives this self-organizing process. On the other hand, in all vertebrates, an accurate retinotectal map can be established with little or no visual experience at the postnatal stage.

A question that has been of great interest to many modelers is: *How does the brain develop the visual maps given only local information available at the synapses?* Most research in this area has focused on the following two types of maps:

The ocular dominance (OD) map, which consists of alternating stripes or blobs with a regular periodicity, with neurons in each stripe responding preferentially to the stimulus in one eye and with interstripe or interblob regions of binocular neurons.

The orientation preference (OP) map, which exhibits stripes or blobs of neurons selective to the same orientation, with nearby stripes tending to code for nearby orientations but interspersed with point singularities (also called pinwheels) at which orientation tuning is relatively weak.

In the literature, computer scientists have developed numerous computational models for stimulating the visual map formation using *local* and *correlation-based* learning rules, (e.g., [21, 330, 558, 597, 624, 626, 804, 874, 921, 963]; for a review of this research area see [498, 620, 673, 674]). [19]

As an example of a model of OD development (taken from [201]), let $\mathbf{w} = (w_L, w_R)$ denote the synaptic weights of two LGN projections from left and right eyes. The postsynaptic output activity is represented as

$$y = w_L x_L + w_R x_R,$$

where $\mathbf{x} = (x_L, x_R)$ denote two retinal inputs from the left and right eyes. The synaptic connections w_L and w_R are assumed to be nonnegative. Ocular dominance arises when one of the weights is pushed to zero while the other remains positive. Assuming the retinal inputs from two eyes are statistically identical, then the autocorrelation matrix of the input can be represented as

$$\mathbf{C} = \langle \mathbf{x}\mathbf{x}^{\mathrm{T}} \rangle = \begin{pmatrix} \langle x_{\mathrm{L}} x_{\mathrm{L}} \rangle & \langle x_{\mathrm{R}} x_{\mathrm{L}} \rangle \\ \langle x_{\mathrm{L}} x_{\mathrm{R}} \rangle & \langle x_{\mathrm{R}} x_{\mathrm{R}} \rangle \end{pmatrix} \equiv \begin{pmatrix} c_1 & c_2 \\ c_2 & c_1 \end{pmatrix},$$

where $c_1 = \langle x_{\mathrm{L}} x_{\mathrm{L}} \rangle = \langle x_{\mathrm{R}} x_{\mathrm{R}} \rangle$ and $c_2 = \langle x_{\mathrm{L}} x_{\mathrm{R}} \rangle = \langle x_{\mathrm{R}} x_{\mathrm{L}} \rangle$. In this simple case, it is easy to calculate the eigenvectors of matrix $\mathbf{C}$ and their corresponding eigenvalues ($\lambda_{1,2} = c_1 \pm c_2$). In order to search for the dominant eigenvector, it can be shown that using a simple Hebbian rule with proper initial conditions and weight normalization constraints, ocular dominance can arise given sufficient competition between the growth of the left- and right-eye synaptic strengths [201].

Binocular Disparity Selectivity Development. Berns et al. [84] presented a computational correlation-based model for the development of *binocular disparity selectivity* in visual cortex. The model is based on Hebbian plasticity at synapses between geniculate and cortical cells. The model is driven by correlated activities in retinal ganglion cells within each eye before birth and additionally between eyes after birth. It was shown in [84] that with no correlations present between the two eyes the cortical model develops only monocular cells, and adding correlation between the eyes produces binocular neurons that may tune to zero disparity. Specifically, let $w_{x\alpha}^{\mathrm{L}}$ and $w_{x\alpha}^{\mathrm{R}}$ denote the synaptic strengths connecting the cortical position x to the retinal position α in the left and right eyes, respectively; then their synaptic modifications are described by the Hebbian rule as follows:

$$\Delta w_{x\alpha}^{\mathrm{L}} = \eta \sum_y \sum_\beta \mathbf{A}_{xy} (w_{y\beta}^{\mathrm{L}} C_{\alpha\beta}^{\mathrm{LL}} + w_{b\beta}^{\mathrm{R}} C_{\alpha\beta}^{\mathrm{LR}}),$$

$$\Delta w_{x\alpha}^{\mathrm{R}} = \eta \sum_y \sum_\beta \mathbf{A}_{xy} (w_{y\beta}^{\mathrm{L}} C_{\alpha\beta}^{\mathrm{RL}} + w_{b\beta}^{\mathrm{R}} C_{\alpha\beta}^{\mathrm{RR}}),$$

where $\mathbf{A}$ denotes the cortical interaction matrix that is defined by $\mathbf{A} = (\mathbf{I} - \mathbf{B})^{-1}$ and $\mathbf{B}$ is the cortical connection matrix; $\mathbf{C}^{\mathrm{LL}}$ and $\mathbf{C}^{\mathrm{RR}}$ are the autocorrelation matrices that represent the correlations of neuronal activities from the left and right eye inputs, respectively, and the cross-correlation matrices $\mathbf{C}^{\mathrm{LR}}$ and $\mathbf{C}^{\mathrm{RL}}$ represent the correlations of input activities between the left and right eyes. The synapses are further subject to subtractive and multiplicative normalization operations that prevent the synaptic weights from growing infinitely. Specifically, the subtractive term is defined as the average of the synaptic modification among the cells, namely, $(1/\ell) \sum_\alpha (\Delta w_{x\alpha}^{\mathrm{L}} + \Delta w_{x\alpha}^{\mathrm{R}})$, where ℓ denotes the total number of the inputs to the cortical cell.

3.3.11 Elastic Net

The elastic net is a self-organizing model that was originally developed by Durbin and colleagues for combinatorial optimization [234]. Rooted in statistical physics, the elastic net is a *generalized deformable model* well suited for many "hard" optimization problems [993]. Unlike simulated annealing [483], elastic net optimization is deterministic and therefore very efficient. For this reason, it has generated a growing interest as a candidate model of brain-style computing, for example, for simulating visual cortical maps [200, 232, 331, 653]. Moreover, it has proven to be useful in solving hard optimization problems, such as finding shortest paths in graph [233] and protein structure matching.

Without loss of generality, let us describe the elastic net with a "matching-nodes-in-the-graph" formulation. Let $\{\mathbf{x}_i\}_{i=1}^N$ denote the positions of the nodes inside a graph and $\{\mathbf{y}_j\}_{j=1}^M$ denote the "elastic points" (each with the same dimensionality as $\mathbf{x}_i$) to be manipulated and located. From an optimization point of view, the elastic net seeks to minimize an energy function, as described by

$$J(\{\mathbf{y}_j\}, \kappa) = -\alpha\kappa \sum_i \log\left(\sum_j e^{-\|\mathbf{x}_i - \mathbf{y}_j\|^2/2\kappa^2}\right) + \beta \sum_j \|\mathbf{y}_j - \mathbf{y}_{j+1}\|^2, \quad (3.186)$$

where κ is a scalar parameter that is crucial to the landscape of the energy function, α and β are two scalar constants that balance the "fitness error" and the regularized constraint, and the term $\|\mathbf{y}_j - \mathbf{y}_{j+1}\|$ may be viewed as a discretized derivative operator of first order [653]. Searching along the gradient descent direction $\Delta\mathbf{y}_j = -\kappa(\partial J/\partial \mathbf{y}_j)$ yields the update equation for the parameters $\mathbf{y}_j$:

$$\Delta\mathbf{y}_j = \alpha \sum_i w_{ij}(\mathbf{x}_i - \mathbf{y}_j) + \beta\kappa(\mathbf{y}_{j+1} + \mathbf{y}_{j-1} - \mathbf{y}_j), \quad (3.187)$$

where the weight parameter w_{ij} is defined by

$$w_{ij} = \frac{e^{-\|\mathbf{x}_i - \mathbf{y}_j\|^2/2\kappa^2}}{\sum_k e^{-\|\mathbf{x}_i - \mathbf{y}_k\|^2/2\kappa^2}}.$$

Notably, equation (3.187) may be viewed as a generalized Hebbian rule in which the first term of the right-hand side is Hebb-like and the second term forces the neighboring elastic points as close as possible. Without further probing the details of the elastic net, we instead use two representative examples below to illustrate the elegance of the elastic net in the context of brain-style computing.

■ EXAMPLE 3.4

In the first example, the elastic net is used to simulate the OD and OP maps in the primary visual cortex (V1). The experimental setup is taken from [653]. Specifically, a 128×128 array of visual cortical units is simulated, each of whose receptive field is parameterized by a five-dimensional vector $\mathbf{y}_j = [\xi_x, \xi_y, l, r, \theta]^{\mathrm{T}}$, where (ξ_x, ξ_y) is the center of the receptive field in visual space and the polar coordinates (r, θ) encode the preferred orientation $\theta/2$ and degree of orientation tuning r. Random stimuli $\mathbf{x}$ (training set) are drawn (uniformly) from the same five-dimensional space, with a grid of $N_x \times N_y \times N_{\text{OD}} \times 1 \times N_{\text{OP}}$ in the rectangle $[0, 1] \times [0, 1] \times [-l, l] \times [0, r] \times [-\pi/2, \pi/2]$. The experimental configuration is set as $N_x = N_y = 20$, $N_{\text{OD}} = 2$, $N_{\text{OP}} = 12$, $l = 0.09$, and $r = 0.16$ and the parameter setup of the elastic net is as follows: $\alpha = 1$, $\beta = 5$, and κ starts from 0.1 and is gradually annealed down to 0.05. Computer simulation results are illustrated in Figure 3.18. As seen in the figure, the OD and OP maps were successfully produced. Finally, we refer the reader to [653] for further discussion on simulating the visual cortical maps with generalized elastic nets.

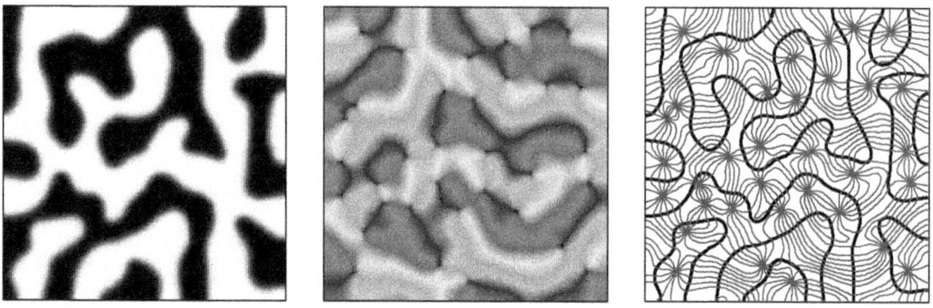

Figure 3.18 Simulated visual maps of OD and OP for an elastic net with derivative order $p = 1$. *Left panel*: ocular dominance map. *Middle panel*: orientation polar map. *Right panel*: contours of ocular dominance and orientation, where the converging singular points represent the pinwheels. (Reproduced, with permission, from (653). Copyright © 2004, by The American Physiological Society.)

Figure 3.19 The TSP for the 15 cities in the United States.

■ EXAMPLE 3.5

In the second example, the elastic net is applied to solve an N-city traveling salesman problem (TSP). This problem is known to be NP hard.[20] Here we only consider a small scale of the TSP with $N = 15$. Specifically, a salesman is supposed to travel over 15 given cities across the United States (Figure 3.19) and is required to visit each city once and only once. Graphically, the TSP is to find the shortest path to complete the whole itinerary given the nodes. In light of the studies in [233, 234], the elastic net parameters used in this experiment are $\alpha = 0.15 \sim 0.2$, $\beta = 1.0$, and κ is initially set as 0.08 and gradually annealed (reduced 10% every five iterations) within 100–200 iterations; a total of 45 "elastic points" are used in the elastic net to link the travel tour. We performed a number of Monte Carlo simulations. Two typical solutions found by the elastic net are shown in Figure 3.20 and the distance matrix is given in Table 3.3.

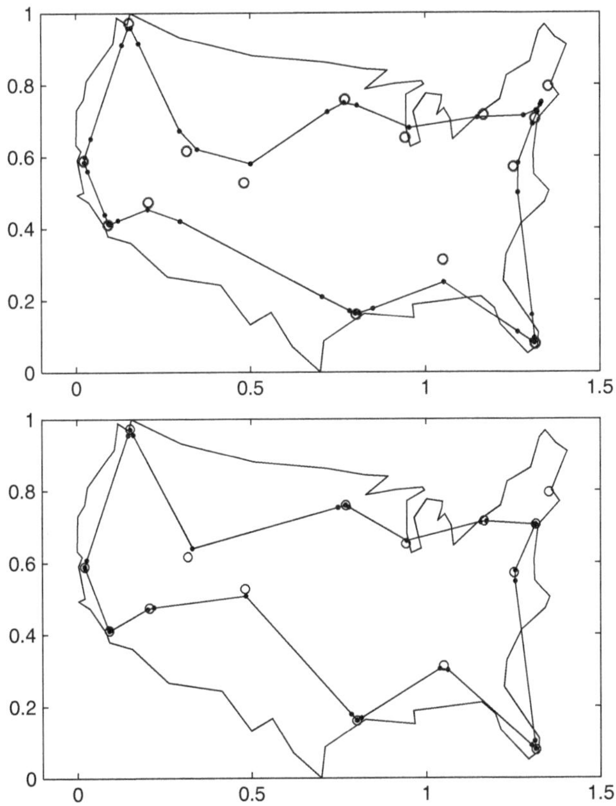

Figure 3.20 Two typical solutions found by the elastic net, with the total tour length 4.3741 (top panel) and 4.3857 (bottom panel). The open circles indicate the 15 cities and the black dots represent the elastic points.

Table 3.3 Symmetric Distance Matrix for 15-City TSP

01	02	03	04	05	06	07	08	09	10	11	12	13	14	15
0	0.097	0.148	0.147	0.377	0.474	0.627	1.189	1.298	1.256	1.128	0.999	0.851	0.547	0.748
	0	0.203	0.245	0.436	0.569	0.717	1.211	1.347	1.317	1.188	1.049	0.911	0.584	0.841
		0	0.167	0.234	0.418	0.653	1.044	1.151	1.115	0.987	0.853	0.709	0.399	0.663
			0	0.321	0.328	0.497	1.168	1.231	1.170	1.047	0.934	0.771	0.517	0.610
				0	0.356	0.684	0.849	0.921	0.881	0.753	0.623	0.476	0.202	0.511
					0	0.353	1.112	1.064	0.962	0.855	0.792	0.607	0.527	0.292
						0	1.464	1.389	1.256	1.172	1.131	0.945	0.871	0.520
							0	0.407	0.566	0.503	0.393	0.553	0.651	1.038
								0	0.192	0.221	0.298	0.465	0.767	0.889
									0	0.132	0.305	0.406	0.761	0.751
										0	0.180	0.277	0.629	0.669
											0	0.186	0.473	0.663
												0	0.370	0.486
													0	0.599
														0

Note: 01: New York City; 02: Boston; 03: Buffalo; 04: Washington, DC; 05: Chicago; 06: Atlanta; 07: Miami; 08: Seattle; 09: San Francisco; 10: Los Angeles; 11: Las Vegas; 12: Salk Lake City; 13: Denver; 14: Minneapolis; 15: Houston.

3.3.12 CMAC and Motor Learning

The cerebellum is a structure located at the back of the brain that is central to motor learning, receiving numerous input projections from sensory and motor systems. Neuroanatomical evidence suggests that the cerebellum is responsible for tuning of motor control for precise actions. Roughly speaking, the inputs of the cerebellar cortex (see Figure 3.21*a*) mainly consist of projections from *mossy fibers* (MFs) and *climbing fibers* (CFs), and its output is mostly in the *cerebellar nuclei* that project the signal to motor control areas. The MFs project to numerous *granule cells* (GCs), and each GC receives synapses of a few randomly connected MF inputs. The GCs also project axons to form *parallel fibers* (PFs), and each *Purkinje cell* (PC) is often connected with a large number of PFs. The role of the CF input is to fire the PC unconditionally, therefore reinforcing PF synapses active at the time of CF discharge. The synapses between the PF and PC are commonly believed to involve an associative learning process that relates the sensory input patterns and an active motor output response. In particular, Marr [590] suggested the main role of the cerebellum is to act as a *pattern recognizer* and a *sparse associative memory*, where the sparsity is achieved by mapping the sensory input space to a high-dimensional state space (i.e., the virtual memory).

Following Marr's study on cerebellar cortex [590], Albus [13] subsequently proposed the CMAC (cerebellar model articulation controller) model that aims to serve as a computational prototype of the cerebellar cortex of mammals. Later, the CMAC model was also used in motor learning and robotics [14, 406]. A schematic of the CMAC network is illustrated in Figure 3.21*b*. In the CMAC, the GCs may be viewed as a set of hard-wired feature detectors that perform feature extraction of specific sensory patterns; the PCs are often modeled as linear neurons that compute a linear weighted sum of the incoming PF inputs.

In the Marr–Albus cerebellum model, motor learning is mediated by a plasticity mechanism known as the LTD of PF synapses onto PCs, and LTD is controlled by an instructive CF signal [432, 433]. Specifically, Albus envisioned a generalized Hebbian learning rule that induces the LTD phenomenon, which would occur only in the presence of a *three-way coincidence* between a CF input (training signal), PC firing (postsynaptic output), and PF synaptic activity (presynaptic input). Such a synaptic plasticity rule bears a strong resemblance to the supervised perceptron learning rule using an error signal provided by the CF. Let θ_i denote the synaptic weight of the ith PF–PC cell synapse; then the synaptic plasticity rule may be generally written as a generic function of three terms, as shown by

$$\Delta\theta_i \propto F(e_{\mathrm{CF}}, x_{\mathrm{PF}}^i, y_{\mathrm{PC}}),$$

where the postsynaptic output is often approximated by $y_{\mathrm{PC}} \approx \sum_i \theta_i x_{\mathrm{PF}}^i$. Applying gradient descent to minimize the error signal [such that $\theta_i \propto -(\partial e_{\mathrm{CF}}^2/\partial\theta_i)$] would yield the perceptron-like learning rule $\Delta\theta_i \propto -x_{\mathrm{PF}}^i e_{\mathrm{CF}}$ (e.g., [599]).

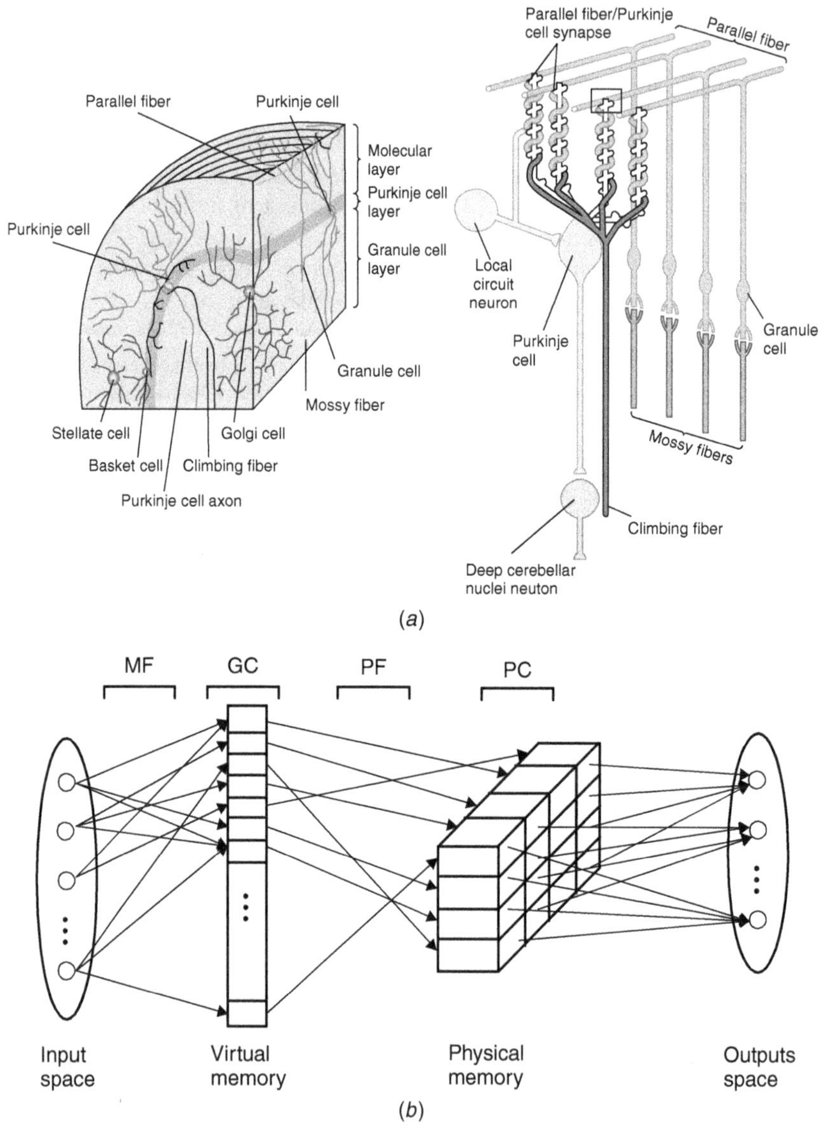

Figure 3.21 A schematic of the (*a*) cerebellum anatomical slice and (*b*) the CMAC network (MF: mossy fibers; GC: granule cells; PF: parallel fibers; PC: Purkinje cells).

Although the Marr–Albus cerebellum model still remains controversial, it has motivated researchers to develop a series of more sophisticated models (e.g., [109, 432, 468]). Notably, Fujita [300] proposed a simple anti-Hebbian learning rule that takes into account the dynamic and temporal characteristics of sensorimotor integration. Accordingly, the change of synaptic efficacy of a single PF synapse is

Table 3.4 Summary of Correlation-Based Computational Neural Models

Computational Models	Role	Operation
Correlation memory matrix	Associative memory	Auto- and cross-correlation
Hopfield network	Content-addressable memory	Recurrent attractor
BSB model	Dynamic associative memory	Recurrent attractor
Autoencoder network	PCA, autoassociation	Self-reconstruction
Binding	Feature conjunction, grouping	Synchronized firing
Synfire chain	Feature conjunction, grouping	Synchronized firing
Correlated oscillator	Sensory segmentation	Synchronized firing
Topographic map	Dimensionality reduction	Correlation-based learning
CMAC	Motor learning	Correlation-based learning

described by the following rule [468]:

$$\Delta\theta_i = -\eta x_{\mathrm{PF}}^i (e_{\mathrm{CF}} - e_{\mathrm{spont}}), \tag{3.188}$$

where x_{PF}^i denotes the firing rate of PF, e_{CF} denotes the firing rate of the CF input, and e_{spont} denotes its spontaneous level. The simple learning rule (3.188) reproduces both the LTD and LTP phenomena in PCs [783]. Equation (3.188) can be viewed as a gradient descent rule, where the error function is defined as the squared distance between e_{CF} and e_{spont} (i.e., $|e_{\mathrm{CF}} - e_{\mathrm{spont}}|^2$).

3.3.13 Summarizing Remarks

To summarize and close this section, we have discussed many computational neuronal models for modeling various brain functions, such as memory, auditory perception, vision, and motor learning. Throughout the discussions, we have witnessed again, as highlighted in Chapter 1, the fundamental role of correlation:

- On the neuron level, correlation serves the basic mechanism of modeling the mutual interactions and synchrony between populations of neurons.
- On the cortex level, correlation allows correlation-based adaptations for forming and reshaping cortical functions.
- On the system level, correlation establishes and consolidates the links between different subregions or modalities.

Finally, correlation-based computational neural models and their characteristics are summarized in Table 3.4.

APPENDIX 3A: MATHEMATICAL ANALYSIS OF HEBBIAN LEARNING*

Mathematically, we can write Hebb's postulate in terms of a differential equation

$$\frac{d\theta_{ij}}{dt} = F(\theta_{ij}; x_i, y_j), \qquad (3.A.1)$$

where θ_{ij} denotes the synapse that connects the ith presynaptic neuron and the jth postsynaptic neuron and F is a yet-unknown function [499, 818]. If we expand the function F around $x_i = y_j = 0$, the resulting expansion up to second order yields

$$\frac{d\theta_{ij}}{dt} \approx c_2^{\text{corr}}(\theta_{ij})x_i y_j + c_2^{\text{post}}(\theta_{ij})x_i^2 + c_2^{\text{pre}}(\theta_{ij})y_j^2$$
$$+c_1^{\text{pre}}(\theta_{ij})y_j + c_1^{\text{post}}(\theta_{ij})x_i + c_0(\theta_{ij}) + \mathcal{O}(\xi^3), \qquad (3.A.2)$$

where $c_k(\theta_{ij})$ $(k = 0, 1, 2)$ denote the k-th order expansion coefficients and $\mathcal{O}(\xi^3)$ denotes the higher-than-2-order term. Note that the first term of (3.A.2) essentially states Hebb's postulate. In the simpler form of (3.A.2), we set all coefficients but $c_2^{\text{corr}}(\theta_{ij})$ to zero; then we obtain

$$\frac{d\theta_{ij}}{dt} = c_2^{\text{corr}}(\theta_{ij})x_i y_j. \qquad (3.A.3)$$

When $c_2^{\text{corr}} > 0$, (3.A.3) is a form of Hebbian learning; when $c_2^{\text{corr}} < 0$, (3.A.3) becomes a form of anti-Hebbian learning.

Suppose we model the postsynaptic neuron by a linear combiner as described by the equation

$$y_j(t) = \sum_k \theta_{ik} x_k(t). \qquad (3.A.4)$$

Then substituting (3.A.4) into (3.A.3) and taking a time average, we obtain

$$\langle \frac{d\theta_{ij}}{dt} \rangle = c_2^{\text{corr}} \sum_k \theta_{ik} \langle x_i(t) x_k(t) \rangle, \qquad (3.A.5)$$

which is in line with the direction of the *principal eigenvector* of the autocorrelation matrix $\mathbf{C} = \{C_{ik}\}$, where

$$C_{ik} = \langle x_i(t) x_k(t) \rangle. \qquad (3.A.6)$$

Thus, we can rewrite (3.A.6) in matrix form

$$\langle \frac{d\boldsymbol{\theta}_i}{dt} \rangle = c_2^{\text{corr}} \mathbf{C} \boldsymbol{\theta}_i. \qquad (3.A.7)$$

*The material presented in this appendix is adapted from [319, 320].

If we discretize time and approximate (3.A.3) with a difference equation, then in light of (3.A.4) we have

$$\Delta\boldsymbol{\theta}_i(t) = \eta\mathbf{x}(t)y(t) = \eta\mathbf{x}(t)\mathbf{x}^{\mathrm{T}}(t)\boldsymbol{\theta}_i(t). \tag{3.A.8}$$

Given an initial estimate $\boldsymbol{\theta}_i(0)$, applying (3.A.8) repeatedly with ℓ data points can yield the "gross" weight change:

$$\Delta\boldsymbol{\theta}_i = \eta\left(\sum_{t=1}^{\ell}\mathbf{x}(t)\mathbf{x}^{\mathrm{T}}(t)\right)\boldsymbol{\theta}_i(0). \tag{3.A.9}$$

Taking an average of both sides of (3.A.9) yields

$$\begin{aligned}\langle\Delta\boldsymbol{\theta}_i\rangle &= \eta\langle\mathbf{x}(t)\mathbf{x}^{\mathrm{T}}(t)\rangle\boldsymbol{\theta}_i(0)\\ &= \eta\mathbf{C}\boldsymbol{\theta}_i(0),\end{aligned} \tag{3.A.10}$$

which has a similar form as the counterpart for the differential equation (3.A.7).

APPENDIX 3B: NECESSITY AND CONVERGENCE OF ANTI-HEBBIAN LEARNING

Following the early discussion of Hebbian learning, let us assume that the postsynaptic activity is represented as a linear sum of presynaptic terms. Written in vector form, we have

$$\begin{aligned}\boldsymbol{\theta}(t+1) &= \boldsymbol{\theta}(t) + \eta\mathbf{x}(t)y(t)\\ &= \boldsymbol{\theta}(t) + \eta\mathbf{x}(t)\mathbf{x}^{\mathrm{T}}(t)\boldsymbol{\theta}(t).\end{aligned} \tag{3.B.1}$$

Taking the expectation of both sides in the above equation yields

$$\boldsymbol{\theta}(t+1) = (\mathbf{I} + \eta\mathbf{C})\boldsymbol{\theta}(t), \tag{3.B.2}$$

where $\mathbf{I}$ denotes the identity matrix. Since the autocorrelation matrix $\mathbf{C}$ is positive definite, so is the sum of $\mathbf{I} + \eta\mathbf{C}$. In other words, all the characteristic roots of the iterative equation (3.B.2) are positive; thus the iterations will lead to divergence with any positive value of η. In conclusion, *pure* Hebbian learning is unstable; equation (3.B.2) will lead $\boldsymbol{\theta}$ to an infinite magnitude, with a direction equal to that of the eigenvector of the correlation matrix $\mathbf{C}$ with the largest eigenvalue.[21]

In contrast, for anti-Hebbian learning, we may similarly derive

$$\boldsymbol{\theta}(t+1) = (\mathbf{I} - \eta\mathbf{C})\boldsymbol{\theta}(t), \tag{3.B.3}$$

and the stability of (3.B.3) is determined by the characteristic roots of $(\mathbf{I} - \eta\mathbf{C})$. In order to assure stability (or asymptotic convergence), we need to confine the learning-rate parameter η to lie within a certain range. Specifically, let $\lambda_{\max}$ denote the maximum eigenvalue of the correlation matrix $\mathbf{C}$; then we may establish the *necessary condition* for the convergence of anti-Hebbian learning [737]:

> If the positive learning-rate parameter satisfies the condition that $0 < \eta < 2/\lambda_{\max}$, where $\lambda_{\max}$ is the maximum eigenvalue of the correlation matrix $\mathbf{C}$, then the anti-Hebbian learning rule (3.B.3) will asymptotically converge.

Thus far, we have restricted ourselves to a linear neuron, namely $y = \mathbf{x}^{\mathrm{T}}\boldsymbol{\theta}$. More generally, the postsynaptic neuron can be modeled by a nonlinear function such as

$$y = f\left(\sum_{i=1}^{m} x_i\theta_i\right) = f(\mathbf{x}^{\mathrm{T}}\boldsymbol{\theta}). \tag{3.B.4}$$

In statistics, the model described by (3.B.4) is known as the *generalized linear model* (GLM), and $f^{-1}(\cdot)$ is called the *canonical link function.* If we expand the nonlinear link function using the Taylor series, we obtain

$$y = f(\xi) = \xi + \frac{1}{2}f'(\xi)\xi^2 + \cdots. \tag{3.B.5}$$

Substituting the linear term $y(t)$ with the expansion terms in (3.B.5) for either Hebbian or anti-Hebbian learning, we then obtain

$$\Delta\boldsymbol{\theta}(t) = \pm\eta\left\{\mathbf{x}(t)\left[\mathbf{x}^{\mathrm{T}}(t)\boldsymbol{\theta}(t) + \frac{1}{2}\alpha\|\mathbf{x}^{\mathrm{T}}(t)\boldsymbol{\theta}(t)\|^2 + \cdots\right]\right\}, \tag{3.B.6}$$

which involves the second-order correlation in the linear Hebbian rule as well as higher order interaction terms. In general, analyzing the convergence of (3.B.6) depends on the specific choice of the nonlinear function f and the order of the approximation; its stability analysis is therefore much more complicated than the linear case.

APPENDIX 3C: LINK BETWEEN HEBBIAN RULE AND GRADIENT DESCENT

For supervised learning, a popular objective function for optimization is the MSE between the target signal $d(t)$ and its estimate $y(t)$. Specifically, we can decompose the expected empirical risk function as

$$\begin{aligned} J &= \mathbb{E}\left[(d(t) - y(t))^2 \middle| \mathbf{x}\right] \\ &= \mathbb{E}\left[(d(t) - \mathbb{E}[d(t)|\mathbf{x}] + \mathbb{E}[d(t)|\mathbf{x}] - y(t))^2 \middle| \mathbf{x}\right] \\ &= \mathbb{E}\left[(d(t) - \mathbb{E}[d(t)|\mathbf{x}])^2 \middle| \mathbf{x}\right] + (\mathbb{E}[d(t)|\mathbf{x}] - y(t))^2 \\ &= \mathrm{var}[d(t)|\mathbf{x}] + (\mathbb{E}[d(t)|\mathbf{x}] - y(t))^2, \end{aligned} \tag{3.C.1}$$

which is known as the *bias-variance* decomposition [313]. When the desired output signal $d(t)$ is a zero-mean random noise process, we then have $\mathbb{E}[d(t)|\mathbf{x}] = 0$, and (3.C.1) reduces to

$$J = \text{var}[d(t)|\mathbf{x}] + y^2(t), \tag{3.C.2}$$

where the first term of the right-hand side is a constant that is independent of the adjusted weight parameters. Let $y(t) = \boldsymbol{\theta}^{\mathrm{T}}\mathbf{x}(t)$; then taking the negative gradient of the second term yields

$$-\frac{\partial J}{\partial \boldsymbol{\theta}} = -\frac{\partial y^2(t)}{\partial \boldsymbol{\theta}} = -2y(t)\mathbf{x}(t). \tag{3.C.3}$$

Hence, minimizing the cost function J in (3.C.2) is equivalent to minimizing the energy (or power) of the output, and performing gradient descent yields a stochastic gradient descent rule that takes the form of an anti-Hebbian rule

$$\boldsymbol{\theta}(t+1) = \boldsymbol{\theta}(t) - \eta\mathbf{x}(t)y(t). \tag{3.C.4}$$

More generally, when the desired output $d(t)$ has a nonzero mean, minimizing the MSE would yield two combining terms that appear in the LMS rule: one being anti-Hebbian and the other being a *forced* Hebbian term.

APPENDIX 3D: RECONSTRUCTION ERROR IN LINEAR AND QUADRATIC PCA

The goal of PCA learning is to estimate the dominant eigenvector(s). In terms of Oja's one-unit PCA model (i.e., $y = \boldsymbol{\theta}^{\mathrm{T}}\mathbf{x}$), the criterion can be rewritten as minimizing the MSE between the original input and the reconstructed input

$$J = \mathbb{E}\left[\|\mathbf{x} - \hat{\mathbf{x}}\|^2\right], \tag{3.D.1}$$

where the reconstructed input $\hat{\mathbf{x}}$ is represented by

$$\hat{\mathbf{x}} = \mathbf{u}y. \tag{3.D.2}$$

Substituting (3.D.2) into (3.D.1) yields

$$J = \mathbb{E}\left[\left\|(\mathbf{I} - \mathbf{u}\boldsymbol{\theta}^{\mathrm{T}})\mathbf{x}\right\|^2\right]. \tag{3.D.3}$$

Minimizing the cost function J (with respect to) $\mathbf{u}$ for a given $\boldsymbol{\theta}$ would yield the solution

$$\mathbf{u}_{\text{opt}} = \arg\min_{\mathbf{u}} J = \frac{\mathbf{C}_{xx}\boldsymbol{\theta}}{\boldsymbol{\theta}^{\mathrm{T}}\mathbf{C}_{xx}\boldsymbol{\theta}}, \tag{3.D.4}$$

and then the reconstructed input can be written as

$$\hat{\mathbf{x}} = \mathbf{u}_{\mathrm{opt}} y = \frac{\mathbf{C}_{xx}\boldsymbol{\theta}\boldsymbol{\theta}^{\mathrm{T}}}{\boldsymbol{\theta}^{\mathrm{T}}\mathbf{C}_{xx}\boldsymbol{\theta}}\mathbf{x} \equiv \mathbf{P}\mathbf{x}, \tag{3.D.5}$$

where the matrix

$$\mathbf{P} = \frac{\mathbf{C}_{xx}\boldsymbol{\theta}\boldsymbol{\theta}^{\mathrm{T}}}{\boldsymbol{\theta}^{\mathrm{T}}\mathbf{C}_{xx}\boldsymbol{\theta}} \tag{3.D.6}$$

defines a *projection operator* which satisfies the property $\mathbf{P}^2 = \mathbf{P}$. If the matrix $\mathbf{C}_{xx}$ is positive definite, then the minimum of the reconstruction error is attained when $\boldsymbol{\theta}$ is the principal eigenvector of $\mathbf{C}_{xx}$; in other words, PCA minimizes the mean-squared reconstruction error.

In the case of multiple-output PCA, let $\mathbf{y} = \mathbf{W}\mathbf{x}$, and let $\hat{\mathbf{x}} = \mathbf{V}^{\mathrm{T}}\mathbf{y} = \mathbf{V}^{\mathrm{T}}\mathbf{W}\mathbf{x}$ be the reconstructed input, and the reconstruction error is given by equation (3.178). Similarly, it can be proved that, by minimizing the reconstruction error, $\hat{\mathbf{x}}$ can be represented by

$$\hat{\mathbf{x}} = \mathbf{P}\mathbf{x},$$

where the matrix $\mathbf{P}$ specifies the subspace defined by an orthogonal projection

$$\mathbf{P} = \mathbf{W}(\mathbf{W}^{\mathrm{T}}\mathbf{W})^{-1}\mathbf{W}^{\mathrm{T}}. \tag{3.D.7}$$

For quadratic PCA, the reconstruction process is nonlinear and a little more complicated [877]. Let $\hat{x}_i$ denote the ith element of the reconstructed input vector $\hat{\mathbf{x}}$; then its optimal reconstruction takes the following form

$$\hat{x}_i^{\alpha} = y^{\alpha} v_i + \sum_{\beta} y^{\alpha} y^{\beta} v_i^{\beta}, \tag{3.D.8}$$

where α denotes the pattern label. Let

$$\mathbf{Y}^{\alpha} = \left(y^{\alpha}, y^{\alpha} y^{\beta}\right)^{\mathrm{T}},$$
$$\mathbf{V} = \left(\mathbf{v}, \mathbf{v}^{\beta}\right)^{\mathrm{T}},$$

then it follows that

$$\hat{\mathbf{x}}^{\alpha} = \mathbf{V} \cdot \mathbf{Y}^{\alpha}, \tag{3.D.9}$$

where the dot product is taken over the index β. The reconstruction error for the quadratic PCA is then written as

$$\begin{aligned} J &= \mathbb{E}\left[\|\mathbf{x} - \mathbf{V} \cdot \mathbf{Y}\|^2\right] \\ &= \sum_{\alpha,i}\left[x_i^{\alpha} - \sum_{j_n,\beta_n} y^{\alpha} y^{\beta_n} a_{ij_n} x_{j_n}^{\beta_n}\right]^2. \end{aligned} \tag{3.D.10}$$

Minimizing (3.D.10) w.r.t. the unknown parameters in $\mathbf{V}$ leads to the matrix equation as follows [877]

$$a_{ij}\Lambda_{jk} = \Lambda_{ik}, \tag{3.D.11}$$

where $\boldsymbol{j}$ and $\boldsymbol{k}$ denotes the combinations of indices $j_1, j_2, \ldots, j_n$ and $k_1, k_2, \ldots, k_n$, respectively, and a_{ij} denotes the combinations among $\{a_{ij_1}, a_{ij_2}, \ldots, a_{ij_n}\}$, and

$$\Lambda_{ik} = \left(\tilde{\mathbf{C}}\tilde{\boldsymbol{\theta}}\right)_i \left(\tilde{\mathbf{C}}\tilde{\boldsymbol{\theta}}\right)_{\boldsymbol{k}}, \tag{3.D.12}$$

$$\Lambda_{\boldsymbol{jk}} = (\tilde{\boldsymbol{\theta}}^{\mathrm{T}}\tilde{\mathbf{C}}\tilde{\boldsymbol{\theta}}) \left(\tilde{\mathbf{C}}\tilde{\boldsymbol{\theta}}\right)_{\boldsymbol{j}} \left(\tilde{\mathbf{C}}\tilde{\boldsymbol{\theta}}\right)_{\boldsymbol{k}}, \tag{3.D.13}$$

where

$$\left(\tilde{\mathbf{C}}\tilde{\boldsymbol{\theta}}\right)_{\boldsymbol{k}} = \prod_{r=1}^{n} \left(\tilde{\mathbf{C}}\tilde{\boldsymbol{\theta}}\right)_r. \tag{3.D.14}$$

Finally, the solution to the minimum of (3.D.10) may be given by [877]

$$a_{ik} = \frac{\left(\tilde{\mathbf{C}}\tilde{\boldsymbol{\theta}}\right)_i \left(\tilde{\mathbf{C}}\tilde{\boldsymbol{\theta}}\right)_{\boldsymbol{k}}}{(\tilde{\boldsymbol{\theta}}^{\mathrm{T}}\tilde{\mathbf{C}}\tilde{\boldsymbol{\theta}}) \left[\sum_{\boldsymbol{k}} (\tilde{\boldsymbol{\theta}}^{\mathrm{T}}\tilde{\mathbf{C}}^2\tilde{\boldsymbol{\theta}})^{|\boldsymbol{k}|}\right]}, \tag{3.D.15}$$

and the minimum error obtained from this solution is

$$J_{\min} = \frac{1}{2}\left[\sum_{\alpha} \|\mathbf{x}^{\alpha}\|^2 - \frac{\tilde{\boldsymbol{\theta}}^{\mathrm{T}}\tilde{\mathbf{C}}^2\tilde{\boldsymbol{\theta}}}{\tilde{\boldsymbol{\theta}}^{\mathrm{T}}\tilde{\mathbf{C}}\tilde{\boldsymbol{\theta}}}\right], \tag{3.D.16}$$

which is obtained when $\tilde{\boldsymbol{\theta}}$ is chosen as the principal eigenvector of the matrix $\tilde{\mathbf{C}}$.

For a general discussion of reconstruction error in nonlinear PCA, see [877].

BIBLIOGRAPHICAL NOTES

Correlation-based learning or generalized Hebbian learning has a long history in the neural computation and computational neuroscience literature (see, e.g., [21, 23, 342, 498, 921, 963]). The biological and physiological roles of Hebb's synapse were reviewed in [122, 472]. Variants of Hebbian learning were also reviewed by [818]. State-of-the-art spike-timing Hebbian synaptic plasticity rules were reviewed in [89]. In contrast to Hebb's synapse, von der Malsburg [922] also proposed a new computational framework that uses temporal information for neuronal coding, which was referred to as Malsburg's synapse by Francis Crick [193].

As instantiations of correlation-based learning, many computational learning algorithms can be unified and categorized. During the 1960s and 1970s, Stephen

Grossberg presented various Hebb-type learning rules as part of his early studies of *embedding fields*, including instar and outstar learning, competitive learning, and self-organizing learning.

The local PCA learning rule was first developed by Oja [676], and the BCM learning rule was originally invented by Bienenstock, Cooper, and Munro [93]. Both of them are arguably biologically plausible. An excellent reference on BCM learning and its application in modeling cortical plasticity is found in [186]. The PCA and MCA were widely used in statistical signal processing applications, such as image compression, noise reduction, curve and surface fitting [982], and beamforming [279]. A unified algorithm was developed in [156, 939] for the PCA and MCA extraction. For textbook discussions of the PCA and MCA learning rules, the reader is referred to [172].

For supervised learning, the LMS learning rule was invented by Widrow and Hoff [951] at Stanford University when they first designed the "Adaline." The generalized delta rule for multilayer network was independently invented by Amari [18], Werbos [948], Parker [703], LeCun [533], and Rumelhart, Hinton, and Williams [780]. The Boltzmann learning and back propagation learning rules were described and popularized in the two-volume PDP (Parallel Distributed Processing) books [781]. The LMS rule was widely studied in adaptive filter theory [369, 376, 953]. Theoretical analysis of the learning mechanism of the LMS filter is given in [655] in light of the law of large numbers.

With reference to reinforcement learning, the TD learning algorithm was first described by Sutton [865] in studying animal classical conditioning. This powerful idea was later generalized to other reinforcement learning paradigms. Excellent resources on reinforcement learning can be found in the textbooks [85, 868].

The earliest reference to the BSS problem is the paper by Jutten and Herault [454], which was motivated by Hebb's postulate of learning. This paper was followed by Comon's ICA paper [180] and that of Bell and Sejnowski [78]. It seems fair to say that in their own individual ways Comon's 1994 paper and the 1995 paper by Bell and Sejnowski have been the major catalysts for the literature in ICA theory, algorithms, and novel applications. Subsequently, the research in BSS and ICA has been extended in various applications. In fact, the literature is so large and diverse that in the course of 10 years ICA has established itself as an indispensable part of the ever-expanding discipline of statistical signal processing and has had a great impact on neuroscience. For textbook treatment of the BSS and ICA theory, see [172, 428]. Information-theoretic learning has been well discussed in the literature [207, 877]. A textbook introduction of information-theoretic learning can be found in [263]. Reviews of unsupervised learning in the context of information-theoretic learning are given in [69, 75, 76, 877]. An excellent source of discussions for Hebbian learning and negative-feedback networks is found in [303].

The role of correlation for associative memory has deep roots in the literature. The learning-matrix network was first invented in 1958 by Karl Steinbuch [851], which uses a binary version of Hebb's rule (i.e., Boolean Hebbian learning) to form associations between pairs of binary patterns; this was elaborated in his classic book *Automat und Mensche* [852]. The storage capacity of the learning-matrix network was later studied by Willshaw, Buneman, and Longuet-Higgins [961] (see also [378]), which further stimulated studies of associative memory [20, 33, 34, 185, 496, 651]. In the 1980s, the associative memory model was extended by the so-called additive model or discrete Hopfield network [178, 402].

Correlation-based learning has also had great success in modeling self-organizing feature maps. Good resources on the SOM can be found in [499, 674, 762]. Detailed studies of correlation-based learning for forming visual maps can be found in [624–626].

Good resources on the binding problem may be found in the review articles of Singer [836, 837] and von der Malsburg [924, 926] and the special issue of the 1999 *Neuron* that includes the most complete bibliographies. The idea of temporal synchrony and oscillatory correlation was originated by von der Malsburg [922], followed by a variety of publications in auditory and visual perception [927, 928, 935]. The neural oscillator model [937] is an extension of such an idea to computational auditory scene analysis.

NOTES

1. An FIR filter whose impulse response has coefficients equal to the elements of an eigenvector is called an *eigenfilter* [369]. The maximum eigenfilter refers to the one associated with the largest eigenvalue of the correlation matrix of the signal component in the filter input; the maximum eigenfilter is an optimum filter in that it produces the maximum SNR at the filter output.
2. For simplicity, we confine our discussion to the single-factor analysis model; however, the single-factor analysis model can be easily extended to multiple factors.
3. For further discussion on the relationship between temporal Hebbian learning and the theory of classic conditioning, attention, and gated dipole theory, the interested reader is referred to the book by Levine [546].
4. In classic conditioning animal learning experiments, the input $\mathbf{x}(t) = [x(t), x(t-1), \ldots, x(t-\tau)]^{\mathrm{T}}$ can be viewed as a vector of binary variables, with each of its components representing the presence or absence of a given stimulus at a specific time.
5. For more neurobiological background and discussion on neuronal coding of prediction errors and rewards, the reader is referred to the review articles [202, 805, 806]; see also [650] for discussion on dopamine neurons representing context-dependent prediction error.
6. The mutual information between two discrete random variables X and Y is defined as

$$I(X, Y) = \sum_{x \in X} \sum_{y \in Y} p(x, y) \log\left(\frac{p(x, y)}{p(x)p(y)}\right).$$

If X and Y are continuous random variables, the mutual information is defined as

$$I(X, Y) = \int_X \int_Y p(x, y) \log\left(\frac{p(x, y)}{p(x)p(y)}\right) dx\, dy.$$

Another useful quantitative information measure between two random variables is the so-called *mean-square contingency*, which is defined as

$$\begin{aligned} C(X,Y)^2 &= \int_X \int_Y \left(\frac{p(x,y)}{p(x)p(y)} - 1 \right)^2 p(x)p(y)\,dx\,dy \\ &= \int_X \int_Y \left(\frac{p(x,y)}{p(x)p(y)} - 1 \right) p(x,y)\,dx\,dy. \end{aligned}$$

It can be proved that the mean-square contingency is lower bounded by mutual information in that $I(X,Y) \leq C(X,Y)^2$ [which is obvious from the fact that $\log(z-1) \leq z$].

7. The equation $\mathbf{y} = \mathbf{W}\mathbf{x}$ can be realized by either a feedforward linear neural network with connection weight matrix $\mathbf{W}$ or a fully recurrent neural network described as $\mathbf{y} = \mathbf{x} - \mathbf{V}\mathbf{y}$ with feedback connection matrix $\mathbf{V}$; these two linear networks are equivalent when $\mathbf{W} = (\mathbf{I} + \mathbf{V})^{-1}$, where $\mathbf{I}$ denotes the identity matrix.
8. Higher order statistics are often characterized by moment or cumulant statistics. The second-, third-, and fourth-order cumulants for a zero-mean random vector $\mathbf{x}$ are defined by

$$\begin{aligned} \mathrm{cum}(x_i, x_j) &= \mathbb{E}[x_i x_j], \\ \mathrm{cum}(x_i, x_j, x_k) &= \mathbb{E}[x_i x_j x_k], \\ \mathrm{cum}(x_i, x_j, x_k, x_l) &= \mathbb{E}[x_i x_j x_k x_l] - \mathbb{E}[x_i x_j]\mathbb{E}[x_k x_l] \\ &\quad - \mathbb{E}[x_i x_k]\mathbb{E}[x_j x_l] - \mathbb{E}[x_i x_l]\mathbb{E}[x_j x_k]. \end{aligned}$$

   Specifically, the second-order cumulant is equal to the second-order moment $\mathbb{E}[x_i x_j]$, which is defined as the correlation between the variables x_i and x_j; similarly, the third-order cumulant is equal to the third-order moment; however, the fourth-order cumulant differs from the fourth-order moment $\mathbb{E}[x_i x_j x_k x_l]$, which merely specifies a fourth-order correlation.
9. In the context of ICA, let $\mathbf{x} = \mathbf{A}\mathbf{s}$ and $\mathbf{y} = \mathbf{W}\mathbf{x}$ denote, respectively, the mixing and unmixing equations; Comon [180] suggested to maximize an objective function, which has the *contrast property* that it is maximized if and only if its argument $\mathbf{W}$ equals to $\mathbf{A}^{-1}$ up to a left multiplication by a diagonal matrix and a permutation matrix. In the deflation approach of ICA, sources are extracted sequentially one by one instead of by simultaneous extraction; in such a case, non-Gaussianity indices are often used as deflation objective functions, which possess the contrast property that the objective function is maximized if and only if its argument is proportional to the source with the highest non-Gaussianity index among those sources not being extracted. The use of the contrast function for blind separation or deconvolution has been discussed in [145–147, 181, 635, 721].
10. This criterion has also been proven equivalent to several other criteria such as the minimum mutual information (MMI) and maximum likelihood [144, 986].
11. A necessary condition of convergence for the vector $\mathbf{v}$ is that all eigenvalues of the matrix $\mathbf{F}$ are between -1 and 1.
12. As a generalization of ICA, topographic ICA [427] allows higher order dependency (such as correlation of energy) between the components. In topographic ICA, the residual dependence structure is used to define a topographic order for the separated components; specifically, a distance between two components may be defined using their higher order correlations, and the distance is used to create a topographic representation.

13. In Bell and Sejnowski's work [79], only gray-scale images are considered; however, the concept can also be generalized to color and stereo images [410].
14. In light of their different properties (e.g., [727]), neuronal synchrony may be loosely categorized into two classes: (i) *oscillatory (or supercritical, superthreshold) synchronization*, which was first proposed by von der Malsburg [922], mostly refers to idealized (periodic) oscillatory activities for population neurons (for which each neuron is oscillatory); and (ii) *excitable (or subcritical, subthreshold) synchronization*, which often refers to the realistic neuronal excitability within single neurons [434].
15. The cocktail party problem, first described by Colin Cherry [166], is a psychoacoustic phenomenon that refers to the remarkable human ability to selectively attend to and recognize one source of auditory input in a noisy environment, where the hearing interference is produced by competing speech sounds or various other noise sources that are often assumed to be mutually independent; the machine cocktail party problem, on the other hand, refers to the problem of designing a machine that imitates the human's capability in a similar context, with the tools of machine learning and signal processing. See [372, 373] for more detailed discussions.
16. The gamma tone filter bank is often used to simulate the basilar membrane in the auditory system; the bandwidth of the filters are set by a psychoacoustically determined critical band function (defined by the masking properties of the human auditory system) such that the filter bandwidth increases with the center frequency. In the literature, the gamma tone auditory filter impulse response is typically described by

    $$\gamma(t) = at^{n-1}e^{-2\pi bt}\cos(2\pi f_c t + \phi) \qquad (t > 0),$$

    where n denotes the order of the filter, a and b are two constant coefficients, f_c denotes the center frequency, and ϕ denotes the phase shift.
17. Note that the window length has to be longer than the fundamental period of the estimated pitch. The fundamental frequency of an adult speech signal varies from 85 to 255 Hz.
18. where K stands for "Katchalsky", named after Aharon Katchalsky, a pioneer of neurodynamics, who studied the collective behavior of neurons.
19. Alan Turing [895] first proposed the idea that "global order can arise from local interactions." Specifically, Turing showed how order patterns such as a leopard's spots may arise spontaneously from random noise by applying a simple and local rule. Turing ran the simulations on one of the first electronic computers at the University of Manchester to generate spots, dapples, and stripelike patterns.
20. A problem is assigned to the NP (nondeterministic polynomial time) class if it is solvable in polynomial time by a nondeterministic Turing machine. A problem is NP hard if an algorithm for solving it can be translated into one for solving any NP problem. Therefore NP hard means "at least as hard as any NP problem," although it might, in fact, be harder.
21. One way to prevent the instability or divergence of Hebbian learning is to impose a constraint on the synaptic weights, such as the unity norm.

4

CORRELATION-BASED KERNEL LEARNING

4.1 BACKGROUND

In the past decade, kernel learning [799] has produced a revolutionary perspective and generated enormous interests in the machine learning community. Representative examples of successful kernel learning methods include the support vector machine (SVM) and kernel PCA (KPCA) [800]. By virtue of using the so-called kernel trick, researchers can readily extend conventional linear learning methods to kernel-based *nonlinear* methods. This is done by projecting the data to a high- or even infinite-dimensional feature space (with the mapping $\phi : X \mapsto \mathcal{F}$), whereas the inner product of the feature space is induced by a positive-definite kernel.

Definition 4.1 *A Hilbert space*[1] *of functions on a set* $\mathcal{X}$ *is said to be a reproducing kernel Hilbert space (RKHS) if there is a kernel function* $K(\mathbf{x}, \mathbf{x}')$ *defined on* $\mathcal{X} \times \chi$ *having the following properties:*

- *For each* $\mathbf{x}' \in \mathcal{X}$, $K(\mathbf{x}, \mathbf{x}')$ *is a function in Hilbert space.*
- *For each* f *in Hilbert space and* $\mathbf{x}'$ *in* X, *it holds that* $\langle f, K(\cdot, \mathbf{x}')\rangle = f(\mathbf{x}')$.

The kernel function $K(\mathbf{x}, \mathbf{x}')$ *that satisfies such conditions is called a reproducing kernel in the Hilbert space.*

Correlative Learning: A Basis for Brain and Adaptive Systems, by Zhe Chen, Simon Haykin, Jos J. Eggermont, and Suzanna Becker

For every positive-definite kernel function K on $\mathcal{X} \times \mathcal{X}$, it is known [45, 930] that there is a unique RKHS on $\mathcal{X}$ with K as its reproducing kernel.

The basic idea of kernel learning is to construct a kernel that measures the *similarity* or *distance* between pairwise variables [798]; once the kernel is chosen, the feature space is automatically determined. Specifically, the kernel defines the inner product between pairs of data points in the feature space in accordance with

$$K(\mathbf{x}_i, \mathbf{x}_j) = \langle K(\cdot, \mathbf{x}_i), K(\cdot, \mathbf{x}_j)\rangle = \langle \phi(\mathbf{x}_i), \phi(\mathbf{x}_j)\rangle, \tag{4.1}$$

where $\phi(\mathbf{x}) = K(\cdot, \mathbf{x})$ denotes the nonlinear mapping from the input space into the RKHS. Equation (4.1) is often referred to as the "kernel trick." In contrast to second-order similarity measures such as the correlation coefficient or degree of angle [defined by (C.4) in Appendix C], the kernel function implicitly takes into account higher order interactions among the random variables because of the nonlinear nature of ϕ (see Figure 4.1 for an illustration). Given ℓ data points, we can therefore construct an $\ell \times \ell$ kernel matrix (or Gram matrix): $\mathbf{K} = \{K_{ij}\} = \{K(\mathbf{x}_i, \mathbf{x}_j)\}$.

In addition, with proper normalization assumptions, the inner product (or correlation) can be viewed as a special form of pairwise distance measure. For instance, in

$$C = \begin{pmatrix} 1.0000 & 0.7352 & 0.4429 & 0.7057 \\ 0.7352 & 1.0000 & 0.4347 & 0.6298 \\ 0.4429 & 0.4347 & 1.0000 & 0.3824 \\ 0.7057 & 0.6298 & 0.3824 & 1.0000 \end{pmatrix} \qquad K = \begin{pmatrix} 1.0000 & 0.2236 & 0.0746 & 0.2161 \\ 0.2236 & 1.0000 & 0.0709 & 0.1330 \\ 0.0746 & 0.0709 & 1.0000 & 0.0672 \\ 0.2161 & 0.1330 & 0.0672 & 1.0000 \end{pmatrix}$$

$$\cos\angle(X_i, X_j) = \frac{\langle X_i, X_j\rangle}{\|X\|_i \cdot \|X_j\|}$$

$$k(X_i, X_j) = \exp\left(\frac{-\|X_i - X_j\|^2}{2\sigma^2}\right) = \frac{\exp(X_i \cdot X_j/\sigma^2)}{\exp(\|X_i\|^2/2\sigma^2)\exp(\|X_j\|^2/2\sigma^2)}$$

$$= \frac{\sum_{k=0}^{\infty}\left(\frac{1}{k!}\right)(X_i/\sigma \cdot X_j/\sigma)^k}{\exp(\|X_i\|^2/2\sigma^2)\exp(\|X_j\|^2/2\sigma^2)} = \langle \phi(X_i), \phi(X_j)\rangle$$

Figure 4.1 Illustration of two similarity measures. The top row shows the face images of the four coauthors of this book. In the bottom, the matrix **C** for cosine angle and the normalized Gaussian kernel matrix **K** ($\sigma = 30$) are shown, which correspond to the similarity measures in the original data space ($\mathbb{R}^{90\times 90}$) and infinite-dimensional feature space, respectively. The feature map of the Gaussian kernel can be expanded in this case.

the original input data space, the expected Euclidean distance can be represented by

$$\begin{aligned}\mathbb{E}[\|\mathbf{x}_i - \mathbf{x}_j\|^2] &= \mathbb{E}[\|\mathbf{x}_i\|^2] + \mathbb{E}[\|\mathbf{x}_j\|^2] - 2\mathbb{E}[\mathbf{x}_i^{\mathrm{T}}\mathbf{x}_j] \\ &= \text{const} - 2\langle \mathbf{x}_i, \mathbf{x}_j \rangle, \end{aligned} \tag{4.2}$$

where the last term denotes the *negative* inner product between $\mathbf{x}_i$ and $\mathbf{x}_j$. Accordingly, in the feature space, it can be shown that

$$\begin{aligned}\left[\phi(\mathbf{x}_i) - \phi(\mathbf{x}_j)\right]^2 &= \langle \phi(\mathbf{x}_i), \phi(\mathbf{x}_i) \rangle + \langle \phi(\mathbf{x}_j), \phi(\mathbf{x}_j) \rangle - 2\langle \phi(\mathbf{x}_i), \phi(\mathbf{x}_j) \rangle \\ &= K(\mathbf{x}_i, \mathbf{x}_i) + K(\mathbf{x}_j, \mathbf{x}_j) - 2K(\mathbf{x}_i, \mathbf{x}_j), \end{aligned} \tag{4.3}$$

in which the distance or correlation can be calculated efficiently by the kernel function. In fact, equation (4.3) defines the RKHS norm induced by the kernel K, namely, $\|\mathbf{x}_i - \mathbf{x}_j\|_K$.

An important class of kernel functions is the so-called Mercer kernel (e.g., [799]).

Definition 4.2 *Let* $K \in L_2(\mathcal{X}^2)$ *be a symmetric real-valued function such that the integral operator*

$$\mathbf{T}_K : L_2(\mathcal{X}) \to L_2(\mathcal{X})$$

$$(\mathbf{T}_K)(\mathbf{x}) = \int_{\mathcal{X}} K(\mathbf{x}, \mathbf{x}') f(\mathbf{x}')\, d\mu(\mathbf{x}')$$

is positive definite; that is, for all $f(\mathbf{x}) \in L_2(\mathcal{X})$ *(i.e., the square integrable function), we have*

$$\int_{\mathcal{X}^2} K(\mathbf{x}, \mathbf{x}') f(\mathbf{x}) f(\mathbf{x}')\, d\mu(\mathbf{x})\, d\mu(\mathbf{x}') \geq 0.$$

A kernel that satisfies Mercer's condition is called the Mercer kernel or "admissible" kernel.

Two of the most popular Mercer kernels are:

- The polynomial kernel [730]: $K(\mathbf{x}, \mathbf{y}) = (r + \mathbf{x} \cdot \mathbf{y})^d$, where $r > 0$, $d \in \mathbb{N}$.
- The translation-invariant kernel [930]: $K(\mathbf{x}, \mathbf{y}) = K(\mathbf{x} - \mathbf{y})$. In the case of a Gaussian kernel $K(\mathbf{x}, \mathbf{y}) = \exp(-\lambda\|\mathbf{x} - \mathbf{y}\|^2)$ (where $\lambda > 0$), its feature space $\mathcal{F}$ has an infinite dimension and the RKHS can be described by the Fourier theory.[2]

There are many ways to construct new kernel functions. For instance, any convex combination of Mercer kernels is also a Mercer kernel. We can also design atypical kernel functions (such as the locally stationary kernel, nonstationary kernel, or reducible kernel) according to the specific problem under study; see [314, 827] for a detailed discussion of this issue.

4.2 KERNEL PCA AND KERNELIZED GHA

In a similar way to linear PCA, KPCA aims to solve the eigenvalue equation

$$\lambda \mathbf{v} = \mathbf{C}\mathbf{v}, \tag{4.4}$$

where λ and $\mathbf{v}$ are respectively the eigenvalues and eigenvectors of the (positive-semidefinite) covariance matrix $\mathbf{C}$, which is defined for the samples $\{\mathbf{x}_1, \ldots, \mathbf{x}_\ell\}$ in the feature space as

$$\mathbf{C} = \frac{1}{\ell}\sum_{i=1}^{\ell} \phi(\mathbf{x}_i)\phi^{\mathrm{T}}(\mathbf{x}_i), \tag{4.5}$$

where we have assumed the features are centered such that $\sum_{i=1}^{\ell} \phi(\mathbf{x}_i) = 0$; in other words, $\mathbf{C}$ is also a correlation matrix. Note that the matrix $\mathbf{C}$ is defined through the outer product instead of the inner product of the samples. Using the kernel trick [800], we can reformulate the problem to obtain a representation of $\mathbf{v}$ in terms of $\phi(\mathbf{x}_i)$.

Specifically, substituting (4.5) into (4.4) yields

$$\lambda \mathbf{v} = \mathbf{C}\mathbf{v} = \frac{1}{\ell}\sum_{i=1}^{\ell} \phi(\mathbf{x}_i)\phi^{\mathrm{T}}(\mathbf{x}_i)\mathbf{v}, \tag{4.6}$$

which indicates that the eigenvectors can be constructed as a linear combination of the input vectors in the feature space:

$$\mathbf{v} = \sum_{i=1}^{\ell} \phi(\mathbf{x}_i)\alpha_i = \boldsymbol{\Phi}^{\mathrm{T}}\boldsymbol{\alpha}, \tag{4.7}$$

where $\boldsymbol{\alpha}$ is a column vector with the ith component defined by $\alpha_i = \phi^{\mathrm{T}}(\mathbf{x}_i)\mathbf{v}/(\lambda\ell)$. All solutions $\mathbf{v}$ to (4.6) or (4.7) lie in the subspace spanned by all of the training samples in the feature space. In light of (4.6) and (4.7), we can solve the alternative eigenvalue equation

$$\lambda \boldsymbol{\Phi}^{\mathrm{T}}\boldsymbol{\alpha} = \frac{1}{\ell}\boldsymbol{\Phi}^{\mathrm{T}}\boldsymbol{\Phi}\boldsymbol{\Phi}^{\mathrm{T}}\boldsymbol{\alpha}. \tag{4.8}$$

Multiplying both sides of (4.8) by $(\boldsymbol{\Phi}\boldsymbol{\Phi}^{\mathrm{T}})^{-1}\boldsymbol{\Phi}$ (i.e., the pseudoinverse of $\boldsymbol{\Phi}^{\mathrm{T}}$) yields

$$\ell\lambda(\boldsymbol{\Phi}\boldsymbol{\Phi}^{\mathrm{T}})^{-1}\boldsymbol{\Phi}\boldsymbol{\Phi}^{\mathrm{T}}\boldsymbol{\alpha} = (\boldsymbol{\Phi}\boldsymbol{\Phi}^{\mathrm{T}})^{-1}\boldsymbol{\Phi}\boldsymbol{\Phi}^{\mathrm{T}}\boldsymbol{\Phi}\boldsymbol{\Phi}^{\mathrm{T}}\boldsymbol{\alpha}, \tag{4.9}$$

which can be further simplified to

$$\ell\lambda\boldsymbol{\alpha} = \mathbf{K}\boldsymbol{\alpha}, \tag{4.10}$$

which is essentially the eigenvalue equation for the kernel matrix $\mathbf{K}$ with $\mathbf{K}_{ij} = K(\mathbf{x}_i, \mathbf{x}_j)$; the coefficient vector $\boldsymbol{\alpha}$ plays the role of the eigenvector of the kernel matrix $\mathbf{K}$ associated with the eigenvalue $\ell\lambda$, which also contains the expansion coefficients of the eigenvector $\mathbf{v}$ of the covariance matrix $\mathbf{C}$.

As the eigenvalue equation is solved for $\boldsymbol{\alpha}^j$ instead of $\mathbf{v}^j$, we normalize the $\boldsymbol{\alpha}^j$ by $\boldsymbol{\alpha}^j \leftarrow \boldsymbol{\alpha}^j/\sqrt{\lambda_j}$ to assure that the eigenvalues $\mathbf{v}^j$ have unity norm in the feature space, that is, the inner product $(\boldsymbol{\alpha}^j \cdot \boldsymbol{\alpha}^j) = 1$. Therefore, the expansion of any vector $\phi(\mathbf{x})$ in the feature space can be calculated via the kernel:

$$\langle v^j, \phi(\mathbf{x})\rangle = \sum_{i=1}^{\ell} \alpha_i^j \phi^{\mathrm{T}}(\mathbf{x}_i)\phi(\mathbf{x}) = \sum_{i=1}^{\ell} \alpha_i^j K(\mathbf{x}_i, \mathbf{x}), \qquad j = 1, 2, \ldots, m,$$

where m denotes the number of nonzero eigenvalues.

For a testing point $\mathbf{x}'$, its principal component is obtained from computing its high-dimensional feature [i.e., $\phi(\mathbf{x}')$] projections onto the eigenvectors

$$\phi(\mathbf{x}') \cdot \mathbf{v} = \sum_{i=1}^{\ell} \alpha_i \tilde{K}(\mathbf{x}', \mathbf{x}_i), = \tilde{\mathbf{K}}\boldsymbol{\alpha}$$

where $\tilde{\mathbf{K}}$ is the centered version of the new kernel matrix $\mathbf{K}$.[3]

It is clear that KPCA requires solving an EVD problem of size $\ell \times \ell$. To perform feature extraction for a new sample, the optimal feature extractor will be expanded in terms of all training samples in the feature kernel space. In practice, the efficiency of such feature extraction might be low when the number of training samples, ℓ, is extremely large. To overcome this problem, it is possible to construct a reduced set $\{\mathbf{x}'\}_{i=1}^{s}$ (where $s < \ell$) from the complete training set and use this subset for feature extraction. As shown in [983], this is equivalent to solving a generalized eigenvalue problem:

$$\frac{1}{\ell}\mathbf{K}_1\mathbf{K}_1^{\mathrm{T}}\boldsymbol{\beta} = \lambda\mathbf{K}_2\boldsymbol{\beta},$$

where $\boldsymbol{\beta}$ plays the role of the new eigenvector and $\mathbf{K}_1$ and $\mathbf{K}_2$ are two kernel matrices with sizes $s \times \ell$ and $s \times s$, defined respectively as follows:

$$\mathbf{K}_1 = \begin{bmatrix} K(\mathbf{x}_1', \mathbf{x}_1) & K(\mathbf{x}_1', \mathbf{x}_2) & \cdots & K(\mathbf{x}_1', \mathbf{x}_\ell) \\ K(\mathbf{x}_2', \mathbf{x}_1) & K(\mathbf{x}_2', \mathbf{x}_2) & \cdots & K(\mathbf{x}_2', \mathbf{x}_\ell) \\ \vdots & \vdots & & \vdots \\ K(\mathbf{x}_s', \mathbf{x}_1) & K(\mathbf{x}_s', \mathbf{x}_2) & \cdots & K(\mathbf{x}_s', \mathbf{x}_\ell) \end{bmatrix},$$

$$\mathbf{K}_2 = \begin{bmatrix} K(\mathbf{x}_1', \mathbf{x}_1') & K(\mathbf{x}_1', \mathbf{x}_2') & \cdots & K(\mathbf{x}_1', \mathbf{x}_s') \\ K(\mathbf{x}_2', \mathbf{x}_1') & K(\mathbf{x}_2', \mathbf{x}_2') & \cdots & K(\mathbf{x}_2', \mathbf{x}_s') \\ \vdots & \vdots & & \vdots \\ K(\mathbf{x}_s', \mathbf{x}_1') & K(\mathbf{x}_s', \mathbf{x}_2') & \cdots & K(\mathbf{x}_s', \mathbf{x}_s') \end{bmatrix}.$$

■ EXAMPLE 4.1

For the purpose of demonstration, in this example we test and compare the Linear and Kernel PCA approaches for real-life handwritten digits. A small subset of the U.S. Postal Service (USPS) database that consists of 300 handwritten digit images of the number 3 was used to compute the eigenvectors in the linear and kernel spaces. Each example digit 3 is a 16×16 gray-scale image; all of the data points are scaled to lie within the region [0, 1]. For KPCA, two types of kernel functions are considered in the experiment. The first one is a third-order polynomial kernel

$$K(\mathbf{x}, \mathbf{x}_i) = (1 + \mathbf{x}^{\mathrm{T}}\mathbf{x}_i)^3$$

and the second one is an isotropic Gaussian kernel

$$K(\mathbf{x}, \mathbf{x}_i) = \exp\left(-\frac{1}{8}\|\mathbf{x} - \mathbf{x}_i\|^2\right).$$

It is noteworthy that the number of eigenvectors in linear PCA is limited by the dimensionality of each data point (here, $N = 256$), whereas in KPCA it has up to 300 eigenvectors (equal to the number of training samples); this allows KPCA to have more choices in feature extraction and representation. For the purpose of visualization, we have also reconstructed the input space from the kernel eigenvectors with the "preimage" method described in [799], as shown in the second and third rows of Figure 4.2. By comparison, the kernelized eigenmaps are better in characterizing the local features of the digit 3 than the linear eigenmaps; it also seems that the Gaussian kernel performs slightly better than the polynomial kernel in this task.

Note that the above formulation of KPCA is an offline method, which might involve a large-scale ($\ell \times \ell$) matrix decomposition operation. It would be appealing to develop an online method for extracting kernel-based principal components. Motivated by Sanger's online GHA for linear PCA, Kim et al. [479] developed an iterative Hebbian learning rule for KPCA. Specifically, in a manner consistent with the GHA notation, the kernelized GHA (KGHA) is written as

$$\mathbf{W}(t+1) = \mathbf{W}(t) + \eta\left\{\mathbf{y}(t)\Phi^{\mathrm{T}}(\mathbf{x}(t)) - \mathrm{LT}[\mathbf{y}(t)\mathbf{y}^{\mathrm{T}}(t)]\mathbf{W}(t)\right\}, \tag{4.11}$$

where $\mathbf{y}(t) = \mathbf{W}(t)\Phi(\mathbf{x}(t))$ and $\Phi(\cdot)$ is a (high-dimensional) mapping function in the feature space. Here it is assumed that for each index i there exists a function $\mathcal{I}(t)$ that maps t to the index set $i \in \{1, \ldots, \ell\}$ such that $\Phi(\mathbf{x}(t)) \equiv \Phi(\mathbf{x}(\mathcal{I}(t))) = \Phi(\mathbf{x}_i)$.

In light of KPCA, it is known that the row vectors of $\mathbf{W}(t)$, denoted by $\{\boldsymbol{\theta}_i(t)\}$, can be expanded in terms of the mapped data points $\Phi(\mathbf{x}_i)$ ($i = 1, 2, \ldots, \ell$). Therefore, $\mathbf{W}(t)$ can be represented via the linear combination of $\Phi(\mathbf{x}_i)$:

$$\mathbf{W}(t) = \mathbf{A}(t)\boldsymbol{\Phi}, \tag{4.12}$$

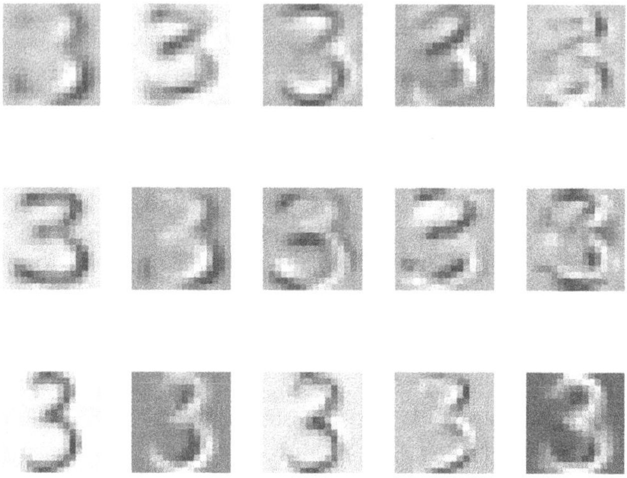

Figure 4.2 Visualization of the eigenvectors or "preimage" patterns calculated from the subset of the USPS handwritten digit 3. *Top row*: the eigenvectors obtained from linear PCA. *Middle and bottom rows*: the preimage patterns obtained from kernel PCA reconstruction using a third-order polynomial kernel (middle row) $K(\mathbf{x}, \mathbf{x}_i) = (1 + \mathbf{x}^{\mathrm{T}}\mathbf{x}_i)^3$ and a Gaussian kernel (bottom row) $K(\mathbf{x}, \mathbf{x}_i) = \exp(-\|\mathbf{x} - \mathbf{x}_i\|^2/8)$. In all cases, the five columns (from left to right) correspond to the associated (1, 2, 4, 8, 16)th eigenvectors.

where $\mathbf{A}(t) = [\mathbf{a}_1^{\mathrm{T}}(t), \ldots, \mathbf{a}_l^{\mathrm{T}}(t)]^{\mathrm{T}}$ is an $\ell \times \ell$ matrix that contains expansion coefficients in the row vectors. Specifically, the ith row vector $\mathbf{a}_i = [a_{i1}, \ldots, a_{i\ell}]$ of $\mathbf{A}(t)$ contains the expansion coefficients of the ith eigenvector of the kernel matrix $\mathbf{K}$, namely,

$$\boldsymbol{\theta}_i(t) = \boldsymbol{\Phi}^{\mathrm{T}}\mathbf{a}_i(t). \tag{4.13}$$

Using the dual representation, the learning rule (4.11) can be reformulated as

$$\mathbf{A}(t+1)\boldsymbol{\Phi} = \mathbf{A}(t)\boldsymbol{\Phi} + \eta\left\{\mathbf{y}(t)\Phi^{\mathrm{T}}(\mathbf{x}(t)) - \mathrm{LT}[\mathbf{y}(t)\mathbf{y}^{\mathrm{T}}(t)]\mathbf{A}(t)\boldsymbol{\Phi}\right\}. \tag{4.14}$$

By introducing a canonical unit ℓ-length column vector $\mathbf{b}(t) = [0, \ldots, 1, \ldots, 0]^{\mathrm{T}}$ [with only the $\mathcal{I}(t)$th element as 1] and by representing the mapped data points as $\Phi(\mathbf{x}(t)) = \boldsymbol{\Phi}^{\mathrm{T}}\mathbf{b}(t)$, the learning rule (4.14) can be written in terms of expansion coefficients as

$$\mathbf{A}(t+1) = \mathbf{A}(t) + \eta\left\{\mathbf{y}(t)\mathbf{b}^{\mathrm{T}}(t) - \mathrm{LT}[\mathbf{y}(t)\mathbf{y}^{\mathrm{T}}(t)]\mathbf{A}(t)\right\}. \tag{4.15}$$

Written in componentwise form, (4.15) is represented as

$$a_{ij}(t+1) = \begin{cases} a_{ij}(t) + \eta y_i(t) - \eta y_i(t)\sum_{k=1}^{i} a_{kj}(t)y_k(t) & \text{if } \mathcal{I}(t) = j, \\ a_{ij}(t) - \eta y_i(t)\sum_{k=1}^{i} a_{kj}(t)y_k(t) & \text{otherwise,} \end{cases} \tag{4.16}$$

where $y_i(t)$ is computed by the kernel matrix followed by the centering operation; that is,

$$y_i(t) = \sum_{i=1}^{\ell} a_{ik}(t)\left[K(\mathbf{x}(t), \mathbf{x}_k) - \overline{K}(\mathbf{x}_k)\right] - \overline{a}_i(t)\sum_{k=1}^{\ell}\left[K(\mathbf{x}(t), \mathbf{x}_k) - \overline{K}(\mathbf{x}_k)\right],$$

with

$$\overline{K}(\mathbf{x}_k) = \frac{1}{\ell}\sum_{m=1}^{\ell} K(\mathbf{x}_m, \mathbf{x}_k), \qquad \overline{a}_i(t) = \frac{1}{\ell}\sum_{m=1}^{\ell} a_{im}(t).$$

In [479], the power of the KGHA was demonstrated in image compression and denoising. Compared to the batch KPCA, the kernelized Hebbian PCA learning algorithm offers advantages in terms of computation and memory efficiency.

As a demonstration, we apply the KGHA to a toy example in which 200 two-dimensional data samples ($\mathbf{x} = [x_1, x_2]^{\mathrm{T}}$) are generated from a nonlinear mapping

$$x_2 = x_1^3 - x_1 + \xi,$$

where x_1 is uniformly distributed in $[-1, 1]$ and ξ denotes additive Gaussian noise with zero mean and variance 0.01. The goal of this task is to extract principal components from the noisy data. In comparison, we also apply KPCA to the same data set. A polynomial kernel with degree 2 was used in the experiment for both algorithms. The experimental results are illustrated in Figure 4.3. As seen from the figure, the results obtained from these two algorithms are almost identical.

4.3 KERNEL CCA AND KERNEL ICA

In a way similar to extending PCA to KPCA, CCA can also be extended to kernel CCA (KCCA). Given two sets of random variables $\{\mathbf{x}_i\}_{i=1}^{\ell} \in \mathbb{R}^p$ and $\{\mathbf{y}_i\}_{i=1}^{\ell} \in \mathbb{R}^q$, KCCA seeks to explore canonical correlation in the high-dimensional feature space. Recalling the formulation of linear CCA in Chapter 2, the conventional correlation matrices are defined in terms of the outer product: $\mathbf{C}_{xx} = \mathbf{X}\mathbf{X}^{\mathrm{T}}$ and $\mathbf{C}_{xy} = \mathbf{X}\mathbf{Y}^{\mathrm{T}}$. By using the kernel trick as in KPCA, we may define the kernel matrices in terms of inner products:

$$\mathbf{K_x} = \Phi(\mathbf{X})^{\mathrm{T}}\Phi(\mathbf{X}), \tag{4.17}$$

$$\mathbf{K_y} = \Phi(\mathbf{Y})^{\mathrm{T}}\Phi(\mathbf{Y}), \tag{4.18}$$

both of which are of size $\ell \times \ell$.

Without going into full mathematical derivation details, it can be shown [52] that KCCA essentially amounts to solving the generalized eigenvalue problem

$$\begin{pmatrix} \mathbf{0} & \mathbf{K_x K_y} \\ \mathbf{K_y K_x} & \mathbf{0} \end{pmatrix}\begin{pmatrix} \xi_1 \\ \xi_2 \end{pmatrix} = \rho \begin{pmatrix} \mathbf{K_x K_x} & \mathbf{0} \\ \mathbf{0} & \mathbf{K_y K_y} \end{pmatrix}\begin{pmatrix} \xi_1 \\ \xi_2 \end{pmatrix}, \tag{4.19}$$

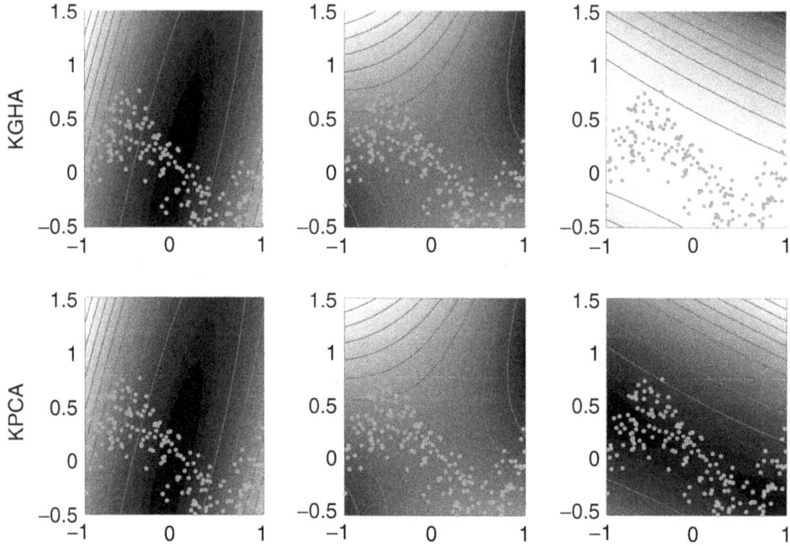

Figure 4.3 Comparison between KGHA and KPCA in learning the principal components of the two-dimensional data samples (shown in red dots). From left to right, the panels show the first three learned principal components visualized with blue contour lines. The KGHA results were obtained after 3000 iterations with a constant learning rate 0.005.

with $\boldsymbol{\xi}_1, \boldsymbol{\xi}_2 \in \mathbb{R}^\ell$; and the canonical correlation of the KCCA can also be defined as

$$\rho = \max_{\boldsymbol{\xi}_1, \boldsymbol{\xi}_2} \frac{\boldsymbol{\xi}_1^{\mathrm{T}} \mathbf{K}_{\mathrm{x}} \mathbf{K}_{\mathrm{y}} \boldsymbol{\xi}_2}{(\boldsymbol{\xi}_1^{\mathrm{T}} \mathbf{K}_{\mathrm{x}} \mathbf{K}_{\mathrm{x}} \boldsymbol{\xi}_1)^{1/2} (\boldsymbol{\xi}_2^{\mathrm{T}} \mathbf{K}_{\mathrm{y}} \mathbf{K}_{\mathrm{y}} \boldsymbol{\xi}_2)^{1/2}}. \tag{4.20}$$

Likewise, KCCA can be generalized for more than two variables. Specifically, given m pairs of multivariate random variables $\{\mathbf{x}_1, \ldots, \mathbf{x}_m\}$, the generalized eigenvalue problem can be written as [52]

$$\begin{pmatrix} \mathbf{K}_1\mathbf{K}_1 & \mathbf{K}_1\mathbf{K}_2 & \cdots & \mathbf{K}_1\mathbf{K}_m \\ \mathbf{K}_2\mathbf{K}_1 & \mathbf{K}_2\mathbf{K}_2 & \cdots & \mathbf{K}_2\mathbf{K}_m \\ \vdots & \vdots & \ddots & \vdots \\ \mathbf{K}_m\mathbf{K}_1 & \mathbf{K}_m\mathbf{K}_2 & \cdots & \mathbf{K}_m\mathbf{K}_m \end{pmatrix} \begin{pmatrix} \boldsymbol{\xi}_1 \\ \boldsymbol{\xi}_2 \\ \vdots \\ \boldsymbol{\xi}_m \end{pmatrix}$$
$$= \lambda \begin{pmatrix} \mathbf{K}_1\mathbf{K}_1 & \mathbf{0} & \cdots & \mathbf{0} \\ \mathbf{0} & \mathbf{K}_2\mathbf{K}_2 & \cdots & \mathbf{0} \\ \vdots & \vdots & \ddots & \vdots \\ \mathbf{0} & \mathbf{0} & \cdots & \mathbf{K}_m\mathbf{K}_m \end{pmatrix} \begin{pmatrix} \boldsymbol{\xi}_1 \\ \boldsymbol{\xi}_2 \\ \vdots \\ \boldsymbol{\xi}_m \end{pmatrix}. \tag{4.21}$$

In short, it is written as $\mathcal{K}\boldsymbol{\xi} = \lambda\mathcal{D}\boldsymbol{\xi}$, where $\mathcal{K}$ is an $m\ell \times m\ell$ matrix with $\mathcal{K}_{ij} = \mathbf{K}_i\mathbf{K}_j$ and $\mathcal{D}$ is an $m\ell \times m\ell$ block-diagonal matrix with $\mathcal{D}_{ii} = \mathbf{K}_i\mathbf{K}_i$. The minimal eigenvalue of equation (4.21), denoted by $\lambda_{\mathcal{F}}(\mathbf{K}_1, \ldots, \mathbf{K}_m)$, is referred to as the *first kernel canonical correlation.*

Similar to the definitions of "generalized variance" and "mutual information" as in linear CCA, the kernel generalized variance (KGV), denoted as $\sigma_{\mathcal{F}}^2$, is defined as [52]

$$\sigma_{\mathcal{F}}^2 = \frac{\det(\mathcal{K})}{\det(\mathcal{D})}. \tag{4.22}$$

Furthermore, the kernelized mutual information is defined by [52]

$$I_{\sigma_{\mathcal{F}}^2}(\mathbf{K}_1, \ldots, \mathbf{K}_m) = -\frac{1}{2}\log\sigma_{\mathcal{F}}^2 = -\frac{1}{2}\log\frac{\det(\mathcal{K})}{\det(\mathcal{D})}. \tag{4.23}$$

Equation (4.23) can be viewed as a natural extension of (2.163) (which is defined originally in the linear input space for the Gaussian variables), which is closely related to the mutual information between the non-Gaussian variables in the input space [52].

Moreover, ICA can also be "kernelized" to yield kernel ICA (KICA). Based on the theoretical framework of KCCA, Bach and Jordan [52] proposed two algorithms to solve the standard ICA problem. Specifically, they proposed the kernel-based contrast function, denoted by $C(\mathbf{W})$ (where $\mathbf{W}$ denotes a demixing matrix in the conventional linear ICA setup, as discussed in Chapter 3), which can be a form of either the kernelized mutual information

$$C(\mathbf{W}) = -\frac{1}{2}\log\sigma_{\mathcal{F}}^2 = -\frac{1}{2}\log\frac{\det(\mathcal{K})}{\det(\mathcal{D})}, \tag{4.24}$$

or

$$C(\mathbf{W}) = -\frac{1}{2}\log\lambda_{\mathcal{F}}(\mathbf{K}_1, \ldots, \mathbf{K}_m). \tag{4.25}$$

Bach and Jordan [52] further proposed several efficient computational algorithms for optimizing the derivative of the above two contrast functions. Specifically, the demixing matrix $\mathbf{W}$ is updated on a Stiefel manifold by the following natural gradient learning rule:

$$\Delta\mathbf{W} = -\eta\left(\frac{\partial C}{\partial\mathbf{W}} - \mathbf{W}\left(\frac{\partial C}{\partial\mathbf{W}}\right)^{\mathrm{T}}\mathbf{W}\right), \tag{4.26}$$

where $\partial C/\partial\mathbf{W}$ denotes the derivative of contrast function $C(\mathbf{W})$ with respect to $\mathbf{W}$. For details of implementation, regularization, and optimization, the reader is referred to [52].

As demonstrated in [52], the KICA algorithm has several advantages that make it superior to the conventional ICA algorithms in practical BSS applications:

- The KICA algorithm is robust to the Gaussianity or near-Gaussianity of the independent sources. In contrast, the performance of many other ICA algorithms often degrades when the sources are close to being Gaussian. This property is appealing since in practice we may not have prior knowledge of the sources.
- The KICA algorithm is very robust to outliers. This property is particularly important because noisy samples and outliers typically exist in practice.

However, as expected, the advantages obtained from KICA also come with a higher computational cost. In general, the convergence of the KICA algorithm is slower than that of the nonkernelized counterparts.

■ EXAMPLE 4.2

In this example, we apply the KICA algorithm (Matlab code available from http://cmm.ensmp.fr/~bach/kernel-ica/) to a simple BSS problem. In this task, the goal is to separate three simulated independent sources. In our experiments, the mixing matrix **A** was randomly generated and the initial demixing matrix **W** was set to be an identity matrix. In order to evaluate the separation performance, we use the so-called Amari distance [29] as the performance index (PI):

$$\mathrm{PI} = \sum_{i=1}^{3}\left(\sum_{j=1}^{3}\frac{|r_{ij}|}{\max_k |r_{ik}|} - 1\right) + \sum_{j=1}^{3}\left(\sum_{i=1}^{3}\frac{|r_{ij}|}{\max_k |r_{kj}|} - 1\right),$$

where $\mathbf{R} = \mathbf{WA} = \{r_{ij}\}$. A total of 100 Monte Carlo experiments were repeated, and the averaged PI was calculated. In the experiments, we always used the standard (default) setup for the KICA algorithm (learning rate 0.001, KGV contrast function, Gaussian kernel with width parameter 0.5). The stopping criterion for (4.26) is set as $\|\mathbf{W}(t+1) - \mathbf{W}(t)\|_F < 0.0001$.

First, we test the robustness of the KICA algorithm to the Gaussianity. In this case, the three mutually independent components (each with 500 data points) contain one Gaussian source (with i.i.d. samples), one near-Gaussian source (95% i.i.d. Gaussian random samples mixed with 5% i.i.d. Laplacian random samples), plus one deterministic sinusoidal signal. The averaged PI obtained from the KICA algorithm upon 100 Monte Carlo runs is 0.08. Figure 4.4 illustrates one separation result. As a comparison, the averaged performance indices from two standard ICA algorithms, Joint Approximate Diagonalization of Eigenmatrices (JADE) [149] (Matlab code available from http://www.tsi.enst.fr/~cardoso/guidesepsou.html) and Infomax with natural gradient [29], are 0.09 and 0.11, respectively. Therefore, in this task the

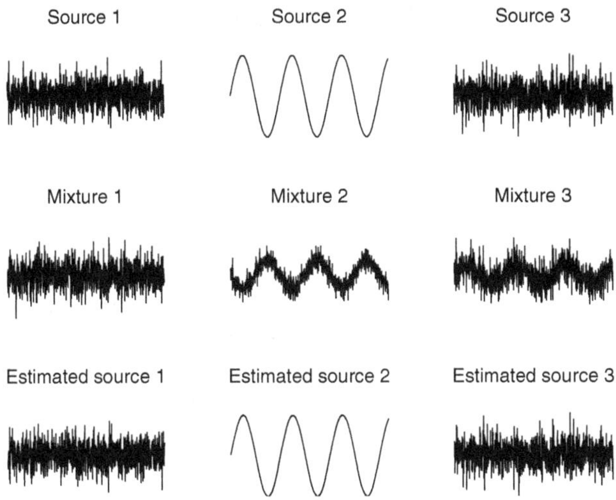

Figure 4.4 One BSS result obtained from the KICA algorithm.

KICA algorithm obtained the best result, while the JADE algorithm slightly outperformed the Infomax algorithm.

Second, we also test the robustness to outliers. In this case, the independent sources (each with 200 i.i.d. samples) are drawn from three probability distributions: Gaussian, uniform (sub-Gaussian), and exponential (super-Gaussian). In this task, we gradually increased the number of outliers (randomly replacing specific source samples with $+5$ or -5 with probability 0.5) and calculated the averaged PI based on 100 Monte Carlo runs. Again, for comparison, the PI statistics of the JADE and Infomax algorithms were also calculated. The performance of the three algorithms in this task is shown in Figure 4.5.

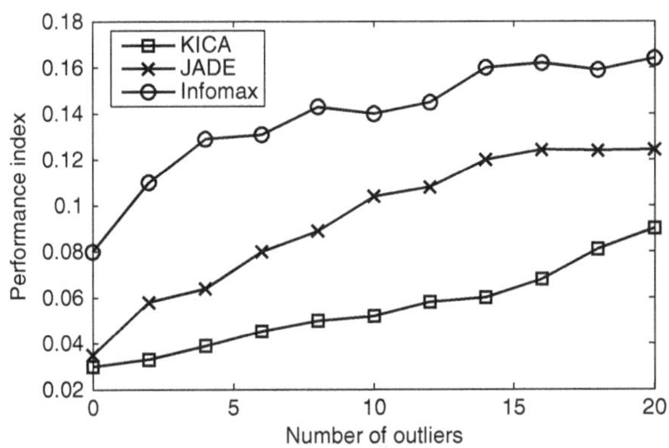

Figure 4.5 The performance of the three algorithms plotted against the number of outliers.

From the curves, we see that the KICA algorithm is much more robust than the other two algorithms.

4.4 KERNEL PRINCIPAL ANGLES

Principal angles are defined as the angles between a pair of vector sets in two linear subspaces, which also relate to the notion of *principal correlation* [329]. Appendix C presents a brief description of principal correlation and principal angles. Recently, Wolf and Shashua [968, 969] extended this concept and derived the so-called kernel principal angles with the kernel trick.

Specifically, let $\mathbf{A} = [\phi(\mathbf{a}_1), \ldots, \phi(\mathbf{a}_\ell)]$ and $\mathbf{B} = [\phi(\mathbf{b}_1), \ldots, \phi(\mathbf{b}_\ell)]$ denote two $N \times \ell$ matrices that both contain ℓ columns, where $\phi(\cdot)$ denotes some mapping from the input space $\mathbb{R}^N$ onto a feature space $\mathcal{F}$; hence, $\mathbf{A}$ and $\mathbf{B}$ represent the nonlinear surfaces in the original input spaces $\{\mathbf{a}_k\}$ and $\{\mathbf{b}_k\}$. The goal of kernel principal angles is to find a *similarity metric*, $f(\mathbf{A}, \mathbf{B})$, which measures the unordered sets of column spaces of $\mathbf{A}$ and $\mathbf{B}$ using the inner product (without the explicit computation of ϕ).

Suppose the columns of $\mathbf{A}$ and $\mathbf{B}$ represent two linear subspaces U_A and U_B in the feature space $\mathcal{F}$ that is induced by a nonlinear mapping ϕ; then the principal angles between the two subspaces, $0 \le \theta_1 \le \cdots \le \theta_\ell \le \pi/2$, are uniquely defined as

$$\cos(\theta_\ell) = \max_{\mathbf{u}\in U_A} \max_{\mathbf{v}\in U_B} \mathbf{u}^{\mathrm{T}}\mathbf{v}$$

$$\text{s.t.} \quad \mathbf{u}^{\mathrm{T}}\mathbf{u} = \mathbf{v}^{\mathrm{T}}\mathbf{v} = 1, \mathbf{u}^{\mathrm{T}}\mathbf{u}_i = \mathbf{v}^{\mathrm{T}}\mathbf{v}_i = 0, \quad i = 1, 2, \ldots, \ell - 1. \tag{4.27}$$

The quantities $\cos(\theta_k)$ are often referred to as *principal correlations* or *canonical correlations* of the matrix pair $(\mathbf{A}, \mathbf{B})$.

Consider the Gram–Schmidt orthogonalization procedure (described in Appendix C) for matrix $\mathbf{A}$, and let $\mathbf{v}_j \in \mathcal{F}$ be defined as

$$\mathbf{v}_j = \phi(\mathbf{a}_j) - \sum_{i=1}^{j-1} \frac{\mathbf{v}_i^{\mathrm{T}}\phi(\mathbf{a}_j)}{\mathbf{v}_i^{\mathrm{T}}\mathbf{v}_i}\mathbf{v}_i. \tag{4.28}$$

Let $\mathbf{V_A} = [\mathbf{v}_1, \ldots, \mathbf{v}_\ell]$ and let

$$\mathbf{s}_j = \left(\frac{\mathbf{v}_1^{\mathrm{T}}\phi(\mathbf{a}_j)}{\mathbf{v}_1^{\mathrm{T}}\mathbf{v}_1}, \ldots, \frac{\mathbf{v}_{j-1}^{\mathrm{T}}\phi(\mathbf{a}_j)}{\mathbf{v}_{j-1}^{\mathrm{T}}\mathbf{v}_{j-1}}, 1, 0, 0, \ldots, 0\right)^{\mathrm{T}}. \tag{4.29}$$

Then $\mathbf{A} = \mathbf{V_A}\mathbf{S_A}$, where $\mathbf{S_A} = [\mathbf{s}_1, \ldots, \mathbf{s}_\ell]$ is an $\ell \times \ell$ upper diagonal matrix. Furthermore, the QR factorization of matrix $\mathbf{A}$ can be rewritten as

$$\mathbf{A} = (\mathbf{V_A}\mathbf{D_A}^{-1})(\mathbf{D_A}\mathbf{S_A}) \equiv \mathbf{Q_A}\mathbf{R_A}, \tag{4.30}$$

where $\mathbf{D_A} = \mathrm{diag}\{\|\mathbf{v}_1\|, \ldots, \|\mathbf{v}_\ell\|\}$ is a diagonal matrix; $\mathbf{R_A} = \mathbf{D_A S_A}$ is upper diagonal, and $\mathbf{Q_A} = \mathbf{A R_A}^{-1}$ is an orthonormal matrix. Repeating the Gram–Schmidt orthogonalization procedure for matrix $\mathbf{B}$, we also obtain $\mathbf{B} = \mathbf{Q_B R_B}$. Finally, the singular values $\{\sigma_1, \ldots, \sigma_\ell\}$ of the matrix $\mathbf{Q}_\mathbf{A}^\mathrm{T}\mathbf{Q_B}$ correspond to the principal correlations $\cos(\theta_k) = \sigma_k$.

Notably, $\mathbf{Q}_\mathbf{A}^\mathrm{T}\mathbf{Q_B} = \mathbf{R}_\mathbf{A}^{-\mathrm{T}}\mathbf{A}^\mathrm{T}\mathbf{B}\mathbf{R}_\mathbf{B}^{-1}$, where $\mathbf{A}^\mathrm{T}\mathbf{B}$ involves only the inner product. Hence, using the kernel trick, the inner product can be computed such that $(\mathbf{A}^\mathrm{T}\mathbf{B})_{ij} = K(\mathbf{a}_i, \mathbf{b}_j)$. Similarly, matrices $\mathbf{D_A}$ and $\mathbf{S_A}$ can be computed by the kernel trick [968]. Since $\mathbf{V_A} = \mathbf{A S}_\mathbf{A}^{-1}$, we can write

$$\mathbf{v}_j = \sum_{i=1}^{j} \alpha_{ij}\phi(\mathbf{a}_i), \tag{4.31}$$

where α_{ij} denotes the ith element of the vector $\boldsymbol{\alpha}_j$ (where $\mathbf{v}_j = \mathbf{A}\boldsymbol{\alpha}_j$). The inner products $\mathbf{v}_j^\mathrm{T}\phi(\mathbf{a}_j)$ and $\mathbf{v}_j^\mathrm{T}\mathbf{v}_j$ can be computed using a kernel as follows:

$$\mathbf{v}_j^\mathrm{T}\phi(\mathbf{a}_i) = \sum_{k=1}^{j} \alpha_{kj} K(\mathbf{a}_i, \mathbf{a}_k), \tag{4.32}$$

$$\mathbf{v}_j^\mathrm{T}\mathbf{v}_j = \sum_{k=1}^{j}\sum_{i=1}^{j} \alpha_{kj}\alpha_{ij} K(\mathbf{a}_k, \mathbf{a}_i). \tag{4.33}$$

Substituting (4.32) and (4.33) into (4.29) leads to the computation of $\mathbf{S_A}$, $\mathbf{D_A}$, and subsequent $\mathbf{R_A}$. A similar procedure can be applied to obtain $\mathbf{S_B}$, $\mathbf{D_B}$, and $\mathbf{R_B}$.

In addition to the above QR-SVD procedure, the kernel principal angles can be alteratively derived from solving a $2\ell \times 2\ell$ generalized eigenvalue problem [969]. Specifically, in the case of nonkernelized principal angles (i.e., ϕ is an identity mapping), the eigenequation is given by

$$\begin{pmatrix} \mathbf{0} & \mathbf{B}^\mathrm{T}\mathbf{A} \\ \mathbf{A}^\mathrm{T}\mathbf{B} & \mathbf{0} \end{pmatrix} \begin{pmatrix} \xi_1 \\ \xi_2 \end{pmatrix} = \lambda \begin{pmatrix} \mathbf{B}^\mathrm{T}\mathbf{B} & \mathbf{0} \\ \mathbf{0} & \mathbf{A}^\mathrm{T}\mathbf{A} \end{pmatrix} \begin{pmatrix} \xi_1 \\ \xi_2 \end{pmatrix}, \tag{4.34}$$

and the generalized eigenvalues $\lambda_1, \ldots, \lambda_{2\ell}$ are related to the principal angles by $\lambda_1 = \cos(\theta_1), \ldots, \lambda_\ell = \cos(\theta_\ell)$ and $\lambda_{\ell+1} = -\cos(\theta_\ell), \ldots, \lambda_{2\ell} = -\cos(\theta_1)$. Since the matrices $\mathbf{A}^\mathrm{T}\mathbf{A}$, $\mathbf{B}^\mathrm{T}\mathbf{B}$, $\mathbf{A}^\mathrm{T}\mathbf{B}$, and $\mathbf{B}^\mathrm{T}\mathbf{A}$ in (4.34) involve only inner products between columns of $\mathbf{A}$ and $\mathbf{B}$, it can be readily kernelized using the kernel trick. In other words, during the inner product computation we can replace $\mathbf{a}_i^\mathrm{T}\mathbf{a}_j$, $\mathbf{b}_i^\mathrm{T}\mathbf{b}_j$, $\mathbf{a}_i^\mathrm{T}\mathbf{b}_j$, and $\mathbf{b}_i^\mathrm{T}\mathbf{a}_j$ by $K(\mathbf{a}_i, \mathbf{a}_j)$, $K(\mathbf{b}_i, \mathbf{b}_j)$, $K(\mathbf{a}_i, \mathbf{b}_j)$, and $K(\mathbf{b}_i, \mathbf{a}_j)$, respectively.

Finally, the similarity metric $f(\mathbf{A}_i, \mathbf{A}_j)$ (for a pair of matrices $\mathbf{A}_i \in \mathbb{R}^{N\times\ell}$ and $\mathbf{A}_j \in \mathbb{R}^{N\times\ell}$) is constructed by the following positive-definite kernel [968, 969]:

$$K(\mathbf{A}_i, \mathbf{A}_j) \equiv f(\mathbf{A}_i, \mathbf{A}_j) = \prod_{k=1}^{\ell} \cos^2(\theta_k), \tag{4.35}$$

where θ_k denotes the principal angles between two linear subspaces. Such a similarity metric can be used for a wide family of kernel learning tools, including classification and clustering.

In [968, 969], the power of kernel principal angles was demonstrated in image/video sequence analysis, with applications in face recognition, irregular motion trajectory detection, and image classification.

4.5 KERNEL DISCRIMINANT ANALYSIS

Analogous to LDA described in Chapter 2, we may extend the idea to feature space, which leads to the method of kernel discriminant analysis. Despite many different formulations (e.g., [65, 621, 799, 983, 987, 1001]) of this problem, the common goal behind them is to optimize the Fisher discrimination ratio in a high-dimensional feature space with the help of a reproducing kernel function, and then the optimization problem is converted into a generalized eigenvalue problem. Here, we use the general multiple classification formulation (from [65]) to illustrate the essential idea.

Consider an N-class discrimination task applied to a data set $X = \{\mathbf{x}_i\}_{i=1}^{\ell}$. We assume the lth class consists of n_l sample points, which is denoted as $X_l = \{\mathbf{x}_k\}_{k=1}^{n_l}$; $X = \bigcup_{l=1}^{N} X_l$. For simplicity, we assume the data points are centered in the feature space. Let $\overline{\phi}_l$ denote the feature mean of the class l:

$$\overline{\phi}_l = \frac{1}{n_l} \sum_{k=1}^{n_l} \phi(\mathbf{x}_{lk}), \tag{4.36}$$

where $\mathbf{x}_{lk}$ is the the kth sample from the class l. Furthermore, let $\mathbf{B}$ denote the covariance matrix of the class centers (i.e., the interclass inertia),

$$\mathbf{B} = \frac{1}{\ell} \sum_{l=1}^{N} n_l \overline{\phi}_l \overline{\phi}_l, \tag{4.37}$$

and let $\mathbf{V}$ denote the total inertia of all the data points in the feature space,

$$\mathbf{V} = \frac{1}{\ell} \sum_{l=1}^{N} \sum_{k=1}^{n_l} \phi_l(\mathbf{x}_{lk}) \phi_l^{\mathrm{T}}(\mathbf{x}_{lk}). \tag{4.38}$$

Similar to the linear LDA, the nonlinear discriminant analysis in feature space can be formulated as a problem of maximizing the interclass inertia while minimizing the intraclass inertia. This is equivalent to solving a generalized eigenvalue problem [65]:

$$\lambda \mathbf{V}\mathbf{u} = \mathbf{B}\mathbf{u} \tag{4.39}$$

or equivalently

$$\lambda \mathbf{u} = \mathbf{V}^{-1}\mathbf{B}\mathbf{u}. \tag{4.40}$$

The largest eigenvalue of (4.40) yields the maximum of the following quotient of the inertia:

$$\lambda = \frac{\mathbf{u}^{\mathrm{T}}\mathbf{B}\mathbf{u}}{\mathbf{u}^{\mathrm{T}}\mathbf{V}\mathbf{u}}, \tag{4.41}$$

which also corresponds to the *Fisher discriminant ratio* in the feature space. Equation (4.41), in turn, by using the kernel trick, is equivalent to the expression

$$\lambda = \frac{\boldsymbol{\alpha}^{\mathrm{T}}\mathbf{K}\mathbf{W}\mathbf{K}\boldsymbol{\alpha}}{\boldsymbol{\alpha}^{\mathrm{T}}\mathbf{K}\mathbf{K}\boldsymbol{\alpha}}, \tag{4.42}$$

where $\boldsymbol{\alpha} = (\alpha_{pq})_{p=1,\dots,N;q=1,\dots,n_p}$ is an $\ell \times 1$ coefficient vector, $\mathbf{W} = (\boldsymbol{W}_l)_{l=1,\dots,N}$ is an $\ell \times \ell$ block-diagonal matrix (in which $\boldsymbol{W}_l$ is an $n_l \times n_l$ matrix with all terms equal to $1/n_l$), and $\mathbf{K} = (\boldsymbol{K}_{pq})_{p=1,\dots,N;q=1,\dots,n_p}$ is an $\ell \times \ell$ symmetric kernel matrix (in which $\boldsymbol{K}_{pq} = \{k_{ij}\}_{i=1,\dots,n_p;j=1,\dots,n_p}$ is an $n_p \times n_p$ matrix).

Applying the eigenvalue decomposition (EVD) to the above kernel matrix $\mathbf{K}$, we have

$$\mathbf{K} = \mathbf{U}\boldsymbol{\Lambda}\mathbf{U}^{\mathrm{T}}. \tag{4.43}$$

Substituting (4.43) into (4.42) yields

$$\lambda = \frac{\boldsymbol{\beta}^{\mathrm{T}}\mathbf{U}^{\mathrm{T}}\mathbf{W}\mathbf{U}\boldsymbol{\beta}}{\boldsymbol{\beta}^{\mathrm{T}}\mathbf{U}^{\mathrm{T}}\mathbf{U}\boldsymbol{\beta}}, \tag{4.44}$$

where $\boldsymbol{\beta} = \boldsymbol{\Lambda}^{1/2}\mathbf{U}^{\mathrm{T}}\boldsymbol{\alpha}$. After simplifying, the equivalent eigenvalue problem is rewritten as

$$\lambda \boldsymbol{\beta} = \mathbf{U}^{\mathrm{T}}\mathbf{W}\mathbf{U}\boldsymbol{\beta}, \tag{4.45}$$

where $\boldsymbol{\beta}$ corresponds to the eigenvector of matrix $\mathbf{U}^{\mathrm{T}}\mathbf{W}\mathbf{U}$.

Upon obtaining $\boldsymbol{\beta}$ and subsequently $\boldsymbol{\alpha}$, the optimal eigenvectors $\mathbf{v}$ can be constructed by

$$\mathbf{v} = \sum_{p=1}^{N}\sum_{q=1}^{n_p} \alpha_{pq}\phi(\mathbf{x}_{pq}). \tag{4.46}$$

After the training phase, it is straightforward to discriminate a new test data point $\mathbf{x}'$ by applying projections of the test point onto the *normalized* eigenvectors $\mathbf{v}$ (s.t. $\mathbf{v}^{\mathrm{T}}\mathbf{v} = \boldsymbol{\alpha}\mathbf{K}\boldsymbol{\alpha} = 1$), namely,

$$\mathbf{v}^{\mathrm{T}}\phi(\mathbf{x}') = \sum_{p=1}^{N}\sum_{q=1}^{n_p} \alpha_{pq} K(\mathbf{x}_{pq}, \mathbf{x}'). \tag{4.47}$$

In [987], it was shown that the kernel Fisher discriminant analysis is essentially equivalent to KPCA plus Fisher LDA. That is, KPCA is first performed and then LDA is used for a second-step feature extraction in the KPCA-transformed subspace. Specifically, it can be proved that maximizing equation (4.42) is equivalent to maximizing a generalized Rayleigh quotient defined as follows:

$$\rho = \frac{\boldsymbol{\beta}^{\mathrm{T}}\mathbf{S}_b\boldsymbol{\beta}}{\boldsymbol{\beta}^{\mathrm{T}}\mathbf{S}_t\boldsymbol{\beta}}, \tag{4.48}$$

where $\mathbf{S}_b = \boldsymbol{\Lambda}^{1/2}\mathbf{U}^{\mathrm{T}}\mathbf{W}\mathbf{U}\boldsymbol{\Lambda}^{1/2}$ and $\mathbf{S}_t = \boldsymbol{\Lambda}$ correspond to, respectively, the between-class and total scatter matrices in the KPCA-transformed space. Finding an optimal value of the vector $\boldsymbol{\beta}$ corresponds to finding the eigenvector associated with the maximum eigenvalue of matrix $\mathbf{S}_t^{-1}\mathbf{S}_b$.

■ EXAMPLE 4.3

In this example, we use two problems to illustrate the kernel discriminant analysis method for two pattern classification tasks that are not linearly separable. The first real-life data set is for a three-way classification problem, while the second synthetic data set is for a two-way classification problem.

The first data set is the iris flower data, a widely used benchmark [65]. The data set contains samples from three collected iris species (each with 50 specimens). Each sample consists of four variables: *sepal length, sepal width, petal length*, and *petal width*. A total of 150 normalized samples (i.e., with zero mean and unit variance) were used in this experiment. It has been known that for this problem one class is linearly separable from the two other and the latter two are not linearly separable from each other. We apply kernel Fisher discriminant analysis (with the same Gaussian kernel setup as [65]) and LDA [i.e., with a linear kernel $K(\mathbf{x}_i, \mathbf{x}_j) = \mathbf{x}_i^{\mathrm{T}}\mathbf{x}_j$] to the same data and project them onto the first two axes (see Figure 4.6). With respect to their decision boundaries, the kernel discriminant analysis has a better discriminant performance in that the three clusters are well separated (Figure 4.6*b*).

The second data set is another widely used benchmark, consisting of the so-called two spirals problem [524]. This synthetic data set consists of two classes of two intertwined spirals with 194 data points. As seen from Figure 4.7*a*, this problem is not linearly separable and in fact has a very complex decision boundary between the two classes. Applying kernel discriminant analysis with a Gaussian kernel to the data, we obtain two well-separated

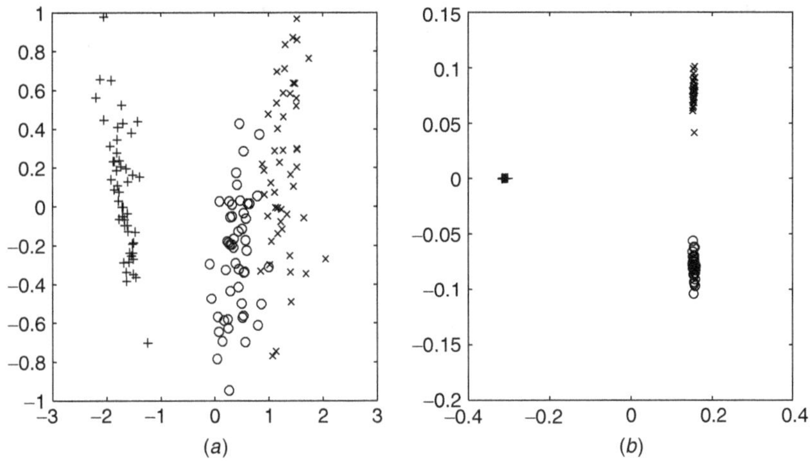

Figure 4.6 Projection of the iris data (three classes labeled by different markers) onto the the first two axes. (*a*) LDA with a linear kernel $K(\mathbf{x}_i, \mathbf{x}_j) = \mathbf{x}_i^T \mathbf{x}_j$. (*b*) Kernel discriminant analysis with a Gaussian kernel $K(\mathbf{x}_i, \mathbf{x}_j) = \exp(-\|\mathbf{x}_i - \mathbf{x}_j\|^2/0.7)$.

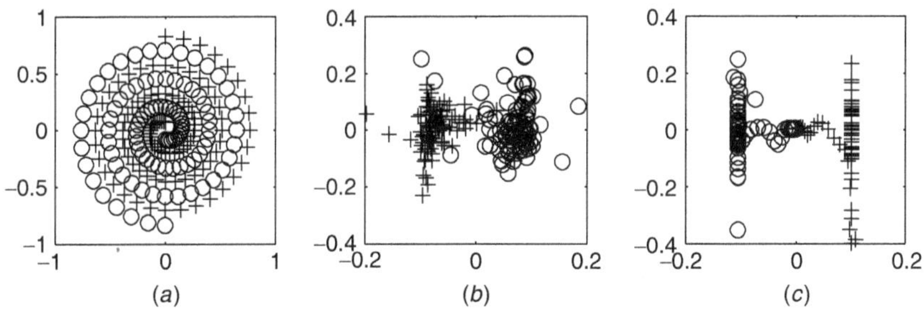

Figure 4.7 (*a*) The two-spirals problem. (*b*) The projection of all data samples on the first two axes using the kernel discriminant analysis with a Gaussian kernel $K(\mathbf{x}_i, \mathbf{x}_j) = \exp(-\|\mathbf{x}_i - \mathbf{x}_j\|^2/0.01)$. (*c*) The projection of the test samples onto the first two axes.

clusters as shown in Figure 4.7*b*. Moreover, we also split the data points (not randomly, but skipping one nearest point along the spiral trajectories) evenly into two groups, one group for training and other group for testing. We then project the 97 testing samples onto the first two axes of the feature space that was learned by the other 97 training samples. Again, we can see the two-class testing data points are well separated (Figure 4.7*c*).

4.6 KERNEL WIENER FILTER

Using the kernel trick again, we can extend linear Wiener filter theory to a nonlinear Wiener filter by invoking kernelization in the RKHS [182, 941, 984].

Recall from Chapter 2 the formulation of the discrete-time linear Wiener filter. Let $\mathbf{x}(t) = [x(t), x(t-1), \ldots, x(t-N+1)]^{\mathrm{T}}$ and $d(t)$ denote the N-dimensional input and scalar output signals, respectively, and suppose the output signal $d(t)$ can be modeled by an FIR filter:

$$\begin{aligned} d(t) &= \sum_{k=0}^{N-1} x(t-k)\theta_k(t) + e(t) \\ &= \mathbf{x}^{\mathrm{T}}(t)\boldsymbol{\theta}(t) + e(t). \end{aligned} \tag{4.49}$$

Multiplying both sides with $\mathbf{x}(t)$ and taking the statistical expectation, by assuming that $\mathbb{E}[\mathbf{x}(t)e(t)] = 0$, we obtain $\mathbb{E}[\mathbf{x}(t)d(t)] = \mathbb{E}[\mathbf{x}(t)\mathbf{x}^{\mathrm{T}}(t)]\boldsymbol{\theta}(t)$. By solving the Wiener–Hopf equation, the Wiener solution is obtained as

$$\boldsymbol{\theta}_o = \mathbf{C}_{xx}^{-1}\mathbf{C}_{xd},$$

where $\mathbf{C}_{xx}$ and $\mathbf{C}_{xd}$ denote the autocorrelation matrix [of $\mathbf{x}(t)$] and cross-correlation vector [between $\mathbf{x}(t)$ and $d(t)$], respectively:

$$\mathbf{C}_{xx} = \mathbb{E}[\mathbf{x}(t)\mathbf{x}^{\mathrm{T}}(t)] \approx \frac{1}{T}\sum_{k=1}^{T}\mathbf{x}(t)\mathbf{x}^{\mathrm{T}}(t),$$

$$\mathbf{C}_{xd} = \mathbb{E}[\mathbf{x}(t)d(t)] \approx \frac{1}{T}\sum_{k=1}^{T}\mathbf{x}(t)d(t).$$

Now, let us formulate the nonlinear Wiener filter in the RKHS. Given the two sequences $\{\mathbf{x}(t)\}_{t=1}^{T}$ and $\{d(t)\}_{t=1}^{T}$, the latter of which defines the span of a subspace in the RKHS, we may construct a *nonlinear* filter with the form

$$y(t) = \phi^{\mathrm{T}}(\mathbf{x}(t))\boldsymbol{\theta}(t), \tag{4.50}$$

where the vector $\boldsymbol{\theta}$ specifies the filter response coefficients and $\phi(\mathbf{x}(t))$ specifies a nonlinear basis function that is defined in a high-dimensional feature space. Similarly, solving the Wiener–Hopf equation

$$\mathbb{E}\left[\phi(\mathbf{x}(t))(d(t) - \phi^{\mathrm{T}}(\mathbf{x}(t))\boldsymbol{\theta}(t)\right] = 0 \tag{4.51}$$

would yield the optimal Wiener solution $\boldsymbol{\theta}_o$, in which we again assume that the feature map and the error signal are uncorrelated, namely $\mathbb{E}[\phi(\mathbf{x}(t))e(t)] = 0$.

In the high-dimensional feature space, using the kernel trick we may avoid the direct calculation of $\phi(\mathbf{x}(t))$ and its outer product $\phi(\mathbf{x})\phi^{\mathrm{T}}(\mathbf{x})$; instead, we calculate the inner product with a Mercer kernel

$$\mathbf{K} = \langle\phi(\mathbf{x}), \phi(\mathbf{x})\rangle,$$

where $\mathbf{K}_{ij} \equiv K(\mathbf{x}_i, \mathbf{x}_j) = \phi^{\mathrm{T}}(\mathbf{x}_i)\phi(\mathbf{x}_j)$, and $\Phi : \mathbf{x} \mapsto K(\mathbf{x}, \cdot)$ denotes the *reproducing kernel map*. For notational convenience, let $\mathbf{d} = [d(1), \ldots, d(T)]^{\mathrm{T}}$ and $\mathbf{k}(t) = \Phi(\mathbf{x}(t)) = [K(\mathbf{x}(t), \mathbf{x}(1)), \ldots, K(\mathbf{x}(t), \mathbf{x}(T))]^{\mathrm{T}}$. We then define the following autocorrelation matrix and cross-correlation vector, respectively, as shown by

$$\mathbf{C}_{\phi\phi} \equiv \mathbb{E}[\phi(\mathbf{x}(t))\phi^{\mathrm{T}}(\mathbf{x}(t))] \approx \frac{1}{T}\sum_{t=1}^{T} \mathbf{k}(t)\mathbf{k}^{\mathrm{T}}(t) = \frac{1}{T}\mathbf{K}^{\mathrm{T}}\mathbf{K}, \tag{4.52}$$

$$\mathbf{C}_{\phi d} \equiv \mathbb{E}[\phi(\mathbf{x}(t))d(t)] \approx \frac{1}{T}\sum_{t=1}^{T} \mathbf{k}(t)d(t) = \frac{1}{T}\mathbf{K}^{\mathrm{T}}\mathbf{d}, \tag{4.53}$$

in which we have used the "reproducing property" of the kernel: $\langle \phi(\mathbf{x}), \phi(\mathbf{x}') \rangle = \langle K(\cdot, \mathbf{x}), K(\cdot, \mathbf{x}') \rangle = K(\mathbf{x}, \mathbf{x}')$. Hence, from the eigenequation $\mathbf{C}_{\phi d} = \boldsymbol{\theta}_o \mathbf{C}_{\phi\phi}$, we have

$$\frac{1}{T}\mathbf{K}^{\mathrm{T}}\mathbf{d} = \boldsymbol{\theta}_o \left(\frac{1}{T}\mathbf{K}^{\mathrm{T}}\mathbf{K} \right), \tag{4.54}$$

and the output of the nonlinear Wiener filter is given by

$$\begin{aligned} y(t) &= \phi^{\mathrm{T}}(\mathbf{x}(t))\boldsymbol{\theta}_0 \\ &= \phi^{\mathrm{T}}(\mathbf{x}(t)) \left(\mathbf{K}^{\mathrm{T}}\mathbf{K}\right)^{-1} \mathbf{K}^{\mathrm{T}}\mathbf{d}. \end{aligned} \tag{4.55}$$

In contrast to (4.50), the dual formulation of the kernel Winer filter can be written as

$$y(t) = \sum_{k=1}^{T} c_k K(\mathbf{x}(t), \mathbf{x}(k)) \quad \text{or} \quad \mathbf{y} = \mathbf{K}\mathbf{c} \tag{4.56}$$

where $\mathbf{y} = [y(1), \ldots, y(T)]^{\mathrm{T}} \in \mathbb{R}^T$, $\mathbf{K} \in \mathbb{R}^{T\times T}$, and the vector $\mathbf{c} = [c_1, \ldots, c_T]^{\mathrm{T}} \in \mathbb{R}^T$ is to be determined in order to minimize the variance of the estimation error. Note that when a d-order polynomial kernel is employed, (4.56) is written as

$$y(t) = \sum_{k=1}^{T} c_k(1 + \mathbf{x}(t) \cdot \mathbf{x}(k))^d,$$

which is also known as a *Volterra filter* of degree d in the literature.[4]

In a manner similar to the linear case, the nonlinear kernel Wiener filter is given by the solution

$$\mathbf{c} = \mathbf{K}^{\dagger}\mathbf{d}, \tag{4.57}$$

where $\mathbf{K}^{\dagger}$ defines the pseudoinverse of the kernel matrix $\mathbf{K}$, which plays the role of the correlation matrix inverse $\mathbf{C}_{xx}^{-1}$. When the matrix $\mathbf{K}$ is square and invertible,

$\mathbf{K}^{\dagger}$ reduces to $\mathbf{K}^{-1}$. In practice, since the signal of interest is often contaminated by noise, it is wise to use a lower rank approximation for the matrix $\mathbf{K}$. Suppose that the signal power is greater than the noise power; then the signal space and noise space can be separated via KPCA [800]. Let $\tilde{\mathbf{K}} = [\mathbf{u}_1, \ldots, \mathbf{u}_m]^{\mathrm{T}} \in \mathbb{R}^{T \times m}$ denote the lower rank kernel that contains the first m dominant $T \times 1$ eigenvectors $\{\mathbf{u}_i\}_{i=1}^{m}$ obtained from diagonalizing $\mathbf{K} = \mathbf{U}\mathbf{\Lambda}\mathbf{U}^{\mathrm{T}}$. Then (4.57) is rewritten as

$$\tilde{\mathbf{c}} = \tilde{\mathbf{K}}^{\dagger}\mathbf{d} = (\tilde{\mathbf{K}}^{\mathrm{T}}\tilde{\mathbf{K}})^{-1}\tilde{\mathbf{K}}^{\mathrm{T}}\mathbf{d}. \tag{4.58}$$

Note that in this case $\tilde{\mathbf{K}}^{\mathrm{T}}\tilde{\mathbf{K}}$ is a diagonal matrix whose entries contain the scaled eigenvalues. Therefore, the matrix inverse is obtained simply by inverting the individual diagonal entries.

Two additional comments are noteworthy:

- Compared to the standard Wiener filter, the kernel Wiener filter is more powerful in characterizing the non-Gaussian nature of a signal or noise because of the incorporation of nonlinearity and higher order correlations. When non-Gaussian signals (such as speech or image) are corrupted by non-Gaussian noise (such as impulsive noise), the kernel Wiener filter typically yields better denoising or restoration performance [182, 941].
- For large-scale problems (in which the number of observations is large), direct matrix inversion may be computationally prohibitive, and a reduced rank representation is therefore desirable. In addition to the EVD, the Cholesky and QR decomposition methods can also be used for this purpose [51]. Moreover, when the matrix $\mathbf{K}$ or $\tilde{\mathbf{K}}$ is ill-conditioned, regularization is required to avoid numerical problems that may arise in computing the matrix inverse or pseudoinverse.

4.7 KERNEL-BASED CORRELATION ANALYSIS: GENERALIZED CORRELATION FUNCTION AND CORRENTROPY

As the correlation function (either autocorrelation or cross-correlation) measures the similarity among the data, this measure can be defined in a similar manner in the feature space. Accordingly, generalized correlation function and correntropy have been proposed for this purpose in the context of kernelization and information-theoretic learning [565, 733, 790].

Definition 4.3 *[790] Let $\{\mathbf{x}_t, t \in \mathcal{T}\}$ be a stochastic process with $\mathcal{T}$ being an index set and $\mathbf{x}_t \in \mathbb{R}^d$. The generalized correlation function $V_{xx}(t_1, t_2)$ is defined as a function from $\mathcal{T} \times \mathcal{T}$ into $\mathbb{R}^+$ given by*

$$\begin{aligned} V_{xx}(t_1, t_2) &= \mathbb{E}\left[\langle \phi(\mathbf{x}_{t_1}), \phi(\mathbf{x}_{t_2})\rangle\right] \\ &= \mathbb{E}\left[K(\mathbf{x}_{t_1}, \mathbf{x}_{t_2})\right] = \mathbb{E}\left[K(\mathbf{x}_{t_1} - \mathbf{x}_{t_2})\right], \end{aligned} \tag{4.59}$$

where $\mathbb{E}[\cdot]$ *denotes the mathematical expectation over the stochastic process* $\mathbf{x}_t$ *and* $K(\cdot,\cdot)$ *is a translation-invariant positive-definite Mercer kernel such as the Gaussian kernel.*

Because of its natural link to the *quadratic Rényi entropy*[5] in the context of Parzen kernel estimation [790], the generalized correlation function is also referred to as *correntropy* [566, 790].

In [790], it is shown that when using a series expansion for the Gaussian kernel (with a width parameter σ), the correntropy function can be written as

$$V_{xx}(t_1, t_2) = \frac{1}{\sqrt{2\pi}\sigma} \sum_{n=0}^{\infty} \frac{(-1)^n}{2^n \sigma^{2n} n!} \mathbb{E}\left[\|\mathbf{x}_{t_1} - \mathbf{x}_{t_2}\|^{2n}\right], \tag{4.60}$$

which involves all the even-order moments of the random variable $\|\mathbf{x}_{t_1} - \mathbf{x}_{t_2}\|$. Specifically, the term corresponding to $n = 1$ in (4.60) is proportional to

$$\mathbb{E}\left[\|\mathbf{x}_{t_1}\|^2\right] + \mathbb{E}\left[\|\mathbf{x}_{t_2}\|^2\right] - 2\mathbb{E}\left[\mathbf{x}_{t_1}^{\mathrm{T}} \mathbf{x}_{t_2}\right], \tag{4.61}$$

where the first two terms correspond to the variance and the third term is similar to the autocorrelation function defined for stochastic processes (except for using the inner product). Hence, the correntropy function defined in the nonlinear feature space generalizes the autocorrelation function defined in the linear space. Note that the definition of the correntropy function assumes wide-sense stationarity [i.e., $V_{xx}(t_1, t_2) = V_{xx}(t_1 - t_2)$], implying that the stochastic process must be strictly stationary on the even moments. On the other hand, the correntropy function also shares many properties with the autocorrelation function, such as symmetry [i.e., $V_{xx}(t_1 - t_2) = V_{xx}(t_2 - t_1)$] and maximum value at the origin [i.e., $V_{xx}(\tau) \le V_{xx}(0) = 1/(\sqrt{2\pi}\sigma)$, $\forall\tau$]. Given a finite set of discrete samples of a stochastic process, the correntropy function can be approximated by

$$V_{xx}(\tau) = \frac{1}{T - \tau + 1} \sum_{t=\tau}^{T} K(\mathbf{x}_t - \mathbf{x}_{t-\tau}). \tag{4.62}$$

In addition, the generalized PSD function is defined similarly to the generalized correlation function, which is referred to as the *correntropy spectral density* (CSD) [790]:

$$S_{xx}(\omega) = \sum_{\tau=-\infty}^{\infty} V_{xx}(\tau) e^{-j\omega\tau}. \tag{4.63}$$

In [733], the correntropy function was used to derive a closed form of the kernel Wiener filter in the feature space. Similar to the preceding discussion, let $\mathbf{x}(t) = [x(t), x(t-1), \ldots, x(t-N+1)]^{\mathrm{T}}$ and $d(t)$ denote the N-dimensional input and scalar output signals, respectively. Also, let $\mathbf{V}$ define an $N \times N$ matrix whose

(i, j)th element is given by $\mathbb{E}[K(x(t-i+1), x(t-j+1))]$. By replacing the autocorrelation function with the correntropy function, in light of the Wiener–Hopf equation, the Wiener solution in the feature space is written as [733]

$$\boldsymbol{\theta}_o = \mathbf{V}^{-1}\mathbb{E}[\phi(\mathbf{x}(t))d(t)] \approx \frac{1}{T}\mathbf{V}^{-1}\sum_{k=1}^{T}\phi(\mathbf{x}(k))d(k). \tag{4.64}$$

With the kernel trick, the output of the kernel Wiener filter is thus given by

$$\begin{aligned} y(t) &= \phi^{\mathrm{T}}(\mathbf{x}(t))\boldsymbol{\theta}_o \\ &\approx \frac{1}{T}\sum_{k=1}^{T} d(k) \sum_{i=0}^{N-1}\sum_{j=0}^{N-1} a_{ij}K(x(t-i), x(k-j)), \end{aligned} \tag{4.65}$$

where a_{ij} denotes the (i, j)th element of the matrix $\mathbf{V}^{-1}$. Notably, equation (4.65) is essentially another way of rewriting (4.56) and (4.57).

More generally, the correntropy function can be defined between two arbitrary random variables $\mathbf{x} \in \mathbb{R}^d$ and $\mathbf{y} \in \mathbb{R}^d$ as

$$V_{xy}(\mathbf{x}, \mathbf{y}) = \mathbb{E}\left[\langle\phi(\mathbf{x}), \phi(\mathbf{y})\rangle\right] = \mathbb{E}\left[K(\mathbf{x} - \mathbf{y})\right], \tag{4.66}$$

which can be viewed as a generalized measure of cross-correlation that evaluates the similarity between two random vectors $\mathbf{x}$ and $\mathbf{y}$. In practice, the joint pdf $p(\mathbf{x}, \mathbf{y})$ is unknown, in which case (4.66) can be approximated by a sample estimator based on a finite number of data points $\{\mathbf{x}_i, \mathbf{y}_i\}_{i=1}^{\ell}$:

$$\hat{V}_{xy}(\mathbf{x}, \mathbf{y}) = \frac{1}{\ell}\sum_{i=1}^{\ell} K(\mathbf{x}_i - \mathbf{y}_i). \tag{4.67}$$

For an in-depth discussion of the mathematical properties and non-Gaussian signal processing applications of the correntropy function (4.66), the reader is referred to [565, 566].

■ EXAMPLE 4.4

In this example, we present a simple experiment (taken from [790]) to illustrate the use of the correntropy function in non-Gaussian signal processing and compare its behavior with the conventional autocorrelation function.

First, we generate three zero-mean white random processes with different distributions: Gaussian, impulsive, and exponential. For each random process, the samples are shifted properly to obtain a zero mean. Because the random processes are white, it is inferred that their autocorrelation functions should be a Dirac delta function. We estimate the autocorrelation function for these three white processes based on 5000 samples, while the correntropy function is estimated from the same 5000 samples, with a chosen kernel width parameter $\sigma = 2$.

Next, we feed the white random processes into a LTI infinite-duration impulse response (IIR) filter, which has the following transfer function in the z-domain:

$$H(z) = \frac{1}{1 - 1.5z^{-1} + 0.8z^{-2}}.$$

We again estimate the autocorrelation and correntropy functions for these three filtered (colored) processes.

The experimental results are shown in Figure 4.8. As seen from the figure, for the white processes, the autocorrelation function is nearly indistinguishable for the three random processes. In contrast, the mean value of the correntropy function varies for different probability distributions (the exponential source ranks the highest, followed by the Gaussian, and then the impulsive). For the filtered process, since linear filtering brings in correlations among the random samples, the shapes of the autocorrelation and correntropy functions will change accordingly. As shown in the figure, the autocorrelation

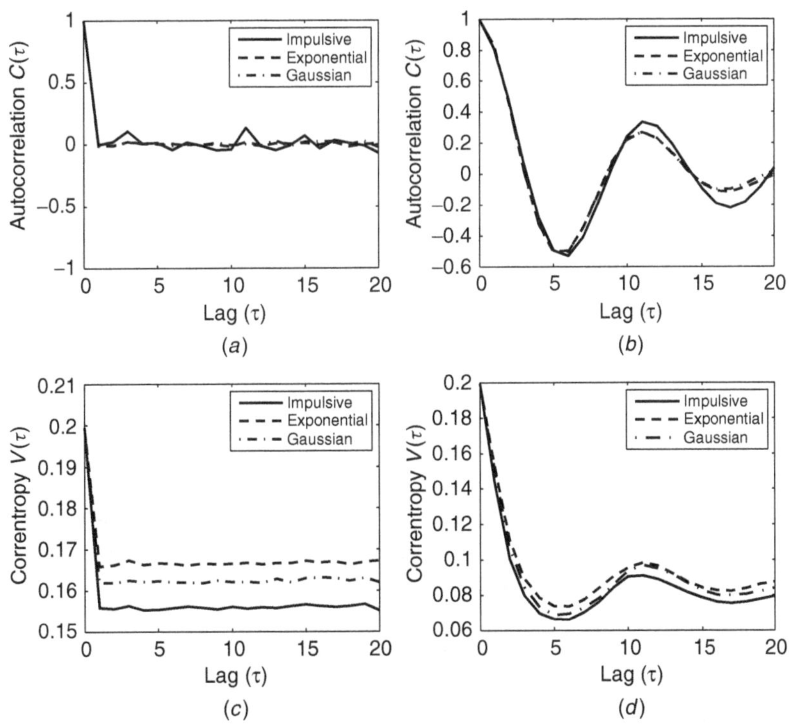

Figure 4.8 Autocorrelation function for the white (*a*) and filtered (*b*) processes. Correntropy function for the white (*c*) and filtered (*d*) processes.

function is again similar for the three filtered processes. However, the correntropy function can distinguish these three filtered processes while preserving their original rankings (namely, exponential source the highest, impulsive source the lowest).

4.8 KERNEL MATCHED FILTER

As discussed in Chapter 2, the matched filter is an optimum filter for signal detection when the target is known at the receiver. In the literature, the so-called *spectral matched filter* has also been designed for hyperspectral target detection, where the linear spectral signal that consists of N spectral bands is modeled as a linear combination of the target spectral signature plus additive noise:

$$\mathbf{x} = a\mathbf{s} + \mathbf{n}, \tag{4.68}$$

where $\mathbf{x}$, $\mathbf{s}$, and $\mathbf{n}$ denote the N-dimensional observation, target, and noise vectors, respectively, and scalar a is an attenuation constant that serves as a target abundance measure: $a = 0$ implies that no target is present and $a > 0$ implies that a target is present. The linear spectral matched filter is designed such that the desired (known) target signal $\mathbf{s}$ is passed through the filter while minimizing the averaged filter output. The optimal filter solution is given by the following impulse response [656]:

$$\mathbf{w}_{\text{opt}} = \frac{\mathbf{C}^{-1}\mathbf{s}^{\text{T}}}{\mathbf{s}^{\text{T}}\mathbf{C}^{-1}\mathbf{s}}, \tag{4.69}$$

where $\mathbf{C}$ denotes the sample covariance matrix of $\mathbf{x}$ based on the observed signal matrix $\mathbf{X} = [\mathbf{x}_1, \ldots, \mathbf{x}_\ell]$. When a new input signal $\mathbf{r}$ is presented to the matched filter, the filter output is given by

$$y_{\mathbf{r}} = \mathbf{w}_{\text{opt}}^{\text{T}}\mathbf{r} = \frac{\mathbf{s}^{\text{T}}\mathbf{C}^{-1}\mathbf{r}}{\mathbf{s}^{\text{T}}\mathbf{C}^{-1}\mathbf{s}}. \tag{4.70}$$

Recently, Nasrabadi and Kwon [517, 656] proposed a kernelized version of the spectral linear filter that exploits the nonlinear correlations between the spectral bands, which are typically ignored in the linear matched filter. Specifically, in line with (4.68), the following nonlinear model was assumed [656]:

$$\Phi(\mathbf{x}) = a_\Phi \Phi(\mathbf{s}) + \mathbf{n}_\Phi, \tag{4.71}$$

where Φ denotes a nonlinear mapping that maps the observed signal inside the linear space into a high-dimensional feature space, a_Φ denotes the corresponding attenuation coefficient in the feature space, and $\mathbf{n}_\Phi$ denotes the noise component in the feature space. Accordingly, the desired matched filter's output for the input $\Phi(\mathbf{r})$ is given by

$$y_{\Phi(\mathbf{r})} = \frac{\Phi(\mathbf{s})^{\mathrm{T}}\mathbf{C}_{\Phi}^{-1}\Phi(\mathbf{r})}{\Phi(\mathbf{s})^{\mathrm{T}}\mathbf{C}_{\Phi}^{-1}\Phi(\mathbf{s})}, \tag{4.72}$$

where $\mathbf{C}_{\Phi}$ denotes the centered covariance matrix in the feature space. Using the kernel trick and KPCA, the following kernelized matched filter can be derived [656]:

$$y_{\mathbf{k}_{\mathbf{r}}} = \frac{\mathbf{k}_{\mathbf{s}}^{\mathrm{T}}\mathbf{K}^{-1}\mathbf{k}_{\mathbf{r}}}{\mathbf{k}_{\mathbf{s}}^{\mathrm{T}}\mathbf{K}^{-1}\mathbf{k}_{\mathbf{s}}}, \tag{4.73}$$

where $\mathbf{K} = K(\mathbf{X}, \mathbf{X})$ denotes an $\ell \times \ell$ Gram matrix calculated from the observation matrix $\mathbf{X}$, with $(\mathbf{K})_{ij} = K(\mathbf{x}_i, \mathbf{x}_j)$, and $\mathbf{k}_{\mathbf{s}} = K(\mathbf{X}, \mathbf{s})$ and $\mathbf{k}_{\mathbf{r}} = K(\mathbf{X}, \mathbf{r})$ denote two $\ell \times 1$ column vectors. Notably, the kernel matrix $\mathbf{K}$ and two empirical kernel maps $\mathbf{k}_{\mathbf{s}}, \mathbf{k}_{\mathbf{r}}$ are required to be properly centered.

In [517, 656], the above-described kernel matched filter was demonstrated to be superior to the standard linear spectral matched filter in terms of reduced detection error.

4.9 DISCUSSION

In this chapter, we have introduced the notion of the kernel for measuring the similarity or distance between pairs of data points in a high-dimensional feature space. By using the kernel trick, we can extend many linear correlation-based statistical algorithms to their kernelized versions, such as KPCA, KCCA, KICA, kernel LDA, and kernel Wiener filter. The concepts of RKHS and reproducing kernel are essential for formulating these kernelized algorithms. The kernelized algorithms can be viewed as the natural nonlinear generalizations of their linear counterparts. Because the kernel function introduces nonlinearity and higher order correlation between variables, the kernelized algorithms often obtain superior performance (in either feature extraction or pattern discrimination) relative to their linear versions. We have presented several toy examples to demonstrate this point in this chapter. Note, however, that the advantages of these kernelized algorithms also come at the expense of higher computational cost. In addition, the linear algorithms often produce results that can be more clearly interpreted.

Recently kernel learning has expanded rapidly and established itself as an important branch of machine learning [799, 827]. This research field is so diverse that it is impossible to cover all important topics here. Nonetheless, we would like to briefly mention several interesting research topics in the context of correlation-based kernel learning.

Gaussian Processes. A stochastic process $\{\mathbf{x}(t)\}$ is called Gaussian if the random variables $\mathbf{x}(t_1), \ldots, \mathbf{x}(t_n)$ are jointly Gaussian for any n and $t_1, \ldots, t_n$. The Gaussian process is the most popular continuous-valued stochastic process that is sufficiently characterized by the mean and covariance functions. Examples of Gaussian processes include the Brownian motion (also called Wiener process [956]) and the Markov Gaussian process (which serves as the basis of the Kalman filter theory [440]). In the context of time series analysis, Parzen [706, 708] showed that the choice of RKHS is equivalent to the choice of a zero-mean stochastic process associated with a correlation kernel function K (which is assumed to be symmetric and positive definite); that is,

$$\mathbb{E}[f(\mathbf{x})] = 0,$$
$$\mathbb{E}[f(\mathbf{x}_i)f(\mathbf{x}_j)] = \sigma^2 K(\mathbf{x}_i, \mathbf{x}_j),$$

where σ^2 denotes the variance of the observed data samples. Essentially, the Gaussian process extends the notion of a set of random variables to random functions, and therefore it provides a tool for probabilistic inference, smoothing, and prediction [755]. When the kernel function K is shift invariant, it gives rise to a stationary stochastic process. For the stationary Gaussian process, the kernel K is an isotropic (i.e., the variances are identical in all directions) Gaussian function

$$K(\mathbf{x}_i, \mathbf{x}_j) = \exp\left(-\frac{\|\mathbf{x}_i - \mathbf{x}_j\|^2}{2\sigma^2}\right). \tag{4.74}$$

From a Bayesian perspective, Gaussian processes are based on the prior assumption that adjacent observations should convey information about each other. The observed variables are Gaussian, and $K(\mathbf{x}_i, \mathbf{x}_j)$ describes the correlation between the observations $f(\mathbf{x}_i)$ and $f(\mathbf{x}_j)$. Provided that two observations $f(\mathbf{x}_1)$ and $f(\mathbf{x}_2)$ are of interest, we can estimate the conditional probability of one given the other as follows:

$$p(f(\mathbf{x}_2)|f(\mathbf{x}_1)) = \frac{p(f(\mathbf{x}_2), f(\mathbf{x}_1))}{p(f(\mathbf{x}_1))}. \tag{4.75}$$

Notably, the marginal probability density $p(f(\mathbf{x}_1))$ and the conditional probability density $p(f(\mathbf{x}_2)|f(\mathbf{x}_1))$ are both Gaussian. Figure 4.9 presents an illustrative example for a simple inference problem (taken from [799]).

The standard Gaussian process has a shift-invariant covariance function, which implicitly assumes stationarity among the data samples. However, it is also possible to introduce nonstationary Gaussian processes for data smoothing [695]. Specifically, Paciorek [695] proposed a non-stationary correlation kernel that has a form of an anisotropic *squared exponential correlation function*:

$$K(\mathbf{x}_i, \mathbf{x}_j) = \sigma^2 |\mathbf{\Sigma}_i|^{1/4} |\mathbf{\Sigma}_j|^{1/4} \left|\frac{\mathbf{\Sigma}_i + \mathbf{\Sigma}_j}{2}\right|^{-1/2} \exp(-Q_{ij}), \tag{4.76}$$

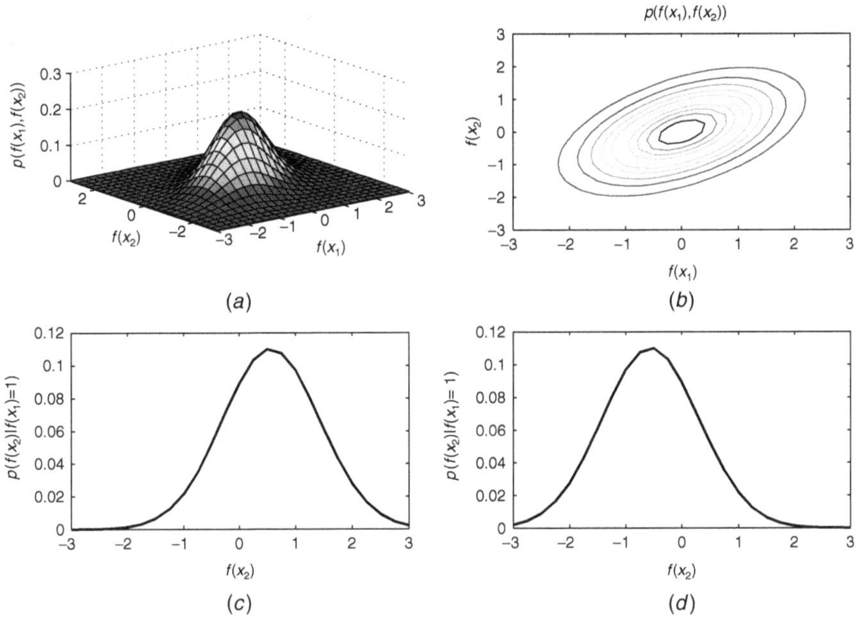

Figure 4.9 (*a*) A two-dimensional joint Gaussian distribution $p(f(\mathbf{x}_1), f(\mathbf{x}_2))$ with zero mean and correlation matrix $\begin{bmatrix} 1 & 0.25 \\ 0.25 & 0.8 \end{bmatrix}$. (*b*) The contour plot of $p(f(\mathbf{x}_1), f(\mathbf{x}_2))$. (*c*) Conditional probability density $p(f(\mathbf{x}_2)|f(\mathbf{x}_1) = 1)$. (*d*) Conditional probability density $p(f(\mathbf{x}_2)|f(\mathbf{x}_1) = -1)$.

where

$$Q_{ij} = (\mathbf{x}_i - \mathbf{x}_j)^{\mathrm{T}} \left(\frac{\boldsymbol{\Sigma}_i + \boldsymbol{\Sigma}_j}{2} \right)^{-1} (\mathbf{x}_i - \mathbf{x}_j), \tag{4.77}$$

in which $\boldsymbol{\Sigma}_i$ and $\boldsymbol{\Sigma}_j$ are two covariance matrices of the Gaussian kernel at data points $\mathbf{x}_i$ and $\mathbf{x}_j$, respectively. If the covariance matrices are constant (i.e., $\boldsymbol{\Sigma}_i = \boldsymbol{\Sigma}_j = \boldsymbol{\Sigma}\ \forall i, j$), then Q_{ij} reduces to the conventional *squared Mahalanobis distance*:

$$Q_{ij} = (\mathbf{x}_i - \mathbf{x}_j)^{\mathrm{T}} \boldsymbol{\Sigma}^{-1} (\mathbf{x}_i - \mathbf{x}_j), \tag{4.78}$$

which is also an anisotropic measure.

From a regularization theory viewpoint [162, 267], choosing a kernel function K is equivalent to assuming a Gaussian prior on the nonlinear functional, with the normalized covariance equal to K. With the stationarity assumption, choosing the covariance kernel is also equivalent to finding the correlation function of the Gaussian process [930]. In addition, the Gaussian process has natural connections to the GLM and the radial basis function (RBF) network [799, 959]. For in-depth discussions of Gaussian processes for regression and classification problems, see [580, 755, 812, 960].

Generalized Correlation Kernel and Sparse Representation. In the context of SVM regression, Papageorgiou et al. [701] proposed a generalized correlation kernel for multiresolution image compression and reconstruction. In general, the covariance kernel is defined as

$$K(\mathbf{x}, \mathbf{y}) = \mathbb{E}\left[(f(\mathbf{x}) - \mu(\mathbf{x}))(f(\mathbf{y}) - \mu(\mathbf{y}))\right], \tag{4.79}$$

where $\mu(\cdot)$ denotes the mean function of the argument. In light of the *spectral theorem*,[6] the correlation kernel can be represented by the sum of a number of basis functions

$$K(\mathbf{x}, \mathbf{y}) = \sum_i \lambda_i \phi_i(\mathbf{x})\phi_i(\mathbf{y}), \tag{4.80}$$

which is essentially the expansion of KPCA. In light of the RKHS theory, the function $f(\mathbf{x})$ can be represented by a reproducing kernel:

$$f(\mathbf{x}) = \sum_{i=1}^{\ell} c_i K(\mathbf{x}, \mathbf{x}_i). \tag{4.81}$$

Motivated by (4.80), Papageorgiou et al. [701] further proposed a *generalized correlation kernel*

$$K_d(\mathbf{x}, \mathbf{y}) = \sum_i (\lambda_i)^d \phi_i(\mathbf{x})\phi_i(\mathbf{y}), \tag{4.82}$$

where the scalar parameter $d \in \mathbb{R}$ controls the *locality* of the kernel: A small d will make $K_d(\mathbf{x}, \mathbf{y})$ look like a Dirac delta function, whereas a large d will make $K_d(\mathbf{x}, \mathbf{y})$ behave smoothly. It has been shown [732] that the linear combination of local correlation kernels is a *sparse representation* for functional approximation that closely relates to the SVM.

BIBLIOGRAPHICAL NOTES

The theoretical foundation of RKHS was established in [45]. The early ideas of applying RKHS to data analysis can be traced back to Kailath, Parzen, and Wahba in the respective fields of time series analysis, signal detection, and data smoothing [456, 708, 930].

The popularity of kernel learning can be partially ascribed to the great successes of the SVM [187] and kernel PCA [800]. Kernel methods have a close relationship to regularization theory, Gaussian process, and statistical learning theory [912]. Their in-depth relationships were reviewed in [267]. Kernel learning has established itself as an important branch of machine learning. An excellent resource for kernel methods is the book by Schökopf and Smola [799]. Extensive references on Gaussian processes for machine learning can be found in [755]. Extensions of CCA to the kernel framework have been addressed by many researchers [52, 352, 516, 520]. Kernel discriminant analysis was first proposed in [621] for the two-class problem, and it was further generalized to the multiple-class problem in [65]. Other variants have also been developed [799, 983, 1001]. The connection between kernel discriminant analysis and KPCA and LDA was established in [987]. Kernel Wiener filters were developed independently by several authors [182, 941, 984, 985]. Motivated by information-theoretic learning [263, 264, 738], kernel-based generalized correlation functions [790] and correntropy function [565, 566] were proposed as similarity measures in feature space based on the quadratic Rényi entropy and Parzen kernel estimator. The correntropy function was also used to derive a closed form for a nonlinear Wiener filter [733].

NOTES

1. A Hilbert space is a complete inner product space which defines an Euclidean space that is *complete, separable*, and generally *infinite dimensional*. Examples of Hilbert space include L_2, $\mathbb{R}^d$, and ℓ^2. However, not every Hilbert space is a RKHS, e.g., $L_2[0, 1]$.
2. Specifically, for a smooth function $f(\mathbf{x}) \in L_2(\chi)$, its RKHS norm associated with kernel K in the feature space $\mathcal{F}$ satisfies the condition [325]

$$\int \frac{|F(\omega)|^2}{\mathcal{K}(\omega)} d\omega < \infty,$$

where $F(\omega)$ and $\mathcal{K}(\omega)$ denote the Fourier transforms of f and K, respectively. Because K is a Mercer kernel, $\mathcal{K}(\omega)$ is real and positive, which implies that the function in the RKHS has a Fourier transform that decays rapidly and $\mathcal{F}$ is a space of smooth functions.
3. The centering operation can be done by computing the kernel matrix $\tilde{\mathbf{K}} = \mathbf{K} - \mathbf{1}_\ell \mathbf{K} - \mathbf{K}\mathbf{1}_\ell + \mathbf{1}_\ell \mathbf{K} \mathbf{1}_\ell$, where $\mathbf{1}_\ell$ denotes an $\ell \times \ell$ matrix with all entries equal to $1/\ell$.
4. The Volterra series expansion is an important way of representing nonlinear functions or nonlinear systems [696, 697, 730]. Consider a continuous and smooth nonlinear mapping $\mathbf{y} = \mathbf{F}(\mathbf{x})$ with $\mathbf{y} \in \mathbb{R}^n$ and $\mathbf{x} \in \mathbb{R}^N$; each output y_k can be expanded in a Taylor series around a fixed point (say, the origin), resulting in

$$y_k = f_k(\mathbf{x}) = a_{0k} + \sum_{i=1}^{m} a_{ik} x_i + \sum_{i=1}^{m} \sum_{j=1}^{m} a_{ijk} x_i x_j + \cdots \qquad (k = 1, 2, \ldots, n),$$

where the coefficients $a_{ik}, a_{ijk}, \ldots$ are obtained from the expansion and $a_{0k} = f_k(0)$. If we let $\mathbf{x}(t) = [x(t), x(t-1), \ldots, x(t-N+1)]^{\mathrm{T}}$, then $\mathbf{y}(t) = \mathbf{f}(\mathbf{x}(t))$ may be viewed as the discrete-time Volterra series expansion. Applications of Volterra series expansion include examples in image modeling [288] and system identification [219].
5. The generalized k-order Rényi entropy $(0 < k \in \mathbb{R})$ is defined as [189]

$$H_k(x) = \frac{1}{1-k} \log\left(\int p(x)^k \, dx \right).$$

When the limit $k \to 1$ is taken, the Rényi entropy reduces to the standard Shannon entropy. When $k = 2$, the Rényi entropy of order 2 is often called the *quadratic Rényi entropy* or *extension entropy*

$$H_2(x) = -\log\left(\int p(x)^2 dx\right).$$

By virtue of the *Jensen inequality* [189], we have $H_2(x) \le H(x)$. In general, Rényi entropy is a nonincreasing function in the sense that $H_k(x) \ge H_r(x)$ for any $r > k$.

6. In linear algebra or functional analysis, the spectral theorem provides conditions under which a matrix or an operator can be diagonalized; the result of the spectral theorem provides a canonical decomposition, also known as spectral decomposition. A representative example is the eigenvalue decomposition of a symmetric or nonsymmetric matrix.

5

CORRELATIVE LEARNING IN A COMPLEX-VALUED DOMAIN

A complex-valued variable comprises a real part and an imaginary part, which uniquely define the *modulus* (or amplitude) and *phase* (or angle) of the complex number.[1] The correlation statistic in a complex-valued domain is similar to that defined in its real-valued counterpart; however, the higher order cumulant statistics and nonlinearity defined for complex-valued variables are more complicated and require special attention. Extensions of correlation to complex random variables, complex random vectors, and complex random processes are well defined in the literature [623].

Complex-valued signals or observations are frequently encountered in practical applications, such as array signal processing, acoustics, imaging, radar, and communications. For instance, data from multiple sensory array are often modeled as a vector of complex random variables in which the phase encodes the spatial information. On the other hand, a real-valued signal in the time domain may also take a complex-valued form in the transform or frequency domain (such as the Fourier transform or Hilbert transform). In engineering, complex-valued neural networks have also been introduced [392, 393] for tackling the complex-valued signals or data. In this chapter, we will extend a number of correlation-based learning algorithms to the complex domain and illustrate their applications in various practical problems in communications, radar, and array signal processing.

Correlative Learning: A Basis for Brain and Adaptive Systems, by Zhe Chen, Simon Haykin, Jos J. Eggermont, and Suzanna Becker

5.1 PRELIMINARIES

A complex random variable $x \in \mathbb{C}$ is defined in the Cartesian form as $x = x_{\text{Re}} + jx_{\text{Im}}$, where $j = \sqrt{-1}$ and the real part $x_{\text{Re}} \in \mathbb{R}$ and imaginary part $x_{\text{Im}} \in \mathbb{R}$ are both real-valued random variables. In most cases, by "complex-valued" variable we mean that the variable is strictly complex if not stated otherwise; that is, the variable's imaginary part is not zero everywhere. For a complex-valued variable $x = x_{\text{Re}} + jx_{\text{Im}}$, its complex conjugate, denoted as x^*, is defined as $x^* = x_{\text{Re}} - jx_{\text{Im}}$. The relationship $x = x^*$ holds if and only if $x_{\text{Im}} = 0$. Alternatively, the complex variable x can also be represented in the polar form as $x = |x|e^{j\theta}$, where $|x| = \sqrt{x_{\text{Re}}^2 + x_{\text{Im}}^2}$ denotes the modulus and $\theta = \arg(x)$ $(0 \le \theta < 2\pi)$ denotes the phase.

The statistical properties of x are characterized by the *joint* probability density function (pdf) of x_{Re} and x_{Im}, $p(x) = p(x_{\text{Re}}, x_{\text{Im}}) \in \mathbb{R}$, provided that it exists. When x_{Re} and x_{Im} are mutually independent, then $p(x) = p(x_{\text{Re}})p(x_{\text{Im}})$. For instance, a complex random variable $x = x_{\text{Re}} + jx_{\text{Im}}$ is called *complex normal* if x_{Re} and x_{Im} are jointly normal (Gaussian); in this case, its pdf is defined as

$$p(x) = p(x_{\text{Re}}, x_{\text{Im}}) = \frac{1}{\sqrt{2\pi \det(\mathbf{\Sigma}_c)}} \exp\left(-\frac{1}{2}[\mathbf{x}_c - \mathbf{m}_c]^{\text{T}} \mathbf{\Sigma}_c^{-1} [\mathbf{x}_c - \mathbf{m}_c]\right), \tag{5.1}$$

where $\mathbf{x}_c = [x_{\text{Re}}, x_{\text{Im}}]^{\text{T}}$ denotes the augmented vector that contains the real and imaginary parts; $\mathbf{m}_c = \mathbb{E}[\mathbf{x}_c]$ and $\mathbf{\Sigma}_c = \mathbb{E}[(\mathbf{x}_c - \mathbf{m}_c)(\mathbf{x}_c - \mathbf{m}_c)^{\text{T}}]$ denote the mean and covariance of $\mathbf{x}_c$, respectively. Consequently, the Shannon entropy of the complex-valued variable x that satisfies (5.1) is given as

$$\begin{aligned} H(x) = H(x_{\text{Re}}, x_{\text{Im}}) &= -\int\int p(x_{\text{Re}}, x_{\text{Im}}) \log p(x_{\text{Re}}, x_{\text{Im}})\, dx_{\text{Re}}\, dx_{\text{Im}} \\ &= \log(2\pi e) + \frac{1}{2}\log\det(\mathbf{\Sigma}_c). \end{aligned}$$

Observe that the entropy $H(x)$ is a quantity that is independent of the mean values $\mathbb{E}[x_{\text{Re}}]$ and $\mathbb{E}[x_{\text{Im}}]$.

Given a complex variable $x = x_{\text{Re}} + jx_{\text{Im}} = |x|e^{j\theta}$, if x_{Re} and x_{Im} are *independent* Gaussian random variables with zero mean and equal variance σ^2, then it is known that the modulus and phase have, respectively, the Rayleigh and uniform distributions given by [702]

$$p_{|x|}(|x|) = \frac{|x|}{\sigma^2}\exp\left(-\frac{|x|^2}{2\sigma^2}\right) \qquad (|x| \ge 0),$$

$$p_\theta(\theta) = \frac{\sec^2\theta}{\pi(\tan^2\theta + 1)} = \begin{cases} 1/\pi, & -\frac{1}{2}\pi < \theta < \frac{1}{2}\pi, \\ 0, & \text{otherwise.} \end{cases}$$

Moment Statistics. Given an appropriate probability metric of random complex variables $x \in \mathbb{C}$, we can identify and calculate the first- and second-order moment/cumulant statistics:

- First-order moment (expected mean):

$$\mathbb{E}[x] = \mathbb{E}[x_{\text{Re}} + jx_{\text{Im}}] = \mathbb{E}[x_{\text{Re}}] + j\mathbb{E}[x_{\text{Im}}].$$

- Second-order moment:

$$\mathbb{E}[x^2] = \mathbb{E}[x_{\text{Re}}^2 - x_{\text{Im}}^2 + 2jx_{\text{Re}}x_{\text{Im}}] = \mathbb{E}[x_{\text{Re}}^2] - \mathbb{E}[x_{\text{Im}}^2] + 2j\mathbb{E}[x_{\text{Re}}x_{\text{Im}}].$$

- Second-order cumulant (variance):

$$\text{var}[x] = \mathbb{E}[|x - \mathbb{E}[x]|^2] = \mathbb{E}[|x|^2] - |\mathbb{E}[x]|^2 .$$

- For two complex random variables x_i and x_j, their covariance is defined as

$$\begin{aligned} C_{ij} &= \mathbb{E}\left[(x_i - \mathbb{E}[x_i])(x_j^* - \mathbb{E}[x_j^*])\right] \\ &= \mathbb{E}[x_i x_j^*] - \mathbb{E}[x_i]\mathbb{E}[x_j^*]. \end{aligned}$$

  Two complex random variables x_i and x_j $(j \neq i)$ are said to be mutually uncorrelated if $C_{ij} = 0$ or $\mathbb{E}[x_i x_j^*] = \mathbb{E}[x_i]\mathbb{E}[x_j^*]$.

In a similar way, we can define higher order cumulant statistics for complex-valued random variables. For instance, for a zero-mean complex-valued random variable x, the third- and fourth-order cumulant statistics (skewness and kurtosis) are defined as [660]

$$\text{skewness}(x) = \frac{\mathbb{E}[|x|^3]}{\left(\mathbb{E}[|x|^2]\right)^{3/2}}, \tag{5.2}$$

$$\text{kurtosis}(x) = \mathbb{E}[|x|^4] - 2\left(\mathbb{E}[|x|^2]\right)^2 - \left|\mathbb{E}[x^2]\right|^2 . \tag{5.3}$$

When the real and imaginary parts of x are mutually uncorrelated and have equal variance $\frac{1}{2}$, then $\mathbb{E}[x^2] = 0$, $\mathbb{E}[|x|^2] = 1$, and (5.2) and (5.3) are simplified to, respectively, $\text{skewness}(x) = \mathbb{E}[|x|^3]$ and $\text{kurtosis}(x) = \mathbb{E}[|x|^4] - 2$.

To extend the analysis from the scalar to the vector case, let $\mathbf{x} = [x_1, \ldots, x_n]$ be a complex-valued random vector and let $\mathbf{x}^{\text{H}} = [x_1^*, \ldots, x_n^*]^{\text{T}} \equiv (\mathbf{x}^*)^{\text{T}}$ denote its Hermitian transpose. The norm and the Hermitian inner product of $\mathbf{x}$ are defined as

$$\|\mathbf{x}\| = \langle \mathbf{x}, \mathbf{x} \rangle^{1/2} = \sqrt{\mathbf{x}^{\text{H}}\mathbf{x}} = \left(\|\mathbf{x}_{\text{Re}}\|^2 + \|\mathbf{x}_{\text{Im}}\|^2\right)^{1/2} = \|\mathbf{x}^*\|, \tag{5.4}$$

$$\langle \mathbf{x}, \mathbf{y} \rangle = \mathbf{x}^{\text{H}}\mathbf{y} = \mathbf{x}_{\text{Re}}^{\text{T}}\mathbf{y}_{\text{Re}} + \mathbf{x}_{\text{Im}}^{\text{T}}\mathbf{y}_{\text{Im}} + j(\mathbf{x}_{\text{Re}}^{\text{T}}\mathbf{y}_{\text{Im}} - \mathbf{x}_{\text{Im}}^{\text{T}}\mathbf{y}_{\text{Re}}) = (\langle \mathbf{y}, \mathbf{x} \rangle)^* . \tag{5.5}$$

It is noted that the inner product is Hermitian and the norm is nonnegative (i.e., $\|\mathbf{x}\| \geq 0$ and the equality holds if and only if $\mathbf{x} = \mathbf{0}$). A complex vector space endowed with the inner product operator is called a *complex inner product space* or *unitary space.*

For a complex-valued vector $\mathbf{x} \in \mathbb{C}^n$, its mean and autocorrelation matrix are defined, respectively, as

$$\mathbb{E}[\mathbf{x}] = (\mathbb{E}[x_1], \ldots, \mathbb{E}[x_n]), \tag{5.6}$$

$$\mathbb{E}[\mathbf{x}\mathbf{x}^{\mathrm{H}}] = \begin{bmatrix} C_{11} & \cdots & C_{1n} \\ \vdots & \ddots & \vdots \\ C_{n1} & \cdots & C_{nn} \end{bmatrix}, \tag{5.7}$$

where $C_{ij} = \mathbb{E}[x_i x_j^*]$. It is noted that the correlation matrix uses the Hermitian transpose instead of the conventional transpose operation, hence (5.7) is different from the so-called *pseudocorrelation matrix*:

$$\mathbb{E}[\mathbf{x}\mathbf{x}^{\mathrm{T}}] = \begin{bmatrix} C'_{11} & \cdots & C'_{1n} \\ \vdots & \ddots & \vdots \\ C'_{n1} & \cdots & C'_{nn} \end{bmatrix}, \tag{5.8}$$

where $C'_{ij} = \mathbb{E}[x_i x_j]$; note that $\mathbb{E}[\mathbf{x}\mathbf{x}^{\mathrm{H}}] = \mathbb{E}[\mathbf{x}^*\mathbf{x}^{\mathrm{T}}]$.

According to the common terminology of the literature (e.g., [723–725]), the complex-valued random vector $\mathbf{x}$ is called *second-order circular* or *strictly proper* if its pseudocovariance matrix is a null matrix, namely, $\mathrm{Pcov}[\mathbf{x}] \equiv \mathbb{E}\left[(\mathbf{x} - \mathbb{E}[\mathbf{x}])(\mathbf{x} - \mathbb{E}[\mathbf{x}])^{\mathrm{T}}\right] = \mathbf{0}$. If its covariance matrix $\mathrm{cov}[\mathbf{x}] \equiv \mathbb{E}\left[(\mathbf{x} - \mathbb{E}[\mathbf{x}])(\mathbf{x} - \mathbb{E}[\mathbf{x}])^{\mathrm{H}}\right]$, is positive definite, then the complex-valued random vector $\mathbf{x}$ is called *full*. If $\mathbb{E}[\mathbf{x}\mathbf{x}^{\mathrm{H}}]$ is diagonal, then we say the random vector $\mathbf{x}$ has *uncorrelated* components; the random vector $\mathbf{x}$ is said to have *strongly uncorrelated* components if $\mathbb{E}[\mathbf{x}\mathbf{x}^{\mathrm{H}}]$ and $\mathbb{E}[\mathbf{x}\mathbf{x}^{\mathrm{T}}]$ are both diagonal. When the real and imaginary parts of $\mathbf{x}$ have equal variance, $\mathbf{x}$ is often said to be *symmetric*. If $\mathbf{x} = [x_1, \ldots, x_n]$ is nonsymmetric, then its *circularity coefficient*, denoted by $\{\lambda_i\}_{i=1}^n$, is defined by the variance difference between the real and imaginary components:

$$\lambda_i = |\mathrm{var}[\mathrm{Re}\{x_i\}] - \mathrm{var}[\mathrm{Im}\{x_i\}]| \qquad (i = 1, \ldots, n). \tag{5.9}$$

Two complex-valued random vectors $\mathbf{x}_1$ and $\mathbf{x}_2$ are said to be uncorrelated if and only if $\mathrm{cov}[\mathbf{x}_1, \mathbf{x}_2] = \mathrm{Pcov}[\mathbf{x}_1, \mathbf{x}_2] = \mathbf{0}$.

Definition 5.1 [723] *A complex random variable x is said to be "circular" if, for any real-valued α, the probability density functions $p(x)$ and $p(e^{j\alpha}x)$ are the same (i.e., rotation invariant).*

Note that the circularity of x implies that $\mathbb{E}[x_{\mathrm{Re}} x_{\mathrm{Im}}] = 0$, but not vice versa.

Given a circular complex-valued variable x, for all $p, q \in \mathbb{N}$, we have

$$\mathbb{E}\left[x^p (x^*)^q\right] = 0 \qquad (p \neq q).$$

For a zero-mean complex random variable x, the second-order circularity implies that $\mathbb{E}[x^2] = 0$, and the real and imaginary parts of x are uncorrelated and have equal variances. For an n-dimensional *circular complex Gaussian* random vector $\mathbf{x} = \mathbf{x}_{\text{Re}} + j\mathbf{x}_{\text{Im}}$, its pdf can be characterized in a compact way [623, 723, 974]:

$$p_{\mathbf{x}}(\mathbf{x}) = \frac{1}{\pi^n \det(\mathbf{\Sigma})} \exp\left[-(\mathbf{x}-\mathbf{m})^{\text{H}} \mathbf{\Sigma}^{-1} (\mathbf{x}-\mathbf{m})\right], \tag{5.10}$$

where $\mathbf{m} = \mathbb{E}[\mathbf{x}]$ and $\mathbf{\Sigma} = \text{cov}[\mathbf{x}]$ denote, respectively, the mean vector and covariance matrix of the n-dimensional complex random variable $\mathbf{x}$. The representation of the pdf (5.10) is more economical than the one that splits the real and imaginary parts and construct a $2n$-dimensional real-valued vector for the *generalized complex Gaussian* pdf, in which $[\mathbf{x}_{\text{Re}}^{\text{T}}, \mathbf{x}_{\text{Im}}^{\text{T}}]^{\text{T}}$ is jointly Gaussian. For a zero-mean complex Gaussian random variable $\mathbf{x} \in \mathbb{C}^n$ with circularity coefficients $\lambda_i \neq 1$ $(i = 1, \ldots, n)$, its Shannon entropy is given by [265]

$$H(\mathbf{x}) = n \log(\pi e) + \log\det(\mathbf{\Sigma}) + \frac{1}{2}\sum_{i=1}^{n} \log(1-\lambda_i^2). \tag{5.11}$$

Note that when the random variable $\mathbf{x}$ is additionally circular the third term on the right-hand side of the above equation vanishes to zero. Because the third term is always nonpositive, it also follows that the entropy of a complex Gaussian random variable is maximized when its pseudocovariance matrix is a null matrix.

Remark: Note that, although the probabilistic property or structure of the complex random variable can be described by its real and imaginary parts, the operational structure cannot; this is due to the fact that the n-dimensional complex space is not equivalent to the $2n$-dimensional real space as an inner product space and they use different algebras [265].

Nonlinearity. Typically, functions of complex variables have rather different mathematical properties (such as convergence, continuity, differentiability, and integrability) from those of real variables [547]. A function whose range is in the complex domain is said to be a complex function, or a complex-valued function.

Definition 5.2 *A complex function is said to be analytic*[2] *on a real plane $\mathbb{R}$ if it is complex differentiable at every point in $\mathbb{R}$.*

Definition 5.3 *A complex function $f(x)$ is analytic on a complex plane if the following two conditions are fulfilled: (i) $f(x)$ is derivable at x; and (ii) there exists a neighborhood $\aleph$ of $x \in \mathbb{C}$ such that $f(\cdot)$ is derivable at every point of $\aleph$. A function that is analytic on the whole complex plane is called an entire function.*

Theorem 5.1 Liouville's theorem [369, 547] *If $f(z)$ is analytic and bounded on the complex plane, then $f(z)$ is a constant.*

To state more precisely, for any complex-valued variable, every bounded [i.e., there exists a real number M such that $|f(x)| = M$ for all $x \in \mathbb{C}$] entire function must be constant.

Because of Liouville's theorem, we know that there is a trade-off between *boundedness* and *analyticity* in the choice of nonlinearity in the complex domain. Namely, if one defines a fully complex and analytic nonlinear function, then it loses the boundedness; on the other hand, if we require the function be bounded, then we suffer from the loss of analyticity since the *Cauchy–Riemann equations* do not hold.[3] In the literature, there are three options for solving this dilemma:

- Choose a nonlinear function $f(\cdot) : \mathbb{R} \to \mathbb{R}$ which only processes the modulus of the complex-valued variable and ignores the phase information; namely $f(x) = f(|x|)$. This is particularly useful when the complex-valued data are *circular*; namely, the pdf of the random variable is rotation invariant in the complex plane.
- Choose a "split" nonanalytic nonlinear function $f(\cdot) : \mathbb{R} \to \mathbb{R}$ such that the real and imaginary parts are processed separately: $f(x) = f(x_{\text{Re}}) + jf(x_{\text{Im}})$. In this case, the function f may satisfy the boundedness condition.
- Choose a fully complex nonlinear function $f(\cdot) : \mathbb{C} \to \mathbb{C}$ such that the property of analyticity is preserved.

As an example, Figure 5.1 illustrates the difference between a split-complex bounded hyperbolic tangent function $\tanh(x) = \tanh(x_{\text{Re}}) + j\tanh(x_{\text{Im}})$ and a fully complex analytic hyperbolic tangent function $\tanh(x)$.

A complex variable $x \in \mathbb{C}$ and its conjugate x^* can be treated as independent variables; therefore, a complex variable and its conjugate are viewed as the result of applying an invertible linear transformation to the variable's real and imaginary parts. Such a treatment may somewhat simplify the complex analysis, especially when encountering the differentiability issue. For instance, the function $f(x) = (|x|)^2$ is *not* a differentiable function on the complex plane [because the function $f(x) = x^*$ is not analytic with respect to x]. However, by treating real and imaginary parts of x and x^* as independent variables, we obtain

$$\frac{\partial |x|^2}{\partial x} = x^* \quad \text{and} \quad \frac{\partial |x|^2}{\partial x^*} = x. \tag{5.12}$$

Gradient and Hessian. The learning-and-optimization procedure often requires the estimation of the gradient or Hessian information, for which it is desirable to derive the complex-valued versions of the gradient and Hessian operators [369, 905].

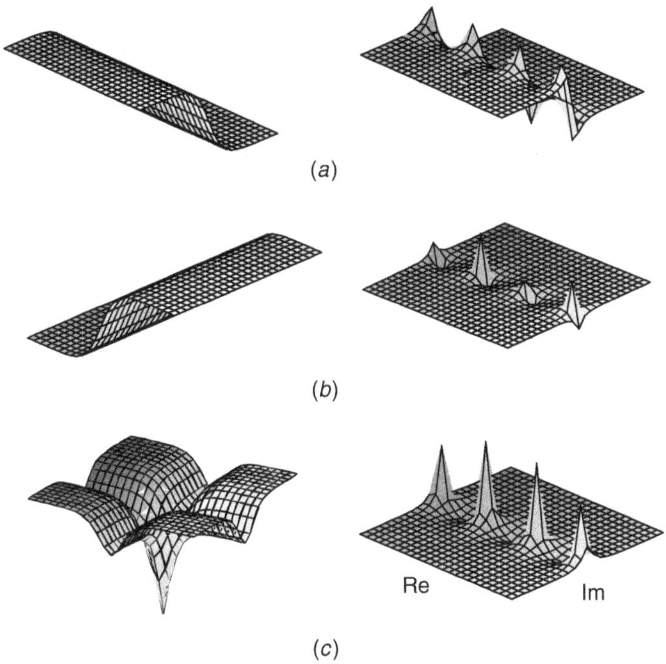

Figure 5.1 Comparison of a split-complex tanh function (left column) and a fully complex analytic tanh function (right column) in terms of (*a*) the real part, (*b*) the imaginary part, and (*c*) the modulus.

Suppose the goal is to optimize a real-valued cost function $J(x)$ $(x \in \mathbb{C})$. The natural way is to calculate its derivative and set it to zero. However, if the cost function $J(x)$ is nonanalytic (and thus nondifferentiable with respect to x), we have to treat x and x^* as two independent variables for optimization; namely, $\mathrm{d}x/\mathrm{d}x = 1, \mathrm{d}x^*/\mathrm{d}x = \mathrm{d}x/\mathrm{d}x^* = 0$. In particular, the following theorem holds:

Theorem 5.2 *If the function $J(x)$ $(x \in \mathbb{C})$ is real valued and analytic with respect to x and x^*, all stationary points can be found by setting the derivatives with respect to either x or x^* to zero.*

Next, let us further consider the problem of optimizing a real-valued, bounded cost function $J(\mathbf{x})$ that has a complex-valued argument $\mathbf{x} \in \mathbb{C}^n$. Since $J(\mathbf{x})$ is nonanalytic (because of its boundedness assumption), its derivative has to be calculated based on real-valued functions. Without loss of generality, we assume $J(\mathbf{x})$ can be decomposed into the form of two real-valued functions $U(\mathbf{x})$ and $V(\mathbf{x})$ as follows:

$$J(\mathbf{x}) = |U(\mathbf{x}) + jV(\mathbf{x})|^2 = U^2(a,b) + V^2(a,b), \tag{5.13}$$

where a and b denote, respectively, the real and imaginary parts of the associated complex-valued variables. Then the partial derivative of $J(\mathbf{x})$ with respect to the real and imaginary parts of $\mathbf{x} \in \mathbb{C}^n$ can be calculated separately as follows:

$$\frac{\partial J(\mathbf{x})}{\partial \mathbf{x}_{\text{Re}}} = 2U\left(\frac{\partial U(a,b)}{\partial a}\frac{\partial a}{\partial \mathbf{x}_{\text{Re}}} + \frac{\partial U(a,b)}{\partial b}\frac{\partial b}{\partial \mathbf{x}_{\text{Re}}}\right) + 2V\left(\frac{\partial V(a,b)}{\partial a}\frac{\partial a}{\partial \mathbf{x}_{\text{Re}}} + \frac{\partial V(a,b)}{\partial b}\frac{\partial b}{\partial \mathbf{x}_{\text{Re}}}\right), \tag{5.14}$$

$$\frac{\partial J(\mathbf{x})}{\partial \mathbf{x}_{\text{Im}}} = 2U\left(\frac{\partial U(a,b)}{\partial a}\frac{\partial a}{\partial \mathbf{x}_{\text{Im}}} + \frac{\partial U(a,b)}{\partial b}\frac{\partial b}{\partial \mathbf{x}_{\text{Im}}}\right) + 2V\left(\frac{\partial V(a,b)}{\partial a}\frac{\partial a}{\partial \mathbf{x}_{\text{Im}}} + \frac{\partial V(a,b)}{\partial b}\frac{\partial b}{\partial \mathbf{x}_{\text{Im}}}\right). \tag{5.15}$$

In light of the Cauchy–Riemann equations, we can rewrite the derivative of $J(\mathbf{x})$ with respect to $\mathbf{x} \in \mathbb{C}^n$ as

$$\frac{\partial J(\mathbf{x})}{\partial \mathbf{x}} = \frac{1}{2}\left(\frac{\partial J(\mathbf{x})}{\partial \mathbf{x}_{\text{Re}}} - j\frac{\partial J(\mathbf{x})}{\partial \mathbf{x}_{\text{Im}}}\right). \tag{5.16}$$

Similarly, we also have

$$\frac{\partial J(\mathbf{x})}{\partial \mathbf{x}^*} = \frac{1}{2}\left(\frac{\partial J(\mathbf{x})}{\partial \mathbf{x}_{\text{Re}}} + j\frac{\partial J(\mathbf{x})}{\partial \mathbf{x}_{\text{Im}}}\right). \tag{5.17}$$

To find the stationary points of $J(\mathbf{x})$ for the complex-valued vector $\mathbf{x} \in \mathbb{C}^n$, we need to solve the equation $\partial J(\mathbf{x})/\partial \mathbf{x} = \mathbf{0}$ or $\partial J(\mathbf{x})/\partial \mathbf{x}^* = \mathbf{0}$. Typically, we define the gradient operator as [369]

$$\nabla J = \frac{\partial J(\mathbf{x})}{\partial \mathbf{x}^*}. \tag{5.18}$$

The stationary point is described by $\nabla J = \mathbf{0}$, which also implies that at a stationary point $\partial J(\mathbf{x})/\partial \mathbf{x}_{\text{Re}} = \partial J(\mathbf{x})/\partial \mathbf{x}_{\text{Im}} = \mathbf{0}$.

Definition 5.4 *A real-valued function $J(\mathbf{x})$ (where $\mathbf{x} \in \mathbb{C}^n$) is said to be convex in the complex plane if*

$$J(\lambda \mathbf{z}_1 + (1-\lambda)\mathbf{z}_2) \leq \lambda J(\mathbf{z}_1) + (1-\lambda) J(\mathbf{z}_2)$$

for all $\mathbf{z}_1, \mathbf{z}_2 \in \mathbb{C}^n$ and $0 \leq \lambda \leq 1$.

Likewise, assuming $J(\mathbf{x}) \in \mathbb{R}$ is is twice differentiable with respect to $\mathbf{x} \in \mathbb{C}^n$, then the Hessian is defined by the second-order derivative

$$\mathbf{H} = \frac{\partial^2 J(\mathbf{x})}{\partial \mathbf{x}\mathbf{x}^{\mathrm{H}}} = \begin{bmatrix} \dfrac{\partial^2 J}{\partial x_1\, \partial x_1^*} & \dfrac{\partial^2 J}{\partial x_1\, \partial x_2^*} & \cdots & \dfrac{\partial^2 J}{\partial x_1\, \partial x_n^*} \\ \dfrac{\partial^2 J}{\partial x_2\, \partial x_1^*} & \dfrac{\partial^2 J}{\partial x_2\, \partial x_2^*} & \cdots & \dfrac{\partial^2 J}{\partial x_2\, \partial x_n^*} \\ \vdots & \vdots & \ddots & \vdots \\ \dfrac{\partial^2 J}{\partial x_n\, \partial x_1^*} & \dfrac{\partial^2 J}{\partial x_n\, \partial x_2^*} & \cdots & \dfrac{\partial^2 J}{\partial x_n\, \partial x_n^*} \end{bmatrix}. \tag{5.19}$$

If the Hessian matrix $\mathbf{H}$ is positive semidefinite (i.e., with nonnegative real eigenvalues), then $J(\mathbf{x})$ is said to be a convex function.[4]

5.2 COMPLEX-VALUED EXTENSIONS OF CORRELATION-BASED LEARNING

5.2.1 Complex-Valued Associative Memory

Analogous to the real-valued, bipolar, discrete Hopfield network [399], the complex-valued Hopfield network can also be developed for multistate associative memory [438, 641, 664]. Specifically, given a complex-valued state vector $\mathbf{x} \in \mathbb{C}^N$, the Lyapunov energy function can be constructed as follows:

$$J(\mathbf{x}) = -\frac{1}{2}\mathbf{x}^{\mathrm{H}}\mathbf{W}\mathbf{x} = -\frac{1}{2}\sum_{i=1}^{N}\sum_{k=1}^{N} w_{ik}x_i^* x_k, \tag{5.20}$$

where $\mathbf{W}$ is a Hermitian matrix with nonnegative diagonal entries (i.e., $w_{ii} \geq 0$) and the synaptic weight matrix that stores the state prototypes is learned from the complex-valued generalization of Hebb's rule [664]:

$$\mathbf{W} = \frac{1}{N}\sum_{l=1}^{\ell} \mathbf{x}_l \mathbf{x}_l^{\mathrm{H}}, \tag{5.21}$$

where $\mathbf{x}_l \mathbf{x}_l^{\mathrm{H}}$ is the instantaneous autocorrelation of $\mathbf{x}_l \in \mathbb{C}^N$. In this case, the complex-valued couplings represent the phase shifts due to finite propagation delays of hidden variables $\mathbf{x} = [x_1, \ldots, x_N]^{\mathrm{T}}$. At each time index t, the neuron's state is updated by the asynchronous rule [664]:

$$x_k(t+1) = \operatorname{csign}_N \left(e^{j(\pi/N)} \sum_i w_{ki} x_i(t) \right), \tag{5.22}$$

where $\mathrm{csign}_N(\cdot)$ is a complex signum function that defines an N-stage phase quantizer for complex numbers as follows:

$$\mathrm{csign}_N(z) = \begin{cases} e^0, & 0 \le \arg(z) \le \frac{2\pi}{N}, \\ e^{j2\pi/N}, & \frac{2\pi}{N} \le \arg(z) \le \frac{4\pi}{N}, \\ \vdots & \vdots \\ e^{j2\pi(N-1)/N}, & \frac{2\pi(N-1)}{N} \le \arg(z) \le 2\pi, \end{cases}$$

where the resolution factor N divides evenly the complex unit circle into N separate sectors and each of them has an angle $2\pi/N$. Notably, when $N = 2$, it is functionally equivalent to the real-valued discrete Hopfield network, in which all neuron states are bipolar real values (i.e., ± 1); the only difference is that the standard Hopfield network does not permit complex-valued connections. Theoretical analysis of such complex-valued neural associative memories can be found in [154, 540, 618].

Similarly, continuous complex-valued associative memories may also be developed [512, 513]. Specifically, a complex-valued continuous Hopfield network may be described by the following differential equations [512]:

$$\tau \frac{du_j(t)}{dt} = -u_j(t) + \sum_{k=1}^{N} w_{jk}^* x_k(t), \tag{5.23}$$

$$x_j(t) = f(u_j(t)), \tag{5.24}$$

where $\tau > 0$ denotes the time constant and $f(\cdot)$ in (5.24) is a complex activation function defined by

$$f(z) = \frac{\lambda z}{\lambda - 1 + |z|}, \qquad z \in \mathbb{C}, \tag{5.25}$$

where λ is a real number that is greater than 1 (i.e., $\lambda - 1 > 0$). Such an activation function is nonanalytic but bounded and it has continuous partial derivatives. The synaptic weights w_{jk} is constructed by the autocorrelation rule (5.21) as in the discrete Hopfield network.

Because of the use of complex number, the storage capacity of the complex-valued Hopfield network depends on the number of states N. Theoretical analysis of the storage capacity of the complex Hopfield network is referred to [165].

5.2.2 Complex-Valued Boltzmann Machine

In parallel to the development in the real domain, the idea of extending the Hopfield network to the Boltzmann machine can be pursued in the complex domain. Specifically, Zemel et al. [997] proposed a complex-valued Boltzmann machine with directional units in order to enhance the representation power of the conventional binary Boltzmann machine. Similar to the complex-valued Hopfield network, the state of each directional unit is described by a complex variable, where the phase component specifies the direction. The energy function is the same as (5.20), and the probability for determining the state of a directional unit $x_i = a_i e^{j\theta_i}$ is described by the so-called *von Mises* (or circular normal) distribution

$$p(X_i = x_i) \propto e^{\beta a_i \cos(\tau_i - \theta_i)}, \qquad x_i \in \mathbb{C}, \qquad a_i > 0, \qquad \theta_i \in (0, 2\pi], \quad (5.26)$$

where $\beta = 1/T$ denotes the reciprocal of the temperature parameter and

$$p(\tau; \overline{\tau}, m) = \frac{1}{2\pi I_0(m)} e^{m\cos(\tau - \overline{\tau})} \quad (5.27)$$

denotes the pdf of the circular normal distribution, in which $\overline{\tau} \in (0, 2\pi]$ specifies the mean direction, $m > 0$ behaves like the reciprocal of the variance parameter of a Gaussian distribution, and $I_0(m) = (1/\pi) \int_0^{\pi} e^{\cos\xi}\, d\xi$ is the modified zero-order Bessel function of the first kind [588]. Given (5.26) and (5.27), the mean of the state is defined by

$$\langle x_i \rangle = r_i e^{j\gamma_i} \quad (5.28)$$

with the mean direction parameter $\gamma_i = \overline{\tau}_i$ and the mean modulus parameter

$$r_i = \frac{I_1(m_i)}{I_0(m_i)} = \frac{I_1(\beta a_i)}{I_0(\beta a_i)}, \quad (5.29)$$

where $I_1(m) = (1/\pi) \int_0^{\pi} e^{m\cos\xi} \cos\xi\, d\xi$ is the modified first-order Bessel function of the first kind.

Analogous to the mean-field approximation for a deterministic binary Boltzmann machine [383, 715], Zemel et al. [997] also developed a mean-field approximation algorithm which allows one to learn the unknown parameters $w_{ki} = b_{ki} e^{j\alpha_{ki}}$ with the following generalized Hebb's rule:

$$\Delta b_{ki} \propto r_k r_i \cos(\gamma_k - \gamma_i + \alpha_{ki}), \quad (5.30)$$

$$\Delta \alpha_{ki} \propto -r_k r_i b_{ki} \sin(\gamma_k - \gamma_i + \alpha_{ki}), \quad (5.31)$$

where $\{r_k, \gamma_k\}$ and $\{r_i, \gamma_i\}$ denote the expected means of the modulus and phase for the directional units k and i, respectively.

5.2.3 Complex-Valued LMS Rule

Let us consider a multidimensional regression model $\mathbf{y} = \mathbf{W}\mathbf{x}$, where $\mathbf{x} \in \mathbb{C}^N$ and $\mathbf{y} \in \mathbb{C}^M$ denote the complex-valued multidimensional input and multidimensional output signals, respectively, and $\mathbf{W} \in \mathbb{C}^{M \times N}$ denotes the complex-valued connection weight matrix. Given the desired (supervised) signals $\mathbf{d}(t)$, the goal of online regression is to seek the optimal $\mathbf{W}$ that minimizes the cost function

$$J(t) = \frac{1}{2}\|\mathbf{e}(t)\|^2 = \frac{1}{2}\|\mathbf{d}(t) - \mathbf{y}(t)\|^2 = \frac{1}{2}\left[\mathbf{d}(t) - \mathbf{y}(t)\right]^{\mathrm{H}}\left[\mathbf{d}(t) - \mathbf{y}(t)\right], \quad (5.32)$$

where $\mathbf{e}(t) = \mathbf{d}(t) - \mathbf{y}(t)$ denotes the estimation error between the desired output $\mathbf{d}(t)$ and the estimated output $\mathbf{y}(t)$. Similar to the real-valued case, the complex-valued LMS learning rule [369, 952] can be derived by stochastic gradient descent $\Delta\mathbf{W} \propto -\partial J/\partial\mathbf{W}$:

$$\Delta\mathbf{W}(t+1) = \eta\mathbf{x}(t)\mathbf{e}^{\mathrm{H}}(t), \tag{5.33}$$

or in scalar form

$$\Delta w_{ij}(t+1) = \eta x_j(t)e_i^*(t), \qquad i = 1, \ldots, M, \qquad j = 1, \ldots, N. \tag{5.34}$$

The complex-valued LMS rule has been widely used in array signal processing and communications [369]. Equation (5.34) can be further generalized to complex-valued backpropagation for a nonlinear multilayer network [83, 328, 369, 545].

■ EXAMPLE 5.1

In this example, we follow [311, 415] and derive a complex-valued multichannel LMS (MCLMS) algorithm for a single input–multiple output (SIMO) blind channel identification problem.

In a SIMO system (see Figure 5.2), a signal $s(t)$ passes through a noisy multipath environment and is collected by an array of sensors at the receiver side. The signal received at the lth sensor is represented as

$$x_l(t) = \mathbf{h}_l^{\mathrm{H}}\mathbf{s}(t) + n_l(t), \qquad l = 1, \ldots, M, \tag{5.35}$$

where $\mathbf{h}_l = [h_{l,0}, h_{l,1}, \ldots, h_{l,L-1}]^{\mathrm{T}} \in \mathbb{C}^L$ denotes the L-tap impulse response of the channel between the source transmitter and the lth sensor; $\mathbf{s}(t) = [s(t), s(t-1), \ldots, s(t-L+1)]^{\mathrm{T}} \in \mathbb{C}^L$ denotes the source signal vector and $n_l(t)$ denotes the additive measurement noise at the lth sensor. Let $\hat{\mathbf{h}}_l = [\hat{h}_{l,0}, \hat{h}_{l,1}, \ldots, \hat{h}_{l,L-1}]^{\mathrm{T}} \in \mathbb{C}^L$ denote the parameter vector of an FIR filter (assuming the order L is known a priori). The goal of blind system identification is to estimate all $\mathbf{h}_l$ using only the observations $x_l(t)$ ($l = 1, \ldots, M$).

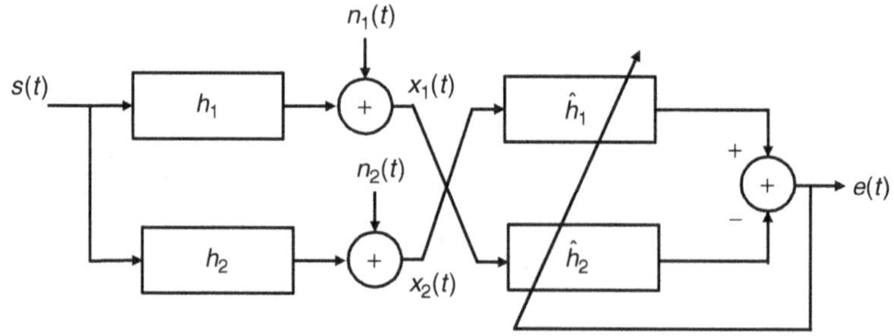

Figure 5.2 Block diagram of SIMO blind channel identification (here $M = 2$).

Here, we assume the following identifiability conditions are satisfied [979]: (i) the channels do not share any common zeros and (ii) the autocorrelation matrix of the source signal is of full rank.

The basic idea of the MCLMS algorithm derived in [415] was based on the cross-relation between two channels [979]: $x_1 * h_2 = s * h_1 * h_2 = x_2 * h_1$. In the noise-free condition, we have

$$\mathbf{x}_l^{\mathrm{H}}(t)\mathbf{h}_m = \mathbf{x}_m^{\mathrm{H}}(t)\mathbf{h}_l, \qquad l, m = 1, 2, \ldots, M, \tag{5.36}$$

where $\mathbf{x}_l(t) = [x_l(t), x_l(t-1), \ldots, x_l(t-L+1)]^{\mathrm{T}}$ denotes the tap-delay vector of observations at the lth sensor at time t. In the presence of noise, the complex error function can be defined as [311]

$$\chi(t) = \sum_{l=1}^{M-1} \sum_{m=l+1}^{M} |e_{lm}(t)|^2, \tag{5.37}$$

where $e_{lm}(t) = \mathbf{x}_l^{\mathrm{H}}(t)\hat{\mathbf{h}}_m - \mathbf{x}_m^{\mathrm{H}}(t)\hat{\mathbf{h}}_l$.

Let $\hat{\mathbf{h}} = [\hat{\mathbf{h}}_1^{\mathrm{T}}, \hat{\mathbf{h}}_2^{\mathrm{T}}, \ldots, \hat{\mathbf{h}}_M^{\mathrm{T}}]^{\mathrm{T}} \in \mathbb{C}^{ML \times 1}$ be a vector of the concatenated M channel estimates; then the optimal estimate of channel responses can be found by solving a constrained optimization problem [415]:

$$\hat{\mathbf{h}}_{\mathrm{opt}} = \arg\min_{\hat{\mathbf{h}}} \mathbb{E}[\chi(t)] \qquad \text{subject to} \quad \|\hat{\mathbf{h}}\| = 1, \tag{5.38}$$

where the unit norm constraint is introduced to avoid the degenerate solution $\hat{\mathbf{h}} = \mathbf{0}$. Alternatively, we can minimize a normalized cost function as follows:

$$J(t) = \frac{\chi(t)}{\|\hat{\mathbf{h}}\|}. \tag{5.39}$$

Applying the stochastic gradient descent with respect to $\hat{\mathbf{h}}$, we obtain [311, 415]

$$\begin{aligned} \hat{\mathbf{h}}(t+1) &= \hat{\mathbf{h}}(t) - \eta\, \nabla J(t) \\ &= \hat{\mathbf{h}}(t) - \frac{1}{\|\hat{\mathbf{h}}\|^2}\left[2\mathbf{R}^*(t)\hat{\mathbf{h}}(t) - 2J(t)\hat{\mathbf{h}}(t)\right], \end{aligned} \tag{5.40}$$

with

$$\mathbf{R}(t) = \begin{bmatrix} \sum_{l\neq 1} \mathbf{R}_{x_l x_l}(t) & -\mathbf{R}_{x_2 x_1}(t) & \cdots & -\mathbf{R}_{x_M x_1}(t) \\ -\mathbf{R}_{x_1 x_2}(t) & \sum_{l\neq 2} \mathbf{R}_{x_l x_l}(t) & \cdots & -\mathbf{R}_{x_M x_2}(t) \\ \vdots & \vdots & \ddots & \vdots \\ -\mathbf{R}_{x_1 x_M}(t) & -\mathbf{R}_{x_2 x_M}(t) & \cdots & \sum_{l\neq M} \mathbf{R}_{x_l x_l}(t) \end{bmatrix}, \tag{5.41}$$

where $\mathbf{R}_{x_l x_m}(t) = \mathbf{x}_l(t)\mathbf{x}_m^{\mathrm{H}}(t) \in \mathbb{C}^{L\times L}$ denotes the cross-correlation matrix between $\mathbf{x}_l(t)$ and $\mathbf{x}_m(t)$ and $\mathbf{R}(t) \in \mathbb{C}^{ML\times ML}$ is a concatenated matrix.

Finally, if the channel estimate is always normalized after each iteration, then the update equation for the complex MCLMS algorithm can be derived as [311]

$$\hat{\mathbf{h}}(t+1) = \frac{\hat{\mathbf{h}}(t) - 2\eta[\mathbf{R}^*(t)\hat{\mathbf{h}}(t) - \chi(t)\hat{\mathbf{h}}(t)]}{\left\|\hat{\mathbf{h}}(t) - 2\eta[\mathbf{R}^*(t)\hat{\mathbf{h}}(t) - \chi(t)\hat{\mathbf{h}}(t)]\right\|}. \tag{5.42}$$

Note that, in this example, the unknown channel impulse responses are identified up to an arbitrary complex-valued gain factor (i.e., with both modulus and phase ambiguity) [311].

5.2.4 Complex-Valued PCA Learning

Complex-Valued Hermitian Eigenvalue Problem. Let $\mathbf{C} = \mathbb{E}[\mathbf{x}\mathbf{x}^{\mathrm{H}}] \in \mathbb{C}^{N\times N}$ denote the correlation matrix of a complex-valued random vector $\mathbf{x} \in \mathbb{C}^N$; the Hermitian eigenvalue problem is

$$\mathbf{C}\mathbf{v} = \lambda\mathbf{v}, \tag{5.43}$$

where λ denotes the real eigenvalue of the complex Hermitian matrix $\mathbf{C}$. Applying the EVD to matrix $\mathbf{C}$ would yield[5]

$$\mathbf{C} = \mathbf{U}\mathbf{\Lambda}\mathbf{U}^{\mathrm{H}}, \tag{5.44}$$

where $\mathbf{U}$ is a unitary matrix such that $\mathbf{U}\mathbf{U}^{\mathrm{H}} = \mathbf{I}$ and $\mathbf{\Lambda}$ is a diagonal matrix with eigenvalues $\{\lambda_i\}_{i=1}^N$ as entries. The spectral radius of matrix $\mathbf{C}$, denoted as $\rho(\mathbf{C})$, is defined as

$$\rho(\mathbf{C}) = \max_{i=1,\ldots,N} |\lambda_i|. \tag{5.45}$$

Let $\mathbf{C} = \mathbf{C}_{\mathrm{Re}} + j\mathbf{C}_{\mathrm{Im}}$ and $\mathbf{v} = \mathbf{v}_{\mathrm{Re}} + j\mathbf{v}_{\mathrm{Im}}$; then (5.43) can be rewritten as

$$(\mathbf{C}_{\mathrm{Re}} + j\mathbf{C}_{\mathrm{Im}})(\mathbf{v}_{\mathrm{Re}} + j\mathbf{v}_{\mathrm{Im}}) = \lambda(\mathbf{v}_{\mathrm{Re}} + j\mathbf{v}_{\mathrm{Im}}), \tag{5.46}$$

and rearranging the terms yields

$$(\mathbf{C}_{\mathrm{Re}}\mathbf{v}_{\mathrm{Re}} - \mathbf{C}_{\mathrm{Im}}\mathbf{v}_{\mathrm{Im}}) + j(\mathbf{C}_{\mathrm{Re}}\mathbf{v}_{\mathrm{Im}} + \mathbf{C}_{\mathrm{Im}}\mathbf{v}_{\mathrm{Re}}) = \lambda\mathbf{v}_{\mathrm{Re}} + j\lambda\mathbf{v}_{\mathrm{Im}}. \tag{5.47}$$

Let us further introduce an augmented real-valued vector $\mathbf{x}_c \in \mathbb{R}^{2N}$,

$$\mathbf{x}_c = \begin{pmatrix} \mathbf{x}_{\mathrm{Re}} \\ \mathbf{x}_{\mathrm{Im}} \end{pmatrix},$$

and its corresponding augmented real-valued correlation matrix $\mathbf{C}_c \in \mathbb{R}^{2N\times 2N}$,

$$\mathbf{C}_c = \mathbb{E}\left[\begin{pmatrix} \mathbf{x}_{\mathrm{Re}} & -\mathbf{x}_{\mathrm{Im}} \\ \mathbf{x}_{\mathrm{Im}} & \mathbf{x}_{\mathrm{Re}} \end{pmatrix}\begin{pmatrix} \mathbf{x}_{\mathrm{Re}}^{\mathrm{T}} & \mathbf{x}_{\mathrm{Im}}^{\mathrm{T}} \\ -\mathbf{x}_{\mathrm{Im}}^{\mathrm{T}} & \mathbf{x}_{\mathrm{Re}}^{\mathrm{T}} \end{pmatrix}\right]$$
$$= \begin{pmatrix} \mathbb{E}[\mathbf{x}_{\mathrm{Re}}\mathbf{x}_{\mathrm{Re}}^{\mathrm{T}} + \mathbf{x}_{\mathrm{Im}}\mathbf{x}_{\mathrm{Im}}^{\mathrm{T}}] & -\mathbb{E}[\mathbf{x}_{\mathrm{Im}}\mathbf{x}_{\mathrm{Re}}^{\mathrm{T}} - \mathbf{x}_{\mathrm{Re}}\mathbf{x}_{\mathrm{Im}}^{\mathrm{T}}] \\ \mathbb{E}[\mathbf{x}_{\mathrm{Im}}\mathbf{x}_{\mathrm{Re}}^{\mathrm{T}} - \mathbf{x}_{\mathrm{Re}}\mathbf{x}_{\mathrm{Im}}^{\mathrm{T}}] & \mathbb{E}[\mathbf{x}_{\mathrm{Re}}\mathbf{x}_{\mathrm{Re}}^{\mathrm{T}} + \mathbf{x}_{\mathrm{Im}}\mathbf{x}_{\mathrm{Im}}^{\mathrm{T}}] \end{pmatrix}.$$

Notably, the matrix $\mathbf{C}_c$ is always positive semidefinite. With these newly introduced notations, we can reformulate (5.43) as an equivalent eigenvalue problem

$$\mathbf{C}_c\mathbf{v}_c = \lambda\mathbf{v}_c, \tag{5.48}$$

where

$$\mathbf{C}_c = \begin{pmatrix} \mathbf{C}_{\mathrm{Re}} & -\mathbf{C}_{\mathrm{Im}} \\ \mathbf{C}_{\mathrm{Im}} & \mathbf{C}_{\mathrm{Re}} \end{pmatrix} \quad \text{and} \quad \mathbf{v}_c = \begin{pmatrix} \mathbf{v}_{\mathrm{Re}} \\ \mathbf{v}_{\mathrm{Im}} \end{pmatrix}. \tag{5.49}$$

Indeed, the eigenvalue from the reformulated eigenequation (5.48) and that from the original eigenequation (5.43) are related by the following theorem:

Theorem 5.3 [265] *Let $\mathbf{C} = \mathbf{C}_{Re} + j\mathbf{C}_{Im}$ (where $\mathbf{C}_{Re} \in \mathbb{R}^{N\times N}$, $\mathbf{C}_{Im} \in \mathbb{R}^{N\times N}$) be a complex Hermitian matrix and define $\mathbf{C}_c$ as a real-valued $2N \times 2N$ matrix according to (5.49). If λ is an eigenvalue of the matrix $\mathbf{C}$, then the matrix $\mathbf{C}_c$ has two eigenvalues as λ.*

Solving a Hermitian eigenvalue problem is computationally expensive, especially when the size of the matrix, N, is large. Preferably, we would like to develop adaptive learning algorithms with lower complexity that extract single or multiple eigenvectors in an efficient fashion. As we will see below, many correlation-based learning algorithms can be developed for complex-valued PCA.

Complex-Valued Oja's Learning Rule. Oja's local PCA learning rule (see Chapter 3) is a simple yet powerful Hebbian learning algorithm for extracting the (single) dominant eigenvector. Similar to the real-valued setting, we consider a MISO linear neuron model $y = \boldsymbol{\theta}^{\mathrm{H}}\mathbf{x}$, where $\mathbf{x} \in \mathbb{C}^N$ denotes the complex-valued N-dimensional input and y denotes the complex-valued scalar output. The one-unit complex-valued PCA learning rule, as an extension of Oja's rule, is given by

$$\boldsymbol{\theta}(t+1) = \boldsymbol{\theta}(t) + \eta y(t)[\mathbf{x}^*(t) - \boldsymbol{\theta}(t)y^*(t)]$$
$$= \boldsymbol{\theta}(t) + \eta\left[y(t)\mathbf{x}^*(t) - |y(t)|^2\boldsymbol{\theta}(t)\right]. \tag{5.50}$$

With a proper choice of learning rate η, after a sufficient number of learning steps, $\boldsymbol{\theta}$ will converge to the principal eigenvector up to an arbitrary angle rotation (i.e., with phase ambiguity).

To analyze the convergence of the one-unit complex PCA learning rule, we rewrite (5.50) in terms of a differential equation

$$\frac{d\boldsymbol{\theta}}{dt} = y(t)\mathbf{x}^*(t) - |y(t)|^2\boldsymbol{\theta}. \tag{5.51}$$

By defining the Hermitian correlation matrix $\mathbf{C} = \mathbb{E}[\mathbf{x}(t)\mathbf{x}^{\mathrm{H}}(t)] = \mathbb{E}[\mathbf{x}^*(t)\mathbf{x}^{\mathrm{T}}(t)]$, taking the expectation of the right-hand side of (5.51) yields

$$\begin{aligned}\frac{d\boldsymbol{\theta}}{dt} &= \mathbb{E}[y\mathbf{x}^* - |y|^2\boldsymbol{\theta}] \\ &= \mathbf{C}\boldsymbol{\theta} - (\boldsymbol{\theta}^{\mathrm{H}}\mathbf{C}\boldsymbol{\theta})\boldsymbol{\theta} = \left(\mathbf{C} - \boldsymbol{\theta}^{\mathrm{H}}\mathbf{C}\boldsymbol{\theta}\right)\boldsymbol{\theta}.\end{aligned} \tag{5.52}$$

The stationary point of (5.52) is determined by the eigenvector $\boldsymbol{\theta}$ by solving a complex-valued eigenvalue problem as follows:

$$\mathbf{C}\boldsymbol{\theta} = \lambda\boldsymbol{\theta} \qquad (\boldsymbol{\theta} \in \mathbb{C}^N), \tag{5.53}$$

where $\lambda = \boldsymbol{\theta}^{\mathrm{H}}\mathbf{C}\boldsymbol{\theta}$ corresponds to the eigenvalue.

In a similar vein to the analysis of the real-valued version of Oja's learning rule [679], the convergence of the one-unit complex PCA learning rule can be stated as follows [280]:

Theorem 5.4 *Suppose* $\mathbf{C} \in \mathbb{C}^{N\times N}$ *is Hermitian with N pairs of eigenvalues and eigenvectors,* $(\sigma_1, \mathbf{q}_1), (\sigma_2, \mathbf{q}_2), \ldots, (\sigma_N, \mathbf{q}_N)$*, and suppose that the eigenvalues are distinct and arranged in a descending order and the eigenvectors are normalized so that* $\mathbf{q}_k^{\mathrm{H}}\mathbf{q}_k = 1$ *and* $\boldsymbol{\theta}^{\mathrm{H}}(0)\mathbf{q}_1 \neq 0$*. Then it holds for equation (5.52) that*

$$\lim_{t\to\infty} \boldsymbol{\theta}(t) = \mathbf{q}_1 e^{j\alpha},$$

where $\alpha \in [0, 2\pi)$ *is an arbitrary real-valued constant.*

To extend PCA to MIMO neurons, let $\mathbf{y} = \mathbf{W}^{\mathrm{H}}\mathbf{x}$ (where $\mathbf{x} \in \mathbb{C}^N$, $\mathbf{y} \in \mathbb{C}^m$, and $\mathbf{W} \in \mathbb{C}^{N\times m}$). The general complex-valued version of Oja's rule can be derived as

$$\Delta\mathbf{W}(t) = \eta\left(\mathbf{x}(t)\mathbf{y}^{\mathrm{H}}(t) - \mathbf{W}(t)\mathbf{y}(t)\mathbf{y}^{\mathrm{H}}(t)\right). \tag{5.54}$$

Written in the form of a differential equation, (5.54) can be formulated by

$$\frac{d\mathbf{W}}{dt} = \mathbf{C}_{xx}\mathbf{W} - \mathbf{W}\mathbf{W}^{\mathrm{H}}\mathbf{C}_{xx}\mathbf{W}, \tag{5.55}$$

where $\mathbf{C}_{xx} = \mathbb{E}[\mathbf{x}(t)\mathbf{x}^{\mathrm{H}}(t)]$ denotes the correlation matrix of $\mathbf{x}$.

Because the above version of Oja's learning rule (5.54) only tracks the principal subspace instead of the principal components of $\mathbf{x}$, it is sometimes referred to as the *principal subspace rule*. To impose more structural constraints on $\mathbf{W}$, Sanger's learning rule can be used for extracting multiple principal components.

Complex-Valued Sanger's Learning Rule. In a similar vein to the real-valued GHA (see Chapter 3), Sanger's learning rule can be reformulated for complex-valued data, which is referred to as the complex-valued GHA rule [1000]:

$$\Delta\mathbf{W}(t) = \eta\left(\mathbf{x}(t)\mathbf{y}^{\mathrm{H}}(t) - \mathbf{W}(t)\mathrm{UT}[\mathbf{y}(t)\mathbf{y}^{\mathrm{H}}(t)]\right). \tag{5.56}$$

Alternatively, if we write $\mathbf{y} = \mathbf{W}\mathbf{x}$ with $\mathbf{W} \in \mathbb{C}^{m\times N}$, then (5.56) is rewritten as

$$\Delta\mathbf{W}(t) = \eta\left(\mathbf{y}(t)\mathbf{x}^{\mathrm{H}}(t) - \mathrm{LT}[\mathbf{y}(t)\mathbf{y}^{\mathrm{H}}(t)]\mathbf{W}(t)\right). \tag{5.57}$$

The notations $\mathrm{UT}[\cdot]$ and $\mathrm{LT}[\cdot]$ denote the operators that return, respectively, the upper triangular and lower triangular parts of the matrix contained within. In particular, equation (5.57) is a complex counterpart of (3.21) in the real domain. The convergence of the complex-valued GHA rule was discussed in [999].

Complex-Valued Brockett's Learning Rule. It is also possible to extend Brockett's generalized subspace learning rule [115] to the complex domain (e.g., [172]). Specifically, in Brockett's subspace learning rule, the network output, denoted by $\mathbf{y} \in \mathbb{C}^m$, is represented as

$$\mathbf{y} = \mathbf{D}\mathbf{W}^{\mathrm{H}}\mathbf{x},$$

where $\mathbf{W} \in \mathbb{C}^{N\times m}$, $\mathbf{x} \in \mathbb{C}^N$, and $\mathbf{D} \in \mathbb{C}^{m\times m}$ is a diagonal matrix with positive and strictly decreasing real-valued entries

$$\mathbf{D} = \mathrm{diag}\{d_1, d_2, \ldots, d_m\},$$

where $d_1 > d_2 > \cdots > d_m > 0$. The purpose of the diagonal matrix $\mathbf{D}$ is to introduce asymmetry between the output units. Brockett's algorithm can be described by a dynamical equation of isopectral flows, and the Brockett flow is obtained from a potential function as the Riemannian gradient flow in the space of all orthogonal matrices [115].

In matrix form, Brockett's complex-valued subspace learning rule is described by [115, 172]:

$$\Delta\mathbf{W}(t) = \boldsymbol{\eta}\left(\mathbf{x}(t)\mathbf{y}^{\mathrm{H}}(t)\mathbf{D} - \mathbf{W}(t)\mathbf{D}\mathbf{y}(t)\mathbf{y}^{\mathrm{H}}(t)\mathbf{D}\right), \tag{5.58}$$

where $\boldsymbol{\eta} = \mathrm{diag}\{\eta_1, \ldots, \eta_m\}$ is a *diagonal* learning-rate matrix typically with different learning-rate parameters for each entry. Two similar versions of (5.58), the so-called *weighted subspace algorithms*, have been proposed in [678],

$$\Delta\mathbf{W}(t) = \boldsymbol{\eta}\left(\mathbf{x}(t)\mathbf{y}^{\mathrm{H}}(t) - \mathbf{W}(t)\mathbf{y}(t)\mathbf{y}^{\mathrm{H}}(t)\mathbf{D}\right), \tag{5.59}$$

as well as in [980],

$$\Delta\mathbf{W}(t) = \boldsymbol{\eta}\left(\mathbf{x}(t)\mathbf{y}^{\mathrm{H}}(t)\mathbf{D}^{-1} - \mathbf{W}(t)\mathbf{D}^{-1}\mathbf{y}(t)\mathbf{y}^{\mathrm{H}}(t)\right). \tag{5.60}$$

In addition, a number of other stochastic adaptive algorithms have been developed for extracting either principal/minor components or the principal subspace. Unified mathematical treatments of these learning rules were discussed in [156, 875]. Specifically, a generalized weighted subspace learning rule can be written as

$$\Delta \mathbf{W}(t) = \eta \left(\mathbf{x}(t)\mathbf{y}^{\mathrm{H}}(t)\mathbf{D}^{-p} - \mathbf{W}(t)\mathbf{y}(t)\mathbf{y}^{\mathrm{H}}(t)\mathbf{D}^{1-p}\right). \tag{5.61}$$

When $p = 0$ and $p = -1$, equation (5.61) reduces to (5.59) and Brockett's rule (5.58), respectively. Let $p = 0.5$ and $\mathbf{W} \leftarrow \mathbf{WD}^{-1/2}$; then equation (5.60) is recovered as a special case.

Complex-Valued APEX Algorithm. In a similar manner to the extensions of the previous algorithms, the APEX algorithm (see Chapter 3) can be extended to the complex-valued domain [157]. Specifically, given a linear neural network with lateral inhibitory connections, let $\mathbf{W} = [\boldsymbol{\theta}_1, \ldots, \boldsymbol{\theta}_m] \in \mathbb{C}^{N\times m}$ denote the complex-valued feedforward connections, $\mathbf{U} = [\mathbf{u}_1, \ldots, \mathbf{u}_m] \in \mathbb{C}^{m\times m}$ denote the complex-valued lateral connections, and $\mathbf{x} \in \mathbb{C}^N$ and $\mathbf{y} \in \mathbb{C}^m$ denote the complex-valued input and output, respectively. Then the network equation can be represented in matrix form as follows:

$$\begin{aligned} \mathbf{y} &= \mathbf{z} + \mathbf{U}^{\mathrm{H}}\mathbf{y} \\ &= \mathbf{W}^{\mathrm{H}}\mathbf{x} + \mathbf{U}^{\mathrm{H}}\mathbf{y}, \end{aligned} \tag{5.62}$$

where $\mathbf{z} = \mathbf{W}^{\mathrm{H}}\mathbf{x}$ and $\mathbf{U}$ is a strictly upper triangular matrix. Alternatively, the network output can be rewritten as

$$y_k = \boldsymbol{\theta}_k^{\mathrm{H}}\mathbf{x} + \mathbf{u}_k^{\mathrm{H}}\mathbf{y}. \tag{5.63}$$

As in the standard APEX algorithm, the learning rules for complex-valued feedforward and lateral connections are described as follows:

$$\begin{aligned} \boldsymbol{\theta}_k &= -\eta\frac{d\boldsymbol{\theta}_k}{dt}, \qquad k = 1, \ldots, m, \\ \mathbf{u}_k &= -\eta\frac{d\mathbf{u}_k}{dt}, \qquad k = 1, \ldots, m, \end{aligned}$$

where the derivatives can be approximated by the Hebbian and anti-Hebbian terms [157, 280]:

$$\frac{d\boldsymbol{\theta}_k}{dt} = \mathbb{E}[y_k^*(\mathbf{x}_k - y_k\boldsymbol{\theta}_k)], \qquad \frac{d\mathbf{u}_k}{dt} = -\mathbb{E}[y_k^*(\mathbf{y}_{[k]} + y_k\mathbf{u}_k)], \tag{5.64}$$

where $\mathbf{y}_{[k]} \triangleq [y_1, y_2, \ldots, y_{k-1}, 0, \ldots, 0]^{\mathrm{T}} \in \mathbb{C}^m$ for $k > 1$ and $\mathbf{y}_{[1]} = [0, 0, \ldots, 0]^{\mathrm{T}}$. Note that, when $m = 1$, it follows that $y_1^*(\mathbf{y}_{[1]} + y_1\mathbf{u}_1) = y_1^*y_1\mathbf{u}_1 = |y_1|^2\mathbf{u}_1$, and then (5.64) reduces to Oja's first principal-component analyzer in the complex domain.

■ EXAMPLE 5.2

Beamforming is a signal processing technique that performs spatial filtering of a signal source in the presence of spatial noise and other disturbing sources by means of an array of antennas or microphones provided that the DOA of the primary source is known [910, 911]. A beamformer may be realized by a complex-weighted neural unit fed with the Fourier transform of the measured signals, thereby bearing a complex-valued nature. A way to train the beamforming neuron is to force it to solve a minimum eigenvalue problem, which is also known as the MCA problem [279]. Specifically, let $y = \boldsymbol{\theta}^{\mathrm{H}}\mathbf{x}$ denote the complex-valued linear neuron output. Then minimizing the power of the output is equivalent to finding the solution to the equation $\mathbb{E}[|y|^2] = \boldsymbol{\theta}^{\mathrm{H}}\mathbf{C}\boldsymbol{\theta}$, where $\mathbf{C} = \mathbb{E}[\mathbf{x}\mathbf{x}^{\mathrm{H}}]$ denotes the correlation matrix of the input, which corresponds to the discrete Fourier transform of the sampled signals coming from the sensors.

In a simple beamforming setup, we consider three sensors that have a geometric layout illustrated in Figure 5.3*a*, where the source is located in the center. For simplicity, all sensors are assumed to be omnidirectional or panoramic. We further assume that the sensor noise is spatially white with unit variance such that the *spectral correlation matrix* of the array input signal $\mathbf{x}$ is decomposed into signal and noise components by

$$\mathbf{C} = \sigma_s^2 \mathbf{a}\mathbf{a}^{\mathrm{H}} + \sigma_n^2 \mathbf{I}, \tag{5.65}$$

where $\mathbf{a} \equiv \mathbf{a}(\alpha)$ denotes a complex-valued steering vector (or DOA vector) that is defined as the vector of phase delays needed to align the array outputs for a plane wave coming from the direction α (see Figure 5.3*b* for illustration). The ratio σ_s^2/σ_n^2 denotes the spectral SNR averaged over all the sensors, and the array gain $G(\alpha)$, which represents the beamforming improvement of

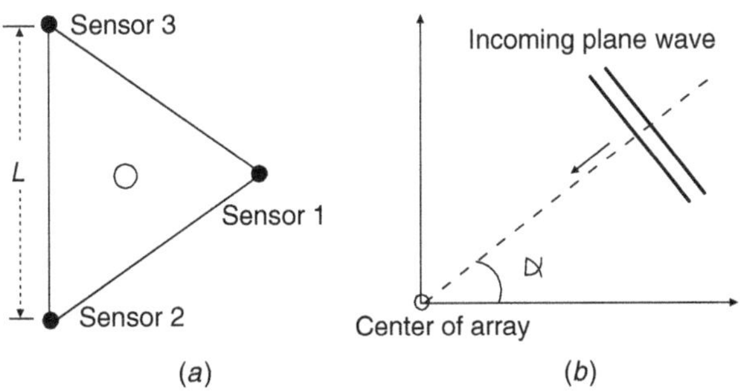

Figure 5.3 (*a*) Sensor array geometry: three sensors are located in the corners of the equilateral triangle, and the transmitter or the loudspeaker is positioned in the center of the triangle. (*b*) Array signal propagation diagram (α denotes the angle between the axis of the linear array and the direction of the desired signal source).

SNR along direction α, is defined by

$$G(\alpha) = \frac{|\boldsymbol{\theta}^{\mathrm{H}}\mathbf{a}|^2}{\boldsymbol{\theta}^{\mathrm{H}}\boldsymbol{\theta}}. \tag{5.66}$$

Specifically, the beamforming problem is reduced to a constrained optimization problem in the complex domain [279]:

$$\min_{\boldsymbol{\theta}} \boldsymbol{\theta}^{\mathrm{H}}\mathbf{C}\boldsymbol{\theta} \qquad \text{s.t.} \quad \boldsymbol{\theta}^{\mathrm{H}}\mathbf{a} = 1 \quad \text{and} \quad \boldsymbol{\theta}^{\mathrm{H}}\boldsymbol{\theta} = \delta^{-2},$$

where the first constraint $\boldsymbol{\theta}^{\mathrm{H}}\mathbf{a} = 1$ forces the unit boresight response, whereas the second constraint $\boldsymbol{\theta}^{\mathrm{H}}\boldsymbol{\theta} = \delta^{-2}$, when combined with the first one, imposes a white-noise gain in the steering direction such that $G(\alpha_s) = \delta^2$, where α_s denotes the DOA of the primary source. Generally, a large value of δ implies small sensitivity to the white noise and thereby better robustness of the beamformer. Notably, if only the first constraint is imposed, then using the method of Lagrange multipliers, we can find that the optimum solution to the constrained optimization problem is [577]

$$\boldsymbol{\theta}_{\mathrm{opt}} = \frac{\mathbf{C}^{-1}\mathbf{a}^*}{\mathbf{a}^{\mathrm{T}}\mathbf{C}^{-1}\mathbf{a}^*},$$

which requires the computation of the matrix inverse $\mathbf{C}^{-1}$. In order to conduct adaptive beamforming, the stochastic adaptive learning rule for updating the weight vector $\boldsymbol{\theta}$ is described by [279]:

$$\Delta\boldsymbol{\theta} = \eta\left[\mathbf{x}y^* - \delta^2|y|^2\boldsymbol{\theta} + \sigma(\|\boldsymbol{\theta}\|^2 - \delta^{-2})\boldsymbol{\theta}\right], \tag{5.67}$$

where σ is a constant that is chosen to be smaller than the power of the incoming input signal. In our experimental scenario, the steering vector is

$$\mathbf{a}^{\mathrm{H}}(\alpha) = \left[\exp\left(\frac{2j\pi r}{\sqrt{3}}\sin\alpha\right),\right.$$
$$\left.\exp\left(\frac{j\pi r}{\sqrt{3}}(\sqrt{3}\cos\alpha - \sin\alpha)\right), \exp\left(-\frac{j\pi r}{\sqrt{3}}(\sqrt{3}\cos\alpha + \sin\alpha)\right)\right],$$

where $r = L/\lambda$, L denotes the distance between the microphones, and λ denotes the wavelength corresponding to the frequency bin that the array is accorded to. The parameter setup in the experiment is $\sigma_s^2 = \sigma_n^2 = 1$ (0 dB), $r = 0.4$, $\eta = 0.0002$, $\delta = 1.5$, and $\sigma = 2$. The experimental performance is shown in Figure 5.4. As seen in the figure, the array beam pattern looks reasonably good, with a strong main lobe around the DOA of the primary signal and significant attenuation in the other directions (with appearance of only a small side lobe).

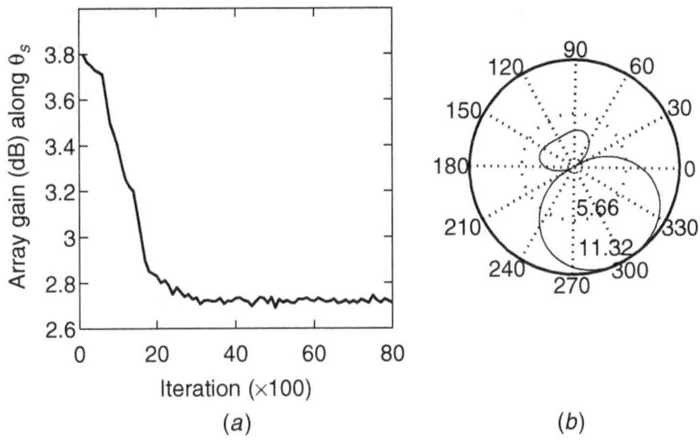

Figure 5.4 The beamformer performance for (*a*) array gain and (*b*) array beam pattern (values in decibels).

5.2.5 Complex-Valued ICA Learning

In a similar vein to the real-valued ICA, we will further consider a complex version of the ICA model: $\mathbf{x} = \mathbf{A}\mathbf{s}$, where $\mathbf{s} \in \mathbb{C}^m$ denotes the m-dimensional complex-valued, elementwise-independent source vector, $\mathbf{x} \in \mathbb{C}^m$ denotes the m-dimensional complex-valued vector of mixture signals, and $\mathbf{A} \in \mathbb{C}^{m \times m}$ denotes a complex-valued mixing matrix. In the complex-valued ICA problem, there are three types of indeterminacies: (i) sign and scaling indeterminacy, (ii) permutation indeterminacy, and (iii) phase indeterminacy. The first two indeterminacies are shared with the real-valued ICA problem, whereas the phase ambiguity arises from the inherent nature of complex-valued variables.

To characterize the identifiability of the complex-valued ICA model, the complex analogs of the well-known *Cramer theorem* and *Darmois–Skitovich theorem*, which are fundamental to the concept of ICA [180], are stated here:

Theorem 5.5 Complex Cramer Theorem [265] *If s_1 and s_2 are independent random variables such that $s_1 + s_2$ is a complex normal random variable, then s_1 and s_2 are both complex normal.*

Theorem 5.6 Complex Darmois–Skitovich Theorem [265] *Let $s_1, \ldots, s_n$ be n mutually independent complex random variables. For $\alpha_i, \beta_i \in \mathbb{C}$ $(i = 1, \ldots, n)$, if the linear forms $x_1 = \sum_{i=1}^{n} \alpha_i s_i$ and $x_2 = \sum_{i=1}^{n} \beta_i s_i$ are independent, then random variables $\{s_i\}$ for which $\alpha_i \beta_i \neq 0$ are complex Gaussian.*

There are several routes for solving the complex ICA problem:

- Complex ICA Based on *Eigenvalue Decomposition*: In this approach, generalization from the real to the complex domain is relatively straightforward by replacing the symmetric covariance matrix with a Hermitian covariance matrix. Examples of this kind include the AMUSE, SOBI, FOBI, and JADE algorithms, which were partially reviewed in Chapter 3 (see also [172]).

- Complex ICA Based on *Strongly Uncorrelating Transformation*: In this approach, second-order statistics (covariance and pseudocovariance) of complex random variables are fully exploited to separate either circular or non circular sources [265, 266].
- Complex ICA Based on *Higher Order Statistics*: In this approach, nonlinearity is used to produce higher order decorrelation. Examples of this kind include adaptive algorithms such as the complex FastICA [94] and complex Infomax [7, 42, 137].

To take a specific case, we can separate the independent sources by imposing nonlinear decorrelation via adaptive anti-Hebbian learning, which is employed in the Infomax or natural gradient algorithm [29, 78]. Let $\mathbf{W} \in \mathbb{C}^{m \times m}$ be a demixing matrix and $\mathbf{y} = \mathbf{W}\mathbf{x} \in \mathbb{C}^m$ be the separated complex signal vector. Then the complex-valued version of the natural gradient learning rule is described by [137]

$$\Delta \mathbf{W} = \eta[\mathbf{I} - \psi(\mathbf{y})\mathbf{y}^{\mathrm{H}}]\mathbf{W}, \tag{5.68}$$

which bears a close resemblance to its real-valued counterpart (3.138). The nonlinear activation function $\psi(\cdot)$ is called the *complex score function* [164, 266].[6] In practice, for the purpose of generating higher order statistics, $\psi(\cdot)$ is chosen to be either a split-complex bounded but nonanalytic function [42] or a fully complex analytic function [7, 137].

For the learning rule (5.68), a stationary point of the solution implies that $\mathbb{E}[\Delta w_{kj}] = 0$, or equivalently

$$\mathbb{E}[\psi(y_k)y_i^*] = \begin{cases} 0, & k \neq i, \\ 1, & k = i, \end{cases} \tag{5.69}$$

which says that $\psi(y_k)$ and y_i are *nonlinearly* uncorrelated. In this ideal case, the output of $\psi(\mathbf{y})$ approximates a uniform distribution to achieve the maximum information transfer and maximum entropy [7].

■ EXAMPLE 5.3

In this example, we study a MIMO blind equalization problem where the goal is to equalize or separate different independent complex-valued transmitted signals in communication with the employed constellation scheme as M-PSK (phase shift keying) and quadrature amplitude modulation (QAM). Here, the source signals include three types of modulated signals—8-PSK, 4-QAM, and 16-QAM—plus the uniformly distributed complex-valued noise that is strongly uncorrelated. Among them, 8-PSK and 4-QAM are noncircular complex sources with constant modulus, while 16-QAM is neither circular nor constant modulus. For each source, 500 i.i.d. samples were generated. The source signals were then mixed by a 4×4 complex random mixing matrix. The complex-valued JADE algorithm [143] was used here for the purpose of source separation. The JADE algorithm is an offline (or batch) ICA algorithm based on joint diagonalization of a set of cumulant matrices with all second- and fourth-order cumulants. Because it involves no nonlinearity but requires solving the eigenvalue problem, it is well suited for both real and complex BSS problems [149]. The experimental results are illustrated in Figure 5.5.

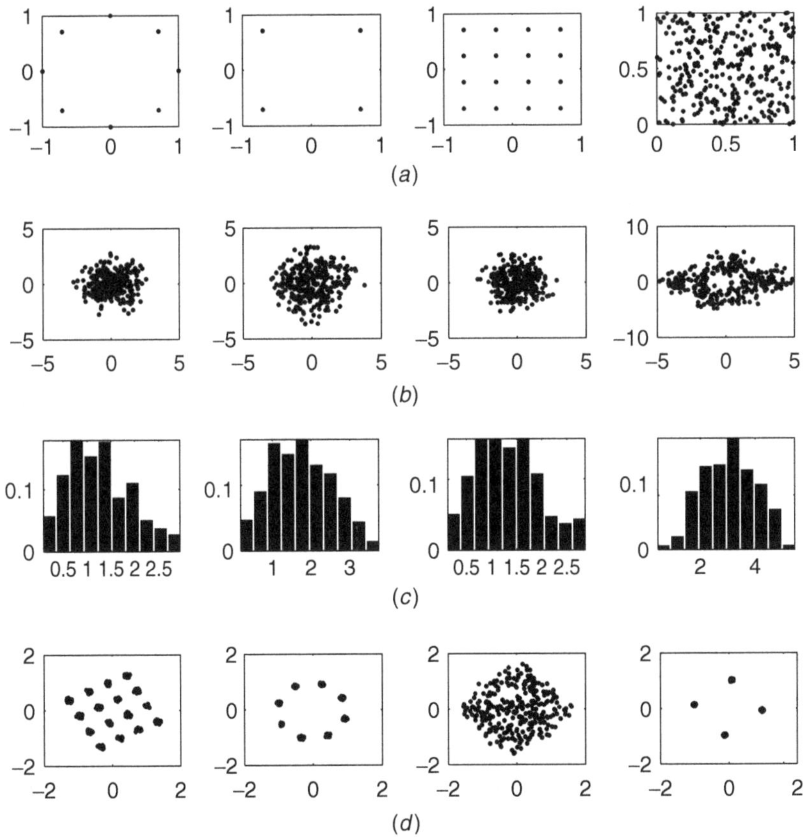

Figure 5.5 (*a*) Constellation of three types of modulated signals (first three columns: 8-PSK, 4-QAM, and 16-QAM) and the scatter plot (real vs. imaginary) of the complex-valued noise (last column). (*b*) Scatter plots (real vs. imaginary) of four observed complex-valued signals (mixed by random complex-valued mixing matrix). (*c*) Histogram of the modulus of the observed signals. (*d*) Scatter plots (real vs. imaginary) of the separated complex-valued signals.

■ EXAMPLE 5.4

A natural application of the complex ICA algorithm is to solve the BSS problem in the frequency domain (e.g., [42, 46, 645]). In a general setting, a convolutive mixture of N source signals $s_i(t)$ can be described as

$$x_j(t) = \sum_{i=1}^{N} \sum_{p=1}^{P} h_{ji}(p) s_i(t - p + 1) \qquad (j = 1, \ldots, m), \tag{5.70}$$

where h_{ji} denotes the impulse response from source i to sensor j. In the frequency domain, using a T-point STFT, we have

$$\mathbf{x}(\omega, n) = \mathbf{H}(\omega)\mathbf{s}(\omega, n), \tag{5.71}$$

where ω denotes the frequency, n represents the time dependence of the STFT, and the mixing matrix $\mathbf{H}(\omega)$ is assumed to be square ($m = N$) and invertible and its entries $H_{ji}(\omega) \neq 0$ ($\forall i, j$). The source separation process at the frequency ω is then formulated as

$$\mathbf{y}(\omega, n) = \mathbf{W}(\omega)\mathbf{x}(\omega, n). \tag{5.72}$$

The learning rule for $\mathbf{W}(\omega)$, similar to the time domain, follows the iterative equation

$$\Delta\mathbf{W}(\omega) = \eta \left[\text{diag} \left\{ \langle \psi(\mathbf{y}(\omega))\mathbf{y}^{\mathrm{H}}(\omega) \rangle \right\} - \langle \psi(\mathbf{y}(\omega))\mathbf{y}^{\mathrm{H}}(\omega) \rangle \right] \mathbf{W}(\omega), \tag{5.73}$$

where the score function used here is a split-complex hyperbolic tangent function

$$\psi(\mathbf{y}) = \tanh(\mathbf{y}_{\mathrm{Re}}) + j \tanh(\mathbf{y}_{\mathrm{Im}}).$$

In the example, the source signals are two male speech signals sampled at 8 kHz in a room environment. Given the 8 kHz sampling frequency, the room impulse response is assumed to have a length of 150 ms (that corresponds to $P = 1200$ taps) and a window length $T = 2500 > 2P = 2400$ was chosen.[7]

The two speech signals were convolved with the room impulse response in the virtual room environment and were then treated as input signals $x_1(t)$ and $x_2(t)$. They were then processed by STFT with a window length of 312.5 ms. The learning-rate parameter was chosen to be a small scalar (with an initial value 0.001 and then gradually decreased after 1000 iterations). Upon the convergence of the frequency-domain ICA learning rule, the original signals were recovered by the inverse STFT. The scaling and permutation problems may be solved by a method proposed in [645] that computes the correlation of the envelopes of the spectrograms (i.e., the interfrequency spectral envelope correlation) or the improved method proposed in [46] based on interfrequency coherency. The experimental flowchart is illustrated in Figure 5.6. The two

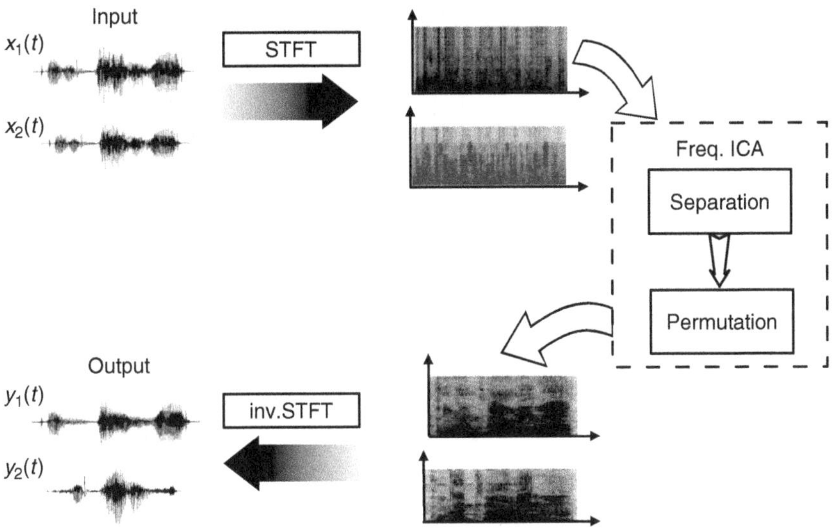

Figure 5.6 The experimental flowchart for frequency-domain BSS of two speech signals.

estimated time-domain speech signals $y_1(t)$ and $y_2(t)$ will be evaluated in SNR as compared with the convolutive mixtures $x_1(t)$ and $x_2(t)$, in which the SNR was calculated in terms of the signal amplitude after proper amplitude scaling. After 10,000 iterations of the learning rule, the averaged SNRs obtained in this experiment are 18.5 and 17.8 dB for two output signals.

5.2.6 Constant-Modulus Algorithm

The *constant-modulus algorithm* (CMA) is an adaptive learning algorithm proposed for blind equalization [218, 363, 365, 369, 446]; it exploits the constant or nearly constant modulus property of most modulated signals used in wireless communication, such as M-PSK or QAM. For simplicity, consider a single input–single output (SISO) system in which the source symbols $\{s(t)\}$ are transmitted through the channel, and we denote the input $\mathbf{x}(t) \in \mathbb{C}^N$ by a sequence of modulated complex-valued symbols $\mathbf{x}(t) = [s(t), s(t-1), \ldots, s(t-N+1)]^{\mathrm{T}}$. The equalizer is an adaptive FIR filter, denoted by the unknown parameter vector $\boldsymbol{\theta} = [\theta_0, \theta_1, \ldots, \theta_{N-1}]^{\mathrm{T}} \in \mathbb{C}^N$, which produces an output signal $y(t) = \boldsymbol{\theta}^{\mathrm{H}}\mathbf{x}(t)$, and the final equalized output corresponds to the approximate transmitted symbol such that $y(t) = \hat{s}(t)$. The goal of the equalizer is to minimize the error signal [denoted by $e(t)$] between the equalized output and the desired output in either blind or, semiblind mode.[8]

Consider the blind equalization problem for a communication channel; the signal processing operation is a form of blind deconvolution as illustrate in Figure 5.7. The equalizer contains an FIR filter and a zero-memory nonlinearity, and the error

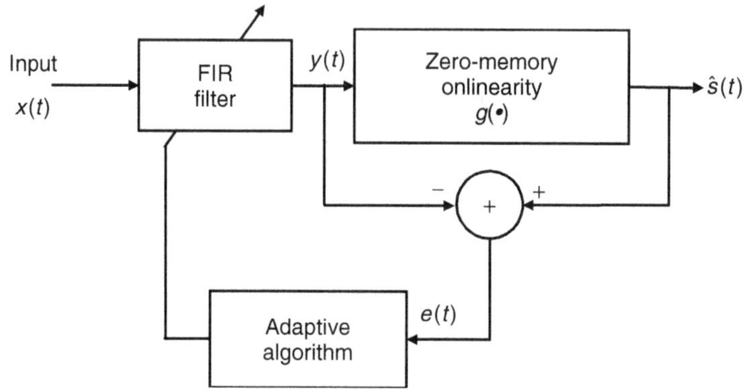

Figure 5.7 Block diagram of blind equalization using the Bussgang-type algorithm.

signal $e(t)$ can be modeled by

$$e(t) = \hat{s}(t) - y(t) = g(y(t)) - y(t), \tag{5.74}$$

where $g(\cdot)$ is a memoryless nonlinear function. Such an operation for blind equalization was known as the "Bussgang" algorithm [126], and the Bussgang-type algorithm approaches the equilibrium when the equalizer satisfies the condition

$$\mathbb{E}[y(t)g(y(t-k))] = \mathbb{E}[y(t)y(t-k)]. \tag{5.75}$$

In other words, a Bussgang process has the property that its autocorrelation function is equal to the cross-correlation between that process and the output of a zero-memory nonlinearity produced by that process. The Bussgang family of unsupervised adaptive filters include the *decision-directed algorithm* [575], the *Sato algorithm* [792], and the CMA for blind equalization [327, 888].

Specifically, in order to exploit the *constant-modulus* (CM) property, Godard [327] proposed to minimize the so-called dispersion cost function:

$$J_{\mathrm{CM}} = \mathbb{E}\left[(|y(t)|^p - \gamma_p)^2\right] = \mathbb{E}\left[(|\boldsymbol{\theta}^{\mathrm{H}}\mathbf{x}(t)|^p - \gamma_p)^2\right], \tag{5.76}$$

where the real-valued constant γ_p is chosen as a function of the source alphabet and of the integer p:

$$\gamma_p = \frac{\mathbb{E}[|s(t)|^{2p}]}{\mathbb{E}[|s(t)|^p]}. \tag{5.77}$$

Specifically:

- When $p = 1$, $\gamma_1 = \mathbb{E}[|s(t)|^2]/\mathbb{E}[|s(t)|]$, we have $J_{\mathrm{CM}} = \mathbb{E}\left[(|y(t)| - \gamma_1)^2\right]$. This case can be viewed as a modification of the Sato algorithm.

- When $p = 2$, $\gamma_2 = \mathbb{E}[|s(t)|^4]/\mathbb{E}[|s(t)|^2]$, we have $J_{\text{CM}} = \mathbb{E}\left[(|y(t)|^2 - \gamma_2)^2\right]$. This case is often referred to as the CMA in the literature.

By applying the gradient descent $\Delta\boldsymbol{\theta}(t) \propto -\partial J_{\text{CM}}/\partial\boldsymbol{\theta}$, the CMA can be described by a complex-valued version of the generalized Hebbian rule

$$\Delta\boldsymbol{\theta}(t) = \eta\mathbf{x}(t)e^*(t), \tag{5.78}$$

where $e(t)$ denotes the error signal. In general, the error signal is given by $e(t) = y(t)|y(t)|^{p-2}(\gamma_p - |y(t)|^p)$; when $p = 2$, it reduces to $e(t) = y(t)(\gamma_2 - |y(t)|^2)$. The Godard algorithm is considered to be the most successful among the Bussgang family. Remarkably, the CMA is very robust and also works reasonably well for non-CM sources [218]. In addition, Godard [327] showed that the MSE performance of the CMA is close to that of the Wiener equalizer.

If the learning-rate parameter η is sufficiently small, the stochastic gradient-based CMA rule (5.78) will converge to the optimal solution (when the global minimum of the cost function function is attained, we have $|y(t)|^2 = \mathbb{E}[|s(t)|^4]/\mathbb{E}[|s(t)|^2]$ and zero intersymbol interference). However, the convergence of the CMA is not guaranteed because the cost function is nonconvex and therefore has many local minima. To better illustrate this point, let us consider a simple example (taken from [218]) where the binary phase shift keying (BPSK) signals (i.e., binary symbols ± 1) are transmitted through a noise-free baseband channel. The channel follows an AR(1) model, in which the source symbol $s(t)$ (channel input) and the observed signal $x(t)$ (channel output) satisfy

$$x(t) + 0.6x(t-1) = s(t), \quad \text{where} \quad \Pr(s(t) = \pm 1) = 0.5, \tag{5.79}$$

and the two-tap equalizer parameter vector is $\boldsymbol{\theta} = [\theta_0, \theta_1]^{\text{T}}$. In this case, $s(t)$ has a constant modulus (i.e., $|s(t)| = 1$), and the CM cost function for the BPSK source is given by

$$J_{\text{CM}} = \mathbb{E}\left[(|y(t)|^2 - 1)^2\right]. \tag{5.80}$$

The ideal equilibria (i.e., global minima) for the CMA in this case are $\pm[1, 0.6]^{\text{T}}$, and the spurious equilibria (i.e., local minima) that are undesirable are $\pm[0, 0.5575]^{\text{T}}$. In addition, there are an extra four saddle points and one maximum (at the origin); hence, there are nine equilibria in total. Figure 5.8 presents an illustration of the three-dimensional plot of the error surface as well as its contour plot.

■ EXAMPLE 5.5

We further consider a SISO blind equalization example with the CMA. In this example, we assume a linear baseband real channel whose impulse response is given by equation (2.91) (Example 2.2). The number of taps of the equalizer is $N = 11$. The channel output SNR is 20 dB, and we employ two

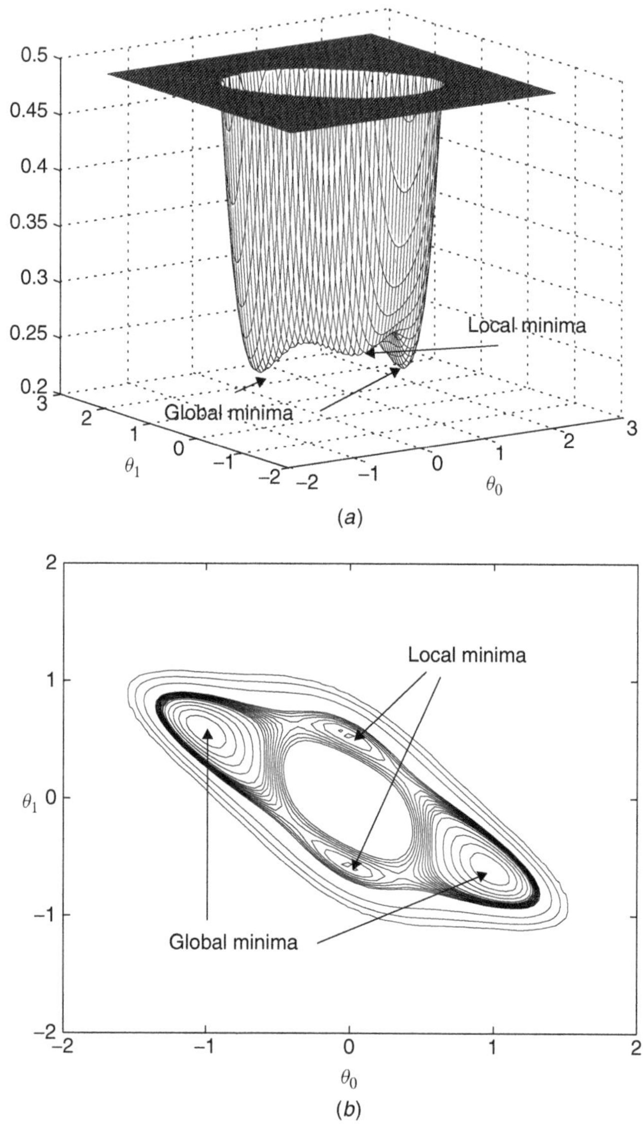

Figure 5.8 (*a*) Three-dimensional plot of the CMA cost function $J_{\rm CM}(\theta_0, \theta_1)$ and (*b*) its contour plot, assuming binary transmission in a noise-free channel. (Reproduced with permission. Copyright © 2001 by Marcel Dekker, Inc.)

constellation schemes: BPSK and quadrature phase-shift keying (QPSK). After randomly generating 4000 binary BPSK symbols or complex-valued QPSK symbols, we run the CMA rule (5.78) with an initial learning rate $\eta = 0.005$ (gradually annealed down to 0.0005). Note that, in the case of

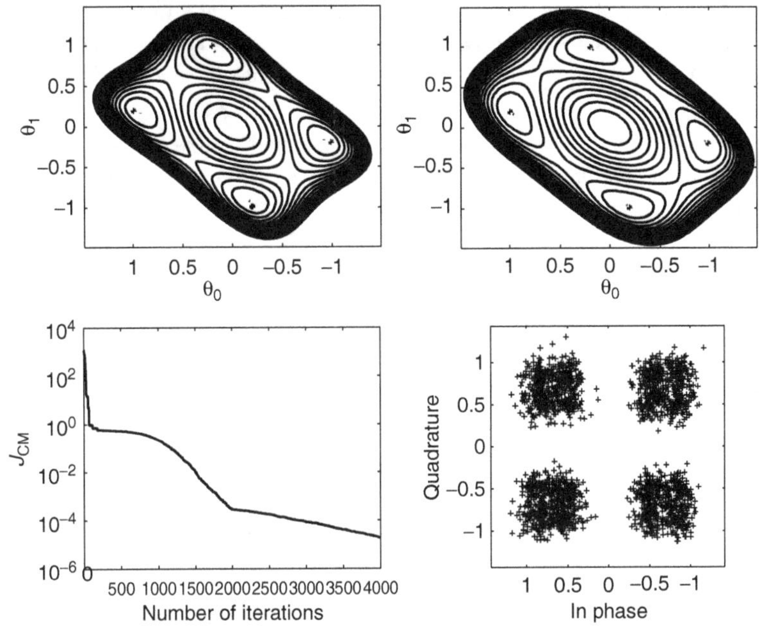

Figure 5.9 *Top two panels*: the CMA error surface contours projected on a two-dimensional space (where asterisks indicate the global minima), assuming BPSK (left) and QPSK (right) transmission and 20 dB SNR. *Bottom left panel*: the learning curve of a successful trial obtained from the CMA. *Bottom right panel*: the equalized QPSK output.

BPSK, $\gamma_2 = 1$; in the case of QPSK, $\gamma_2 = 0.5$, and the memoryless nonlinear function is

$$g(y(t)) = y_{\text{Re}}(1 + \gamma_2 - |y_{\text{Re}}|^2) + j y_{\text{Im}}(1 + \gamma_2 - |y_{\text{Im}}|^2).$$

The experimental results are shown in Figure 5.9. As seen from the figure, with a sufficiently small learning rate, the CMA might be able to escape from local minima and converge to optimal (or suboptimal) solution. However, in general, the convergence speed of the unsupervised CMA (for blind equalization) is slower than that of the supervised adaptive filtering (such as the LMS filter, see Example 2.2).

5.3 KERNEL METHODS FOR COMPLEX-VALUED DATA

5.3.1 Reproducing Kernels in the Complex Domain

Similar to the real vector space $\mathbb{R}^N$, the complex vector space $\mathbb{C}^N$ is also a finite-dimensional Hilbert space, with associated definitions of inner product and norm defined in the preceding section. A finite-dimensional Hilbert space always has a reproducing kernel; hence a unique reproducing kernel can always be found in the complex vector space. Most properties of the reproducing kernel in the real domain also hold in the complex domain. Here we only point out several differences.

Lemma 5.1 *Let $\{u_n : 1 \leq n \leq N\}$ be an orthonormal basis in a RKHS, where N is either finite or infinite; the reproducing kernel $K(\mathbf{x}', \mathbf{x})$ in the complex domain is given by*

$$K(\mathbf{x}', \mathbf{x}) = \sum_{n=1}^{N} u_n(\mathbf{x}) u_n^*(\mathbf{x}'),$$

where $u_n^*(\mathbf{x}')$ denotes the complex conjugate of $u_n(\mathbf{x}')$.

Lemma 5.2 *For a reproducing kernel $K(\mathbf{x}', \mathbf{x})$, the following equations hold:*

$$K(\mathbf{x}', \mathbf{x}) = K^*(\mathbf{x}, \mathbf{x}'),$$
$$\|K(\cdot, \mathbf{x})\|^2 = K(\mathbf{x}, \mathbf{x}) \geq 0,$$

where $K^*(\mathbf{x}, \mathbf{x}')$ denotes the complex conjugate of $K(\mathbf{x}, \mathbf{x}')$.

The reproducing kernel matrix $\mathbf{K} = \{K_{ij}\} \equiv \{K(\mathbf{x}_i, \mathbf{x}_j)\}$ is also called the *Gram matrix*, which is Hermitian (namely, $K_{ij} = K_{ji}^*$) in the complex domain. A complex Hermitian matrix $\mathbf{K}$ is positive definite since, for all $c_i \in \mathbb{C}$,

$$\sum_{i,j} c_i c_j^* K(\mathbf{x}_i, \mathbf{x}_j) = \langle \sum_i c_i \phi(\mathbf{x}_i), \sum_j c_j \phi(\mathbf{x}_j) \rangle = \left\| \sum_i c_i \phi(\mathbf{x}_i) \right\|^2 \geq 0,$$

where $\phi(\cdot)$ is a nonlinear function defined in the high-dimensional complex-valued feature space and its inner product defines the kernel $K(\mathbf{x}_i, \mathbf{x}_j) = \langle \phi(\mathbf{x}_i), \phi(\mathbf{x}_j) \rangle$. It is noted that

$$\begin{aligned}
\|\phi(\mathbf{x}_i) - \phi(\mathbf{x}_j)\|^2 &= \left[\phi(\mathbf{x}_i) - \phi(\mathbf{x}_j)\right]^{\mathrm{H}} \left[\phi(\mathbf{x}_i) - \phi(\mathbf{x}_j)\right] \\
&= K(\mathbf{x}_i, \mathbf{x}_i) + K(\mathbf{x}_j, \mathbf{x}_j) - K(\mathbf{x}_i, \mathbf{x}_j) - K(\mathbf{x}_j, \mathbf{x}_i) \\
&= K(\mathbf{x}_i, \mathbf{x}_i) + K(\mathbf{x}_j, \mathbf{x}_j) - 2\,\mathrm{Re}\left\{K(\mathbf{x}_i, \mathbf{x}_j)\right\}.
\end{aligned}$$

In terms of choosing kernels, two classes of kernel functions can be considered for complex-valued data: (i) The first class is the Hermitian kernel, which is Hermitian symmetric and complex valued in off-diagonal elements; the Hermitian kernel can be viewed as being induced by the *complex inner product* in the feature space. Examples of this kind include the d-order polynomial kernel

$$K(\mathbf{x}_i, \mathbf{x}_j) = (1 + \mathbf{x}_i^{\mathrm{H}} \mathbf{x}_j)^d, \qquad \mathbf{x}_i, \mathbf{x}_j \in \mathbb{C}^N, \qquad d \in \mathbb{N},$$

and the trigonometric kernel

$$K(\mathbf{x}_i, \mathbf{x}_j) = \cos \angle(\mathbf{x}_i, \mathbf{x}_j) = \frac{\mathbf{x}_i^{\mathrm{H}} \mathbf{x}_j}{\|\mathbf{x}_i\| \cdot \|\mathbf{x}_j\|}, \qquad \mathbf{x}_i, \mathbf{x}_j \in \mathbb{C}^N.$$

The second class is the real-valued symmetric kernel that takes the same form as in the real domain; such a real-valued kernel can be viewed as being induced by the *distance* or *probability* metric between two complex-valued variables. For instance, the Gaussian kernel belongs to this kind:

$$K(\mathbf{x}_i, \mathbf{x}_j) = \exp\left[-\frac{(\mathbf{x}_i - \mathbf{x}_j)^{\mathrm{H}}(\mathbf{x}_i - \mathbf{x}_j)}{\sigma^2}\right], \qquad \mathbf{x}_i, \mathbf{x}_j \in \mathbb{C}^N, \qquad \sigma \in \mathbb{R}.$$

The real-valued symmetric kernel is also a special case of the Hermitian kernel when all imaginary components vanish or remain zeros with probability 1. Notably, these two classes of kernel functions are both positive definite kernels.

5.3.2 Complex-Valued Kernel PCA

In Chapter 4, we derived the KPCA algorithm in the real domain. Without too much difficulty, the complex-valued version of KPCA can also be derived, which seeks to solve a kernelized Hermitian eigenvalue problem.

Define the Hermitian correlation matrix

$$\mathbf{C} = \frac{1}{\ell} \sum_{i=1}^{\ell} \phi(\mathbf{x}_i) \phi^{\mathrm{H}}(\mathbf{x}_i). \tag{5.81}$$

Then the Hermitian eigenvalue problem in the RKHS is rewritten as

$$\lambda \mathbf{v} = \mathbf{C}\mathbf{v} = \frac{1}{\ell} \sum_{i=1}^{\ell} \phi(\mathbf{x}_i) \phi^{\mathrm{H}}(\mathbf{x}_i) \mathbf{v}, \qquad \lambda \in \mathbb{R}, \tag{5.82}$$

which indicates that the eigenvectors can be constructed as a linear combination of the input vectors in the feature space:

$$\mathbf{v} = \sum_{i=1}^{\ell} \alpha_i \phi(\mathbf{x}_i), \tag{5.83}$$

where $\boldsymbol{\alpha}$ is a complex-valued column vector with the ith component defined as $\alpha_i = \phi^{\mathrm{H}}(\mathbf{x}_i)\mathbf{v}/(\lambda\ell)$. As shown previously in Chapter 4, we can reformulate a dual eigenvalue problem using the kernel representation

$$\ell \lambda \boldsymbol{\alpha} = \mathbf{K} \boldsymbol{\alpha}, \tag{5.84}$$

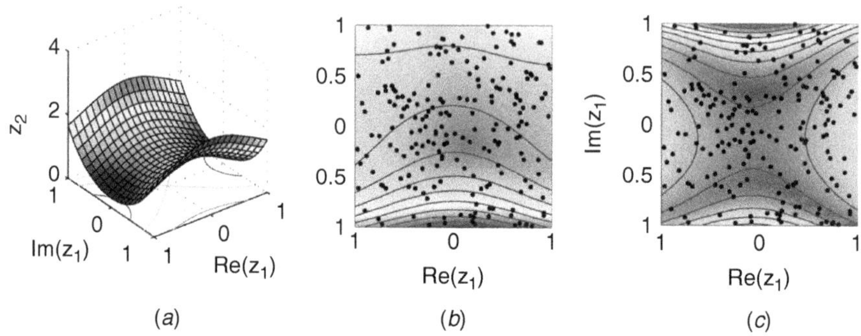

Figure 5.10 (*a*) Functional mapping. (*b,c*) The projections of the first (*b*) and second (*c*) eigenvectors in the feature space (training samples shown in black dots).

where the complex-valued coefficient vector $\boldsymbol{\alpha}$ plays the role of the eigenvector of the Hermitian kernel matrix $\mathbf{K}$ associated with the real eigenvalue $\ell\lambda$.

To illustrate the complex KPCA, we consider a simple toy example that has the following functional mapping:

$$z_1 = z_1^{\mathrm{Re}} + jz_1^{\mathrm{Im}}, \quad z_1^{\mathrm{Re}}, z_1^{im} \in [-1, 1]$$

$$z_2 = |\cos^2(z_1)| + \xi, \quad \xi \in \mathcal{N}(0, 0.05).$$

A total of 400 random samples $\mathbf{z}_i = [z_1, z_2]^{\mathrm{T}} \in \mathbb{C}^2 (i = 1, \ldots, 400)$ are generated as the training set. After learning the eigenvectors with 3rd-order polynomial kernel, we project the testing points onto the first two dominant eigenvectors in the feature space. The results are shown in Figure 5.10.

Likewise, many other correlation-based kernelized algorithms can be generalized to the complex domain. We will not repeat them here simply due to the close resemblance.

5.4 DISCUSSION

In comparison with real numbers, complex numbers offer an additional representation power that is appealing for directional data with orientation or phase attribute. Such data are frequently found, such as wind speed, magnetic field, or optical flow. Complex-valued signals also arise in many real-life applications, such as communications, array signal processing, remote sensing, and imaging. Therefore, how to extend the idea of correlative learning to the complex domain is an interesting research topic. In this chapter, we considered complex-valued Hebbian learning and complex-valued neural networks. As discussed throughout the chapter, we have observed similarity between the development of the complex-valued correlation-based learning paradigms and that of their real counterparts. On the other hand, complex-valued correlative learning also poses some challenges in computational neural coding and pattern recognition (e.g., [640, 944]).

BIBLIOGRAPHICAL NOTES

Complex numbers and complex analysis have a long history in mathematics. Extending correlation-based statistical analysis or adaptive algorithms to the complex domain is useful for complex-valued data encountered in array signal processing, imaging, remote sensing, radar, and communications. Second-order correlation statistics have again played important roles in complex-valued signal processing. The mathematical treatment of second-order complex random vectors and the circular and noncircular complex Gaussian distributions are discussed in [723, 724].

Analogous to their real-valued counterparts, complex-valued neural networks have many unique properties and deserve special research attention [392, 393]. In the literature, many versions of complex-valued neural networks have been proposed, such as the complex-valued Hopfield network [641], complex-valued SOM [351], and complex-valued MLP. In-depth discussion of the complex-valued LMS and backpropagation algorithms was given in [369]. A complex-valued real-time recurrent learning (RTRL) algorithm was also developed in [328] for recurrent neural networks.

The complex-valued PCA theory was first developed to analyze two-dimensional vector fields such as winds and currents or the complex-valued data induced by the Fourier or Hilbert transform of the real-valued data [403]. Applications of complex-valued principal- or minor-component analysis were reviewed and discussed in [279, 577], which are useful in array signal processing, beamforming, and teleconferencing. Extensions of complex-valued nonlinear PCA were also discussed in [278, 280, 756].

Complex-valued ICA algorithms have been developed from several different roots, such as the complex JADE [143], the complex FastICA for both circular sources [94] and general sources [226], the complex Infomax or natural gradient [7, 42, 137], and many other variants [164, 265, 266, 279, 280]. However, a complete theoretical understanding of the complex ICA problem somewhat remains missing in the literature. Complex ICA algorithms have also been applied to neurophysiological data, such as functional magnetic resonance imaging (fMRI) [138] and electroencephalography (EEG) [42].

The blind equalization problem arises in wired and wireless communications with the goal of reducing the intersymbol interference among the transmission. The very first idea of a blind equalization algorithm, bearing a form of unsupervised filter, was introduced by Bussgang in his 1952 technical report at MIT. A modern rediscovery of such an idea was independently found in the publications of Godard [327] and Treichler and Agee [888]. In fact, the Bussgang family of unsupervised adaptive filters includes the decision-directed algorithm [575], the Sato algorithm [792], as well as the CMA. Just as the LMS algorithm has established itself as the workhorse for supervised linear adaptive filtering, the CMA has become the workhorse for blind channel equalization. A review of the CMA in the context of blind equalization is given in [446]. For detailed treatments of blind equalization and blind deconvolution, see [218, 363, 365].

NOTES

1. For more discussions on the properties, history, and applications of complex numbers, the interested reader is referred to the online URL source http://en.wikipedia.org/wiki/Complex_number.
2. The terms *holomorphic function*, *differentiable function*, and *complex differentiable function* are sometimes used interchangeably with "analytic function."
3. Cauchy–Riemann equations state that the partial derivatives of a complex function $f(z) = u(x, y) + jv(x, y)$ along the real and imaginary axes should be equal: $\partial u/\partial x = \partial v/\partial y$ and $\partial v/\partial x = -\partial u/\partial y$.
4. If a complex-valued function $J(\mathbf{x}) : \mathbb{C}^n \to \mathbb{C}$ is twice differentiable and the complex Hessian matrix is positive semidefinite, then it is said that the function $J(\mathbf{x})$ at every point $\mathbf{x}$ is *plurisubharmonic*; since $J(\mathbf{x})$ is continuous, it is also called a *pseudoconvex function* [507]. Note that this is different from the real-valued case, where a twice continuously differentiable real-valued function with a positive-semidefinite real Hessian matrix at every point is convex.
5. In contrast, the *Takagi factorization* [404] seeks to factorize the complex symmetric matrix $\mathbf{C}$ (such as the pseudocovariance matrix) into the form $\mathbf{C} = \mathbf{U}\mathbf{\Lambda}\mathbf{U}^{\mathrm{T}}$, where $\mathbf{U}$ is a unitary matrix and $\mathbf{\Lambda}$ is the diagonal singular-value matrix.
6. In the real-valued case, the activation function $\psi(u)$ is often chosen to match the score function associated with the pdf of the sources, which is defined as $\psi(u) = -\mathrm{d}\log p(u)/\mathrm{d}u$. However, when complex-valued functions are employed to generate the nonlinearities, direct interpretation of $\psi(\cdot)$ in the context of the cumulative distribution function is lost [7].
7. The reason that the time frame window size T must be longer than P is threefold [43]: (i) Linear convolution can be approximated by a circular convolution if $T > 2P$; (ii) if we need to estimate the inverse of a system with impulse response P taps long, the length of the impulse response of the inverse system must be longer than P; and (iii) provided a noise canceler is used, the FIR filter's length must also be longer than P.
8. There are many linear equalizer algorithms in the literature; a MMSE solution is given by the optimum Wiener equalizer. In the case of semiblind equalization, at the first-stage of the *training phase*, the error signal is produced by the difference between the estimate and a supervised pilot signal $e(t) = d(t) - y(t) \equiv s(t) - y(t)$: at the second stage of the *decision-directed phase*, the error signal is given by $e(t) = \hat{s}(t) - y(t)$, where $\hat{s}(t)$ is the symbol estimate generated by the (hard or soft) decision device.

6

ALOPEX: A CORRELATION-BASED LEARNING PARADIGM

6.1 BACKGROUND

ALOPEX, short for *ALgorithm Of Pattern EXtraction*, was originally designed in the 1970s as an optimization procedure for pattern extraction in the visual system [355, 900]. In its first appearance, ALOPEX was developed for extracting visual receptive fields,[1] in which the *response feedback* was used to construct visual patterns that optimize the neurons' responses. The underlying assumption in the ALOPEX procedure is that, apart from noise fluctuations, the response of a neuron in the visual pathway increases as the stimulus approaches some optimal pattern, that is, one that matches its receptive field. In principle, any visual event or sequence of events displayed on the retina may match the receptive field of a neuron (or population of neurons). Such neurons act as *detectors* of the specific sensory trigger features defined by their receptive fields. In particular, when the detectors' generated patterns (starting with a random pattern) match the desired receptive field (i.e., they are highly correlated), the neuron is likely to produce a high response (i.e., with high firing rate). In [355], the ALOPEX process takes the feedback of the neurons' firing responses and further optimizes its produced patterns until the correlations between the ALOPEX's output patterns and the neuronal receptive fields' patterns are sufficiently high; by then coincidence detection is accomplished with a trial-and-error stimulus pattern-matching process. A mathematical analysis of the ALOPEX process for the model described in [355] was given by Amari [22] (see Appendix 6A).

Correlative Learning: A Basis for Brain and Adaptive Systems, by Zhe Chen, Simon Haykin, Jos J. Eggermont, and Suzanna Becker

Since its first appearance, ALOPEX has been widely used for modeling the dynamical aspects of the visual system, particularly its use of feedback. A classic study of reciprocal pathways in visual circuits was presented in [356]; another example is to use ALOPEX for modeling visual attention [437] with feedback pathways. Nowadays, the name ALOPEX has gradually gone beyond its original meaning. ALOPEX has also been used to model other neural structures, going beyond visual cortex. For instance, the ALOPEX process was suggested to play a critical role in the thalamus via the thalamocortical (feedforward) and corticothalamic (feedback) loops [356, 644]. In Chapter 7, we will also present one example of using ALOPEX for modeling sensory systems.

Another application of ALOPEX is its use as a universal gradient-free nonlinear optimization procedure for various optimization problems [354], such as training neural networks [901], control [914], and combinatorial optimization [354]. In particular, ALOPEX was popularized and introduced to the neural computation community by Unnikrishnan and Venugopal [902]. Bia [90] also proposed a quasi-deterministic version of ALOPEX, which was termed ALOPEX-B. ALOPEX-B was developed to overcome some of the limitations of the original algorithm in [902]. Recently, some sophisticated versions of ALOPEX have also been developed [163, 374, 791]. In this chapter, we will present an in-depth overview of these algorithms that use the correlation-based paradigm for learning or optimization.

6.2 THE BASIC ALOPEX RULE

Heuristics. Before presenting a rigorous mathematical derivation, we give a heuristic illustration of the key ideas underlying the development of the ALOPEX procedure.

Without loss of generality, let us first consider a one-dimensional example. Suppose that the goal is to minimize or maximize an objective function $J(\theta)$, where θ is the parameter to be optimized. By definition, the gradient of $J(\theta)$ is given by the following equation[2]:

$$\frac{\partial J(\theta)}{\partial \theta} = \lim_{\delta\theta \to 0} \frac{J(\theta + \delta\theta) - J(\theta)}{\delta\theta} = \lim_{\delta\theta \to 0} \frac{\delta J}{\delta\theta} \approx \frac{\Delta J}{\Delta\theta},$$

where the approximation is valid when $\Delta\theta$ is sufficiently small and therefore approximates the infinitesimal perturbation $\delta\theta$. Note that the algebraic sign of the gradient remains unchanged if we substitute $\Delta J/\Delta\theta$ with the product form $\Delta\theta$ ΔJ; in other words, they only differ in quantity. When the unknown parameter is multidimensional (i.e., the scalar θ is replaced by a vector $\boldsymbol{\theta}$), using $\Delta\boldsymbol{\theta}\,\Delta J$ as a gradient estimate will allow one to find the nearest local minimum/maximum, but multidimensional optimization methods based on gradient search all suffer from the problem of becoming trapped in poor local optima. In order to circumvent this limitation, we need to introduce noise to allow some probability of escape from local optima. How to control the amount of the noise is the key in the ALOPEX procedure. In the next section, we will discuss this issue in detail and finally lead to the appealing features of this correlation-based learning paradigm.

Mathematical Derivation. By analogy to the correlative form of Hebbian learning, we will derive a simple correlative form of the ALOPEX learning rule. We do so by relating an incremental *continuous-time* perturbation in the weight vector, $\delta\boldsymbol{\theta}$, to the correlation between a *discrete-time* change in the weight vector, $\Delta\boldsymbol{\theta}$, and the corresponding incremental continuous-time perturbation in the objective function $\delta J = J(\boldsymbol{\theta} + \delta\boldsymbol{\theta}) - J(\boldsymbol{\theta}) \approx J(\boldsymbol{\theta} + \Delta\boldsymbol{\theta}) - J(\boldsymbol{\theta})$, defined as [301]

$$\delta\boldsymbol{\theta} \propto \langle \Delta\boldsymbol{\theta}, \delta J \rangle, \tag{6.1}$$

where the time-average operator $\langle x, y \rangle$ accounts for temporally local correlations between two variables x and y. Moreover, invoking the first-order Taylor series, we may approximate δJ due to discrete-time changes in the individual elements of the N-dimensional weight vector $\boldsymbol{\theta}$ as

$$\delta J \approx \sum_{j=1}^{N} \left. \frac{\partial J}{\partial \theta_j} \right|_{\boldsymbol{\theta}} \Delta\theta_j.$$

Correspondingly, we may write

$$\langle \Delta\theta_i, \delta J \rangle \approx \sum_{j=1}^{N} \left. \frac{\partial J}{\partial \theta_j} \right|_{\boldsymbol{\theta}} \langle \Delta\theta_i, \Delta\theta_j \rangle, \qquad i = 1, \ldots, N. \tag{6.2}$$

Assuming that the Euclidean norm $\|\Delta\boldsymbol{\theta}\| \ll 1$ and that the "averaged" individual element changes $\Delta\theta_i$ $(i = 1, \ldots, N)$ are independent of each other (locally in time), we may approximate the cross-correlation term on the right-hand side of (6.2) as

$$\langle \Delta\theta_i, \Delta\theta_j \rangle \approx \eta\, \Delta\theta_i^2 \delta_{ij},$$

where η is a small-valued positive constant, and

$$\delta_{ij} = \begin{cases} 0, & i \neq j, \\ 1, & i = j, \end{cases}$$

is the Kronecker delta. Accordingly, we may further approximate (6.2) as

$$\begin{aligned} \langle \Delta\theta_i, \delta J \rangle &\approx \eta \left. \frac{\partial J}{\partial \theta_i} \right|_{\boldsymbol{\theta}} \Delta\theta_i^2 \\ &\approx \eta\, \Delta J\, \Delta\theta_i, \qquad i = 1, \ldots, N. \end{aligned}$$

In vector form, we thus have the compact relation

$$\Delta\boldsymbol{\theta}(t+1) \propto \eta\, \Delta\boldsymbol{\theta}(t)\, \Delta J(t), \tag{6.3}$$

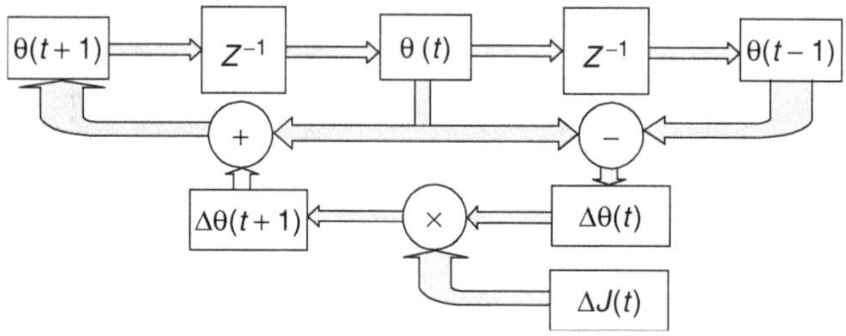

Figure 6.1 Signal-flow graph representation of the ALOPEX procedure.

where

$$\Delta\boldsymbol{\theta}(t) = \boldsymbol{\theta}(t) - \boldsymbol{\theta}(t-1), \tag{6.4}$$

$$\Delta J(t) = J(t) - J(t-1). \tag{6.5}$$

Stated in words, the correction in the update formula (6.3) is proportional to the *instantaneous* correlation or product between the weight modification $\Delta\boldsymbol{\theta}(t)$ in two consecutive time steps and the corresponding objective function change $\Delta J(t)$, where the algebraic sign (positive or negative) on the right-hand side of (6.3) depends on whether the objective function is to be maximized or minimized (see Figure 6.1 for the signal-flow illustration). The algorithm for the weight changes given by (6.3) forms the basis for ALOPEX as discussed below, which additionally incorporates a stochastic decision rule for determining the direction of weight change.

6.3 VARIANTS OF ALOPEX

6.3.1 Unnikrishnan and Venugopal's ALOPEX

Without loss of generality, let us assume the optimization goal is to minimize a generic objective function $J(t)$ which is assumed to be a bounded, continuous or piecewise continuous (but not necessarily differentiable) function of some unknown parameters.

In the context of training neural networks, ALOPEX was introduced by Unnikrishnan and Venugopal [901, 902] as a correlation-based, gradient-free learning procedure. Specifically, let $\boldsymbol{\theta}$ denote the weight vector that includes all unknown parameters. The learning rule is described as

$$\boldsymbol{\theta}(t+1) = \boldsymbol{\theta}(t) + \eta\boldsymbol{\xi}(t), \tag{6.6}$$

where η is the learning-rate parameter. The vector $\boldsymbol{\xi}(t)$ is a random vector with its jth entry determined elementwise by

$$\xi_j(t) = \text{sgn}(u_j - p_j(t)), \qquad u_j \sim \mathcal{U}(0, 1), \tag{6.7}$$

$$p_j(t) = \phi\left(\frac{c_j(t)}{T(t)}\right) = \frac{1}{1 + \exp\left[-c_j(t)/T(t)\right]}, \tag{6.8}$$

$$c_j(t) = \Delta\theta_j(t)\,\Delta J(t), \tag{6.9}$$

where u_j is a uniformly distributed random variable drawn from region (0, 1), $\text{sgn}(\cdot)$ is the signum function, and $\phi(\cdot)$ is the logistic sigmoid function. The key term is $c_j(t)$, which correlates changes in the cost function with parameter vector changes; it is the scalar version of equation (6.3). At each time step, the ALOPEX procedure updates $\theta_j(t)$ by $\pm\eta$ with probability $p_j(t)$ (Boltzmann distribution) or $1 - p_j(t)$. The change of the cost function $\Delta J(t) > 0$ [or $\Delta J(t) < 0$] will make the probability of moving each $\theta_j(t)$ in the same (or opposite) direction greater than 0.5, which thereby favors the changes to decrease the cost function $J(t)$. In addition, $T(t)$ is a time-varying annealing parameter that plays a similar role to "temperature" in simulated annealing [483]. Specifically, $T(t)$ can be updated every T_0 (where $T_0 > 1$ is a predefined integer) iterations as follows:

$$T(t) = \begin{cases} T(t-1) & \text{if } t \text{ is not a multiple of } T_0, \\ \frac{\eta}{T_0} \sum\limits_{k=t-T_0}^{t} |\Delta J(k)| & \text{otherwise.} \end{cases} \tag{6.10}$$

The temperature parameter is critical in that it determines how sharply the probability $p_j(t)$ is pushed towards 0 or 1 with increasing magnitude of the correlation $c_j(t)$. The annealing schedule given in equation (6.10) implies that ALOPEX has a self-scaling property in that the determination of $p_j(t)$ relies on the comparison of current $\Delta J(t)$ and the average of recent past values.

In the optimization procedure, the ALOPEX rule starts with a randomly initialized parameter vector $\boldsymbol{\theta}(0)$ and stops when the cost function $J(t)$ is sufficiently small. The stochastic component $\boldsymbol{\xi}(t)$, being a random force with certain acceptance probability, is included to help the algorithm escape from local minima. Another point to make here is that in the ALOPEX procedure the parameter vector $\{\boldsymbol{\theta}(t), t \geq 0\}$ is not first-order Markovian, since $\boldsymbol{\theta}(t)$ depends on both $\boldsymbol{\theta}(t-1)$ and $\boldsymbol{\theta}(t-2)$. By introducing another auxiliary variable vector $\mathbf{z}(t) = [\boldsymbol{\theta}(t), \boldsymbol{\theta}(t-1)]$, $\mathbf{z}(t)$ becomes a finite-state ergodic Markov chain under regular conditions [791].

6.3.2 Bia's ALOPEX-B

A major feature of the ALOPEX proposed by Unnikrishnan and Venugopal is the use of an annealing schedule that was motivated by simulated annealing [483].

Despite its physical insight, such an annealing schedule often suffers from slow convergence in optimization. To improve this problem, Bia [90] developed a quasi-deterministic version of ALOPEX, which was called ALOPEX-B. Unlike the ALOPEX described by equations (6.6)–(6.10), ALOPEX-B does not employ any annealing scheme and uses fewer tuning parameters, thereby exhibiting a simpler implementation and reportedly faster convergence. Consistent with the preceding notation, ALOPEX-B proceeds as follows:

$$\boldsymbol{\theta}(t+1) = \boldsymbol{\theta}(t) + \eta\boldsymbol{\xi}(t), \tag{6.11}$$

$$\xi_j(t) = \text{sgn}(u_j - p_j(t)), \qquad u_j \sim \mathcal{U}(0, 1), \tag{6.12}$$

$$p_j(t) = \phi(C_j(t)), \tag{6.13}$$

$$C_j(t) = \frac{\text{sgn}(\Delta\theta_j(t))\,\Delta J(t)}{\sum_{k=2}^{t} \lambda(\lambda - 1)^{t-k}|\Delta J(k-1)|}, \tag{6.14}$$

where $0 < \lambda < 1$ is a forgetting parameter. An optimal forgetting parameter is often problem specific; a typical value is often chosen within the range $[0.35, 0.7]$ according to some empirical studies.

It is noteworthy that in ALOPEX-B the acceptance probability $C_j(t)$ replaces $c_j(t)/T(t)$ in equation (6.8); in other words, $T_0 = 1$ is always used for each iteration.

6.3.3 Improved Version of ALOPEX-B

In practical experiments [163], it was found that it is more efficient to combine equations (6.11) and (6.3) in a hybrid learning form, which leads to the modified ALOPEX-B:

$$\boldsymbol{\theta}(t+1) = \boldsymbol{\theta}(t) + \eta\boldsymbol{\xi}_t - \gamma\,\Delta\boldsymbol{\theta}(t)\,\Delta J(t), \tag{6.15}$$

where γ is another learning-rate (or step-size) parameter, $\boldsymbol{\xi}_t$ corresponds to the same stochastic term in (6.11) without invoking the temperature annealing, and $\Delta\boldsymbol{\theta}(t)\,\Delta J(t)$ corresponds to the product term on the right-hand side of equation (6.3). The motivation for inclusion of the noise term $\boldsymbol{\xi}_t$ is to introduce a small amount of randomness in the direction of weight change, thereby helping the algorithm escape from local minima.

The modified ALOPEX-B seeks two types of correlation:

- The first kind of correlation takes the form of instantaneous cross-correlation described by the product term $\Delta\boldsymbol{\theta}(t)\,\Delta J(t)$.
- The second kind of correlation appears in the computation of $\boldsymbol{\xi}_t$ as in equations (6.12)–(6.14), which determines the acceptance probability of random perturbation force $\boldsymbol{\xi}_t$.

We note that when the term $\boldsymbol{\xi}(t)$ takes a simplified form of noise, equation (6.15) reduces to the special form described in [898, 899]:

$$\boldsymbol{\theta}(t+1) = \boldsymbol{\theta}(t) - \eta\, \Delta\boldsymbol{\theta}(t)\, \Delta J(t) + \boldsymbol{u}(t), \tag{6.16}$$

where $\boldsymbol{u}(t)$ denotes a Gaussian noise vector. The additive noise term $\boldsymbol{u}(t)$ differs from $\boldsymbol{\xi}(t)$ in that it ignores the correlation information that is used to determine the noise amount in either equations (6.8) and (6.9) or equations (6.13) and (6.14).

6.3.4 Two-Timescale ALOPEX

Motivated by the two-timescale stochastic approximation method (e.g., [104]), Sastry et al. [791] proposed a two-timescale version of ALOPEX which was called 2t-ALOPEX. The key feature of 2t-ALOPEX is to recursively update the acceptance probability $p_j(t)$ that appears in (6.8). Specifically, the iterative update rule is given by

$$\begin{aligned} p_j(t) &= (1-\lambda)p_j(t-1) + \lambda\zeta_j(t) \\ &= p_j(t-1) + \lambda(\zeta_j(t) - p_j(t-1)), \end{aligned} \tag{6.17}$$

where $0 < \lambda < 1$ and $\zeta_j(t)$ is defined as

$$\zeta_j(t) = \phi\left(\xi_j(t-1)\frac{J(\boldsymbol{\theta}(t)) - J(\boldsymbol{\theta}(t) - \eta\boldsymbol{\xi}(t-1))}{\eta T(t)}\right), \tag{6.18}$$

with $\phi(\cdot)$ being a logistic sigmoid function and $T(t)$ the temperature parameter appearing in (6.10). The motivation for this modification (to Unnikrishnan and Venugopal's ALOPEX) is to incorporate a heuristic approximation of the first-order Taylor series. Specifically, let $\Delta J(t) = J(\boldsymbol{\theta}(t)) - J(\boldsymbol{\theta}(t) - \eta\boldsymbol{\xi}(t-1))$, and in light of (6.9), the correlation term $c_j(t)$ can be approximated by [791]

$$\begin{aligned} c_j(t) &\approx \eta\xi_j(t-1)\left[\eta\sum_{k=1}^{N}\frac{\partial J(\boldsymbol{\theta}(t))}{\partial\theta_k}\xi_k(t-1)\right] \\ &= \eta^2\frac{\partial J(\boldsymbol{\theta}(t))}{\partial\theta_j} + \eta^2\sum_{k\neq j}\frac{\partial J(\boldsymbol{\theta}(t))}{\partial\theta_k}\xi_j(t-1)\xi_k(t-1). \end{aligned} \tag{6.19}$$

When η is small, the second term on the right-hand side of (6.19) is expected to be very small in magnitude due to the terms $\xi_j\xi_k$ averaging close to zero. Therefore, $c_j(t)$ would be primarily determined by the jth partial derivative of the cost function J, thus providing (with a high probability) the correct descent direction for (6.6). In 2t-ALOPEX, λ is chosen to be much greater than η; thus the dynamics of $p_j(t)$ is also much faster than that of $\boldsymbol{\theta}(t)$. The theoretical analysis of 2t-ALOPEX is presented in Appendix 6B.

6.3.5 Other Types of Correlation Mechanisms

Three different types of correlational structure can be incorporated into ALOPEX-type learnig procedures. The first is the *time-averaged correlation*:

$$\theta_j(t+1) = \theta_j(t) - \eta R_j(t) + u_j, \tag{6.20}$$

$$R_j(t) = \lambda R_j(t-1) + \Delta J(t)\, \Delta\theta_j(t), \tag{6.21}$$

where $0 < \lambda < 1$ and the instantaneous correlation is substituted by a window-averaged correlation estimate. Note that by this change the current parameter is influenced by the errors in previous steps (i.e., penalizing temporal trajectories), and the learning rule is forced to search for a locally smooth solution in the parameter space.

The second type of correlational structure is the *inverse correlation*:

$$\theta_j(t+1) = \theta_j(t) - \eta \frac{\Delta J(t)}{\Delta\theta_j(t)} + u_j, \tag{6.22}$$

where the instantaneous value $\Delta J(t)/\Delta\theta_j(t)$ replaces its product value. The inverse correlation, however, has the disadvantage that the crosstalk noise amplifies as $\Delta\theta_j(t)$ becomes small in comparison with $\Delta J(t)$, since $\Delta J(t)$ might include the change caused by other $\theta_k(t)$ for $k \neq j$ [301]. In addition, the inverse correlation often invokes a numerical issue in practice: If $\Delta\theta_j(t)$ is very small, it can cause overflow problems in computer simulations.

Finally, the third type of correlational structure is the *gain-and-loss discriminated correlation*:

$$\theta_j(t+1) = \begin{cases} \theta_j(t) - \eta\, \Delta J(t)\, \Delta\theta_j(t) + u_j & \text{if } \Delta J(t) < 0, \\ \theta_j(t) - \eta \frac{\Delta J(t)}{\Delta\theta_j(t)} + u_j & \text{if } \Delta J(t) > 0, \end{cases} \tag{6.23}$$

which is a form of either *gain-emphasized* correlation [when $\Delta J(t) < 0$] or *loss-suppressed* correlation [when $\Delta J(t) > 0$] [301]. When $\Delta\theta_j$ gives rise to a desired gain [i.e., $\Delta J(t) < 0$], $\Delta J(t)$ is multiplied by $\Delta\theta_j(t)$, the gain is further used to bring in a bigger change of θ_j, and thus a lower potential of J at a farther point is an attractor. When $\Delta\theta_j$ results in an undesired loss [i.e., $\Delta J(t) > 0$], $\Delta J(t)$ is divided by $\Delta\theta_j(t)$, and the loss moves θ_j according to the approximate gradient direction. The motivation of such discriminated correlations is to change the parameters via the *attractive force* of the global minimum and the *repulsive force* of the local gradient.

6.4 DISCUSSION

Summarization of Features. Thus far, we have discussed several different versions of ALOPEX. Despite some implementation differences, they do share many common features, as summarized below:

- The ALOPEX learning rule (6.3) can be viewed as a generalized form of the *differential Hebbian rule* as discussed earlier in Chapter 3.

- The ALOPEX optimization procedure is gradient free and is independent of the objective function and network (model) architecture.
- The optimization is *synchronous* in the sense that all parameters are updated in parallel, thereby sharing the features of algorithmic simplicity and ease of hardware implementation.
- The optimization relies on noise, whose main role is to control the search direction, while usually taking steps in the optimal direction but occasionally allowing steps in the (locally) suboptimal direction. This allows the algorithm to escape from the local minima or maxima by introducing randomness into the search procedure.
- The basic principle of the ALOPEX algorithm is a *trial-and-error* process, similar in spirit to the "weight perturbation" method (also called "MIT rule") in the control literature.
- The ALOPEX rule only invokes either a Hebbian or an anti-Hebbian term [depending on the objective function $J(t)$ to be maximized or minimized] but not both together; in the simplest Hebbian form without constraints (such as weight normalization), it might be potentially unstable.

Comparison with Hebbian Synaptic Plasticity. Despite the fact that ALOPEX and Hebb's original rule are both correlative learning algorithms by nature, ALOPEX distinguishes itself from Hebb's rule in a number of ways. First, Hebb's rule is restricted to using information locally available to a single neuron,[3] whereas ALOPEX is a very general optimization procedure that may potentially incorporate a global cost function. Second, Hebb's rule only characterizes the synaptic plasticity between individual pairs of neurons, whereas the ALOPEX rule is potentially applicable to modeling the synaptic plasticity within a population of neurons.

In using APLOEX for modeling brain functions, it is worth pointing out several important neurobiological considerations:

- ALOPEX is characterized by a temporally asymmetric synaptic plasticity process, implying causality between weight changes and subsequent cost function changes (in the sense that the action $\Delta\boldsymbol{\theta}$ yields either a reward or a penalty measured by ΔJ). The issue of which works best, a quantitative real-valued error signal or a bipolar signal (success or failure), is still under debate.
- The convergence properties of the ALOPEX learning procedure depend upon adding a certain amount of noise. In neurobiological systems, noise may come into play in a number of ways, for example, at the level of synaptic transmission or in the generation of an action potential at the cell body, any of which would lead to randomness in neural plasticity.
- ALOPEX optimizes a global objective function with respect to the adjustable synaptic weights. Thus, the underlying philosophy behind equation (6.3) could be characterized by "*think globally, act locally and synchronously*." In biological systems, it is unclear how a global objective function could be communicated [68]. The best candidate mechanism for such a process is the TD

error signal, which may be communicated via the firing pattern of dopamine neurons [805].

Hindsight. Interestingly, a description of a learning procedure strikingly similar to ALOPEX was discussed in Marvin Minsky's illuminating review paper "Steps Towards Artificial Intelligence" in 1961 [629]:

> Multiple simultaneous optimizers search for a (local) maximum value of some function $J(x_1, \ldots, x_n)$ of several parameters. Each unit u_i independently "jitters" its parameter x_i, perhaps randomly, by adding a variation $d_i(t)$ to a current mean value $m_i(t)$. The changes in the quantities x_i and J [namely, Δx_i and ΔJ] are correlated, and the result is used to slowly change m_i. The filters are to remove DC components. This technique, a form of coherent detection, usually has an advantage over methods dealing separately and sequentially with each parameter. Cf. the discussion of "informative feedback" in Wiener [1948, p. 133]. A great variety of hill-climbing systems have been studied under the names of "adaptive" or "self-optimizing" servomechanisms.

It can readily be seen that the above statement is indeed a description of the idea underlying the stochastic correlative learning algorithms discussed in this chapter.

ALOPEX for Optimization in Complex Domain. In Chapter 5, we discussed complex-valued correlation-based learning and optimization algorithms. ALOPEX can also be used for complex-valued optimization. Moreover, since ALOPEX is gradient free and model independent, the adaptation of its optimization procedure to the complex domain is straightforward and does not require the differentiability of either the cost function or the nonlinear activation function.

Specifically, let $J(\boldsymbol{\theta})$ denote the real-valued scalar cost function to be minimized, and let $\boldsymbol{\theta}$ and $\boldsymbol{\theta}^*$ denote the unknown complex-valued parameter vector and its complex conjugate, respectively; then the complex-valued version of (6.15) can be reformulated as follows:

$$\boldsymbol{\theta}(t+1) = \boldsymbol{\theta}(t) + \eta \boldsymbol{\xi}(t) - \gamma \, \Delta\boldsymbol{\theta}^*(t) \, \Delta J(t), \tag{6.24}$$

where $\Delta\boldsymbol{\theta}^*(t) = \boldsymbol{\theta}^*(t) - \boldsymbol{\theta}^*(t-1)$ and $\Delta J(t) = J(\boldsymbol{\theta}(t)) - J(\boldsymbol{\theta}(t-1))$. It is noted that the product term $\Delta\boldsymbol{\theta}^*(t)\,\Delta J(t)$ is reminiscent of the complex-valued gradient operator $\nabla J_{\theta} = \frac{\partial J(\boldsymbol{\theta})}{\partial \boldsymbol{\theta}^*}$ defined in equation (5.18).

■ EXAMPLE 6.1

Complex-valued neural networks [392, 662] have recently become an important topic of research due to some of their unique properties that are distinct from their real-valued counterparts. Correspondingly, many learning algorithms, such as the complex-valued LMS, complex-valued backpropagation, and complex-valued RTRL algorithm (e.g., [83, 328, 350, 369, 480, 545, 952]), have been developed for optimizing the complex-valued synaptic weights of the networks. One surprising observation reported in [662] is that the simple exclusive-OR (XOR) problem that is unsolvable by the conventional (real-valued) Perceptron with a single layer of weights can be solved with ease in the complex domain using a complex-valued input–output encoding scheme as demonstrated in Tables 6.1 and 6.2.

We now describe a set of simulations on a simple pattern classification problem (see Tables 6.3 and 6.4, taken from [661]) to illustrate the feasibility of using a complex-valued version of ALOPEX for training a complex-valued MLP. Two types of neural networks are used here: (i) a real-valued MLP network net2-4-2 which is trained by the conventional real-valued ALOPEX-B

Table 6.1 Real Encoding (of Two Inputs and One Output) for XOR Problem

Input		Output
x_1	x_2	y
0	0	0
0	1	1
1	0	1
1	1	0

Table 6.2 Complex Encoding (of One Input and One Output) for XOR Problem

Input, $x = x_{\text{Re}} + jx_{\text{Im}}$	Output, $y = y_{\text{Re}} + jy_{\text{Im}}$
$-1 - j$	1
$-1 + j$	0
$1 - j$	$1 + j$
$1 + j$	j

Table 6.3 Real Encoding (of Two Inputs and Two Outputs) for Pattern Classification Problem

Input		Output	
x_1	x_2	y_1	y_2
−1	−1	1	1
1	−1	0	1
1	1	0	0
−1	1	1	0

Table 6.4 Complex Encoding (of One Input and One Output) for Pattern Classification Problem

Input, $x = x_{\text{Re}} + jx_{\text{Im}}$	Output, $y = y_{\text{Re}} + jy_{\text{Im}}$
$-1 - j$	$1 + j$
$1 - j$	j
$1 + j$	0
$-1 + j$	1

and (ii) a complex-valued MLP net1-3-1 which is trained by the complex-valued ALOPEX-B. The training procedure is stopped when the MSE is smaller than 0.001. The experimental results based on 20 Monte Carlo random runs are summarized in Table 6.5. As seen, the performance of the complex-valued MLP is much better than its real counterpart in terms of faster convergence speed as well as sharper decision boundaries. The complex decision boundary for the complex-valued MLP is illustrated in Figure 6.2.

Table 6.5 Comparison of Real- and Complex-Valued MLP Networks in Pattern Classification Example

	Real-Valued net2-4-2	Complex-Valued net1-3-1
Number of free parameters	22	20
Average convergence rate (epochs)	1647 ± 909	989 ± 437
Angles of decision boundary	76 ± 16	90 ± 0

Note: Based on 20 Monte Carlo runs with different initial conditions.

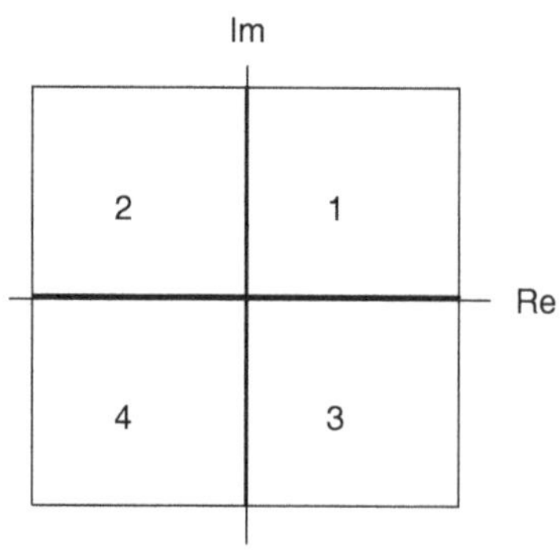

Figure 6.2 The decision boundary.

Notably, the decision boundary for the real part and that for the imaginary part intersect orthogonally [661].

6.5 MONTE CARLO SAMPLING-BASED ALOPEX

In preliminary simulations, it was found that although ALOPEX-B and its improved version often converge more quickly than Unnikrishnan and Venugopal's version of ALOPEX, they also tend to get trapped in local minima more frequently since no annealing scheme is used [163]. This fact motivated the development of the Monte Carlo sampling-based ALOPEX discussed in this section.

The idea of using Monte Carlo methods for optimization is not new; genetic algorithms and simulated annealing [483] are two representative examples. Essentially, sampling-based ALOPEX attempts to combine the advantages of simplicity and fast convergence rate of the improved ALOPEX-B and the robustness of the sequential Monte Carlo sampling technique.

6.5.1 Sequential Monte Carlo Estimation

For our exposition purpose, let us formulate a generic *parameter estimation* problem in the form of a state-space model (SSM):

$$\boldsymbol{\theta}_{t+1} = \boldsymbol{\theta}_t + \boldsymbol{\nu}_t, \tag{6.25a}$$

$$\mathbf{y}_t = f(\boldsymbol{\theta}_t, \mathbf{x}_t) + \mathbf{v}_t, \tag{6.25b}$$

where the nonlinear measurement equation (6.25b), parameterized by $\boldsymbol{\theta}$, determines the mapping $f : X \mapsto Y$, given a number of inputs $\mathbf{x}_t$ and outputs $\mathbf{y}_t$. The additive terms $\boldsymbol{\nu}_t$ and $\mathbf{v}_t$ are process noise and measurement noise, respectively. In general, f can be a neural network or some other parameterized model.

In the sequential Monte Carlo framework, $\boldsymbol{\theta}_t$ is estimated via particle filtering that follows a recursive Bayesian estimation procedure [141, 158, 225]. Simply put, a particle filter uses a number of random samples called "particles" sampled directly from the state space of parameter values to represent the posterior density and updates the posterior density by involving new observations; the "particle system" is properly *located, weighted*, and *propagated* recursively according to Bayes's rule. Among many variations, one of the most popular particle filters is the *sampling–importance–resampling* (SIR) filter. The basic principle of the SIR filter is to use the importance sampling trick

$$\int f(\boldsymbol{\theta})p(\boldsymbol{\theta})\,d\boldsymbol{\theta} = \int f(\boldsymbol{\theta})\frac{p(\boldsymbol{\theta})}{q(\boldsymbol{\theta})}q(\boldsymbol{\theta})\,d\boldsymbol{\theta}, \tag{6.26}$$

where $q(\cdot)$ and $p(\cdot)$ are proposal and target densities, respectively. Given a number of i.i.d. samples $\{\boldsymbol{\theta}^{(i)}\}$ that are drawn from the proposal distribution $q(\boldsymbol{\theta})$, we can estimate the mean of $f(\boldsymbol{\theta})$ as

$$\mathbb{E}_p[f] \approx \frac{1}{N_p} \sum_{i=1}^{N_p} W(\boldsymbol{\theta}^{(i)}) f(\boldsymbol{\theta}^{(i)}) \equiv \hat{f}, \tag{6.27}$$

where the $W(\boldsymbol{\theta}^{(i)}) = p(\boldsymbol{\theta}^{(i)})/q(\boldsymbol{\theta}^{(i)})$ are called the *importance weights*. If the normalizing factor of $p(\boldsymbol{\theta})$ is not known, then $W(\boldsymbol{\theta}^{(i)}) \propto p(\boldsymbol{\theta}^{(i)})/q(\boldsymbol{\theta}^{(i)})$. To ensure that $\sum_{i=1}^{N_p} W(\boldsymbol{\theta}^{(i)}) = 1$, we further calculate

$$\hat{f} = \frac{(1/N_p)\sum_{i=1}^{N_p} W(\boldsymbol{\theta}^{(i)}) f(\boldsymbol{\theta}^{(i)})}{(1/N_p)\sum_{j=1}^{N_p} W(\boldsymbol{\theta}^{(j)})} \equiv \sum_{i=1}^{N_p} \tilde{W}(\boldsymbol{\theta}^{(i)}) f(\boldsymbol{\theta}^{(i)}),$$

where

$$\tilde{W}(\boldsymbol{\theta}^{(i)}) = \frac{W(\boldsymbol{\theta}^{(i)})}{\sum_{j=1}^{N_p} W(\boldsymbol{\theta}^{(j)})}$$

are called the *normalized importance weights*. By choosing a factorized proposal distribution, the importance weights can be updated recursively as follows [225]:

$$W_t^{(i)} = W_{t-1}^{(i)} \frac{p(\mathbf{y}_t|\boldsymbol{\theta}_t^{(i)}, \mathbf{x}_t) p(\boldsymbol{\theta}_t^{(i)}|\boldsymbol{\theta}_{t-1}^{(i)})}{q(\boldsymbol{\theta}_t^{(i)}|\boldsymbol{\theta}_{0:t-1}^{(i)}, \mathbf{y}_t)}, \tag{6.28}$$

where $p(\boldsymbol{\theta}_t^{(i)}|\boldsymbol{\theta}_{t-1}^{(i)})$ is called the transition prior that corresponds to the process equation (6.25a) and $p(\mathbf{y}_t|\boldsymbol{\theta}_t^{(i)}, \mathbf{x}_t)$ is called the likelihood model that corresponds to the measurement equation (6.25b).

When the proposal $q(\boldsymbol{\theta}_t^{(i)}|\boldsymbol{\theta}_{0:t-1}^{(i)}, \mathbf{y}_t)$ is taken as the transition prior, the importance weights turn out to be proportional to the likelihood. It is well known that the SIR filter suffers from an intrinsic problem: As time increases, the distribution of the importance weights becomes more and more skewed; after a few iterations, only very few particles have nonzero importance weights. This phenomenon is often called the *weight degeneracy* or *sample impoverishment* problem. One empirical measure of sample efficiency is the variance of the importance weights (e.g., [225]):

$$\hat{N}_{\text{eff}} = \frac{1}{\sum_{i=1}^{N_p} (\tilde{W}_t^{(i)})^2}. \tag{6.29}$$

We may also suggest another empirical efficiency measure, namely, the KL divergence between the proposal and target densities, denoted by $D(q\|p)$. Given N_p

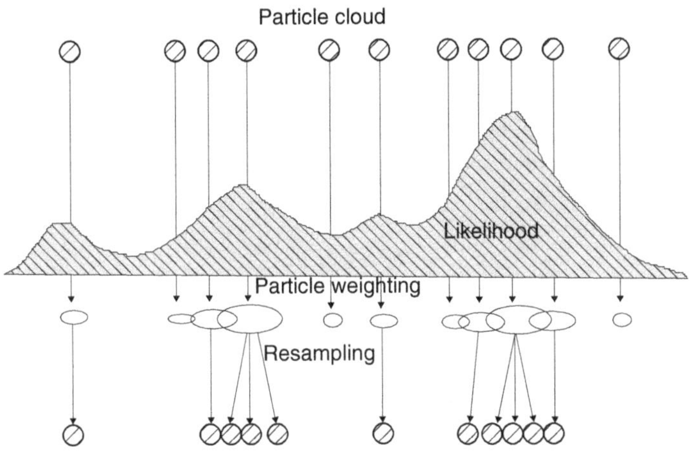

Figure 6.3 A graphical illustration of sequential SIR.

samples drawn from the proposal q, the KL divergence $D(q\|p)$ is approximated by

$$D(q\|p) = \mathbb{E}_q\left[\log\frac{q(\boldsymbol{\theta})}{p(\boldsymbol{\theta})}\right] \approx \frac{1}{N_p}\sum_{i=1}^{N_p}\log\frac{q(\boldsymbol{\theta}^{(i)})}{p(\boldsymbol{\theta}^{(i)})}$$

$$= -\frac{1}{N_p}\sum_{i=1}^{N_p}\log\left[W(\boldsymbol{\theta}^{(i)})\right], \qquad (6.30)$$

where $\{\boldsymbol{\theta}^{(i)}\}$ are drawn from $q(\boldsymbol{\theta})$. When $q = p$ and $W(\boldsymbol{\theta}^{(i)}) = 1$ for all i, $D(q\|p) = 0$. Since $D(q\|p) \geq 0$, $-\log\left[W(\boldsymbol{\theta}^{(i)})\right]$ should be nonnegative. In practice, we instead calculate the logarithm of the normalized importance weights $\hat{N}_{\text{KL}} = -(1/N_p)\sum_{i=1}^{N_p}\log(\tilde{W}\left[\boldsymbol{\theta}^{(i)}\right)]$, which achieves the minimum value $\hat{N}_{\text{KL}}^{\min} = \log(N_p)$ when all $\tilde{W}(\boldsymbol{\theta}^{(i)}) = 1/N_p$. Our previous studies have confirmed that $\hat{N}_{\text{KL}}$ is a good measure that is also consistent with $\hat{N}_{\text{eff}}$: When $\hat{N}_{\text{KL}}$ is small, $\hat{N}_{\text{eff}}$ is usually large and vice versa.

The improvement scheme for the sample impoverishment problem is to introduce a resampling step [225, 332]. Basically, the resampling step is to multiply the particles with high normalized importance weights and discard the particles with low normalized importance weights. Intuitively, more importance weights are imposed on the high-likelihood region. (see Figure 6.3 for an illustration) Resampling can be understood as a sort of selection/reproduction scheme similar to the genetic algorithm. On the other hand, resampling also brings in correlation within the samples, which is called the *loss of diversity*. It has been suggested that the insertion of a Markov chain Monte Carlo (MCMC) step after resampling may help increase the diversity of the samples (see e.g., [225]).

6.5.2 Sampling-Based ALOPEX

The following two sampling-based ALOPEX procedures naturally integrate the features of the ALOPEX and particle filter; they are recursive and fall under the Bayesian estimation framework. Like other ALOPEX procedures, they are gradient free and suitable for either online (sequential) or offline (batch) learning.

In order to avoid the "blind" random-walk behavior, we use a "relaxation" model in place of (6.25a):

$$\boldsymbol{\theta}_{t+1}^{(i)} = \boldsymbol{\mu}_t + \alpha(\boldsymbol{\theta}_t^{(i)} - \boldsymbol{\mu}_t) + \sqrt{1-\alpha^2}\sigma \boldsymbol{\nu}_t, \tag{6.31}$$

where $\boldsymbol{\mu}_t = \sum_{i=1}^{N_p} \tilde{W}_t^{(i)} \boldsymbol{\theta}_t^{(i)}$ denotes a weighted mean; the noise vector $\boldsymbol{\nu}_t$ is standard Gaussian distributed, $\boldsymbol{\nu}_t \sim \mathcal{N}(0, \mathbf{I})$; and σ is the standard deviation controlling the degree of variation in $\boldsymbol{\theta}$, which often requires some prior knowledge of the problem. The relaxing parameter $\alpha \in [-1, 1]$ controls the degree of overrelaxation (or underrelaxation):

- When $\alpha = -1$, (6.31) reduces to an extreme overrelaxation $\boldsymbol{\theta}_{t+1}^{(i)} = 2\boldsymbol{\mu}_t - \boldsymbol{\theta}_t^{(i)}$.
- When $\alpha = 0$, (6.31) reduces to a random walk $\boldsymbol{\theta}_{t+1}^{(i)} = \boldsymbol{\mu}_t + \sigma \boldsymbol{\nu}_t$.
- When $0 < \alpha < 1$, (6.31) is an underrelaxation model.
- When $\alpha = 1$, (6.31) reduces to a stationary point $\boldsymbol{\theta}_{t+1}^{(i)} = \boldsymbol{\theta}_t^{(i)}$.

In summary, our first sampling-based ALOPEX (termed Algorithm 1 hereafter) proceeds as follows:

1. For $i = 1, \ldots, N_p$, initialize $\boldsymbol{\theta}_0^{(i)} \sim p(\boldsymbol{\theta}_0)$, and set $W_0^{(i)} = 1/N_p$.
2. Predict $\boldsymbol{\theta}_t^{(i)}$ from (6.31).
3. Update the samples $\boldsymbol{\theta}_t^{(i)}$ via the modified ALOPEX-B procedure (6.12)–(6.15).
4. Evaluate the importance weights $W_t^{(i)} = W_{t-1}^{(i)} p(\mathbf{y}_t | \boldsymbol{\theta}_t^{(i)}, \mathbf{x}_t)$ and $\tilde{W}_t^{(i)} = (W_t^{(i)} / (\sum_{j=1}^{N_p} W_t^{(j)}))$.
5. Calculate $\hat{N}_{\text{eff}}$ and $\hat{N}_{\text{KL}}$; if $\hat{N}_{\text{eff}} < 0.8N_p$ or $\hat{N}_{\text{KL}} > 3 \log(N_p)$, go to step 6; otherwise go to step 7.
6. Resampling: Generate a new particle set $\{\boldsymbol{\theta}_t^{(j)}\}$ and reset the weights $\tilde{W}_t^{(j)} = 1/N_p$.
7. Repeat steps 2–5.

Note that when $N_p = 1$ Algorithm 1 reduces to a generalized form of ALOPEX-B, which involves an additional randomness through (6.31). In addition, there is no reason why we cannot use specific $\alpha^{(i)}$ for different $\boldsymbol{\theta}^{(i)}$; α can also be time varying, but we have not investigated these issues here. We fixed α for each specific problem in the experiments reported later, but the optimal α often varies from one problem to another.

It is of interest to compare our algorithm with other sampling-based optimization algorithms (e.g., Fisher scoring [112] and HySIR [209]), for training neural networks. The complexity of our algorithm [$\mathcal{O}(N_p N)$] is much smaller than these two algorithms [$\mathcal{O}(N_p N^2)$] simply because of avoiding the calculation of the Jacobian matrix. Our algorithm is also much simpler than another sampling-based gradient-free estimation technique: the unscented particle filter [906, 933], which is typically of $\mathcal{O}(N_p N^3)$ complexity.

In what follows, we propose another Monte Carlo sampling-based ALOPEX procedure (hereafter termed Algorithm 2) that is motivated by the hybrid Monte Carlo (HMC) method [230, 579]. The idea of HMC is to augment the state space $\boldsymbol{\theta}$ with a momentum variable $\boldsymbol{\rho}$. The energy-conserving Hamiltonian dynamics is defined as

$$\mathcal{H}(\boldsymbol{\theta}, \boldsymbol{\rho}) = E(\boldsymbol{\theta}) + \mathcal{K}(\boldsymbol{\rho}), \tag{6.32}$$

where $E(\boldsymbol{\theta})$ is the *potential energy* function,[4] whereas $\mathcal{K}(\boldsymbol{\rho}) = \boldsymbol{\rho}^{\mathrm{T}}\boldsymbol{\rho}/2$ is the *kinetic energy*. The samples are drawn from the joint distribution

$$\begin{aligned} p_{\mathcal{H}}(\boldsymbol{\theta}, \boldsymbol{\rho}) &= \frac{1}{Z} \exp\left[-\mathcal{H}(\boldsymbol{\theta}, \boldsymbol{\rho})\right] \\ &= \frac{1}{Z} \exp[-E(\boldsymbol{\theta})] \exp\left[-\mathcal{K}(\boldsymbol{\rho})\right], \end{aligned} \tag{6.33}$$

where Z is a normalizing constant. Note that the term $\exp[-E(\boldsymbol{\theta})]$ is essentially the likelihood up to a normalizing factor. The momentum dynamics can be approximated by the ensuing difference equations

$$\boldsymbol{\rho}_t = \nabla\boldsymbol{\theta}_t \approx \Delta\boldsymbol{\theta}_t, \tag{6.34a}$$

$$\nabla\boldsymbol{\rho}_t = -\frac{\partial E(\boldsymbol{\theta}_t)}{\partial \boldsymbol{\theta}_t} \approx -\frac{\Delta E(\boldsymbol{\theta}_t)}{\Delta\boldsymbol{\theta}_t}, \tag{6.34b}$$

where, obviously, all of the terms are intermediate results obtained from the ALOPEX-like algorithm without additional computing overhead. By doing so, the posterior of $\boldsymbol{\theta}_{t+1}$ is proportional to $p(\boldsymbol{\theta}_{t+1}|\boldsymbol{\theta}_t) p_{\mathcal{H}}(\boldsymbol{\theta}_t, \boldsymbol{\rho}_t) = p(\boldsymbol{\theta}_{t+1}|\boldsymbol{\theta}_t)$ $\exp(-\frac{1}{2}\boldsymbol{\rho}_t^{\mathrm{T}}\boldsymbol{\rho}_t) p(\mathbf{y}_t|\boldsymbol{\theta}_t)$. Equivalently, while keeping the importance weights proportional to the likelihood, (6.31) is substituted by

$$\begin{aligned} \tilde{\boldsymbol{\theta}}_{t+1}^{(i)} &= \boldsymbol{\theta}_{t+1}^{(i)} + \beta\, \Delta\boldsymbol{\theta}_t^{(i)} \\ &= \boldsymbol{\mu}_t + \alpha(\boldsymbol{\theta}_t^{(i)} - \boldsymbol{\mu}_t) + \beta\, \Delta\boldsymbol{\theta}_t^{(i)} + \sqrt{1-\alpha^2}\sigma \mathbf{v}_t, \end{aligned} \tag{6.35}$$

where β is a momentum coefficient. Equation (6.35) essentially describes a second-order AR model compared to the first-order models (6.25a) and (6.31); it also implies that $p(\tilde{\boldsymbol{\theta}}_{t+1}|\boldsymbol{\theta}_t, \Delta\boldsymbol{\theta}_t) \propto p(\boldsymbol{\theta}_{t+1}|\boldsymbol{\theta}_t)\ \exp(-\Delta\boldsymbol{\theta}_t^{\mathrm{T}}\, \Delta\boldsymbol{\theta}_t)$. Algorithm 2 differs from Algorithm 1 only in the second step where (6.31) is replaced by (6.35).

Thus far, formulations of Monte Carlo sampling-based ALOPEX are discussed in a supervised learning framework. However, they can readily be used for unsupervised learning in which the log-likelihood function $\mathcal{L}(\mathbf{x})$ is related to the potential energy function: $\mathcal{L}(\mathbf{x}) = -E(\mathbf{x}, \boldsymbol{\theta})$.

■ EXAMPLE 6.2

Suppose we are given a fourth-order discrete-time linear system characterized by the transfer function [791]

$$H(z) = \frac{0.05 - 0.4z^{-1}}{1 - 1.1314z^{-1} + 0.25z^{-2}}, \tag{6.36}$$

which has one zero at 8 and two poles at 0.8303 and 0.3011, with a gain of 0.05. Taking the inverse z-transform of $H(z)$ yields the impulse response for this ARMA(2, 2) (autoregressive moving-average) model:

$$\mathbf{h} = [0.0500, \quad -0.3434, \quad -0.4011, \quad -0.3679]^{\mathrm{T}}.$$

The task of system identification is to estimate the transfer function (or impulse response) given some observed input–output data. The input data are generated as a white Gaussian noise sequence with zero mean and unit variance, and output data are obtained by passing the input data through the desired transfer function subject to additional Gaussian noise corruption with resultant 10 dB SNR. For simplicity, we assume the order of the system is available or can be estimated in advance; then the identification problem reduces to seeking an "optimal" model

$$H(z) = \frac{b_0 + b_1 z^{-1}}{1 + a_1 z^{-1} + a_2 z^{-2}} \tag{6.37}$$

which is parameterized by four parameters: b_0, b_1, a_1, and a_2. The optimization problem is then to find the optimal values of these four parameters in order to minimize the MSE. During the learning process, we also monitor the norm between the true and estimated impulse responses, $\|\mathbf{h} - \boldsymbol{\theta}(t)\|$.

For the purpose of comparing the convergence and performance of the *iterative* gradient-based and gradient-free learning methods, we have employed three representative algorithms for this simple task: LMS, ALOPEX-B, and sampling-based ALOPEX. Given the same initial conditions, their learning curves are shown in Figure 6.4. The LMS learning rule is sequential and updates at each time step; with learning-rate parameter $\eta = 0.01$, it converges to the Wiener solution within 1000 steps. In contrast, the ALOPEX learning rules are run in batch mode and updated at each epoch (by scanning all data); ALOPEX-B and sampling-based ALOPEX also converge to the Wiener solution within about 200 and 100 epochs, respectively. In other words, sampling-based ALOPEX (with $N_p = 5$) converges at about twice the rate of ALOPEX-B. The experimental parameters for ALOPEX are $\eta = 0.05$, $\gamma = -0.01$, $\lambda = 0.5$, $\sigma = 0.02$, $\alpha = -0.5$, and $\beta = 0.005$.

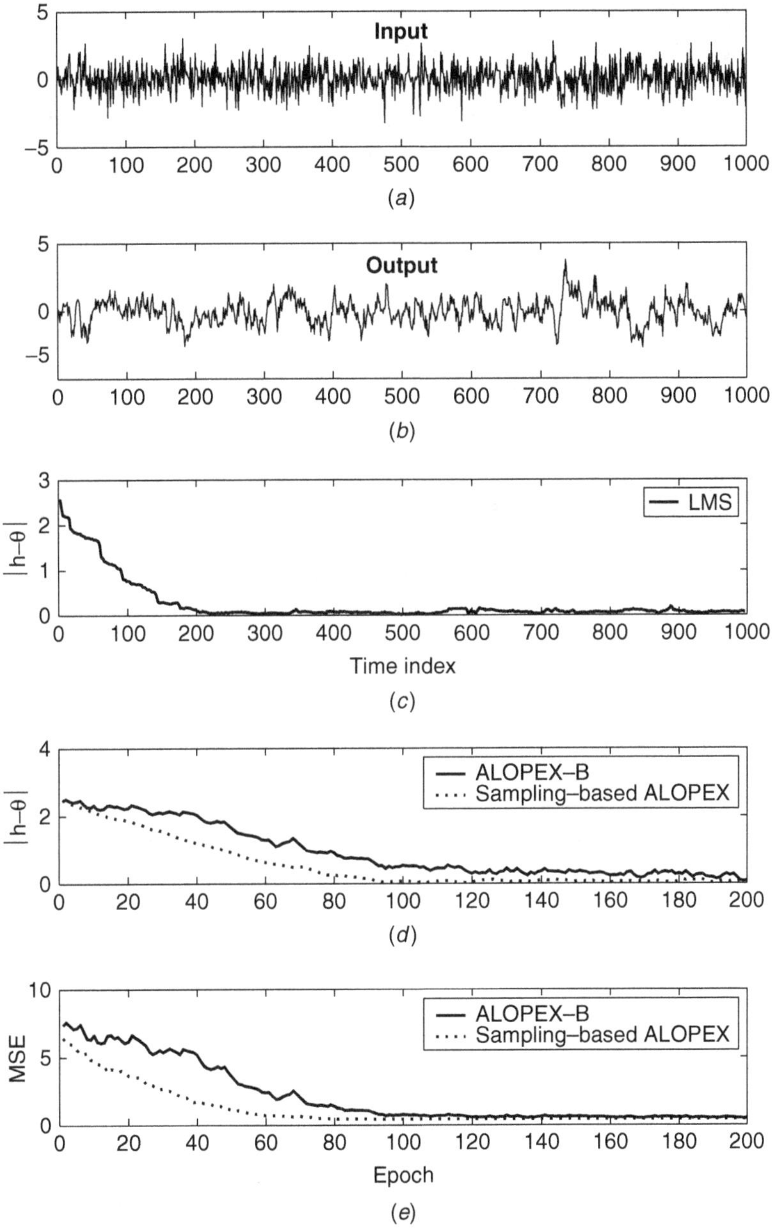

Figure 6.4 (*a*) White Gaussian noise sequence with 1000 input data points. (*b*) Noisy output data (with 10 dB SNR). (*c*) The norm between the true and estimated impulse responses, $\|\mathbf{h} - \theta(t)\|$, from the sequential LMS learning process. (*d*) The $\|\mathbf{h} - \theta(t)\|$ curves from the batch ALOPEX learning process. (*e*) The MSE learning curves.

6.5.3 Remarks

Tricks of the Trade. It is noted that there are many hand-tuned parameters involved in the above-described Monte Carlo sampling-based ALOPEX procedures. In practice, finding these optimal parameters can be time-consuming and difficult. In light of our empirical experiments, we summarize some rules of thumb for selecting those free parameters:

- Learning-rate and step-size parameters: For ALOPEX-B, η is often chosen in the range $[0.05, 0.1]$ and γ is fixed to be 0.01 in most of our experiments.
- Forgetting parameter: In ALOPEX-B, λ is often taken from the region $[0.35, 0.7]$; the smaller the λ, the less influence is induced by previous error estimates. For online learning (on sequential data), λ is usually set to a small value.
- Relaxing parameter: α is taken from the region $[-1, 1]$. When $\alpha > 0$, it corresponds to overrelaxation, and when $\alpha < 0$, it corresponds to underrelaxation. In the initial training, α can be set positive to accelerate the initial convergence; as the error surface becomes more hilly, we can switch to underrelaxation. In our experiments, α is always set to a negative value for online learning.
- Momentum coefficient: By analogy to a physical particle system, gradient-type optimization can be imagined as moving a massless particle (i.e., $\boldsymbol{\theta}$) toward the bottom of a potential well [739]. Imagining the massless particle as a particle with a quantitative mass, we know from Newtonian mechanics that the greater the mass, the greater is the momentum. Since the normalized importance weights are directly related to the likelihood values, ideally it is hoped that the "important" particles (with higher likelihood) are more active. Therefore we assign greater momentum values to them and smaller momentum values to the "idle" particles. Heuristically, for the ith particle, we may set $\beta^{(i)} = \tilde{W}^{(i)}\beta_0$, where $\beta_0 = 1 - \eta$ is a constant. Besides this more sophisticated version, an alternative, simpler setup can be used: $\beta = \eta/10$.
- Diffusion coefficient: σ is initially set to a small constant (depending on the region of the parameter $\boldsymbol{\theta}$); as batch learning progresses, this parameter can be reduced according to an annealing schedule after 1000 iterations $\sigma = \sigma_0/\log(t)$. In online learning, σ remains constant.
- If parameter $\boldsymbol{\theta}$ is subject to a positive constraint (e.g., the width parameter of the radial basis function), one can introduce a surrogate parameter, $\boldsymbol{\vartheta} \equiv \ln \boldsymbol{\theta}$ or $\boldsymbol{\theta} \equiv \exp(\boldsymbol{\vartheta})$, and then use the ALOPEX procedure to update the surrogate parameter $\boldsymbol{\vartheta}$ (with a different prior, of course).

Statistical Physics Interpretation. It is noted that Unnikrishan and Venugopal's ALOPEX procedure has its origins in statistical physics, similar to the Metropolis algorithm [617] and simulated annealing [483]. It is therefore befitting that we explore a statistical physics interpretation of the sampling-based ALOPEX procedures in terms of an *interacting particle system* (IPS). The IPS [555] can be regarded as a dynamic interactive system with a collection of many particles interacting according to simple and local rules. The IPS has been successfully utilized to model such diverse phenomena as magnetism, population growth, and propagation of information and opinions.

Imagine sampling-based ALOPEX as an interactive dynamical composition system. On the one hand, the elements in the system are *spatially independent* (i.i.d. samples) and *temporally correlated* (correlative learning rule). On the other hand, the elements are *globally correlated* (from the correlation learning rule, the change of each element is influenced by others) but also *locally independent.* Finally, the system is not only *cooperative* in parameter space, because every element contributes to the same energy function, but also *competitive* in sample space, because different samples try to find the minimum energy, so the one that finds a locally minimal energy has the highest likelihood. In light of these observations, sampling-based ALOPEX provides a simulation analog for systems with combined cooperative and competitive behavior, which is likely to be a feature of the human brain.

APPENDIX 6A: ASYMPTOTIC ANALYSIS OF ALOPEX PROCESS

In the original presentation, ALOPEX was used as an optimization method for determining the visual receptive field of a single neuron. Visual patterns presented to an experimental subject are successively modified by the feedback of the response of a neuron such that they finally converge to the receptive field pattern of the neuron. Amari [22] has given a detailed mathematical analysis of this process. We briefly highlight the results here.

Let $\mathbf{x}$ be a pattern vector on the retina and $\mathbf{y} = \mathbf{x} + \mathbf{n}$ be a noisy version of $\mathbf{x}$, with $\mathbf{n}$ being an additive and independent noise pattern, and let $J = f(\mathbf{y})$ be the response of a single neuron for the stimulus pattern $\mathbf{y}$. Then the ALOPEX process is described by the following difference equation:

$$\mathbf{x}(t+1) = (1-\eta)\mathbf{x}(t) + \eta\,[J(t) - J(t-1)]\left[\mathbf{y}(t) - \mathbf{y}(t-1)\right], \quad (6.A.1)$$

where $0 < \eta < 1$ is a small learning-rate parameter.

It was proved in [22] that, upon convergence, $\mathbf{x}(t)$ reaches the final equilibrium point $\tilde{\mathbf{x}}$, which satisfies the equation

$$\tilde{\mathbf{x}} = 2\mathbb{E}\left[\mathbf{n} f(\tilde{\mathbf{x}} + \mathbf{n})\right]. \quad (6.A.2)$$

Namely, the equilibrium point is equal to the cross-correlation between the noise pattern and the estimated neuronal response. Specifically, Amari [22] also showed that:

- When the receptive field response is linear, namely $f(\mathbf{x}) = \mathbf{x}^{\mathrm{T}}\boldsymbol{\theta}$ (where $\boldsymbol{\theta}$ denotes the receptive field parameter vector), $\mathbf{x}(t)$ converges to a constant multiple of the receptive field vector; then equation (6.A.2) is simplified to

$$\begin{aligned}\tilde{\mathbf{x}} &= 2\mathbb{E}\left[(\tilde{\mathbf{x}}+\mathbf{n})^{\mathrm{T}}\boldsymbol{\theta}\mathbf{n}\right]\\ &= 2\mathbb{E}\left[\tilde{\mathbf{x}}^{\mathrm{T}}\mathbf{n}\right]\boldsymbol{\theta} + 2\mathbb{E}\left[\|\mathbf{n}\|^2\right]\boldsymbol{\theta} \propto \boldsymbol{\theta},\end{aligned}$$

where the last line holds because $\mathbb{E}\left[\tilde{\mathbf{x}}^{\mathrm{T}}\mathbf{n}\right] = 0$ and $\mathbb{E}\left[\|\mathbf{n}\|^2\right]$ is a constant.

- When the receptive field response is nonlinear, under certain regular conditions, equation (6.A.2) still remains valid and the learning process is stable.

APPENDIX 6B: ASYMPTOTIC CONVERGENCE ANALYSIS OF 2T-ALOPEX

The asymptotic convergence analysis of the 2t-ALOPEX presented here is excerpted from [791]. The theoretical analysis is established using the tools of ordinary differential equations (ODEs) and two-timescale stochastic approximation [104].

Suppose that a constant temperature parameter $T(t) = T$ is used during the learning process. Denote $\boldsymbol{p}(t) = [p_1(t), \ldots, p_N(t)]^{\mathrm{T}}$ and $\boldsymbol{\zeta}(t) = [\zeta_1(t), \ldots, \zeta_N(t)]^{\mathrm{T}}$. The 2t-ALOPEX algorithm can be rewritten in vector form as follows:

$$\boldsymbol{\theta}(t+1) = \boldsymbol{\theta}(t) + \eta[F(\boldsymbol{\theta}(t), \boldsymbol{p}(t)) + \mathbf{w}(t)], \tag{6.B.1}$$

$$\boldsymbol{p}(t+1) = \boldsymbol{p}(t) + \mu[G(\boldsymbol{\theta}(t), \boldsymbol{p}(t)) + \mathbf{v}(t)], \tag{6.B.2}$$

where

$$F(\boldsymbol{\theta}, \boldsymbol{p}) = \mathbb{E}\left[\boldsymbol{\xi}(t)\middle|\boldsymbol{\theta}(t) = \boldsymbol{\theta}, \boldsymbol{p}(t) = \boldsymbol{p}\right], \tag{6.B.3}$$

$$G(\boldsymbol{\theta}, \boldsymbol{p}) = \mathbb{E}\left[\boldsymbol{\zeta}(t) - \boldsymbol{p}(t)\middle|\boldsymbol{\theta}(t) = \boldsymbol{\theta}, \boldsymbol{p}(t) = \boldsymbol{p}\right], \tag{6.B.4}$$

where $\mathbb{E}[\cdot]$ denotes the expectation and $\mathbf{w}(t)$ and $\mathbf{v}(t)$ are two zero-mean i.i.d. noise sequences

$$\mathbf{w}(t) = \boldsymbol{\xi}(t) - F(\boldsymbol{\theta}(t), \boldsymbol{p}(t)), \tag{6.B.5}$$

$$\mathbf{v}(t) = [\boldsymbol{\zeta}(t) - \boldsymbol{p}(t)] - G(\boldsymbol{\theta}(t), \boldsymbol{p}(t)). \tag{6.B.6}$$

Under the assumption that the learning-rate parameter η is an order of magnitude smaller than λ, the dynamics of $\boldsymbol{p}(t)$ evolves much faster than that of $\boldsymbol{\theta}(t)$. Equations (6.B.1) and (6.B.2) correspond to the "almost equilibriated" [for process $\boldsymbol{p}(t)$] and "almost constant" [for process $\boldsymbol{\theta}(t)$] dynamics in light of the two-timescale stochastic approximation theory [104].

In (6.B.2), by fixing $\boldsymbol{\theta}(t) = \boldsymbol{\theta}$ and with a sufficiently small μ, the asymptotic behavior of a suitably interpolated continuous-time version of the process $\boldsymbol{p}(t)$, denoted by $\boldsymbol{p}_{\theta}(t)$, can be approximated by the solution of the following ODE:

$$\dot{\boldsymbol{p}}_{\theta} = G(\boldsymbol{\theta}, \boldsymbol{p}_{\theta}), \quad \boldsymbol{p}_{\theta}(0) = \boldsymbol{p}(0), \tag{6.B.7}$$

where $G(\boldsymbol{\theta}, \boldsymbol{p}_{\boldsymbol{\theta}})$ is given by the limit on the right-hand side of (6.B.4) as $\mu \to 0$. Suppose the ODE (6.B.7) has a globally asymptotically stable equilibrium point, denoted by $\tilde{\boldsymbol{p}}(\boldsymbol{\theta})$. Replacing $\boldsymbol{p}(t)$ with $\tilde{\boldsymbol{p}}(\boldsymbol{\theta}(t))$ in the slowly evolving process in (6.B.3), it follows that a suitably interpolated continuous-time version of the process $\boldsymbol{\theta}(t)$, denoted by $\overline{\boldsymbol{\theta}}(t)$, would be approximated by the following ODE (with a sufficiently small η):

$$\frac{d\overline{\boldsymbol{\theta}}}{dt}(t) = F(\overline{\boldsymbol{\theta}}, \tilde{\boldsymbol{p}}(\overline{\boldsymbol{\theta}}(t))), \qquad \overline{\boldsymbol{\theta}}(0) = \boldsymbol{\theta}(0). \tag{6.B.8}$$

If the ODE (6.B.8) has a globally asymptotically stable solution for each $\boldsymbol{\theta}$, then the asymptotic behavior of $\boldsymbol{\theta}(t)$ is well approximated by the solution of (6.B.8) with almost sure (a.s.) sense convergence. In fact, the jth component of the vector $F(\boldsymbol{\theta}, \tilde{\boldsymbol{p}}(\boldsymbol{\theta}))$ has the same algebraic sign as $-(\partial J(\boldsymbol{\theta})/\partial\theta_j)$ for all $\boldsymbol{\theta} \in \mathbb{R}^N$, which would lead to the conclusion that 2t-ALOPEX results in a local minimum of the cost function $J(\boldsymbol{\theta})$. The interested reader is referred to [791] for detailed mathematical proof.

BIBLIOGRAPHICAL NOTES

The name of ALOPEX first appeared in the literature in 1974 for its use in extracting visual receptive fields [355] followed by related papers in vision research [437, 900]. Later, ALOPEX was used as an optimization tool for modeling attention and perception systems, especially in biology and neuroscience [356, 437, 898]. The idea behind ALOPEX is extremely simple, and discussion of it actually appeared in Minsky's review paper [629]. Mathematical analysis of the ALOPEX process for determination of visual receptive fields was given in Amari [22]. Since the 1990s, variants of ALOPEX were developed for training multilayer neural networks [901, 902] as a substitute for backpropagation. Most variants of ALOPEX were developed in the past few years, including Bia's ALOPEX-B [90] and the two-timescale ALOPEX [791]. The Monte Carlo sampling-based ALOPEX was first described in [163] and then published in [374]. Thus far, ALOPEX has been applied in numerous applications, including control [914], symplectic nonlinear component analysis, [705], biomedicine [198], auditory stimuli optimization [41], resource allocation [699], learning decision trees [821], figure–ground segregation [159], model-based hearing-aid design [101, 160], and even brain–machine interface design. A collected volume on ALOPEX-related research work can be found in the book edited by Tzanakou [899].

NOTES

1. Harth and Tzanakou [355] defined the receptive field as that spatiotemporal stimulus pattern which maximally affects the firing rate of a given neuron.

2. This is known as the *finite forward-difference* approximation in optimization theory [281]. For greater accuracy, one can replace the "forward-difference" term with the "central-difference" term:

$$\frac{\partial J(\theta)}{\partial \theta} \approx \frac{J(\theta + \delta\theta) - J(\theta - \delta\theta)}{2\delta\,\theta} \qquad (|\delta\theta| \to 0).$$

However, the forward-difference approximation is simpler from the implementation perspective.
3. A major criticism of Hebbian synaptic plasticity lies in its neglect of feedback, which brings a difficulty in modeling realistically structured neural circuits. Ramón y Cajal's postulated "dynamic polarization" law stipulates that dendrites and somas are the only receptive areas for the synaptic input, and the resulting output pulses are transmitted unidirectionally along the axon to its target. This postulate assumes that no signals travel backward along the dendrites. However, as reviewed in [493], recent studies have showed that this is not the complete story. Instead, signal action potentials can propagate not only forward from their initiation site along the axon but also backward into the dendritic tree (a phenomenon known as *antidromic spike propagation*). Koch [493] suggested that the backpropagating action potentials be viewed as a sort of "acknowledgment" feedback. According to this theory, a Hebbian synapse is strengthened if a presynaptic spike coincides with the postsynaptic spike that is generated close to the soma and spreads back along the dendritic tree to the synapse.
4. Generally, the quadratic cost function $J(t)$ can be viewed as a potential energy function (up to some scaling factor); when the cost function is nonquadratic, it cannot always be viewed as a potential function unless it is nonnegative and bounded. Sometimes, it is possible to convert an objective function to a potential energy function via functional transformation. For instance, if the objective function is the likelihood function, then the potential energy function may be represented by a scaled version of the negative log-likelihood function.

7

CASE STUDIES

In this chapter, we present several case studies that reflect the nature of this book. The case studies are in three categories: (i) modeling the correlative brain, (ii) applying correlative learning for modeling perceptual functions of the brain, and (iii) applying correlative learning for engineering applications. Each case study is independent and stands alone; the interested reader can select to read any of these according to his or her interests.

The four case studies are:

Case 1: A neurophysiological study of auditory cortical map reorganization.

Case 2: Learning neurocompensator—a model-based hearing compensation design.

Case 3: Online learning of neural networks.

Case 4: Kalman filtering in computational neural modeling—learning shape and motion from image sequences.

Notably, these four case studies are partially excerpted or adapted from the following previously published articles with permission of the corresponding copyright holders:

- J. J. Eggermont. Temporal modulation transfer functions in cat primary auditory cortex: separating stimulus effects from neural mechanisms. *Journal of*

Correlative Learning: A Basis for Brain and Adaptive Systems, by Zhe Chen, Simon Haykin, Jos J. Eggermont, and Suzanna Becker

Neurophysiology, Vol. 87, pp. 305–321. Copyright © 2002 by The American Physiological Society, reprinted with permission.

- J. J. Eggermont. Properties of correlated neural activity clusters in cat auditory cortex resemble those of neural assemblies. *Journal of Neurophysiology*, Vol. 96, pp. 746–764. Copyright © 2006 by The American Physiological Society, reprinted with permission.
- A. J. Noreña and J. J. Eggermont. Comparison between local field potentials and unit cluster activity in primary auditory cortex and anterior auditory field in the cat. *Hearing Research*, Vol. 166, pp. 202–213. Copyright © 2002 by Elsevier, reprinted with permission.
- A. J. Noreña, B. Gourévitch, N. Aizawa, and J. J. Eggermont. Spectrally enhanced acoustic environment disrupts frequency representation in cat auditory cortex. *Nature Neuroscience*, Vol. 9, No. 7, pp. 932–939. Copyright © 2006 by Nature Publishing Group, reprinted with permission.
- Z. Chen, S. Becker, J. Bondy, I. Bruce, and S. Haykin. A novel model-based hearing compensation design using a gradient-free optimization method. *Neural Computation*, Vol. 17, No. 12, pp. 2648–2671. Copyright © 2005 by MIT Press, reprinted with permission.
- S. Haykin, Z. Chen, and S. Becker. Stochastic correlative learning algorithms. *IEEE Transactions on Signal Processing*, Vol. 52, No. 8, pp. 2200–2209. Copyright © 2004 by IEEE, reprinted with permission.
- S. Haykin. Kalman filtering and its neural implications. In M. A. Arbib, Ed., *Handbook of Brain Theory and Neural Networks*, 2nd ed., pp. 590–594. Copyright © 2002 by MIT Press, reprinted with permission.
- G. Patel, S. Becker, and R. Racine. Learning shape and motion from image sequences. In S. Haykin, Ed., *Kalman Filtering and Neural Networks*, pp. 69–81 Copyright © 2001 by Wiley, reprinted with permission.

7.1 HEBBIAN COMPETITION AS BASIS FOR CORTICAL MAP REORGANIZATION?

Background on Auditory Tonotopic Maps. Adult cortex is known to be plastic, that is it changes its organization to suit particular demands imposed by the environment. The process of reorganization can be called *learning*. It can also be an adaptive response to changing conditions, for example, as a result of aging; in some cases it can lead to maladaptive consequences, as in tinnitus (a perceived ringing, hissing, or buzzing sound in the absence of an external stimulus) [253]. The organizational changes that are most easily quantified are those that are expressed in the form of topographic maps. In the auditory cortex an example of such a map is the continuous representation of acoustic frequency versus cortical location, which is known as the *tonotopic map*; it is a map of the one-dimensional receptor surface in the inner ear, with frequency varying along one dimension and other features such as intensity level varying in a patchy fashion along the

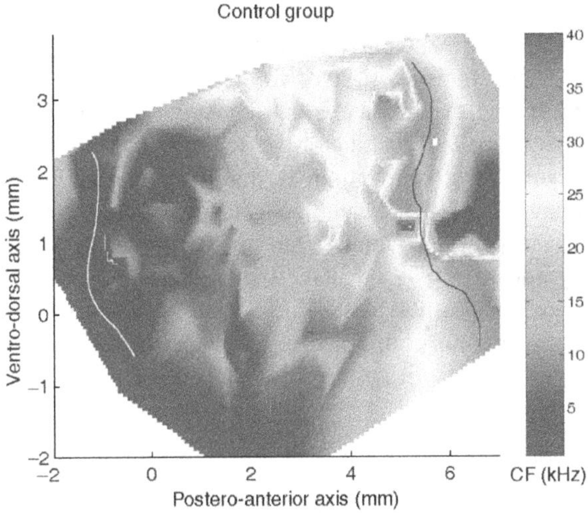

Figure 7.1 False color map of the tonotopic organization in the cat's auditory cortex. The color bar indicates the CF in kilohertz. The (0,0) coordinate represents the tip of the PES (posterior ectosylvian sulcus). The horizontal axis runs parallel to the midline from posterior to anterior. The vertical axis indicates ventral to dorsal distance. (From data presented in (667)).

other dimension (Figure 7.1). In Figure 7.1, the normal tonotopic map shows a progression of characteristic frequencies (CFs) from left bottom to right top in primary auditory cortex (A1). Then a reversal of the frequency gradient takes place and marks the border with anterior auditory field (AAF). The boundary of A1 with AAF is indicated by the black line, and that between A1 and posterior areas by the white line. Perpendicular to the frequency gradient we observe sheets (going through all cortical layers) of locations with similar CFs, that is, the isofrequency sheets. The boundary line between A1 and AAF is indeed such a sheet with a CF of approximately 40 kHz.

Neural Connections. The nerve cells that provide the output of the auditory cortex are the pyramidal cells. They process sound-evoked inputs from the inner ear via the brainstem and midbrain and activity of the thalamocortical afferent fibers that synapse predominantly in cell layers III and IV onto the pyramidal cells (see Chapter 1). Besides transmitting neural activity to other cortical areas, there is also a more localized output from the pyramidal cells through so-called horizontal fibers that are found predominantly in layer III. These horizontal fibers extend for several millimeters within the isofrequency sheets on either side of the cell, but also, albeit less frequently, perpendicular to those sheets thereby providing heterotopic connectivity between cells with vastly different CFs [537].

Thus in a simplified scheme, neglecting for a moment the inhibitory inputs to pyramidal cells, the pyramidal cell receives inputs from thalamic cells with a diverse range of CFs (see Chapter 1) and from other pyramidal cells of even greater range of frequency preferences. Both sets of inputs are excitatory, and under normal conditions the thalamocortical inputs dominate despite that they form only 10–15% of the synapses. Their efficiency derives from the correlations between the input spike times from several thalamic cells that converge on the same pyramidal cell [124] and their relatively fast conduction velocity (3.3 m/s [784]). In contrast, the horizontal fibers are slower conducting (0.5 m/s) and the inputs they provide are likely less synchronized [4]. As a result, the synaptic coupling between the thalamic outputs and the pyramidal cells may be stronger than that between the horizontal fibers and the pyramidal cells as thalamocortical fibers are much more likely to fire a pyramidal cell than a horizontal fiber, a simple consequence of a Hebbian synapse. Of course, inhibitory inputs to pyramidal cells are important in shaping both the spectral and temporal response properties of pyramidal cells [665].

Input and Output Tuning of Pyramidal Cells. The wide frequency range of inputs from thalamic neurons causes the excitatory postsynaptic potentials (EPSPs) to be much wider tuned than the spikes [873], that is, the inputs to the pyramidal cells are much broader tuned than their outputs. The narrower tuning at the output stage is thought to be caused by inhibitory activity. The tuning for extracellularly recorded local field potentials (LFPs) is similar to that for EPSPs [467].

Figure 7.2 shows, for typical sets of recordings, dot rasters for multiunit (MU) spikes (red dots) and LFP triggers (black dots). The upper panel (Figure 7.2*a*) represents a recording site in AAF and the two other panels represent recording sites in A1. The LFP triggers often display repeated activity, with a period of 25–40 ms depending on the recording. This represents repeated triggers for the same multiphasic LFP waveform [254]. This oscillatory behavior is most pronounced at high intensity levels (45–75 dB) and close to the CF of the recording site, that is, when the LFP amplitude is largest (Figure 7.2*a*). A feature of the LFP triggers is that they can also occur randomly produced by spontaneous EEG spindles. These spindles are present when the stimulus is not strong enough, for example, when the frequency is outside the response area, to synchronize the spindles with stimulus onset into an LFP. In general, the latency of the LFP triggers is slightly shorter than that for MU spikes; visual detection thresholds are very similar (Figures 7.2*b*,*c*) or slightly lower (Figure 7.2*a*) for LFP triggers and MU spikes. What is most obvious is that the range of frequencies evoking LFP triggers is much larger than the range evoking MU activities.

Figure 7.3 show examples of frequency-tuning curves for LFP (red lines) and MU (shaded areas) for four different recording sites. Specifically, MU tuning curves could consist of two disjointed areas located within one broad LFP tuning curve (Figure 7.3*d*). The LFP tuning curves represent the input from thalamocortical fibers indicating the wide CF range of the input neurons. Generally, the MU tuning curves, reflecting the pyramidal cell output, are contained fully within the LFP tuning curve boundaries but are much narrower as a result of intracortical inhibition.

Feedforward inhibition from thalamic neurons via an inhibitory interneuron causes the responses of the pyramidal cells to be terminated by postactivation

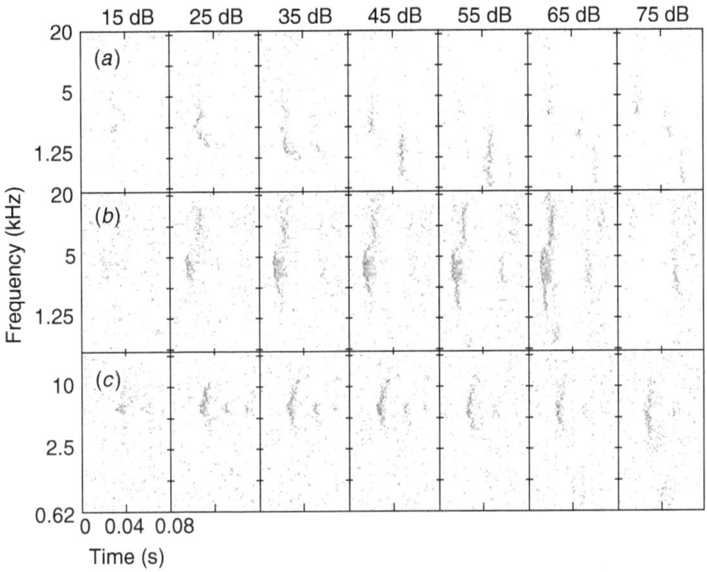

Figure 7.2 Three sets of seven dot rasters showing spectral and temporal response properties of LFP and MU activity. Each dot raster is obtained at a fixed intensity level; the intensity level ranged between 15 and 75 dB- SPL (indicated above the upper panel). MU spikes are shown in red and LFP triggers are shown in black. (*a*) Responses from a recording site in AAF; the MU response intensity function is monotonic and the tuning curve is clearly asymmetric to low frequencies. (*b*) Responses from a recording site in A1; the response intensity function is monotonic and the tuning curve is relatively broad. The tuning curves corresponding to these responses are shown in Figure 7.3*c*. (*c*) Responses from neurons in A1; the MU response intensity function is nonmonotonic and the tuning curve is symmetric and relatively narrow. (Reprinted from *Hearing Research*, Vol. 166, A.J. Noreña and J.J. Eggemont, Comparison between local field potentials and cluster activity in primary auditory cortex and anterior auditory field in the cat, pp. 202--213. Copyright © 2002, with permission from Elsevier.)

suppression (Figure 7.2), especially at high stimulus levels. Horizontal fibers do not have this feature; thus their inputs are more sustained and the output of the pyramidal cell will reflect that.

Synaptic Depression. Central nervous system synapses onto pyramidal cells typically show depression upon repeated stimulation; that is, their transmitter output probability severely declines with each subsequent stimulus until a steady state is reached [50]. In the auditory system the synapses in the brainstem are very precise and reliable and can follow very high input rates without depression [795]. Synapses between the midbrain and the thalamus and also between the thalamus and cortical pyramidal cells are rapidly exhausted by high input rates (Figure 7.4).

Exhausting Thalamocortical Synapses. Having now laid out the basics prerequisites for this case study, let us present a condition in which an animal

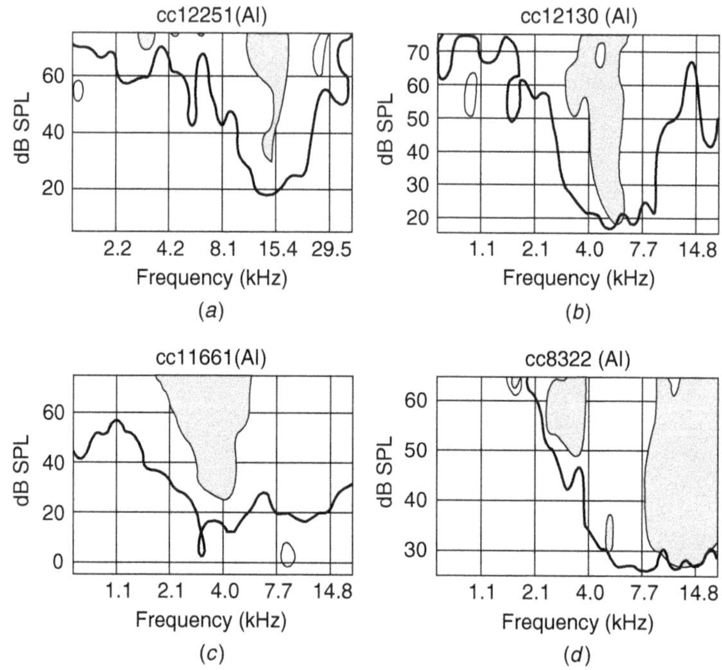

Figure 7.3 Four examples of excitatory frequency-tuning curves for MU (gray shading) and LFP (red lines). The tuning curves are drawn as contour lines at 25% of the maximum response. All the panels show frequency-tuning curves from recording sites located in A1. (*a, b*) Tuning curves for LFP and MU are relatively narrow and symmetric. (*c*) Tuning curves are broad, especially for LFP. (*d*) Tuning curve of the MU is multipeaked. The corresponding dot raster of the tuning curves in (*c*) is shown in Figure 7.2*b*. (Reprinted from *Hearing Research* Vol. 166, A.J. Noreña and J.J. Eggemont, Comparison between local field potentials and unit cluster activity in primary auditory cortex and auterior auditory field in the cat, pp. 202--213. Copyright © 2002, with permission from Elsevier.)

is continuously stimulated with sound at a level that does not cause damage to the ear but that is present 24 h per day, 7 days a week, for several months. The average repetition rate of the tone pips for this sound is 96 Hz, but the sound is not periodic as the 50-ms tones (see Figures 7.4 and 7.5 for the envelope and response of the tone pip) of the frequencies between 4 and 20 kHz are randomly drawn according to uncorrelated Poisson processes with mean rate of 3 Hz for each frequency. Figure 7.5 presents the stimulus envelope, the spectrogram, and the average carrier and modulation spectrum. We can observe the considerable AM of the sound. During the experiment, while the cats passively listened to the sound, they were likely ignoring it as the sound did not have any meaning.

The narrow-band acoustic environment is expected to activate neurons in the 4–20-kHz region of the tonotopic map and not to affect frequency regions below or above. For control animals (Figure 7.6*a* top row) a gradient in activity along the posterior–anterior axis can be observed, reflecting the tonotopic organization. This

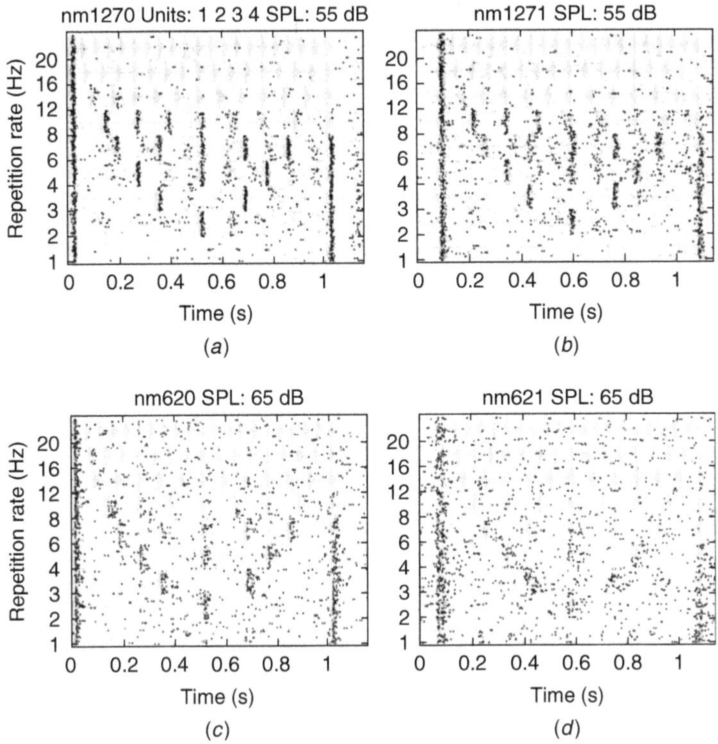

Figure 7.4 (*a, c*) Dot-raster displays for gamma tone trains. (*b, d*) Time-reversed gamma tone trains superimposed on the stimulus envelope. Note that stimulus-following responses cease at repetition rates around 12 Hz. (Reprinted from (249), with permission. Copyright © 2002 by the American Physiological Society.)

is much less clear from the LFPs (Figure 7.6*b* top row) as these are much more broadly tuned as shown previously (Figures 7.2 and 7.3). After the long exposure period the tonotopic maps obtained showed that the percentage of neurons in the designated region of the map that still responded to those frequencies was reduced to 10–15% (Figure 7.6*a* bottom). The remainder of the neurons in this range now responded to frequencies either above 20 kHz or below 4 kHz. A small subset did respond also to their "assigned" frequency and in addition to the high-frequency region, the low-frequency region, or all three frequency regions (Figure 7.6*a*). The LFPs were equally affected in that their amplitudes were greatly reduced for frequencies in the 4–20 kHz range (Figure 7.6*b*). This indicates that the thalamic input to the pyramidal cells was already affected. The spike data indicated that there was additional modification of the cortical tonotopic map over and above that occurring in the thalamus [669].

Horizontal Fibers Take Over. Figure 7.7 shows in some detail individual MU responses across the entire intensity range. The most important cue to the underlying

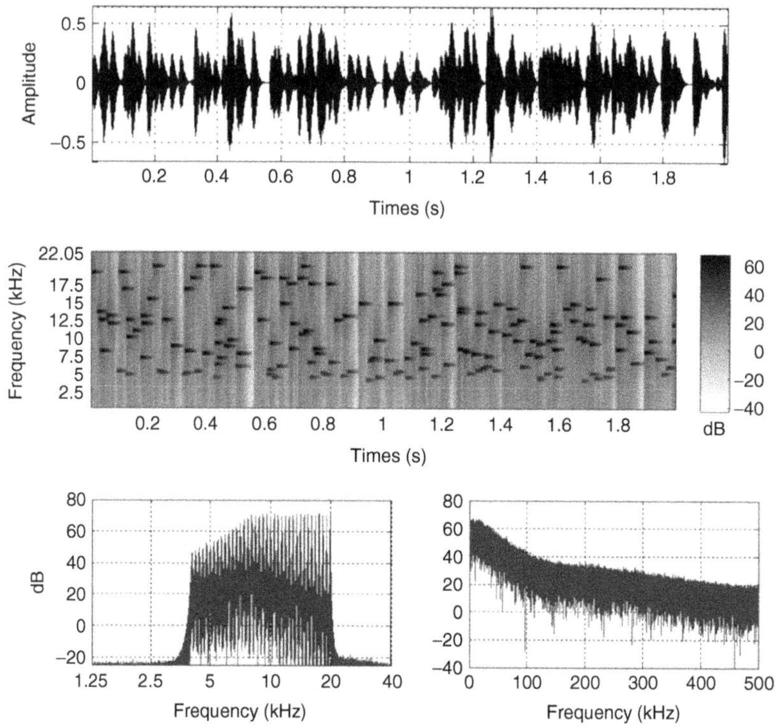

Figure 7.5 Waveform, spectrogram, and average carrier and signal envelope spectra of a 2-s long sequence of the acoustic environment.

changes are found in the raster plots. In the figure, each dot represents an action potential. The dot-raster panels consist of eight subpanels each representing the action potentials as a function of tone pip frequency and time after tone pip onset for a particular intensity (from −5 to 65 dB in 10-dB steps). The standard responses in the normal example (leftmost column of Figure 7.7) are short-latency (< 25 ms), sharp responses that are curtailed by postactivation suppression at higher stimulus level. For lower levels the range of frequencies that causes a response becomes narrower and the response latencies increase. The boundaries of the responses across stimulus levels illustrate the frequency-tuning curve of the neuron. The control example likely has a threshold between 5 and 15 dB with a CF around 15 kHz. The frequency-tuning curves (lower panels) calculated over 0–25 ms and between 25 and 100 ms show essentially the same frequency selectivity.

The examples in columns 2 and 3 of Figure 7.7 show a different picture: The frequency-tuning curves for 0–25 ms show the anticipated tuning for the neurons' locations. Those for longer latencies show the extra low- and high-frequency components. These are also clear in the dot rasters. These low- and high-frequency, longer latency, sustained inputs are likely resulting from horizontal fiber input to the pyramidal cells. The latency increase corresponds to what one expects from

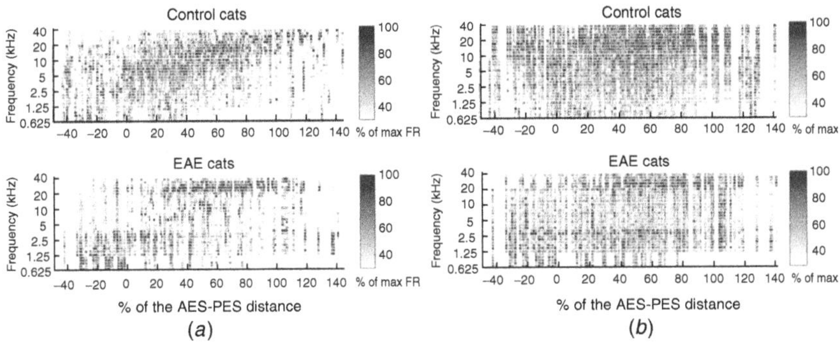

Figure 7.6 Firing rate as a percentage of the maximum firing rate per recording (*a*) and averaged LFP amplitude (*b*) averaged across three intensities (35, 45, and 55 dB SPL) as a function of electrode location along the postero--anterior axis (abscissa) and stimulus frequency (ordinate). Gray-scale bars, percentage of maximum firing rate or maximum amplitude. These data illustrate the dense spatial sampling in the two groups over the postero--anterior axis and the gap in responsiveness in EAE cats for tone frequencies between 4 and 20 kHz.

the slow-conducting horizontal fibers and the distance from the low- or high-CF neurons to the affected frequency region. Examples in columns 4 and 5 of Figure 7.7 show that when the location-based (and $\leq$25-ms) tuning largely disappears (see bottom panels), the responses to low and high frequencies are all sustained (they last at least as long as a tone pip, i.e., $\geq$50 ms) and are of long latency.

Changing Neural Correlation Strengths. The dominance of the inputs to the pyramidal cells from the horizontal fibers is likely the result of a competitive process between the depressed thalamic fiber inputs and the active horizontal fibers originating from cortical pyramidal cells with sensitivities in the low- and high-frequency regions adjacent to the 4–20 kHz region. The continuous stimulation at high rate exhausts the thalamocortical synapses to such an extent that synchronous activation is no longer an option. The fact that even 12 h after the exposure, that is, during the acute recordings, there was no recovery suggests that the synapses are not functioning anymore. This is corroborated by the strong increase in spontaneous spike-timing correlation for distances up to 3 mm away [100% of anterior–posterior ectosylvan sulcus (AES–PES) distance is approximately 8 mm] in the reorganized A1 in exposed animals compared to normal controls (Figure 7.8). In addition to this expansion of the correlated region, the strength of the cross-correlation is also greatly increased. Since the correlation strength was corrected for effect of changes in firing rate, it indicates stronger synapses, more shared branched axons, or both.

Synaptic Competition. Similar competitive processes likely take place after noise-induced hearing loss. It has been known for some time that mechanical damage to a restricted part of the inner ear in adult animals results in clear reorganization of the frequency place map in contralateral A1 [767] and in the auditory

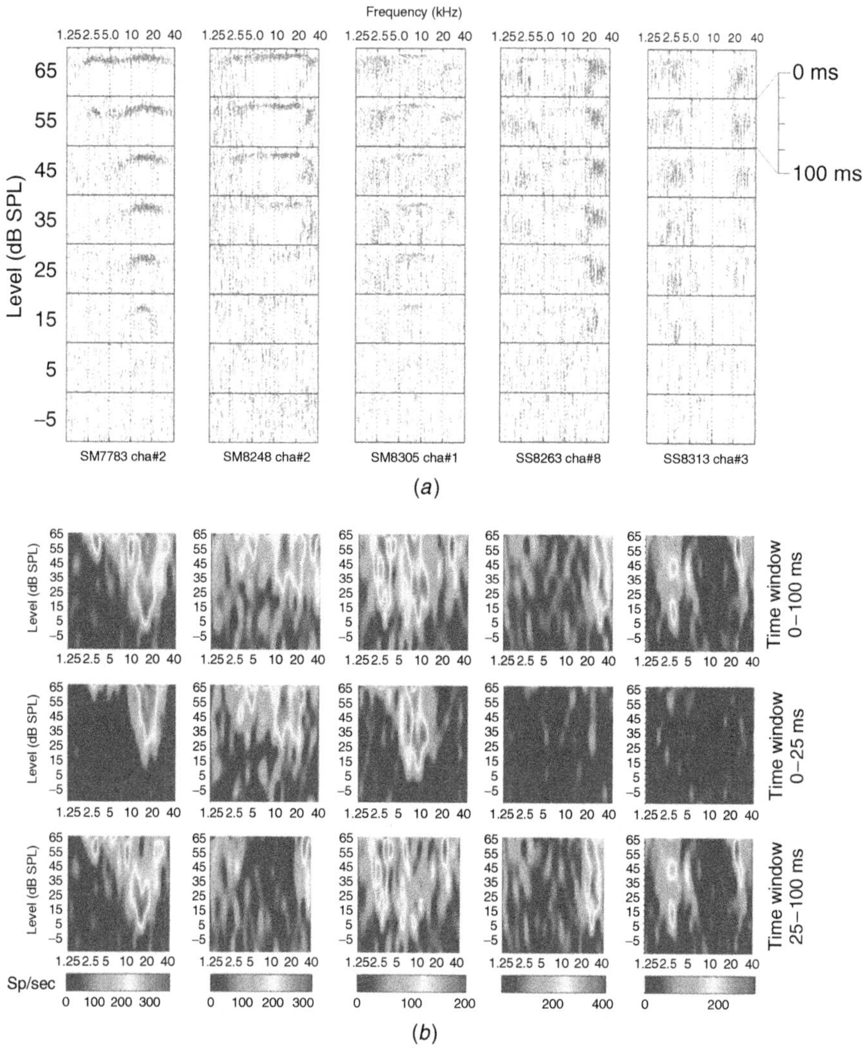

Figure 7.7 Raster plots and tuning curves of selected individual recordings. (*a*) Dot rasters show recorded spikes as a function of frequency and intensity. For each intensity level, the diagram shows a 0--100-ms time window from stimulus onset (0 at top, 100 at bottom). Data are shown for one control cat (first column) and four exposed cats (columns 2--5). (*b*) Rate--frequency--intensity area for MU activity shown in (*a*) (Columns in (b) correspond to columns in (a).) These areas were derived for all spikes (within the time window 0--100 ms), early spikes (within the time window 0--25 ms), and late spikes (within the time window 25--100 ms). Horizontal colored bars, firing rate. (Reprinted from (669) with permission. Copyright © 2006 by the Nature Publishing Group.)

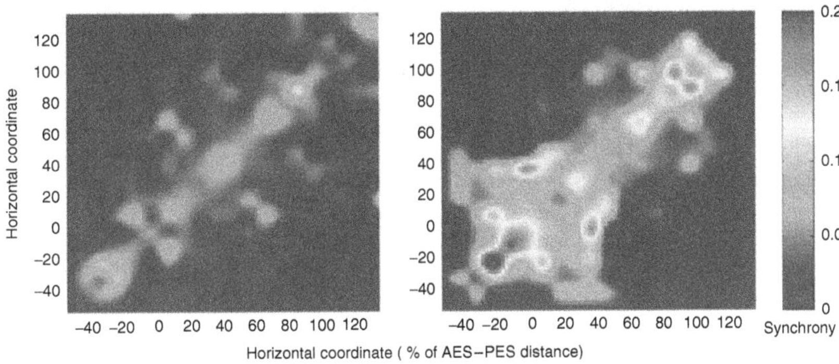

Figure 7.8 Neural synchrony, defined here as the peak strength of the cross-correlogram, is presented as a function of the position of the two recording electrodes along the postero–anterior axis (abscissa) in control (left panel) and exposed cats (right panel). The colored bar indicates the strength of neural synchrony. In control cats, the strongest synchrony was found between neighboring electrodes in the array and most correlations occurred locally. Note the increased synchrony in exposed cats compared to control cats, especially for larger distances between electrodes. This probably signifies the stronger connections over large distances (that is, into the reorganized region) made by horizontal fibers. In these cats, the range of strong correlations is much larger, especially in the −50 to 50% region, which reflects the entire area with characteristic frequencies below 5 kHz but also a substantial part of the 5–20-kHz area. In addition, the area with characteristic frequencies above 20 kHz (70–125%) also showed strongly increased neural synchrony.

thalamus [464]. However, only patchy changes occurred in the auditory midbrain [431] and none whatsoever in the cochlear nucleus [743]. See Figure 1.16 for the organization of the auditory pathways.

After noise trauma [667] that resulted in a sloping hearing loss for frequencies above 8 kHz with maximum loss of about 40 dB at 32 kHz, the tonotopic map changed dramatically and did not contain recording sites in A1 with sensitivity to frequencies above 25 kHz, and borders between cortical areas A1 and AAF can no longer be drawn on the basis of map gradient reversals (Figure 7.9). Noise trauma causes only a partial deafferentation compared to the complete one following mechanical damage to the cochlea in the studies by Irvine and colleagues [431], but nevertheless the changes are considerable. Noise-induced hearing loss is accompanied in the brainstem and midbrain by a reduction in inhibitory activity. This induces disinhibition of excitatory inputs from the thalamus within the LFP tuning areas (Figures 7.2 and 7.3) that span the normal hearing frequency range (i.e., below 8 kHz) and allow a shift in the tuning of the pyramidal cell to lower CFs. For large distances from the normal hearing frequency edge, the horizontal fibers will carry the dominant input to the partially deafferented pyramidal cells.

The map reorganization thus results at least in part from strengthening of the horizontal connections from pyramidal cells at the edge of the hearing loss (CFs in the 8-kHz range). These edge neurons synapse with the pyramidal cells in the hearing loss range above 16 kHz where the hearing loss was about 30 dB and partially

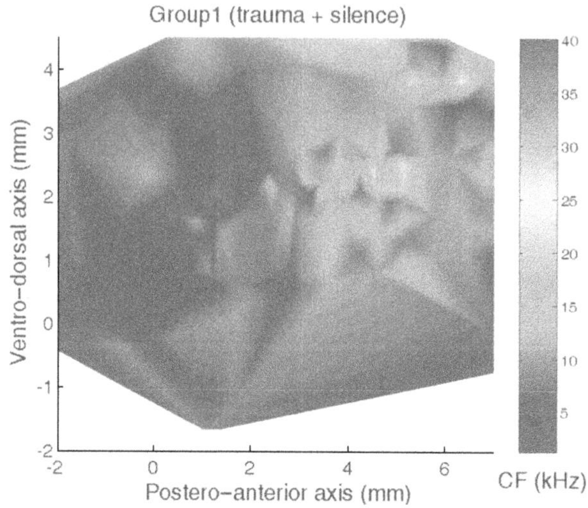

Figure 7.9 Cortical tonotopic map in a group of cats with noise-induced high-frequency hearing loss. Comparison with Figure 7.1 suggests a massive change in the map, especially in the anterior part of the cortex where normally high frequencies are presented (from data presented in (667)).

deprived from thalamic input. Thus it is expected that the normal dependence of the spike–timing correlation with distance (Figure 7.10) will be changed after trauma. As seen from Figure 7.10, in control conditions, the peak cross-correlation coefficient decreases with distance in roughly exponential fashion, with a space constant of about 4 mm.

In the A1 of cats with 5–6 kHz tone-induced hearing loss (Figure 7.11), there is a relative increase in the peak cross-correlation coefficient for distances around 3 mm, corresponding to the distance between the 4–8 kHz region with hearing loss less than 20 dB and the region between 16 and 32 kHz with hearing loss of 30–40 dB. These correlation findings are very similar to those in cortical reorganization following exposure to multifrequency sound without a hearing loss, suggesting that this multifrequency sound produced a functional central lesion in the auditory cortex (and likely also in the thalamus) that is not accompanied by hearing loss. Both the noise-induced hearing loss and the long-term exposure to nondeafening sounds produce changes in auditory tonotopic maps.

Conclusion. In this case study, we show that the changes following long-duration nontraumatizing sound exposure and following noise-induced hearing loss, that is, changes in tonotopic maps and increased neural synchrony both in strength and in spatial extension, are very similar. The tonotopic map changes are likely the result of a synaptic competition between thalamocortical inputs and horizontal fiber inputs; the synaptic adaptation process is referred to as synaptic plasticity or learning. It is highly likely that conditions under which such associative learning takes place will show comparable changes, albeit not on such a large spatial scale

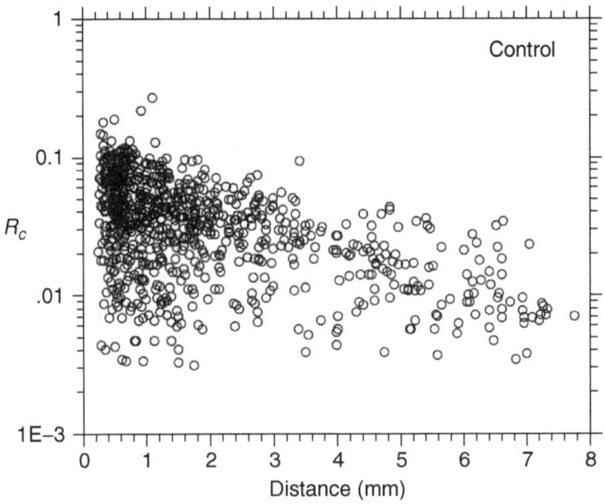

Figure 7.10 Changes in the peak cross-correlation coefficient (R_C) between pairs of spiking neurons in control A1 as a function of distance in the posterior–anterior direction (from data presented in (250)).

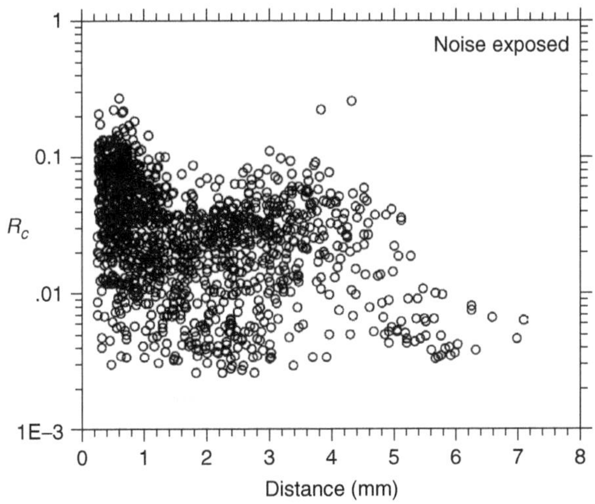

Figure 7.11 Changes in the peak cross-correlation coefficient (R_C) between pairs of spiking neurons in noise-exposed A1 as a function of distance in the posterior–anterior direction.

and most probably not as easily visualized (see Section 1.9). The increased synaptic strengths may not be all between neighboring neurons but could be locally dense and sparse over larger distances such that local clusters of highly correlated neurons [250] are functionally (and anatomically) connected between different cortical areas.

7.2 LEARNING NEUROCOMPENSATOR: MODEL-BASED HEARING COMPENSATION STRATEGY

7.2.1 Background

Current fitting strategies for hearing aids set the amplification in each frequency channel based on the hearing-impaired person's audiogram, which measures pure-tone thresholds for each of a small set of frequencies. However, it is well known that the detection of a sound can be strongly masked in the presence of background noise, competing speech, and so on. It is therefore not surprising that many people with hearing loss end up not wearing their hearing aids. The devices are unhelpful and may even worsen the wearer's ability to hear sounds under noisy listening conditions. Directional microphones and other generic signal processing strategies for noise reduction have resulted in modest benefits in some contexts but not dramatic improvement. Instead, the approach we take here is to treat hearing aid design as a neural coding problem. We start with detailed models of the normal auditory nerve as well as that of a hearing-impaired person. We then search for a signal transformation that, when applied to the input to the impaired model, will result in a neural code that is close to that of the intact model. We refer to this strategy as neural compensation [73]. The signal transformation is highly nonlinear and dynamic and calculates the gain in each frequency channel by combining information across multiple channels rather than using a static set of channel-specific gains. The neurocompensator should therefore be capable of approximating the contrast enhancement function of the normal ear. A schematic of normal/impaired hearing systems as well as the neural compensation is illustrated in Figure 7.12. The goal of the neurocompensator is to restore near-normal firing patterns in the auditory nerve in spite of the hair cell damage in the inner ear; ideally, it attempts to compensate the hearing impairment in the auditory system and match the output of the compensated system as closely as possible to the output of the normal hearing system. In other words, by regarding the outputs of the normal/impaired hearing systems as the neural codes generated by the brain, we attempt to maximize the similarity of the neural codes generated from the models H and $\hat{\mathrm{H}}$ in Figure 7.12.

7.2.2 Biologically Inspired Hearing Compensation Strategy

Overview of System. Given the neurocompensator diagram illustrated in Figure 7.12, the learning of the adaptive hearing system is shown in Figure 7.13. First, the time-domain audio (speech or natural sound) signal is converted into the frequency domain through STFT. The role of the neurocompensator, which is modeled through frequency-dependent gain coefficients for different bands (to be described later in this section), is to conduct spectral enhancement in the frequency domain. Given the normal (H) and impaired ($\hat{\mathrm{H}}$) auditory models, the feedback error is calculated via a probabilistic metric by comparing the spike train images generated by the normal and compensated hearing systems. Furthermore, a gradient-free ALOPEX optimization procedure uses the error for updating the neurocompensator's parameters to minimize the discrepancy between the neural codes generated from the normal and impaired hearing models.

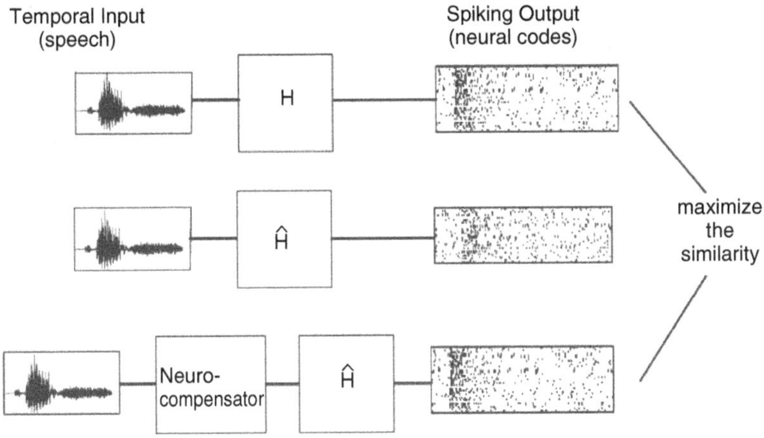

Figure 7.12 A schematic of neurocompensation. *Top*: normal hearing system. *Middle*: impaired hearing system. *Bottom*: neurocompensator followed by the impaired hearing system. The hearing systems map the temporal speech signal input to a spike train map (neural codes) output; the H and Ĥ denote the input–output mappings of the normal and impaired ear models, respectively. The neurocompensator acts as a preprocessor before the impaired ear model in order to produce neural codes similar to as the normal neural codes from the normal ear model. (Reprinted from (160) with permission. Copyright © 2005 by MIT Press.)

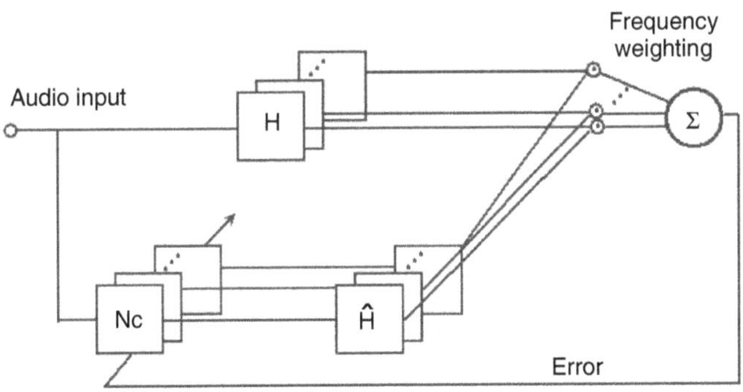

Figure 7.13 Block diagram of algorithm for training the neurocompensator (Nc). The normal (H) and impaired (Ĥ) auditory models' output is a set of spike trains at different best frequencies, which are then subjected to an onset detection process, while the neurocompensator is represented as a preprocessor that calculates gains for each frequency. The error is the KL divergence between the probability distributions of the two models' outputs. (Reprinted from (160) with permission. Copyright © 2005 by MIT Press.)

Experimental Data. The audio data presented to the ear models can be either speech or any other natural sound. In our experiments, the speech data are selected from the TIMIT and the TIDIGITS databases. From the TIMIT database, a total of 10 spoken sentences by different male and female speakers are used for the simulations reported here. In the TIDIGITS database, the data consist of English-spoken digits (in the form of isolated digits or multiple-digit sequences) recorded in a quiet environment. All speech samples were sampled or resampled to 16 kHz before being presented to the auditory models. Some of the speech samples used in the experiments are listed in Table 7.1. Ideally, all of the speech samples are truncated to within the same length.

Auditory Models. The auditory peripheral model used here is based on the earlier work of Bruce and colleagues [123]. In particular, the model consists of a middle-ear filter, time-varying narrow- and wide-band filters, inner and outer hair cell models, synapse model, and spike generator, describing the auditory periphery path from the middle ear to the auditory nerve. More recently, a new middle-ear model and a new saturated exponential synapse gain control have been incorporated into that model. The hearing-impaired version of the model described in detail in [101] simulates a typical steeply sloped high-frequency hearing loss.

With the normal or impaired auditory models [123], the spike train maps can be generated via feeding the temporal audio (speech or natural sound) signal to the system. We further process the auditory representation generated by the auditory nerve models by applying an onset detection procedure [102] consisting of a derivative mask with *rectification* and *thresholding*. This removes much of the noisy spontaneous spiking and high degree of steady-state information in the signal-driven spike trains. The resultant spike train onset map is used here as the basis for comparing the neural codes generated by the normal and impaired models.

Probabilistic Modeling. In order to compare the neural codes of the normal and impaired models, we characterized the spike train onset time–frequency map, which contains a number of two-dimensional data points (represented as black dots in the output image), by its probability density function. To overcome the inherent noisiness of the spike-generating and onset detection processes, we chose a two-dimensional mixture of Gaussians to characterize this distribution, given its spatial smoothing property across the spectral–temporal plane. Suppose that $D_1 \equiv \{\mathbf{x}_i\}_{i=1}^{\ell}$ and $D_2 \equiv \{\mathbf{z}_i\}_{i=1}^{\ell'}$ denote the two-dimensional neural codes (i.e., the onset spike

Table 7.1 Selected Speech Samples used in the Experiments

Speech Sample	Speaker	Content
TIMIT-1	Male	/The emperor had a mean temper./
TIMIT-2	Female	/His scalp was blistered by today's hot sun./
TIMIT-3	Female	/Would a tomboy often play outdoor?/
TIMIT-4	Male	/Almost all of the colleges are now coeducational./
TIDIGITS-1	Male	/one/
TIDIGITS-2	Female	/one, two/
TIDIGITS-3	Female	/nine, five, one/
TIDIGITS-4	Male	/eight, one, o, nine, one/

train binary images) that are calculated from the normal and impaired hearing models [123], respectively.[1] Assume that $p(D_1|M)$ is a probabilistic model that characterizes the data D_1 where M here is represented by a Gaussian mixture model, that is, $M \equiv \{c_j, \boldsymbol{\mu}_j, \boldsymbol{\Sigma}_j\}_{j=1}^{K}$.

Note that $\{\mathbf{x}_i\} \in D_1$ are the data points calculated from the normal ear model (with input–output mapping H) given the audio (speech) data; suppose the data $\{\mathbf{x}_i\} \in \mathbb{R}^d$ are drawn from a two-dimensional ($d = 2$) mixture of Gaussian density:

$$p(\mathbf{x}) = \sum_{j=1}^{K} p(j)p(\mathbf{x}|j)$$

$$= \sum_{j=1}^{K} c_j \frac{1}{\sqrt{(2\pi)^d |\boldsymbol{\Sigma}_j|}} \exp\left(-\frac{1}{2}|\mathbf{x} - \boldsymbol{\mu}_j|^{\mathrm{T}} \boldsymbol{\Sigma}_j^{-1} |\mathbf{x} - \boldsymbol{\mu}_j|\right), \tag{7.1}$$

where c_j is the prior probability for the jth Gaussian component, with mean $\boldsymbol{\mu}_j$ and covariance matrix $\boldsymbol{\Sigma}_j$. Given a total of ℓ data points in the time–frequency spike–train onset map, we can calculate the joint likelihood of the data given the mixture model M:

$$p(D_1|M) = \prod_{i=1}^{\ell} p(\mathbf{x}_i). \tag{7.2}$$

Alternatively, we can calculate the log likelihood

$$\mathcal{L} = \log p(D_1|M) = \sum_{i=1}^{\ell} \log p(\mathbf{x}_i) \tag{7.3}$$

and the associated average log-likelihood $\mathcal{L}_{\text{av}} = \mathcal{L}/\ell$. Here, we have not used any model selection procedure for Gaussian mixture modeling. Nevertheless, it is straightforward to use a penalized maximum-likelihood measure that incorporates a complexity metric such as the Bayesian information criterion (BIC) for model selection. For a K-mixture of Gaussians model, the BIC is defined as

$$\text{BIC}(K) = \sum_{i=1}^{\ell} \log p(\mathbf{x}_i|\boldsymbol{\theta}) - \frac{\ell_K}{2} \log \ell,$$

where $\ell_K = K\left[1 + d + d(d+1)/2\right]$ represents the total number of free parameters in the model. Figure 7.14 shows comparison curves of log-likelihood and BIC as functions of the number of mixtures, K.

The clustering is fitted via a mixture of elliptical Gaussians using the EM algorithm (see Appendix E for details). Based on our empirical observations, the following strategies were used for the probabilistic fitting:

- We rescale the time and frequency ranges for better Gaussian mixture fitting; an optimal *scale ratio* (time vs. frequency) of 0.25 applied to the *normalized*

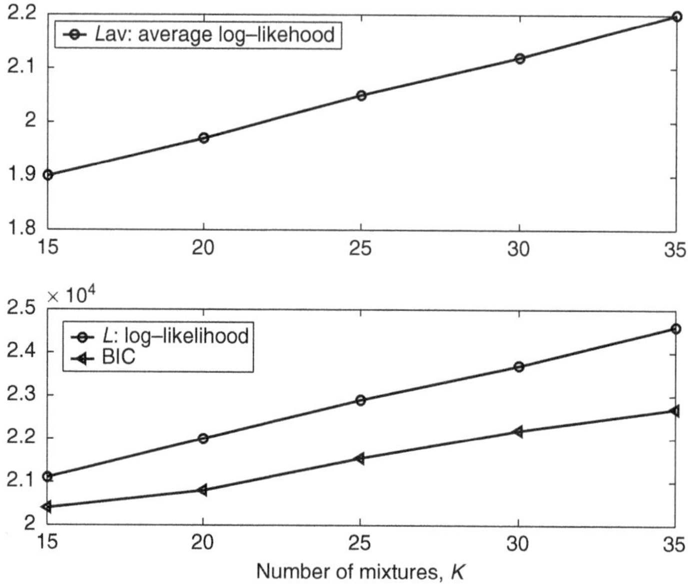

Figure 7.14 The averaged and joint log-likelihood and the BIC parameters against different numbers of mixtures, averaging on different trials for one set of spike train data.

time–frequency coordinate is suggested; namely, the time axis is constrained within the region [0, 1], whereas the frequency axis is within the region [0, 0.25]. This is tantamount to scaling the variance of the coordinates and compressing the data in terms of their distance, which is advantageous for probabilistic fitting (see Figure 7.15 for illustrations).

- For the spike train onset map, a fixed number of 20 mixtures of elliptical Gaussians is used to characterize the data distribution.
- We use the K-means clustering method [231] to initialize the mean parameters to accelerate the convergence. Typically, 10–20 iterations of the batch EM algorithm would produce reasonable fitting results.

Spectral Enhancement. Spectral enhancement is achieved through the neurocompensator. The underlying principle is to control the spectral contrast via the gain coefficients using the idea of *divisive normalization* [811]. In particular, the frequency-dependent gain coefficient G, at the ith frequency band, is calculated as

$$G_i = \frac{\|f_i\|^2}{\sum_j v_{ji}\|f_j\|^2 + \sigma}, \tag{7.4}$$

where i and j represent the indices of the frequency bands; v_{ji} denotes the cross-frequency-effect coefficient; G_i is a nonlinear function of the weighted input (frequency) power, $\|f_i\|^2$, divided by the weighted sum of all the frequencies'

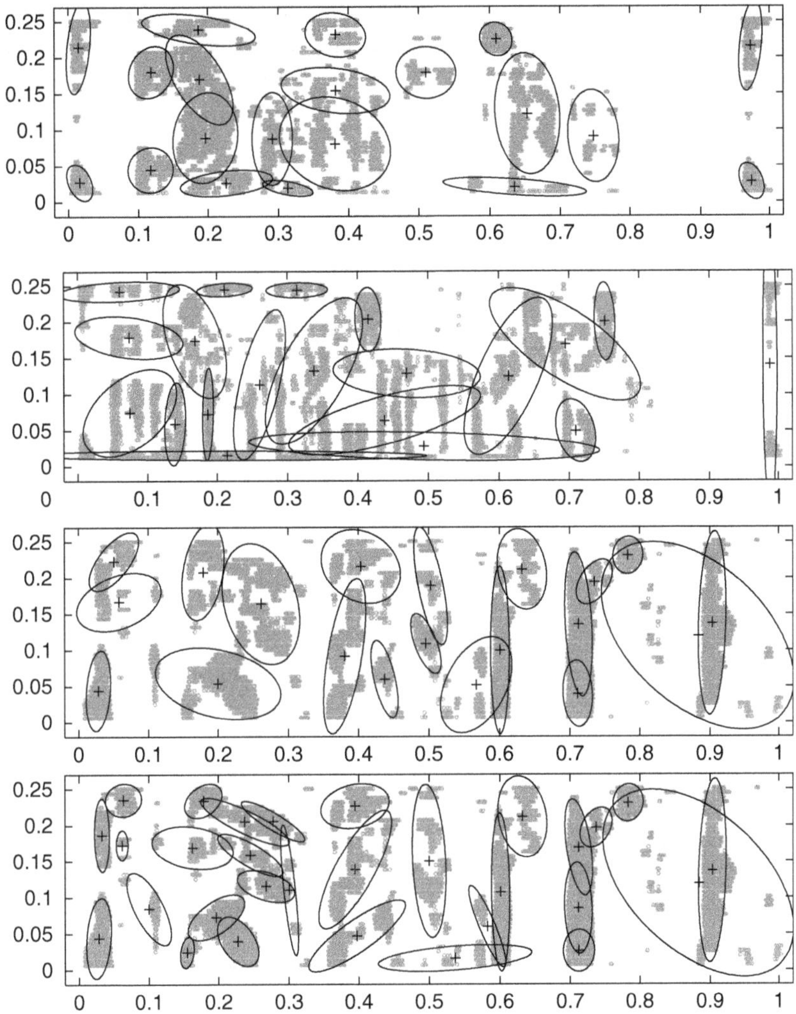

Figure 7.15 Three selected sets of spike train data calculated from the normal hearing model and their probabilistic fittings using 20 (the first three plots) or 30 (the fourth plot) Gaussian mixtures. In these four plots, the horizontal axis represents scaled time and the vertical axis represents scaled frequency, with a frequency–time scale ratio of 0.25. For the third plot, $\mathcal{L} = 22009$, $\mathcal{L}_{\text{av}} = 1.97$, and $\text{BIC}(20) = 20891$; for the fourth plot, $\mathcal{L} = 23942$, $\mathcal{L}_{\text{av}} = 2.14$, and $\text{BIC}(30) = 22264$. It is evident that the fourth plot is a better fit than the third one. (Reprinted from [160] with permission. Copyright © 2005 by MIT Press.)

power; and σ is a regularization constant that ensures that the gain coefficient G_i does not go to infinity. The design of the gain coefficient function is the essence of a neurocompensator. Applying gain coefficients to frequency bands is tantamount to implementing a bank of nonlinear filters, the motivation of which is to mimic the inner hair cells' frequency response. The divisive normalization was originally aimed at suppressing the statistical dependency between the filters' responses [811]. Here, we employ a similar functional form, but rather than adapting the normalization coefficients to optimize information transmission, we adapt the parameters to optimize a measure of the similarity between the neural codes generated by the two models.

For the present purpose, a slightly different version of (7.4) is used:

$$G_i = h\left(\frac{w_i \| f_i \|^2}{\sum_j v_{ji} \| f_j \|^2 + \sigma}\right), \quad \text{where} \quad w_i \propto G_i^{\text{NAL-RP}}, \tag{7.5}$$

where $G_i^{\text{NAL-RP}}$ represents a positive coefficient based on NAL-RP (national acoustics lab-revised profound), a standard hearing aid fitting protocol [131] that can be calculated from the ith frequency band [101], and $h(\cdot)$ is a continuous, smooth (e.g., sigmoid) function that constrains the range of the gains as well as ensures that the gains will vary smoothly in time. When $h(\cdot)$ is linear and $G_i^{\text{NAL-RP}} = 1$, equation (7.5) reduces to (7.4). On the other hand, when all $v_{ji} = 0$ and $h(\cdot)$ is linear, equation (7.5) reduces to the standard, fixed linear gain NAL-RP algorithm. We have chosen w_i to be proportional (in value) to the $G_i^{\text{NAL-RP}}$ that is given by the standard NAL-RP algorithm for calculation of the gains, while assuring that w_i will not be so large or small as to push the sigmoid function into the saturated region where derivatives would be near zero; w_i will be fixed after appropriate scaling. For the hearing aid application, it is appropriate to constrain $G_i \geq 0$. Now, the goal of the learning procedure is to find the optimal parameters $\{v_{ji}\}$ that compensate the hearing impairment or intelligibility according to a certain performance metric. Because these normalization parameters are adapted to compensate for impaired auditory peripheral processing, we expect them to mimic the true neurobiological filter that they are substituting for. For example, for a fixed frequency channel j, v_{ji} might evolve toward an "on-center, off-surround" shape filter. Since the neurocompensator attempts to substitute the role of a real neurobiological filter, it is reasonable to impose biologically realistic constraints on the compensator parameters: The gain coefficients G_i should be nonnegative, bounded, and varying smoothly over a short period of time. It is important to note that, unlike the traditional hearing aid algorithms, the parameters to be optimized are *not* independent, in the sense that the cross-frequency interference may cause modifying one parameter to indirectly affect the optimality of the others. All of these issues make the learning of the neurocompensator a hard optimization problem and the solution might not be unique.

7.2.3 Optimization

Let $\boldsymbol{\theta} \equiv \{v_{ji}\}$ denote the vector that contains all of the parameters to be estimated in the neurocompensator. Let $D_2 = \{\mathbf{z}_i\}$ denote the data calculated from the deficient ear model (with input–output mapping $\hat{\mathrm{H}}$) after preprocessing the speech signal with the neurocompensator parameterized by $\boldsymbol{\theta}$. Let $p(D_2|M, \boldsymbol{\theta})$ be the marginal likelihood of the impaired model's spike trains having been generated by a normal model; then the associated log-likelihood can be written as

$$\mathcal{L}'_{\text{av}} = \frac{1}{\ell'} \log\, p(D_2|M, \boldsymbol{\theta}) = \frac{1}{\ell'} \log\left(\prod_{i=1}^{\ell'} \sum_{k=1}^{K} c_k \mathcal{N}(\boldsymbol{\mu}_k, \boldsymbol{\Sigma}_k; \mathbf{z}_i)\right)$$

$$= \frac{1}{\ell'} \sum_{i=1}^{\ell'} \log\left(\sum_{k=1}^{K} c_k \mathcal{N}(\boldsymbol{\mu}_k, \boldsymbol{\Sigma}_k; \mathbf{z}_i)\right),$$

where M is a Gaussian mixture model fitted to the normal hearing model's output, D_1, by maximizing $\log p(D_1|M)$, which can be optimized offline as a preprocessing step. One way of optimizing the neurocompensator would be to maximize $\mathcal{L}'_{\text{av}}$ with respect to $\boldsymbol{\theta}$; however, directly maximizing it may cause a "saturation" since the number of points in D_2, ℓ', might grow over ℓ. A better objective function that does not suffer this pitfall is the KL divergence between the probability of observing the impaired model's output under the normal versus impaired density function. Unfortunately, calculating the latter is much more costly, because it must be done repeatedly, interleaved with optimization of the neurocompensator parameters $\boldsymbol{\theta}$. We therefore consider a discrete sampling approach to estimate this density which is computationally simpler than fitting a Gaussian mixture model.

Specifically, we quantize or discretize evenly the spike train onset map into a number of bins where each bin contains zero or more of the spikes. To quantitatively measure the discrepancy between the normal spike train and reconstructed spike train maps, we calculate the probability of each bin that covers the spikes; this can be easily done by counting the number of the spikes in the bin and further normalizing by the total number of spikes in the whole spike train map. In particular, the objective function to be *minimized* is a quantized form of the KL divergence:

$$J \equiv \text{KL}(D_2 \| D_1) = \sum_{i}^{\#\text{bins}} p(\text{bin}_i|D_2) \log \frac{p(\text{bin}_i|D_2)}{p(\text{bin}_i|D_1)}, \tag{7.6}$$

where $p(\text{bin}_i|D_1)$ and $p(\text{bin}_i|D_2)$ represent the probabilities of the ith bin that contains the spikes in the normal and reconstructed spike train maps, respectively. Note that $p(\text{bin}_i|D_1)$ can be calculated (only once) in the preprocessing step. In our experiment, we quantize evenly the spike train map into a (40-time) × (10-frequency) mesh grid (see Figure 7.16 for illustration), with a total number of 400 bins.

However, equation (7.6) suffers from two drawbacks: (i) For some bins, the denominator $p(\text{bin}_i|D_1)$ can be zero, thereby causing a numerical problem. (ii) There is no smoothing between two discrete maps; hence it will suffer from the noise in the spiking and/or onset detection processes. Fortunately, since we have the Gaussian mixture probabilistic fitting for D_1 at hand, this can provide a spatial smoothing across the neighboring (time and frequency) bins, thereby counteracting the noise effect. To overcome the above two problems, we therefore

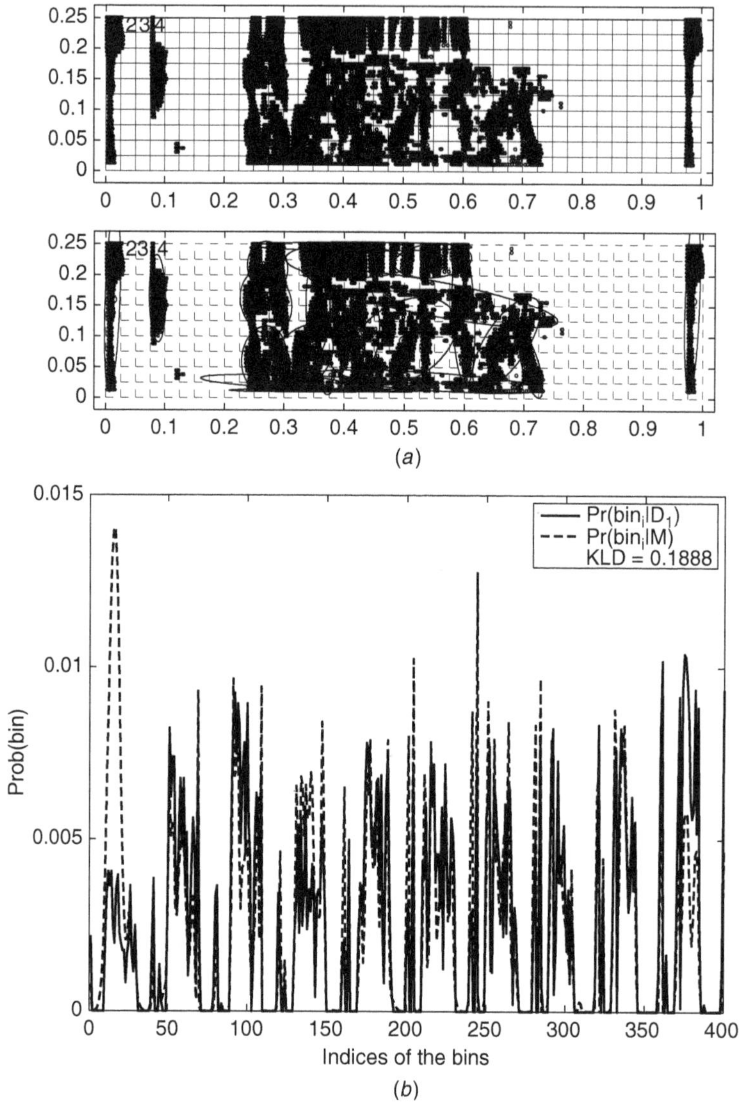

Figure 7.16 (*a*) A grid quantization compared with a Gaussian mixture fitting on the spike train map. Each map contains $40 \times 10 = 400$ bins; the arabic numerals inside the bins indicate their respective indices. (*b*) The approximation comparison between $p_1 = p(\text{bin}_i|D_1)$ and $p_2 = p(\text{bin}_i|M)$ ($i = 1, \ldots, 400$), $\text{KL}(p_1 \| p_2) = 0.1888$. (Reprinted from (160) with permission, Copyright © 2005 by MIT Press.)

substitute $p(\text{bin}_i|D_1)$ (quantized version) with $p(\text{bin}_i|M)$ (continuous version), where $p(\text{bin}_i|M)$ is calculated by fitting the center point in the ith bin with the Gaussian mixture model M divided by a normalization factor $\sum_j p(bin_j|M)$ (see Figure 7.16 for illustration). To do so, we modify (7.6) to obtain our final objective function:

$$J \equiv \mathrm{KL}(D_2 \| M) = \sum_{i}^{\#\text{bins}} p(\text{bin}_i | D_2) \log \frac{p(\text{bin}_i | D_2)}{p(\text{bin}_i | M)}. \tag{7.7}$$

Note that $p(\text{bin}_i|M)$ is usually a nonzero value due to the overlapping Gaussian covering, although it can be very small.[2] As before, $p(\text{bin}_i|M)$ can be calculated in the preprocessing step. When $p(\text{bin}_i|D_2) = p(\text{bin}_i|M)$, it follows that $J = 0$; otherwise J is a nonnegative value given $0 \leq p(\text{bin}_i|D_2) < 1, 0 \leq p(\text{bin}_i|M) < 1$. Since the probability $p(\text{bin}_i|D_2)$ can be zero, we have assumed that $0 \log 0 = 0$.

It is noted that direct calculation of the gradient $\partial J/\partial \boldsymbol{\theta}$ in either (7.6) or (7.7) is inaccessible due to the characteristics of the ear model as well as the form of the objective function; hence we can only resort to gradient-free optimization, which will be discussed below. During the training phase, the gain coefficients are adapted to minimize the discrepancy between the "neurocompensated" and original spike trains.

The optimization algorithm used here is a modified version of ALOPEX-B that is described earlier in Chapter 6. We reorganize the unknown parameters into a vector $\boldsymbol{\theta}$. The algorithm starts with a randomly initialized parameter $\boldsymbol{\theta}(0)$ and stops when the cost function $J(t)$ is sufficiently small or a predefined maximal step is reached. The stochastic component $\boldsymbol{\xi}(t)$, being a random force with certain acceptance probability, is included to help the algorithm escape from local minima.

The entire learning procedure is summarized as follows:

1. Initialize the parameters: $\{v_{ji}\} \in \mathcal{U}(-0.5, 0.5), \sigma = 0.001$; randomly select one speech sample.
2. Load the selected speech data, the associated spike train fitting mixture parameters $M \equiv \{c_i, \boldsymbol{\mu}_i, \boldsymbol{\Sigma}_i\}$, and the probability $p(\text{bin}_i|M)$, the latter two of which are precalculated offline.
3. Apply the STFT to the speech data (128-point FFT with a 64-point overlapping Hamming window); the results of time–frequency analysis then provide the temporal–spectral information across 20 frequency bands.
4. Apply the gain coefficients to the frequency bands according to (7.5); perform inverse Fourier transform to reconstruct the time-domain waveform.
5. Present the reconstructed waveform to the hearing-impaired ear model; produce a neurocompensated spike train map.
6. Using the quantized approximation to the hearing-impaired data probability density and the precalculated Gaussian mixture model, calculate the objective function (7.7).
7. Apply the ALOPEX procedure [described in equations (6.12)–(6.15)] to optimize unknown parameters.
8. Repeat steps 3–7 for a fixed number (say 100) of iterations.
9. Select another speech sample; repeat steps 2–8. Repeat the whole procedure until the convergence criterion is satisfied.

7.2.4 Experimental Results

In general, finding the optimal $\boldsymbol{\theta}$ from normal spike train is an ill-posed inverse problem; hence it is impossible to build a perfect inverse model. However, it is hoped that the reconstructed spike train image from the compensated hearing-impaired model is close to the one from the normal hearing model after the learning of the neurocompensator. Figure 7.17 shows the learning curve of the optimization. Figure 7.18 shows the learned weight coefficients of the Neurocompensator. Figure 7.19 presents the comparison between the *normal, deficient*, and *neurocompensated* spike train maps of the training speech sample.

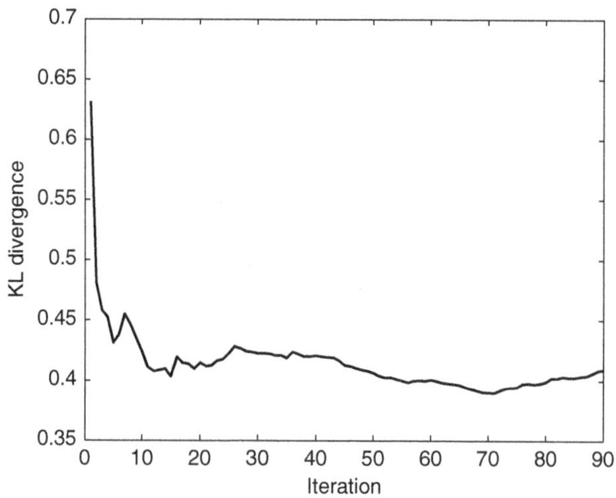

Figure 7.17 Learning curve of one speech sample using synchronous optimization. The KL divergence starts with 0.63 and stays around 0.4 after 90 iterations. (Reprinted from (160) with permission. Copyright © 2005 by MIT Press.)

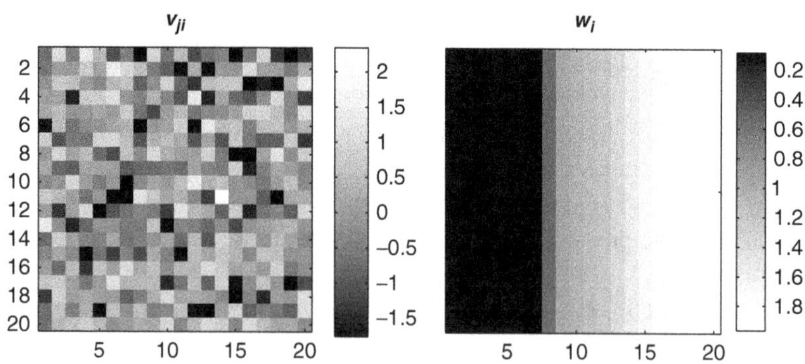

Figure 7.18 Visualization of the learned weights $\{v_{ji}\}$ and fixed weights $\{w_i\}$ of the Neurocompensator. The learned parameters $\{v_{ji}\}$ are displayed in a 20 × 20 matrix, with each column representing the weights associated with the 20 frequency bands.

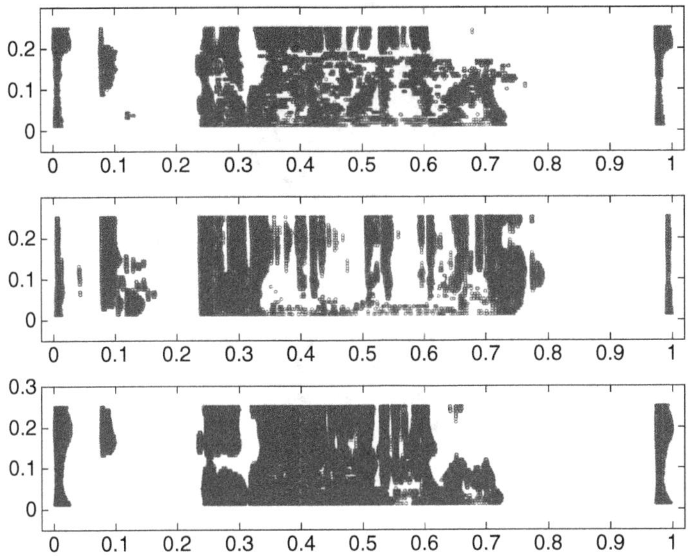

Figure 7.19 Comparisons of normal, deficient, and neurocompensated (respectively from top to bottom panels) spike train onset maps. The deficient spike train map is generated using the hearing-impaired model applied to the deficient waveform (which is produced by preprocessing the signal through the standard NAL-RP algorithm, with all gains set to $G_i \equiv 7G_i^{\text{NAL-RP}}$ for the 20 time-frequency bands and then reconstructing the signal by inverse FFT). The KL divergence between the deficient and normal spike trains is 0.664 before the learning, as opposed to 0.42 between the neurocompensated and normal spike trains after the learning. (Reprinted from (160) with permission, Copyright © 2005 by MIT Press.)

Table 7.2 Training and Testing Results of the Experimental Data in Table 7.1

Speech Sample	$\text{KL}_{\text{init}}(D_2 \| M)$	$\text{KL}_{\text{end}}(D_2 \| M)$	$\text{KL}_{\text{end}}(D_2 \| D_1)$	$\text{KL}(D_1 \| M)$
TIMIT-1	1.2058	**0.4462**	1.2828	0.1885
TIMIT-2	0.6152	0.4697	1.9255	0.2493
TIMIT-3	0.6692	0.6105	1.7367	0.2741
TIMIT-4	0.6477	0.4666	1.8329	0.2743
TIDIGITS-1	1.0626	**0.1798**	0.5591	0.0547
TIDIGITS-2	1.0234	0.4345	1.5918	0.1634
TIDIGITS-3	0.4913	0.2013	0.5759	0.0871
TIDIGITS-4	0.6346	**0.2599**	0.3757	0.1888

Note: The rightmost column $\text{KL}(D_1 \| M)$ indicates the approximation accuracy between the quantized pmf and continuous Gaussian mixture pdf on the neural codes obtained from the normal hearing system; it can be roughly viewed as a lower bound for the values in the third and fourth columns, which are the final values of $\text{KL}(D_2 \| M)$ and $\text{KL}(D_2 \| D_1)$ for the training or testing data after the learning is terminated. The second and third columns show the values of $\text{KL}(D_2 \| M)$ before/after employing the neurocompensator; the numbers in boldface indicate the training results.

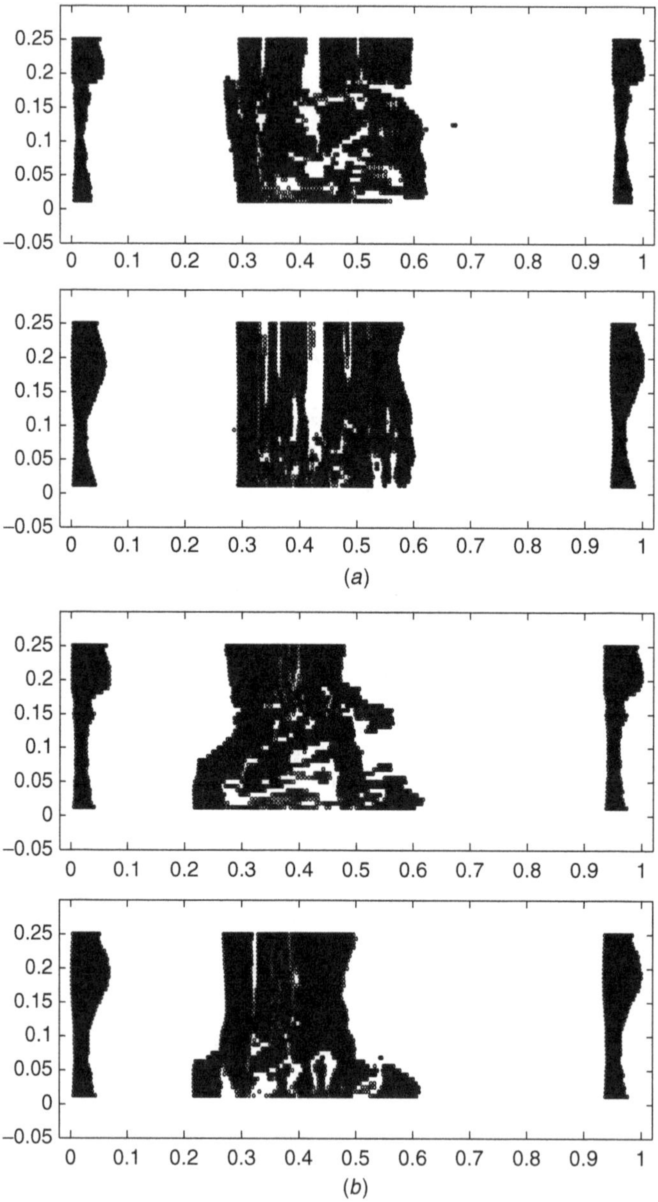

Figure 7.20 Testing results on two untrained continuous speech samples. Comparison is made between the normal and neurocompensated spike train onset maps. The KL divergence of equation (7.7) is 0.2013 between the top two maps (*a*) and 0.5591 between the bottom two maps (*b*). (Reprinted from [160], with permission. Copyright © 2005 by MIT Press.)

Upon completion of the training process, we freeze $\boldsymbol{\theta}$ and further test the neurocompensator on some unseen speech samples. The training and testing KL divergence results of the experimental data are summarized in Table 7.2. Two sets of testing results on two spoken speech signals are shown in Figure 7.20; it is seen that the neurocompensated spike train maps are reasonably close to the normal ones, though not perfect. This is quite encouraging given the fact that we have only used about 3.7 seconds of speech for training; ideally, given sufficient computational power, we should use as many speech samples as possible for training. It is hoped that, by averaging across more speech samples (with different contexts, speakers, spoken speeds, etc.), the learning process can yield a more accurate and robust solution.

7.2.5 Summary

Here, the hearing aid design problem is cast as a neural coding problem, and a neurocompensator is designed to compensate for the hearing loss and enhance the speech. The hearing compensation strategy proposed here allows us to take into account physiological data to design a person-specific hearing aid, that is, one that is tailored to a particular individual's hearing loss profile. An ultimate test of the efficacy of the hearing compensation strategy will be to conduct human hearing tests. The hearing–impaired person(s) will listen to the reconstructed speech waveform yielded from the hearing aid device (i.e., neurocompensator) and compare the intelligibility quality with and without the hearing compensation. Note that once the training is accomplished the hearing test requires no additional computational effort and is easily performed. Furthermore, once the neurocompensator parameters are optimized, the algorithm represented by (7.5) could be straightforwardly and efficiently implemented in a digital hearing aid circuit.

For a detailed discussion and suggested future research, the reader is referred to [160].

7.3 ONLINE TRAINING OF ARTIFICIAL NEURAL NETWORKS

7.3.1 Background

Artificial neural networks have been widely used in various engineering applications, such as pattern recognition, time series prediction, and control. The inherent properties of artificial neural networks, such as nonlinearity, generalization ability, noise tolerance, and robustness, have made them an appealing tool for many "black-box" modeling tasks [671]. Despite its generic nature, a better understanding and close examination of the problem at hand will also help in training the neural networks, including incorporating prior knowledge, regularization, and choosing the network architecture and the objective function.

Different network architectures often require different learning algorithms for optimizing the network parameters. For instance, the feedforward MLP often uses backpropagation, whereas recurrent MLP often uses backpropagation through time

(BPTT) or a RTRL. In general, engineers have to tune their learning procedure according to the network architecture and design the optimal parameter setup via trial and error for specific problems and specific cost functions.

The ALOPEX, as a correlation-based learning paradigm, has been proposed for training feedforward and recurrent networks [90, 902]. As discussed earlier in Chapter 6, different from conventional learning procedures such as backpropagation or the extended Kalman filter (EKF), the ALOPEX-type optimization procedure is independent of either the network architecture or the objective function. Despite its being operationally independent of the selected objective function, the form of the objective function has a direct influence on the optimization or learning performance. In practice, the best choice of objective function often requires specific analysis and prior knowledge of the problem at hand, detailed discussions of which, however, is beyond the focus here.

In what follows, we apply the sampling-based ALOPEX procedures that were described in Chapter 6 to train artificial neural networks for two engineering problems, financial data prediction and system identification, using both real-life and synthetic data. More experimental results for other problems can be found in [163, 374].

7.3.2 Parameter Setup

Given an MLP network, all the unknown parameters (synaptic weights or biases) are put into a parameter vector $\boldsymbol{\theta}$ whose dimensionality is equal to the total number of unknown parameters. In the experiments reported here, the initial parameters of the state vector $\boldsymbol{\theta}_0$ are uniformly distributed inside the region $[-1.5, 1.5]$. Once $\theta_j(0)$ is generated, an initial Gaussian prior $\mathcal{N}(\theta_j(0), 0.5)$ is used for generating the samples $\{\theta_j^{(i)}\}$. The error measure is simply the MSE:

$$J = \frac{1}{2\ell}\sum_{t=1}^{\ell} \|\mathbf{y}_t - \hat{\mathbf{y}}_t\|^2,$$

with ℓ denoting the total number of observations. For sequential data, MSE corresponds to the averaged prediction error. For sampling-based ALOPEX, we only monitor the minimum MSE among all $\{\boldsymbol{\theta}^{(i)}\}$; the one achieving the MMSE is regarded as the *maximum a posteriori* (MAP) estimate. For sequential data, the typical parameter setup is as follows: $\sigma \in [0.5, 1.0]$, $\gamma = 0.01$, $\eta = 0.1$, $\beta = 0.01$, $\lambda = 0.1$; for nonsequential data, $\sigma \in [0.01, 0.02]$, $\gamma = 0.01$, $\eta \in [0.05, 0.1]$, $\beta = \eta/10$, $\lambda = 0.5$. The relaxing parameter is often chosen in the region $\alpha \in [-0.7, 0.5]$. For online learning, we always use the overrelaxation model, namely $\alpha < 0$; the resampling step is performed in every time step.

7.3.3 Online Option Price Prediction

In the past decade, connectionist models such as the MLP and RBF networks have been used successful in financial time series forecasting and analysis (see e.g.,

[420, 421]). The financial data (e.g., stock exchange, interest rate, foreign exchange, etc.) are known to be nonlinear and nonstationary, thus providing a good test bed for neural network modeling and prediction. The real-life experimental data used here consist of five pairs of call and put option contracts on the FTSE100 index (daily close prices from February 1994 to December 1994). The accessible data include *strike prices, call option prices*, and *put option prices*.[3]

In the literature, the classic *Black–Scholes* formula was proposed for the call option price [96]:

$$C = f(S, X, T), \tag{7.8}$$

where C denotes the call option price, T represents the maturity time, and S and X represent the stock (asset) price and the strike (exercise) price of the option, respectively. The form of parametric function f often depends on the specific underlying asset and the market. In reality, the call option price data are inherently generated from complex and stochastic dynamics which rely on a lot of factors that introduce various kinds of noise to the data. Due to this reason, the Black–Scholes parametric model often suffers from violations of the underlying assumptions, such as *lognormality* or *sample-path continuity*; it is also not robust to the colored noise. The nonstationarity of the financial data often necessitates sequential tracking, which requires that the model be updated correspondingly online. This is in contrast to the common approach that uses a fixed-weight neural network for the out-of-the-sample data, assuming a suboptimal network being trained offline given sufficient training data. Our approach here does not impose such a restriction, although a pretrained network (including model selection) with an offline data set will be intuitively helpful.

In the remainder of this section, two different approaches to the problem of option price prediction are presented.

Generic Approach. In a generic approach, we use a *time-varying* nonparametric model (i.e., MLP network) to track the stochastic dynamics. We use strike price X and maturity time T as two inputs (with appropriate normalization preprocessing) feeding an MLP with architecture net2-6-2 (two inputs, two outputs, and six hidden units), where the two outputs correspond to the call option and put option prices. We have tried different option data and compared the sampling-based ALOPEX with the EKF and HySIR algorithms [208]. The specific parameters for this task are $\sigma = 0.8$, $\alpha = -0.7$. Using $N_p = 50$ particles, the Monte Carlo average results are summarized in Table 7.3. Generally, when the number of particles is increased, the prediction performance is also improved. The prediction curves (of one trial) of call and put option prices for the strike price data 3125 and 3325 with Algorithm 2 are shown in Figure 7.21, respectively. As seen from the figure, the sampling-based ALOPEX produces a reasonable tracking trajectory of the highly nonstationary price data, though the exact prediction results are not very accurate.

From Table 7.3, it is observed that the modified ALOPEX-B fails to track the sequential data; the performance of sampling-based ALOPEX is significantly better

Table 7.3 Comparative Experimental Results of Option Pricing Prediction

Algorithm	data 2925	data 3025	data 3125	data 3225	data 3325
ALOPEX-B	0.2891	0.2231	0.1921	0.1837	0.1071
Algorithm 1	0.0403	0.0404	0.0383	0.0352	0.0242
Algorithm 2	0.0399	0.0395	0.0366	0.0310	0.0231
EKF	0.0408	0.0396	0.0401	0.0307	0.0215
HySIR	0.0389	0.0379	0.0369	0.0293	0.0194

Note: The values in the table are averaged one-step-ahead prediction MSE based on 20 Monte Carlo runs with different initial randomseeds.

than ALOPEX-B, close to or slightly better than EKF, and slightly worse than the HySIR algorithm. Under the same conditions, the HySIR algorithm's complexity ($\mathcal{O}(N_p N^2 N_{\text{out}})$, where N_{out} denotes the number of MLP output neurons [208]) and CPU time, however, are much greater than that of the sampling-based ALOPEX [$\mathcal{O}(N_p N)$]. In terms of CPU time, the sampling-based ALOPEX procedures need slightly more time per step than the EKF for this task. Nevertheless, it is expected that, when the size and structural complexity of the neural network are increased, the sampling-based ALOPEX may exhibit a greater computational advantage. It may thus be said that the proposed sampling-based ALOPEX procedures provide a good trade-off between performance and computational complexity for tracking the option price tendency. In addition, they are also amenable to parallel implementation.

Data Driven Approach. In terms of financial data prediction, it is often beneficial to explore the structural properties of the data, even if the data are of limited size. For the financial data at hand, we also investigate another data-driven predictive model. Under certain assumptions, (7.8) can be simplified by normalizing the call option price C and stock price S by the strike price X; in particular, we have[4]

$$\frac{C}{X} = f\left(\frac{S}{X}, T\right). \tag{7.9}$$

The correlation analysis between C/X and S/X and normalized T is shown as a scatter plot in Figure 7.22. In the data-driven approach, we use an MLP net2-4-1 to model the dynamics (7.9) and test the tracking performance of Algorithm 2. Using 50 particles, one prediction curve for the call option prices is shown in Figure 7.23. Compared to the generic approach (see Figure 7.21), the data-driven approach appears to produce more accurate prediction results.

7.3.4 Online System Identification

Next, we test the sampling-based ALOPEX for the system identification problem [568, 839]. The purpose of this experiment is to illustrate the suitability of

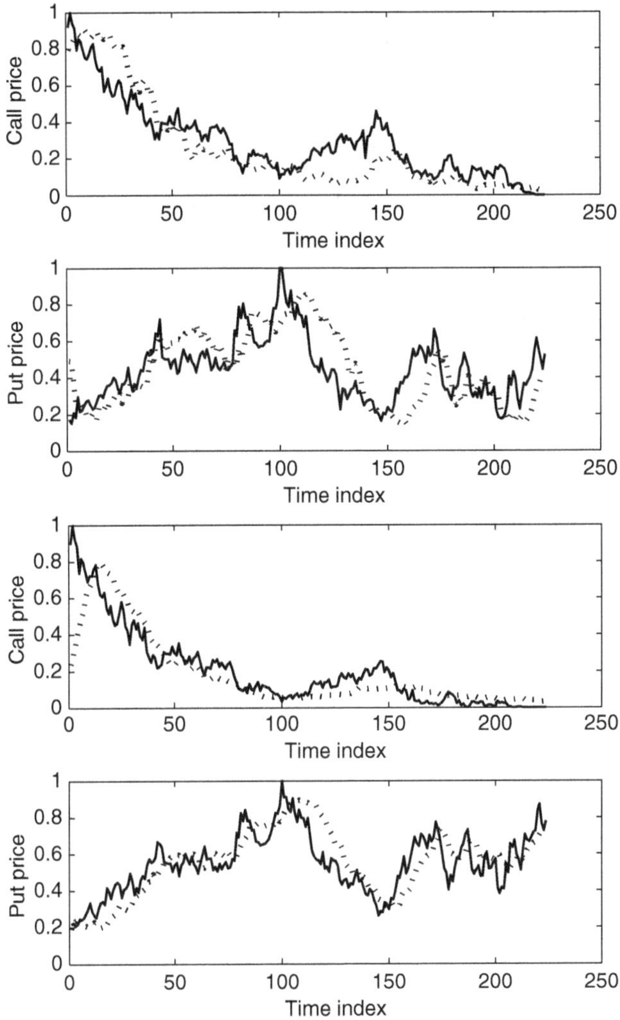

Figure 7.21 Call and put option prices prediction curves (top two panels: strike price data 3125; *bottom two panels:* data 3325) produced by Algorithm 2 in one Monte Carlo run (solid line: true value; dotted line: predicted value). (Copyright © 2004 by IEEE. Adapted, with permission, from S. Haykin, Z. Chen, and S. Becker, Stochastic correlative learning algorithms, *IEEE Transactions on Signal Processing,* Vol. 52, No. 8, pp. 2200–2209, August 2004.)

the proposed sampling-based ALOPEX for an online black-box (neural network) modeling approach. See Figure 7.24 (left panel) for an illustration.

Let us consider a two-link robot arm system the solid and dashed lines in the right panel of Figure 7.24 show the "elbow-up" and "elbow-down" situation, respectively. For a given pair of angles (α_1, α_2), the end-effector position of the

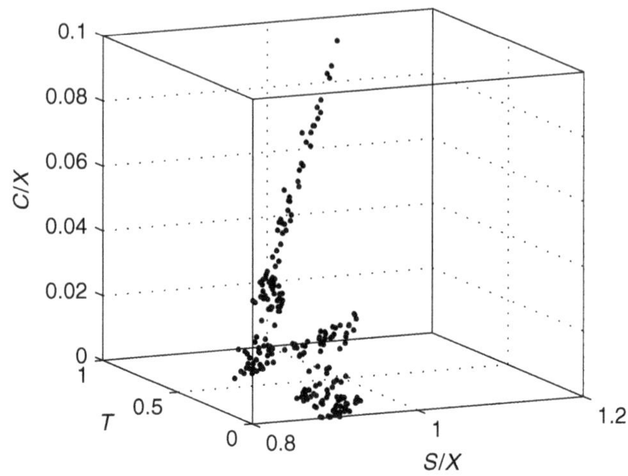

Figure 7.22 Scatter plot of C/X, S/X, and normalized maturity time T for strike price data 3325. (Copyright © 2004 by IEEE. Adapted, with permission, from S. Haykin, Z. Chen, and S. Becker, Stochastic correlative learning algorithms, *IEEE Transactions on Signal Processing*, Vol. 52, No. 8, p. 470, August 2004.)

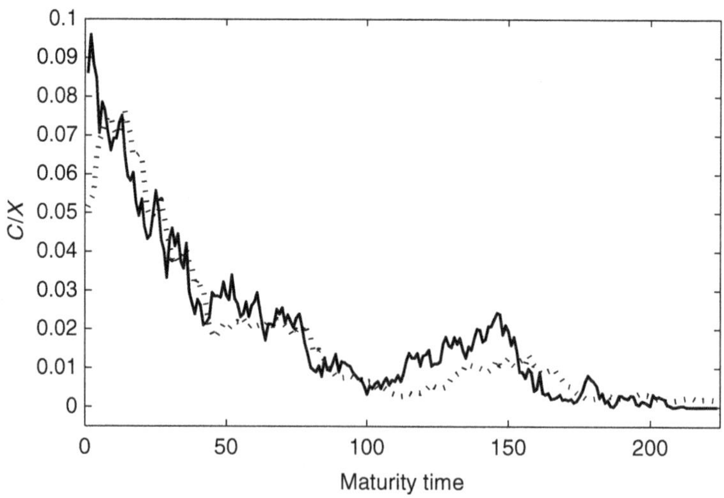

Figure 7.23 The C/X prediction curve (for strike price data 3225) produced by Algorithm 2. Solid line: true value; dotted line: predicted value (Copyright © 2004 by IEEE. Adapted, with permission, from S. Haykin, Z. Chen, and S. Becker, Stochastic correlative learning algorithms, *IEEE Transactions on Signal Processing*, Vol. 52, No. 8, p. 470, August 2004.)

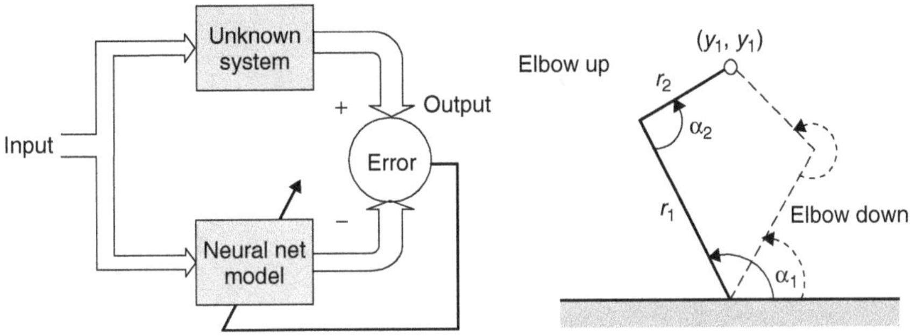

Figure 7.24 *Left panel*: block diagram of system identification using a black-box modeling approach. *Right panel*: two-link robot arm. (Copyright © 2004 by IEEE. Adapted, with permission, from S. Haykin, Z. Chen, and S. Becker, Stochastic correlative learning algorithms, *IEEE Transactions on Signal Processing*, Vol. 52, No. 8, p. 470, August 2004.)

robot arm is determined whose system is described by the Cartesian coordinates

$$\begin{aligned} y_1 &= r_1 \cos(\alpha_1) - r_2 \cos(\alpha_1 + \alpha_2), \\ y_2 &= r_1 \sin(\alpha_1) - r_2 \sin(\alpha_1 + \alpha_2), \end{aligned}$$

where $r_1 = 0.8$, $r_2 = 0.2$, $\alpha_1 \in [0.3, 1.2]$, and $\alpha_2 \in [\pi/2, 3\pi/2]$. Finding the mapping from (α_1, α_2) to (y_1, y_2) is referred to as *forward kinematics*. Reformulating the system dynamics in a state-space form so as to obtain sequential data for the problem at hand, we may write

$$\begin{aligned} \mathbf{x}_{t+1} &= h(\mathbf{x}_t) + \mathbf{w}_t, \\ \mathbf{y}_t &= \begin{bmatrix} \cos(\alpha_{1,t}) & -\cos(\alpha_{1,t} + \alpha_{2,t}) \\ \sin(\alpha_{1,t}) & -\sin(\alpha_{1,t} + \alpha_{2,t}) \end{bmatrix} \begin{bmatrix} r_1 \\ r_2 \end{bmatrix} + \mathbf{v}_t, \end{aligned}$$

where $h(\cdot)$ is a piecewise linear function, $\mathbf{x} = [\alpha_1, \alpha_2]^{\mathrm{T}}$, $\mathbf{y} = [y_1, y_2]^{\mathrm{T}}$, and the noise vectors are chosen as $\mathbf{w}_t \sim \mathcal{N}(\mathbf{0}, \mathrm{diag}\{0.008^2, 0.08^2\})$, $\mathbf{v}_t \sim \mathcal{N}(\mathbf{0}, 0.005 \times \mathbf{I})$. The task of system identification is to train a neural network, given the input–output pairs, to learn the underlying robot arm dynamics and to provide a predictive model for the dynamics. A total set of 630 pairs of input–output data is constructed, where the input sequence follows a piecewise linear dynamics subject to a Gaussian noise perturbation. In order to track the system dynamics, we apply Algorithm 2 to train a two-layer MLP net2-6-2, using 20 particles. The system identification results are shown in Figure 7.25. As shown in the figure, the network quickly tracks the system dynamics, roughly within about 50 iterations.

7.3.5 Summary

In this section, we applied the Monte Carlo sampling-based ALOPEX procedures developed in Chapter 6 for online financial data prediction and system identification problems. As observed in the experiments, the incorporation of a sequential

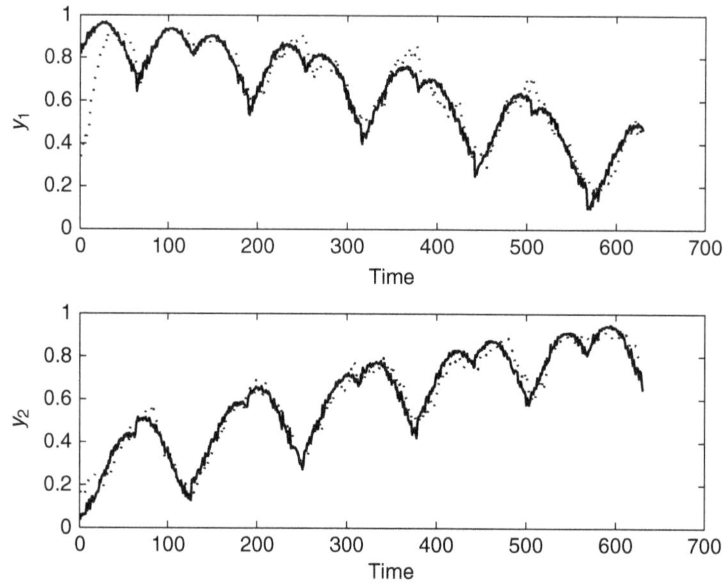

Figure 7.25 Comparison of the predicted (dotted line) and true (solid line) trajectories. (Copyright © 2004 by IEEE. Adapted, with permission, from S. Haykin, Z. Chen, and S. Becker, Stochastic correlative learning algorithms, *IEEE Transactions on Signal Processing*, Vol. 52, No. 8, p. 470, August 2004.)

Monte Carlo simulation (or particle-filtering) procedure allows us to boost the performance of the conventional ALOPEX, in particular for tackling the online (sequential) data. Our Monte Carlo optimization method presents a computational trade-off between complexity and performance (or convergence speed). By combining the gradient-free ALOPEX procedure with sequential Monte Carlo sampling, the proposed algorithms may find their niches in many real-life engineering applications. The simplicity of these algorithms also allows the possibility for a parallel implementation in hardware. Although here we have merely discussed the online learning problem, the sampling-based ALOPEX is also applicable for offline (batch) regression and classification problems [163].

7.4 KALMAN FILTERING IN COMPUTATIONAL NEURAL MODELING

7.4.1 Background

The time-domain description of a system by a State-Space Model (SSM), depicted in Figure 7.26, is of profound importance. The notion of *state* plays a key role in the formulation of this model. The state, denoted by the vector $\mathbf{x}(t)$, is defined as any set of quantities that would be sufficient to uniquely describe the unforced dynamic behavior of the system at discrete time t. The model of Figure 7.26 is

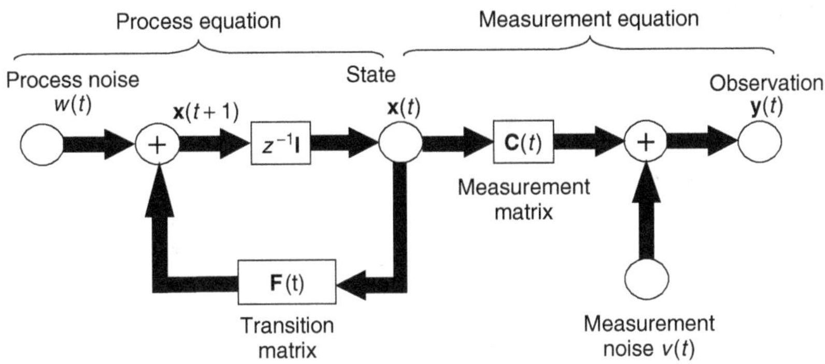

Figure 7.26 Signal-flow graph representation of a linear, discrete-time dynamical system.

not only mathematically convenient but also offers a close relationship to physical/neurobiological reality and a basis for accounting for the statistical behavior of the system.

In a special linear form, the SSM can be described by two basic equations as follows:

- *Process equation*:

$$\mathbf{x}(t+1) = \mathbf{F}(t)\mathbf{x}(t) + \mathbf{w}(t), \tag{7.10}$$

where $\mathbf{F}(t)$ is a transition matrix for the state (from time t to $t+1$) and the vector $\mathbf{w}(t)$ denotes additive dynamic noise.

- *Measurement equation*:

$$\mathbf{y}(t) = \mathbf{C}(t)\mathbf{x}(t) + \mathbf{v}(t), \tag{7.11}$$

where the vector $\mathbf{y}(t)$ denotes the observation, $\mathbf{C}(t)$ is a measurement matrix, and the vector $\mathbf{v}(t)$ denotes additive measurement noise.

According to this model, the state $\mathbf{x}(k)$ is hidden and therefore unknown, and the goal is to estimate it using the sequence of observations $\mathcal{Y}_t = \{\mathbf{y}(1), \ldots, \mathbf{y}(t)\}$. The sequential estimation problem is called *filtering* if $k = t$, *prediction* if $k > t$, and *smoothing* if $1 < k < t$. Unlike smoothing, both filtering and prediction are real-time operations.

In a classic paper, Kalman [461] derived a general solution for the linear filtering problem, and with it the celebrated Kalman filter was born.[5] The essence of Kalman filtering lies in a closed-loop form of a *predictor–corrector*, which contains the *time update* [equations (7.12a) and (7.12b)] and *measurement update* [equations (7.12d) and (7.12e)]:

$$\hat{\mathbf{x}}(t|\mathcal{Y}_{t-1}) = \mathbf{F}(t-1)\hat{\mathbf{x}}(t-1|\mathcal{Y}_{t-1}), \tag{7.12a}$$

$$\mathbf{P}(t|t-1) = \mathbf{F}(t-1)\mathbf{P}(t-1|t-1)\mathbf{F}^{\mathrm{T}}(t-1) + \mathbf{\Sigma}_{\mathbf{w}}, \tag{7.12b}$$

$$\mathbf{G}(t) = \mathbf{P}(t|t-1)\mathbf{C}^{\mathrm{T}}(t)\left[\mathbf{C}(t)\mathbf{P}(t|t-1)\mathbf{C}^{\mathrm{T}}(t) + \mathbf{\Sigma}_{\mathbf{v}}\right]^{-1}, \tag{7.12c}$$

$$\hat{\mathbf{x}}(t|\mathcal{Y}_{t}) = \hat{\mathbf{x}}(t|\mathcal{Y}_{t-1}) + \mathbf{G}(t)\left[\mathbf{y}(t) - \mathbf{C}(t)\hat{\mathbf{x}}(t|\mathcal{Y}_{t-1})\right], \tag{7.12d}$$

$$\mathbf{P}(t|t) = \mathbf{P}(t|t-1) - \mathbf{G}(t)\mathbf{C}(t)\mathbf{P}(t|t-1), \tag{7.12e}$$

where $\mathbf{\Sigma}_{\mathbf{w}}$ and $\mathbf{\Sigma}_{\mathbf{v}}$ are the covariance matrices of the zero-mean dynamic and measurement noise processes, respectively; $\mathbf{P}(t|t-1)$ and $\mathbf{P}(t|t)$ denote error covariance matrices of the predicted and filtered estimates of the state, respectively; $\mathbf{G}(t)$ in (7.12c) is known as the *Kalman gain* that is used for computing the measurement correction; and the error vector $\mathbf{e}(t) = \mathbf{y}(t) - \mathbf{C}(t)\hat{\mathbf{x}}(t|\mathcal{Y}_{t-1})$ is called the *innovation* [457, 461]. Equation (7.12d) can be viewed as an error-correcting learning rule, in which the Kalman gain plays the role of an adaptive modulation factor. Notably, under the assumption that the dynamic noise and measurement noise are uncorrelated, white Gaussian processes, the Kalman filter is a recursive estimator that is optimum in the minimum MSE or, equivalently, maximum-likelihood sense [440].[6]

Because of its mathematical elegance and the recursive estimation nature, the Kalman filter has been widely used in engineering (signal processing, control, communications, etc.), machine learning, as well as computational neuroscience. In what follows, we will give a short overview of the use of the Kalman filter in neuroscience for modeling some brain functions.

7.4.2 Overview of Kalman Filter in Modeling Brain Functions

Dynamic Model of Visual Recognition. As discussed in Chapter 1, the visual cortex contains a hierarchically layered structure (from V1 to V5) and massive interconnections within the cortex and between the cortex and the visual thalamus (i.e., LGN). Specifically, the visual cortex is endowed with two key anatomical properties:

- *Abundant Use of Feedback.* The connections between any two connected areas of the visual cortex are bilateral, thereby accommodating the transmission of forward as well as feedback signals between the interconnected cortical areas.
- *Hierarchical Multiscale Structure.* The RF of lower area cells in the visual cortex span only a small fraction of the visual field, whereas the RFs of higher area cells increase in size until they span almost the entire visual field. It is this constrained network structure that makes it possible for the fully connected visual cortex to perform prediction in a high-dimensional data space with a reduced number of free parameters and therefore in a computationally efficient manner.

In a series of studies, Rao and Ballard [749–751] exploited these two properties of the visual cortex to build a dynamic model of visual recognition, recognizing that vision is fundamentally a nonlinear dynamic process. The Rao–Ballard model of visual recognition is a hierarchically organized neural network with each intermediate level of the hierarchy receiving two kinds of information: bottom-up information from the preceding level and top-down information from the higher level. For its implementation, the model uses a multiscale estimation algorithm that may be viewed as a hierarchical form of the EKF. In particular, the Kalman filter is used to simultaneously learn the feedforward, feedback, and prediction parameters of the model using visual experiences in a dynamic environment. The resulting adaptive processes operate at two different timescales:

- A *fast* dynamic state estimation process, which allows the dynamic model to anticipate incoming stimuli
- A *slow* Hebbian learning process, which provides for synaptic weight adjustments in the model

Specifically, the Rao–Ballard model can be viewed as a neural network implementation of the EKF that employs top-down feedback between layers, which is able to learn the visual RFs for both static images and time-varying image sequences. The dynamic internal model introduced by Rao and Ballard is very appealing in that it is simple, flexible, yet powerful and it allows a Bayesian interpretation of visual perception [490, 541, 754].

Dynamic Model for Sound Stream Segregation. As is well known in the computational neuroscience literature, auditory perception shares many common features with visual perception (e.g., [822]). Specifically, Elhilali [257] addressed the problem of sound stream segregation within the framework of computational auditory scene analysis (CASA). In the computational model therein, the hidden vector contains an internal (abstract) representation of sound streams; the observation is represented by a set of feature vectors or acoustic cues (e.g., pitch, onset) derived from the sound mixture. Since temporal continuity in sound streams is an important clue, it can be used to construct the process equation. The measurement equation describes the cortical filtering process with the cortical model's parameters. The basic component of dynamic sound stream segregation is twofold: First, infer the distribution of sound patterns into a set of streams at each time instant; second, estimate the state of each cluster given the new observations. The second estimation problem is solved by a Kalman-filtering operation, and the first clustering problem may be solved by a Hebb-like competitive learning operation. In a simple figure–ground perception setup, the sound stream of interest is clustered and extracted as the "figure" while the rest of the sound streams all fall into the "background" of the auditory scene.

The dynamic nature of the Kalman filter is important not only for sound stream segregation but also for sound localization and tracking, all of which are regarded as the key ingredients for *active audition* [373].

Dynamic Models for Cerebellum and Motor Learning. The cerebellum has an important role to play in the control and coordination of movements which are ordinarily carried out in a very smooth and almost effortless manner. In the literature, it has been suggested that the cerebellum plays the role of a controller or the neural analog of a dynamic state estimator. The key point in support of the dynamic state estimation hypothesis is embodied in the following statement, the validity of which has been confirmed by decades of work on the design of automatic tracking and guidance systems:

> Any system, be it a biological or artificial system, required to predict and/or control the trajectory of a stochastic multivariate dynamic system, can only do so by using or invoking the essence of Kalman filtering in one way or another.

Building on this key point, Paulin [710] presents several lines of evidence that favor the hypothesis that the cerebellum is a neural analog of a dynamic state estimator. A particular line of evidence presented therein relates to the vestibular–ocular reflex (VOR), which is part of the oculomotor system. The function of the VOR is to maintain visual (i.e., retinal) image stability by making eye rotations that are opposite to head rotations. This function is mediated by a neural network that includes the cerebellar cortex and vestibular nuclei. Now, from modern control theory we know that a Kalman filter is an optimum linear system with minimum variance for predicting the state trajectory of a dynamic system using noisy measurements; it does so by estimating the particular state trajectory that is most likely given an assumed model for the underlying dynamics of the system. A consequence of this strategy is that, when the dynamic system deviates from the assumed model, the Kalman filter produces estimation errors of a predictable kind, which may be attributed to the filter believing in the assumed model rather than the actual sensory data. According to Paulin [710], estimation errors of this kind are observed in the behavior of the VOR.

The human motor system involves various computational tasks such as motor control, motor coordination, control, planning, prediction, and learning (for excellent reviews of computational issues in motor control and learning, the reader is referred to [449, 885, 971]). In modeling the sensorimotor loop, Wolpert and colleagues [972] proposed the Kalman filter for sensorimotor integration. Typically, the hidden state in the motor system involves parameters related to movement, such as the direction of movement, velocity, acceleration, posture, and joint torques. The Kalman filter combines the forward model and the sensory feedback to predict or estimate the state of interest; and the objective of the filter is to compensate for sensorimotor delays and to reduce the uncertainty in the state estimate that arises from the noise inherent in both sensory and motor signals. In addition, by predicting future states and sensory feedback, the model can reduce the effects of feedback delays in sensorimotor loops or can provide a mechanism for determining whether a movement is self-produced or produced externally [971].

Dynamic Model for Hippocampus. In the field of computational neuroscience, an important component of hippocampal function is spatial learning and localization. The hypothesis that the hippocampus represents a cognitive map [682] requires that the place cells of the hippocampus form an integrated neural representation of space, and the plasticity (size and shape) of the place fields allows them to adapt as the position in the environment changes. This is much like a mobile robot navigating in the field that requires continuous map localization.

In [107, 108], Bousquet et al. proposed a computational hippocampal model for animals such as rats which conducts Kalman filtering. Specifically, the state vector was defined to contain the estimated centers of places encountered by the animal that is represented in CA1; the animal's dead-reckoning system, being a system model in the process equation, predicts the new position of the animal based on its previous position estimate and actual animal motion; the measurement equation describes the spatial relationship between the estimated position of the animal and the center of the current place. The *predictor–corrector* framework allows the hippocampus to localize and learn sequentially the spatial positions and associate them with the dead-reckoning estimate, even in the presence of perceptual aliasing.

In an independent study, Lörincz and Buzsáki [571] also suggested the role of Kalman filtering in modeling the entorhinal–hippocampal loop (recall Figure 1.15). Specifically, it was suggested in their computational model that the entorhinal cortex (EC) compares the difference between neocortical representations (primary input) and the feedback information conveyed by the hippocampus (the "reconstructed input"), and the error initiates plastic changes in the hippocampal networks (error compensation), which is achieved by predictive structures, such as the CA3 recurrent network and EC–CA1 connections; alteration of intrahippocampal connections further gives rise to a new hippocampal output; the hippocampus generates separated (independent) outputs that are used to train long-term memory traces in the EC.

To summarize, the "predictor–corrector" nature of the Kalman filter lends itself as a good candidate for *predictive coding* in computational neural modeling, which is a fundamental property for the autonomous brain functions in a dynamic environment. It is also important to note that in the above examples the hypothesis that the neural system (hippocampus, cerebellum, or neocortex) is a neural analog of a Kalman filter is not to be taken to imply that, in physical terms, the neural system resembles a Kalman filter. Rather, in general, biological systems need to do some form of state estimation, and the pertinent neural algorithms may have the general flavor of a Kalman filter. Many brain functions that were discussed here (summarized in Table 7.4) seem to be possible candidates for performing such computations. Moreover, some form of state estimation is quite likely broadly distributed throughout other parts of the central nervous system.

In addition, it is noteworthy that the use of Kalman filter in computational neural modeling is not limited by sequential state estimation; it can also be used for parameter estimation of a model (such as a neural network) or estimation of both [367]. In the following, we will present an example of using a Kalman filter for training a recurrent neural network in a visual recognition application [709].

Table 7.4 Examples of Kalman Filter in Computational Neural Modeling of Brain Functions

	Visual	Auditory	Motor	Hippocampus
State	Visual RFs	Sound patterns	Movement para.	Positions of place field
Observation	Retinal images	Acoustic cues	Sensory inputs	Visual cue of positions
Function	Dynamic vision	Stream segregation	Control	Localization of spatial maps

7.4.3 Kalman Filter for Learning Shape and Motion from Image Sequences

Motivation of Computational Neural Model. The architecture of our computational neural model proposed here is motivated by two key anatomical features of the mammalian neocortex, the extensive use of feedback connections, and the hierarchical multiscale structure.

Feedback is a ubiquitous feature of the brain, both between and within cortical areas. Whenever two cortical areas are interconnected, the connections tend to be bidirectional [274]. Additionally, within every neocortical area, neurons within the superficial layers are richly interconnected laterally via a network of horizontal connections [576]. The dense web of feedback connections within the visual system has been shown to be important in suppressing background stimuli and amplifying salient or foreground stimuli [419]. Feedback is also likely to play an important role in processing sequences. Clearly, we view the world as a continuously varying sequence rather than as a disconnected collection of snapshots. Seeing the world in this way allows recent experience to play a role in the anticipation or prediction of what will come next. The generation of predictions in a perceptual system may serve at least two important functions: first, to the extent that an incoming sensory signal is consistent with expectations, intelligent filtering may be done to increase the SNR and resolve ambiguities using context; second, when the signal violates expectations, an organism can react quickly to such changing or salient conditions by deemphasizing the expected part of the signal and devoting more processing capacity to the unexpected information. Top-down connections between processing layers or lateral connections within layers or both might be used to accomplish this. Lateral connections allow for local constraints about moving contours to guide the expectations.

Prediction in a high-dimensional space is computationally complex in a fully connected network architecture. The problem requires a more constrained network architecture that will reduce the number of free parameters. The visual system has done just that. In the earliest stages of processing, cells' RFs span only a few degrees of visual angle, while in higher visual areas cells' RFs span almost the entire visual field [690]. Consequently, this feature should be taken into account when designing our computational neural model (e.g., [534]).

Model Description. Prediction in a high-dimensional sensory data space, such as a 50-pixel image, using a fully connected recurrent network is not feasible, because the number of connections is typically one or more orders of magnitude larger than the dimensionality of the input, and the so-called *node-decoupled extended Kalman filter* (NDEKF) algorithm [273, 367] requires adapting these unknown parameters for typically hundreds to thousands of iterations. The problem requires a more constrained network architecture that would reduce the number of free parameters. Motivated by the hierarchical architecture of real visual systems, we designed our model network with a similar hierarchical architecture in which the first layer of units was connected to relatively small, local 5×5 pixel regions of the image and a subsequent layer spanned the entire visual field.

A four-layer recurrent network of architecture net100-16-8R-100, as depicted in Figure 7.27, was used in our experiments. Training images of size 10×10 which are arranged in a vector format of size 100×1 were used to form the input to the the networks. As shown in Figure 7.27*a*, the input image is divided into 4 nonoverlapping RFs of size 5×5. Further, the 16 units in the first hidden layer are divided into 4 banks of 4 units each. Each of the 4 units within a bank receive

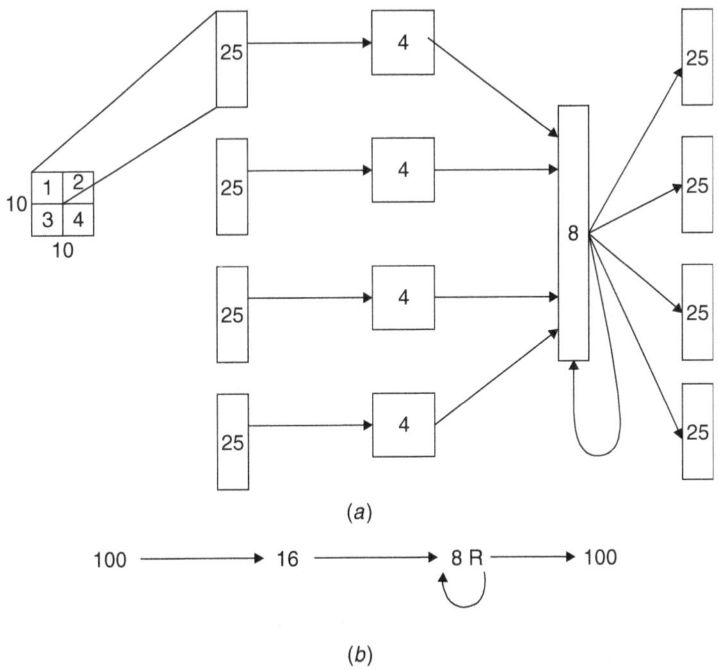

Figure 7.27 Diagram of the recurrent network used in the experiment. The numbers in the boxes indicate the number of units in each layer or module, except in the input layer, where the RFs are numbered 1–4. Local RFs of size 5×5 at the input are fed to the 4 banks of 4 units in the first hidden layer. The second layer of 8 units then combines these local features learned by the first hidden layer. (Reprinted from (709) with permission. Copyright © 2001 by Wiley.)

inputs from one of the 4 RFs. This describes how the 10×10 image is connected to the 16 units in the first hidden layer. Each of these 16 units feeds into a second hidden layer of 8 units. The second hidden layer has recurrent connections (note that recurrence is only within the layer but not between layers).

Thus, the input layer of the network is connected to small and local regions of the image. The first layer processes these local RFs separately in an effort to extract relevant local features. These features are then combined by the second hidden layer to predict the next image in the sequence. The predicted image is represented at the output layer. The prediction error is then used in the EKF equations to update the weights. This process is repeated over several epochs through the training image sequences until a sufficiently small incremental MSE is obtained.

Experiment 1. In the first experiment, the model is trained on images of two different moving shapes, where each shape has its own characteristic movement, namely, shape and direction of movement are perfectly correlated. The sequence of eight 10×10 pixel images in Figure 7.28*a* is used to train a four-layered (100-16-8R-100) network to make one-step predictions of the image sequence. In the first four time steps a circle moves upward within the image, and in the last four time steps a triangle moves downward within the image. At each time step, the network is presented with one of the eight 10×10 images as input (divided into

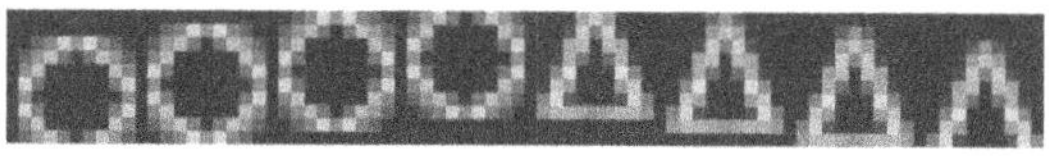

(*a*)

(*b*)

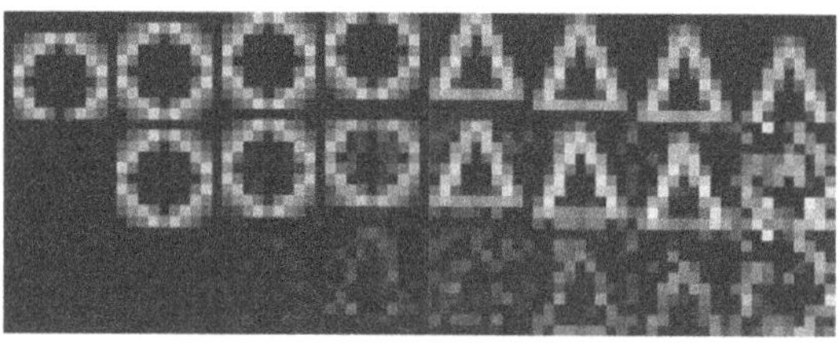

(*c*)

Figure 7.28 Experiment 1: one-step and iterated prediction of image sequence. (*a*) Training sequence. (*b*) One-step prediction. (*c*) Iterated prediction. In (*b*) and (*c*), the three rows correspond to the input, prediction, and error, respectively. (Reprinted from [709] with permission. Copyright © 2001 by Wiley.)

four 5×5 RFs as described above) and generates in its output layer a prediction of the input at the next time step, but it is always given the correct input at the next time step. Training was stopped after 20 epochs through the training sequence. Figure 7.28*b* shows the network operating in one-step prediction mode on the training sequence after training. It makes excellent predictions of the object shape and also its motion. Figure 7.28*c* shows the network operating in an autonomous mode after being shown only the first image of the sequence. In this multistep prediction case, the network is only given external input at the first time step in the sequence. Beyond the first time step, the network is given its prediction from time $t - 1$ as its input at time t, which could potentially lead to a buildup of prediction errors over many time steps. This shows that the network has reconstructed the entire dynamics, to which it was exposed during training, when provided with only the first image. This is indeed a difficult task. It is seen that as the iterative prediction proceeds the residual errors (third row in Figure 7.28c) are amplified at each step.

Experiment 2. Next, a network with the same architecture net100-16-8R-100 used in experiment 1 was trained with three sequences, each consisting of four images, in the following order:

- Circle moving right and up (cru)
- Triangle moving right and down (trd)
- Square moving right and up (sru)

During training, at the beginning of each sequence, the network states were initialized to zero, so that the network would not learn the order of presentation of the sequences. The network was therefore expected to learn the motions associated with each of the three shapes and not the order of presentation of the shapes.

During testing, the order of presentation of the three sequences varied, as shown in Figure 7.29*a*. The trained network does well at the task of one-step prediction, only failing momentarily at transition points where we switch between sequences. It is important to note that one-step prediction, in this case, is a difficult and challenging task because the network has to determine (i) what shape is present and (ii) which direction it is moving in without direct knowledge of inputs some time in the past. In order to make good predictions, it must rely on its recurrent or feedback connections, which play a crucial role in the present model.

We also tested the model on a set of occluded images—images with regions that are intentionally blanked. Remarkably, the network makes correct one-step predictions, even in the presence of occlusions, as shown in Figure 7.29*b*. In addition, the predictions do not contain occlusions, that is, they are correctly filled in, demonstrating the robustness of the model to occlusions. In Figure 7.29*c*, when the network is presented with sequences that it had not been exposed to during training, a larger residual error is obtained, as expected. However, the network is still capable of identifying the shape and motion, although not as accurately as before.

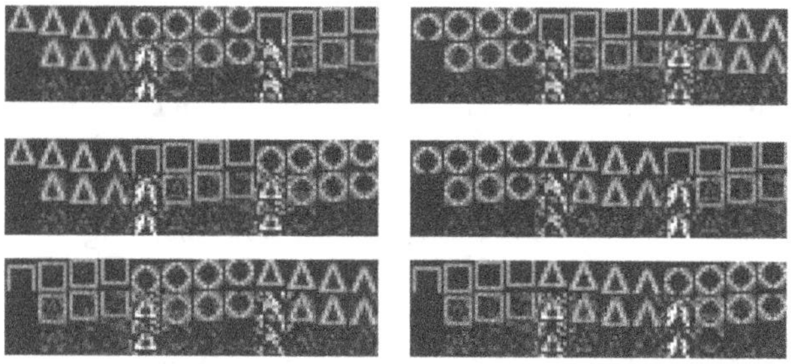

(*a*) Various combinations of sequences used in training

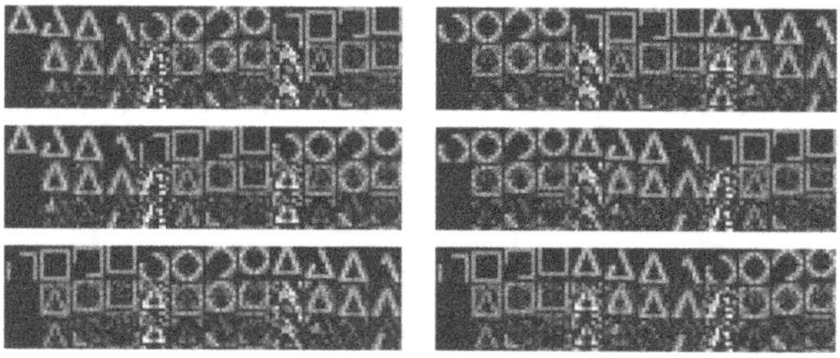

(*b*) Same sequences as in (*a*) but with occlusions

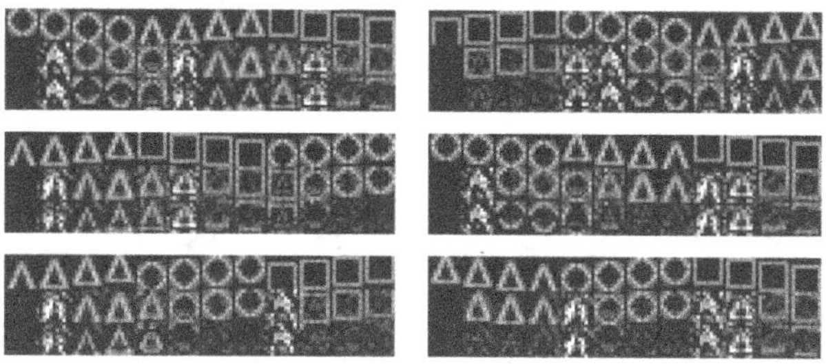

(*c*) Predicition on some sequences not seen during training

Figure 7.29 Experiment 2 one-step prediction of image sequences using the trained recurrent network. The three rows in each image correspond to the input, prediction, and error, respectively. (Reprinted from (709) with permission. Copyright © 2001 by Wiley.)

Experiment 3. In experiment 1, the network was presented with short sequences (four images) of only 2 shapes (circle and triangle), and in experiment two an extra shape (square) was added. In experiment 3, to make the learning task even more challenging, the length of the sequences was increased to 10 and the restriction of one direction of motion per shape was lifted. Specifically, each shape was permitted to move right and either up or down. Thus, the network was exposed to different shapes traveling in similar directions and also the same shape traveling in different directions, increasing the total number of images presented to the network from 8 images in experiment 1 and 12 images in experiment 2 to 100 images in this experiment. In effect, there is a substantial increase in the number of learning patterns and thus a substantial increase in the complexity of the learning task. However, since the number of weights in the network is limited and remains the same as in the other experiments, the network cannot simply memorize the sequences.

A network with the same 100-16-8R-100 architecture was trained on six sequences, each consisting of 10 images (see Figure 7.30) in the following order:

- Circle moving right and up (cru)
- Square moving right and down (srd)
- Triangle moving right and up (tru)
- Circle moving right and down (crd)
- Square moving right and up (sru)
- Triangle moving right and down (trd)

Training was performed in a similar manner as done in experiment 2. During testing, the order of presentation of the six sequences was varied; several examples are shown in Figure 7.31. As in the previous experiments, even with the larger number of training patterns, the network is able to predict the correct motion of the shapes, only failing during transitions between shapes. It is also capable of distinguishing between the same shapes moving in different directions as well as different shapes moving in the same direction using context available via the recurrent connections.

The failure of the model to make accurate predictions at transitions between shapes can also be seen in the residual error that is obtained during prediction. The residual error in the predicted image is quantified by calculating the mean-squared

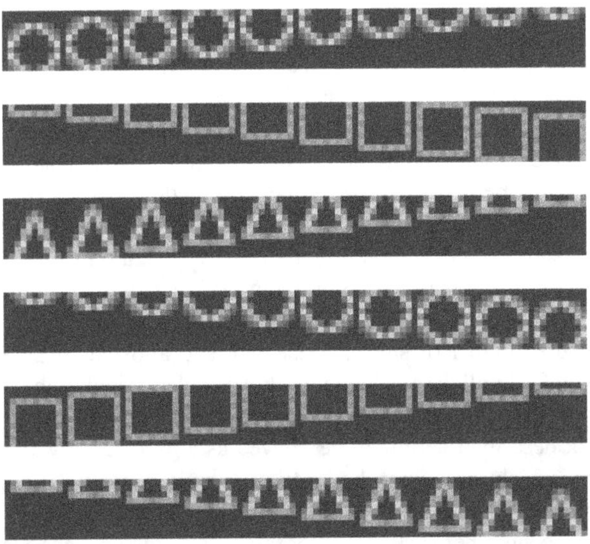

Figure 7.30 Experiment 3: the six image sequences used for training. (Reprinted from (709) with permission. Copyright © 2001 by Wiley.)

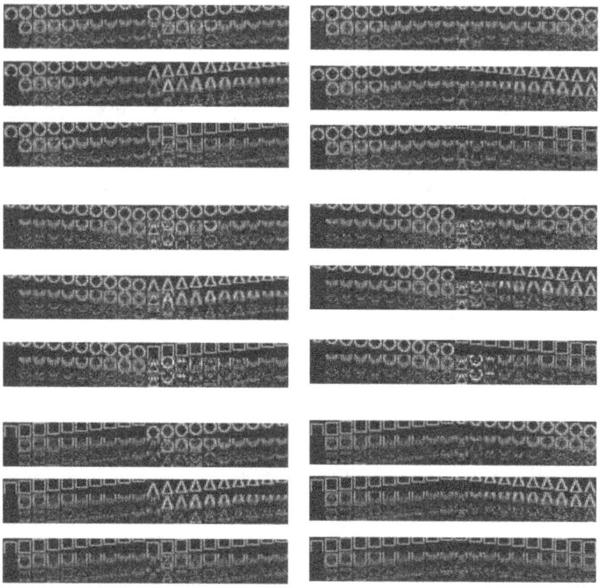

Figure 7.31 Experiment 3: one-step prediction of image sequences using the trained recurrent network. The three rows in each image correspond to the input, prediction, and error, respectively. (Reprinted from (709) with permission. Copyright © 2001 by Wiley.)

prediction error as shown in Figure 7.32. The figure shows how the mean-squared prediction error varies as the prediction continues. Note the transient increase in error at transitions between shapes.

Discussion. In this case study, we have dealt with time series prediction of high-dimensional signals: moving visual images. This situation is much more complicated than a one-dimensional case in that the system has to deal with simultaneous shape and motion prediction. The recurrent neural network model was trained by the EKF method to perform one-step prediction of image sequences in a specific order. In the testing phase, the order of the sequences was varied and the network was asked to predict the correct shape and location of the next image in the sequence. The complexity of the problem was increased from experiment 1 to experiment 3 as we introduced occlusions, increased both the length of the training sequences and the number of shapes presented, and allowed shape and motion to vary independently. In all cases, the network was able to predict the correct motion of the shapes, failing only momentarily at transitions between shapes.

The network described here may be viewed as a first step toward modeling the mechanisms by which the human brain might simultaneously recognize and track moving stimuli. Any attempt to model both shape and motion processing simultaneously within a single network may seem to be at odds with the well-established finding that shape and spatial information are processed in separate pathways of the visual system [631]. An extreme version of this view posits that form-related

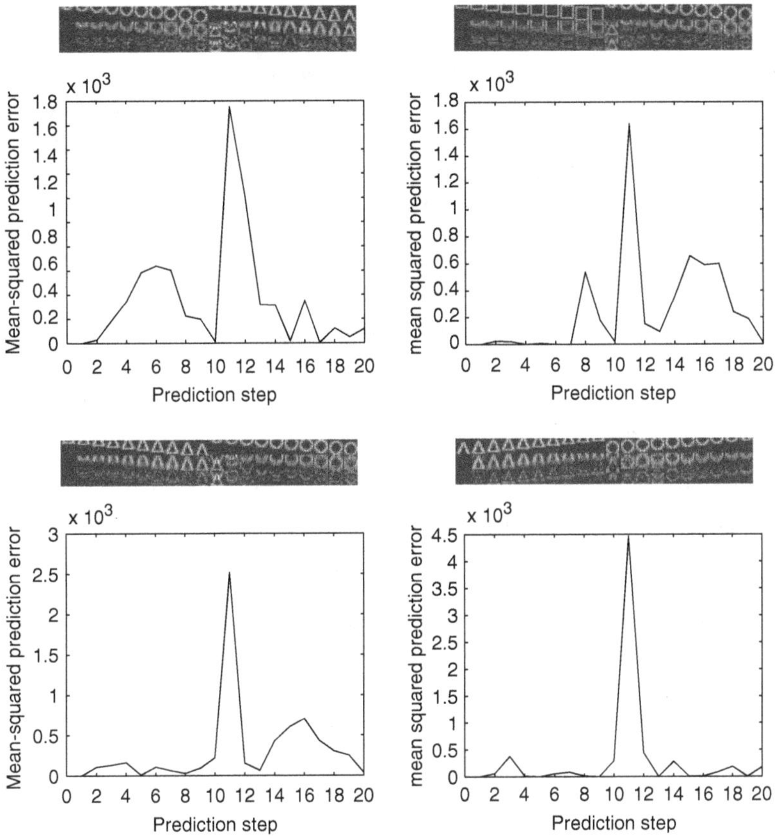

Figure 7.32 Mean-squared prediction error in one-step prediction of image sequences using the trained recurrent network. The three rows in each image correspond to the input, prediction, and error, respectively. The graphs below the images show how the mean-squared prediction error varies as the prediction proceeds. (Reprinted from [709] with permission. Copyright © 2001 by Wiley.)

features are processed strictly by the ventral "what" pathway and motion features are processed strictly by the dorsal "where" pathway. Anatomically, however, there are cross-connections between the two pathways at several points [214]. Furthermore, there is ample behavioral evidence that the processes of shape and motion perception are not completely separate. For example, it has long been established that we are able to infer shape from motion (e.g., [443]). Conversely, under certain conditions object recognition can be shown to drive motion perception [745]. In addition, Stone [858] has shown that viewers are much better at recognizing objects when they are moving in characteristic, familiar trajectories as compared to unfamiliar trajectories. These findings suggest that, when shape and motion are tightly correlated, viewers will learn to use them together to recognize objects. This is exactly what happens in our computational model described here.

To accomplish temporal processing in our computational model, we have incorporated within-layer recurrent connections in the network architecture. Another possibility would be to incorporate top-down recurrent connections. As is well known, a key anatomical feature of the visual system is top-down feedback between visual areas [419]. Top-down connections could allow global expectations about the three-dimensional shape of a moving object to guide predictions. Thus, it would be valuable to extend the model to allow top-down feedback, as suggested in the Rao–Ballard model [750]. Other models of cortical feedback for modeling the generation of expectations have also been proposed (e.g., [356, 643]).

Natural visual systems can deal with an enormous space of possible images under widely varying viewing conditions. It would be useful to extend our computational model to deal with more realistic images. Many additional complexities would arise in natural images that were not present in the artificial image sequences used here. For example, the simultaneous presence of both foreground and background objects may hinder the prediction accuracy. Natural visual systems likely use attentional filtering and binding strategies to alleviate this problem. For example, Moran and Desimone [634] have observed cells that show a suppressed neural response to a preferred stimulus if unattended and in the presence of an attended stimulus. Another simplification of the moving images in our experiments is that shape remained constant for many time frames, whereas for real three-dimensional moving objects the shape projected onto a two-dimensional image may change dramatically over time, because of rotations as well as nonrigid motions (e.g., bending). Humans are able to infer three-dimensional shape from nonrigid motion, even from highly impoverished stimuli such as moving light displays [443]. It is likely that the architecture described here could handle changes in shape provided shape changes predictably and gradually over time.

7.4.4 General Remarks and Implications

As discussed in this section, the Kalman filter (including its variants and nonlinear extensions) is a powerful idea rooted in modern control theory and adaptive signal processing; it has withstood the test of time, having remained highly popular since 1960. Under the ideal conditions of linearity and Gaussianity, Kalman filtering produces an optimal estimate of the hidden state of a dynamic system in either the minimum-variance or maximum-likelihood sense. The state estimation procedure is recursive, which makes it highly amenable to real-time implementation using digital processing.

In the context of neurobiology, the Kalman filter may provide insights into visual recognition [749], motor control [971], and neuronal decoding [976]. One important issue regarding neural implementations of Kalman filtering is its biological plausibility. Specifically, the calculation of Kalman gain involves a matrix-inverse operation, which appears to be an obvious obstacle at the first sight. Then the natural question to ask is how to implement the Kalman filtering operation via local interaction? For an interesting discussion of possible neural implementations of the Kalman filter, the reader is referred to [729]. On the other hand, the brain might not necessarily implement the exact form of Kalman filtering in accordance with equations (7.12a)–(7.12e); rather, there is high likelihood that *approximate* forms of Kalman filtering are performed in certain parts of the brain, with the "predictor–corrector" closed–loop operated recursively.

Finally, with an aim to designing an adaptive system that mimics certain functions of the brain, we are certainly not limited by implausible neurobiological mechanisms; instead, we will build the system by incorporating the strengths of the modern signal processing or machine learning methods. On the one hand, the Kalman filter provides an indispensable tool and an enabling technology for the design of automatic tracking and guidance systems [338]. On the other hand, the Kalman filter can also be used by all means to enhance machine learning (e.g., [870]) or improve the convergence of learning in artificial neural networks (e.g., [367]).

NOTES

1. In general, $\ell \neq \ell'$, where ℓ and ℓ' denote the total number of points in D_1 and D_2, respectively.
2. To avoid the numerical problem in practice, we add a very small value (say, 10^{-16}) to the denominator to prevent overflowing.
3. A *derivative* is a financial instrument whose value replies on some basic cash product. An *option* is a particular type of derivative that gives the holder the right to do something. For example, a *call option* allows the holder to buy a cash product at a specified date in the future. The price at which the option is exercised is known as the *strike price*, while the date at which the option lapses is referred to as the *maturity time*. A *put option* allows the holder to sell the underlying cash product.
4. Theoretically, this normalization is valid at least when the stock returns are independently distributed [420].
5. The continuous-time version of the Kalman filter is also referred to as the Kalman–Bucy filter [462].
6. For details on the Kalman filter and its variants as well as relevant theory, the reader is referred to [338, 369, 459]. Extensions of Kalman filtering to general nonlinear and non-Gaussian scenarios, such as the *unscented Kalman filter* [452] and *particle filter* [225], are discussed in [158, 367].

8

DISCUSSION

> There is no scientific study more vital to man than the study of his own brain. Our entire view of the universe depends on it.
>
> —Francis Crick

8.1 SUMMARY: WHY CORRELATION?

In this monograph, we have proposed that correlative learning constitutes a fundamental basis for both the human brain and adaptive systems. The design and development of the latter are heavily inspired by the efficiency and flexibility of the brain. In describing the essential principles, we have covered a wide range of interdisciplinary topics in computational neuroscience, neural computation, signal processing, and machine learning. Along these lines, we have seen many emergent cross-fertilized ideas and examples motivated from the notion of correlation. Why correlation, and why is it so important? Although it should be clear from the previous chapters, at this point, it is worthwhile to once more summarize the prominent role of correlation; in what follows, our elucidations are structured along three branches: Hebbian plasticity and the correlative brain, correlation-based signal processing, and correlation-based machine learning.

Correlative Learning: A Basis for Brain and Adaptive Systems, by Zhe Chen, Simon Haykin, Jos J. Eggermont, and Suzanna Becker

8.1.1 Hebbian Plasticity and the Correlative Brain

According to Sigmund Freud's philosophy (*Project for a Scientific Psychology*, 1895) [292], a conceptual tenet of modern neuroscience is computation. The computational properties of the brain are a direct consequence of its circuitry, and the computation is carried out within neurons or among the population of neurons through massive numbers of synaptic interconnections. In essence, synaptic plasticity underlies the neuronal mechanism of "learning" or "adaptation" at the microscopic level of the brain. Simply, synaptic plasticity is governed by a correlation-based neuronal mechanism [377]: *When an axon of cell A is near enough to excite a cell B and repeatedly or persistently takes part in firing it, some growth process or metabolic change takes place in one or both cells such that A's efficiency, as one of the cells firing B, is increased.* This now famous conjecture put forward by the McGill University Professor Donald Hebb, now generally known as Hebb's rule, has been cited and modified to appear in countless and diverse publications. More than a half century has elapsed, and it is clear that Hebb's rule has passed the test of time. Simply put, the Hebbian postulate of learning proposes a local correlative rule to adapt the wiring of the neurons that fire together—neurons that fire synchronously acquire accordingly enhanced synaptic strengths. Since his original postulate, Hebb's rule has been repeatedly modified and generalized, as reviewed in Chapter 3. A modern form of Hebbian learning is STDP [89], which was inspired by neurophysiological findings. The temporally asymmetric STDP can yield a differential Hebbian learning rule, where synaptic strengths are changed according to the correlation between the derivatives of the rates instead of the correlation between rates [765, 977]. Temporally asymmetric STDP also connects Hebbian learning with predictive coding within TD learning [752, 753]: If a feature in the synaptic input pattern can reliably predict the occurrence of a postsynaptic spike and seldom comes after a postsynaptic spike, the synapses related to that feature are strengthened, giving that feature more control over the firing of the postsynaptic cell.

At the microscopic level, neuronal synchrony refers to correlated firing among a population of neurons within a short (milliseconds) or long (tens of milliseconds) range. The theories of STDP (e.g., [89, 319, 752, 766]) as well as synfire chains [4] were developed along this line. At the macroscopic level, correlation is a basic computational function exploited by the human brain. Specifically, the brain explores the sensory environment in a multitude of ways and uses the information so gathered to control behavior. More specifically, correlation is used in the formation of topographic maps, detection of events, association of patterns, and recall of memory [241]. The gamma oscillations (30–90 Hz) that were observed in the scalp EEG of various human sensory and cognitive processing tasks (e.g., [235, 503, 836]) clearly indicate precise synchronization of receptive potential generators in the brain (because otherwise the tiny transmembrane currents of the myriads of neurons contributing to the EEG would not summate effectively but would cancel out). The theory of oscillatory correlation has been suggested as a plausible neural basis for feature binding—a central notion in sensory perception, object recognition, attention, and knowledge representation [924, 926], although this is by no means an established fact [774].

Not surprisingly, correlation theory has been applied successfully to model brain functions in sensory (visual, auditory, somatosensory, and olfactory) systems, memory and spatial navigation systems (hippocampus), and motor systems (cerebellum). In Cook's words [183], correlated activities are believed to be prominent at every timescale in the central nervous system: starting from short-term experiences of coincidence detection, novelty detection, perception, learning, and long-term memory to long-term evolution, all of which reflect the ubiquitous nature of correlation in characterizing the intelligence of the human brain.

8.1.2 Correlation-Based Signal Processing

Correlation is a fundamental statistical measure of second-order statistics. In analyzing signals in a dynamic environment, statistical correlation or ensemble correlation characterizes a wide class of (wide-sense) stationary stochastic processes and therefore establishes its nonsubstitutable position in statistical signal processing. In Chapter 2, we have reviewed the roles, both classic and modern, of correlation in signal processing problems, such as spectrum analysis, signal filtering or prediction, matching filters, and correlation detection.

It has been noted that the classic signal processing techniques are often built on the assumptions of stationarity, Gaussianity, and linearity of the studied systems or signals. These assumptions, while sometimes fairly well justified, are frequently violated in reality in the physical world. In order to build a reliable engineering system, insights from mathematics and physics are important [368]; above all, robustness is a central issue. Bearing this goal in mind, modern signal processing techniques will be devoted to developing robust statistical tools for nonstationary, non-Gaussian, and/or nonlinear signals and systems.

Recently, a general research trend in signal processing is to go beyond second-order statistics for statistical estimation or detection. Higher order statistics or information criteria are known for their superior roles in characterizing the statistical dependency between random variables and stochastic processes. Naturally and expectedly, this idea can be universally applied to stochastic filtering, matched filtering, correlation detection, feature extraction, and classification (e.g., [263, 264]).

8.1.3 Correlation-Based Machine Learning

Correlation is essentially a method for seeking the "patterns," while one of the goals of unsupervised learning is to discover the hidden regularity or internal representation of the data, which is characterized by second- or higher order, linear or nonlinear correlations. Many statistical learning algorithms, such as PCA, CCA, SFA, and ICA, are based on this basic principle. Correlation is a measure of *distance* or *similarity* between pairwise random variables; therefore it is naturally used as a quantitative criterion for measuring learning performance. Mutual information can be viewed as a generalized measure of correlation which involves the probability density function and thereby the complete information of moment/cumulant statistics. Information-theoretic learning paradigms are based on optimizing a measure of mutual information, entropy, or information transfer; this class of algorithms is also closely related to the second-order decorrelation-based learning algorithms, which may be viewed as special cases.

Correlation can be viewed as a measure of the inner product between two random variables in a linear space. The kernel method is a powerful tool to extend this concept from linear to nonlinear (potentially high-dimensional) feature space, thereby naturally generalizing the notion of higher order correlation. The essential idea of the kernel methods is to use the so-called kernel trick that calculates the inner product between pairwise data points, thereby sidestepping the direct computation of the outer product in feature space [799]. Kernel methods have intrinsic connections with regularization theory and Gaussian processes, in which a regularization operator and a covariance operator are defined in the functional space, respectively. Unlike other nonlinear correlation-based learning methods, kernel learning implicitly defines the high-dimensional nonlinear features by choosing only a specific kernel function which is free of the risk of overfitting given a small amount of observed samples. We have presented several representative examples in Chapter 4, such as kernel PCA, kernel CCA, kernel discriminant analysis, and kernel Wiener filter, all of which naturally generalize the traditional correlation-based signal processing and statistical analysis tools. It is anticipated that the biologically inspired kernel-based methods (e.g., [801, 831]) will lead to a new realm of signal and pattern analysis in the near future.

8.2 EPILOGUE: WHAT NEXT?

After reading this monograph, we hope the reader will have an appreciation of the importance of correlation and correlative learning in various scientific and engineering fields, especially in the fields of computational neuroscience, signal processing, and machine learning. Now, the next question that naturally arises is: What next, and what will we do about it? Although this is an open-ended question, we would like to pinpoint two important directions for future research.

8.2.1 Generalizing the Correlation Measure

As we refer to correlation throughout the monograph, we mostly constrain ourselves to univariate or multivariate (real- or complex-valued) random variables or random processes; however, the notion of correlation is by no means limited by this assumption. In contrast, it remains challenging to analyze nonvectorial symbols or sequences which nowadays are frequently encountered in many applications, such as texts and webs, biological DNA sequences, and neuronal spike trains. In the meantime, much work still needs to be done for the nontypical discrete-time signals that either have uneven sampling rates or have missing data in the temporal recordings, in which cases conventional correlation analysis has to be modified to accommodate such unfavorable (but quite possible) conditions in practice. It would also be valuable to formulate well-developed measures of correlation or mutual information for random point processes.

On the other hand, as multichannel or cross-modality signal recordings become more popular nowadays, it will be important to address the notion of multifacet correlation, which takes distinct forms across different (e.g., temporal, spatial, and spectral) domains. How to integrate these cross-modality correlations is an important subject of research. It is also desirable to define a multiscale, multitime correlation function [294] that measures the similarity of the event at different time and different scale, for instance, as defined by

$$C_{N,n}^{p,q}(\tau) = \langle x_n^q(0), x_N^p(\tau) \rangle,$$

where n and N denote two scale parameters and p and q denote two order parameters. Such a correlation measure might be important for analyzing fractallike physical or physiological signals, which might also be important for research in computational and traditional neuroscience.

Again, kernel learning theory will continue to play an important role in contributing new insights and tools for analyzing atypical signals and structured data. Essentially, incorporating a priori knowledge into designing problem-specific kernel functions seems like a natural route to pursue. For instance, learning or designing kernel functions to accommodate the nonstationarity is important for temporal signals. Research topics are wide open, especially in an attempt to solve challenging real-life problems in engineering and neuroscience. Above all, the holy grail of researching "learning" is, first, to help human beings understand the observations collected from nature (including the human brain) and, second, to build reliable and efficient machines in practical applications to mimic or outperform human performance.

8.2.2 Deciphering the Correlative Brain

In order to demystify the human brain, we have to understand the language it uses. Whenever neurons are interacting, communicating, or cooperating with each other, the common and unique language they use consists of patterns of spikes or action potentials (i.e., transient electrical discharges), which is often referred to as the "neural code." How do we characterize these correlative neuronal firings, decipher the neural codes within single neurons or populations of neurons, and use mathematical and computational tools to characterize spiking dynamics? Finding the answers to these questions is the key to understanding the correlative brain [118, 504].

A direct method for analyzing spikes is to record the spike trains produced by neurons in vivo. At the cellular level, multielectrode recording is a powerful tool to reveal the internal synchronization of neuronal firing activity. We have discussed this extensively in Chapter 1. Although most studies are restricted to the subcortical and cortical areas of cats or monkeys, there is no strong reason to believe that human neocortex employs an utterly different strategy for information encoding. Nowadays, modern multichannel electrode recording techniques allow one to simultaneously record from more than 100 channels. However, spikes are not recorded directly. Instead, it is the extracelluar voltage potentials that are recorded by electrodes, which can represent, depending on the electrode impedance, the simultaneous electrical activities of a small number of neurons. Therefore, we have to rely on a "spike-sorting" procedure to identify and classify the spike events [241, 551]. The purpose of spike pattern classification is to detect the patterns of spike timing and measure the association and correlation among neural spike trains; these methods provide a way of evaluating higher order (instead of pairwise) neural interactions in the ensemble spike activity [118, 275, 592]. With multielectrode recordings of spike trains from the brain, the goal of neural decoding is to "read" the mind [91, 250, 255, 527, 949].

It is well known that the brain generates oscillatory electrical potentials (also called "brain waves") that are large enough to be detected and recorded by electrodes at the surface of the scalp. The EEG signal is both a consequence and a sign of correlated activities in the brain [183]. As a noninvasive recording technique, EEG is the reflection upon the scalp of the summed synaptic potentials of millions of neurons; the neurons self-organize into transient networks that synchronize in time and space to produce a mixture of short bursts of oscillations that are observable in the EEG recordings. Generally, low-frequency brain waves (such as theta waves, 4–8 Hz, and alpha waves, 8–12 Hz) are found in conditions of sleep or relaxation, and high-frequency gamma waves (30–100 Hz) are more frequently observed during high-level cognitive tasks, which indeed reveal the role of oscillatory synchrony in those active mental processes. Because of its good time resolution, the EEG provides a useful way to investigate brain activities. Another noninvasive multichannel recording technique is MEG, which detects the tiny magnetic fields created as individual neurons synchronize their synaptic currents within the brain; it can pinpoint the active region to within a centimeter and can follow the movement of brain activity as it travels from region to region within the brain; MEG generally has equally good temporal resolution but superior spatial localization compared to EEG, largely because it records activity within smaller distances from the sensor and is not affected by skull impedance and spreading scalp conductance. More recently, many advanced imaging techniques have been developed for studying brain functions. Among the diverse range of imaging tools currently available, one of the most promising is fMRI. Functional MRI uses magnets to detect magnetic molecules within the brain and exploits the changes in the magnetic properties of hemoglobin as it carries oxygen, thereby measuring the so-called *blood-oxygenation-level-dependent* (BOLD) signal [442] (see Figure 8.1 for an illustration). Without making direct measurements of neuronal firing, BOLD fMRI monitors the local changes of blood flow—the phenomena that occur due to regional change of neuronal activity (physically, neuronal activation requires increased oxygen consumption and further results in a local decrease in the concentration of deoxyhemoglobin, which causes an increase in the homogeneity of the static magnetic field and yields an increase in the fMRI signal). There is no doubt

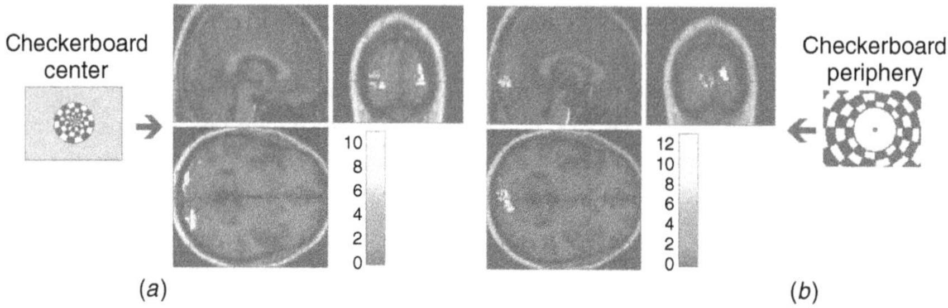

Figure 8.1 Illustration of fMRI for human brain. The imaging activation patterns are compared with two different types of visual stimuli (checkerboard center vs. checkerboard periphery) for one healthy human subject; the warm-colored areas reflect the activated neuronal activities. (Courtesy of Dr. Christine Boucard.)

that, with integration of both direct, invasive recordings (such as the spike trains and local field potentials) and noninvasive measurements (such as those from EEG, MEG, and fMRI), this opens a window for studying brain functions and ultimately leads to a better understanding of the correlative brain.

In helping to decipher the brain with various advanced recording/imaging technologies, numerous emerging signal processing, statistical estimation, and machine learning methods have been developed in the past few decades. With no exception, the correlation-based signal processing and neural/machine learning algorithms that were discussed in this monograph are anticipated to play a prominent role in advancing toward this goal. It is our hope that this book will serve as a useful reference in this odyssey.

APPENDIX A

AUTOCORRELATION AND CROSS-CORRELATION FUNCTIONS

A.1 AUTOCORRELATION FUNCTION

Consider a time-limited (or band-limited) signal $x(t)$,

$$x(t) = \begin{cases} x(t), & 0 \leq t \leq T, \\ 0, & \text{otherwise;} \end{cases} \tag{A.1}$$

its autocorrelation function is defined as

$$\begin{aligned} C_{xx}(t, t+\tau) &= \mathbb{E}[x(t)x(t+\tau)] \\ &\approx \frac{1}{T}\int_0^T x(t)x(t+\tau)\,dt, \end{aligned} \tag{A.2}$$

where the definition equation in the first line is specified for random signals whereas the second line is more general and also applicable for deterministic signals. If the random signal $x(t)$ is drawn from an *ergodic* stochastic process, then the ensemble average can be approximated by the time average by allowing the duration T to approach infinity.

Some important concepts and properties related to the autocorrelation are summarized here:

- If $x(t)$ is drawn from a wide-sense stationary process, then its autocorrelation function is shift invariant, namely,

$$C_{xx}(t, t+\tau) = C_{xx}(\tau). \tag{A.3}$$

- The autocorrelation function is symmetric, namely, $C_{xx}(\tau) = C_{xx}(-\tau)$ and $C_{xx}(\tau) \leq C_{xx}(0) = \sigma_x^2$, where $\sigma_x^2 = \text{var}[x(t)]$ denotes the variance of the $x(t)$.
- The *normalized* autocorrelation function is defined as

$$\overline{C}_{xx}(\tau) = \frac{C_{xx}(\tau)}{C_{xx}(0)}. \tag{A.4}$$

- The decaying rate and the limit of the autocorrelation function can be characterized by [525]

$$|C_{xx}(\tau)| \leq C_{xx}(0) \cos\left(\frac{\pi}{1 + T/\tau}\right) \qquad (\tau < T). \tag{A.5}$$

- If $x(t)$ is wide-sense stationary, its autocorrelation function can be written in terms of spectral representations in light of the *Wiener–Khinchin theorem*

$$C_{xx}(\tau) = \frac{1}{2\pi} \int_{-\infty}^{\infty} S_{xx}(\omega) e^{j\omega\tau} \, d\omega, \tag{A.6}$$

where $S_{xx}(\omega)$ denotes the power spectral density of $x(t)$.
- Let $x_1(t)$ denote the Hilbert transform of $x(t)$:

$$x_1(t) = -\frac{1}{\pi} \int_{-\infty}^{\infty} \frac{x(\tau)}{t - \tau} \, d\tau; \tag{A.7}$$

then it can be proved [525] that the autocorrelation of $x_1(t)$ is equal to that of $x(t)$, namely,

$$C_{x_1 x_1}(\tau) = C_{xx}(\tau), \tag{A.8}$$

whereas $x_1(t)$ is orthogonal (or uncorrelated) to $x(t)$, namely, $\mathbb{E}[x_1(t) x(t)] = 0$.

A.2 CROSS-CORRELATION FUNCTION

For two time-limited signals $x(t)$ and $y(t)$, the cross-correlation function may be defined as

$$\begin{aligned} C_{xy}(t, t+\tau) &= \mathbb{E}[x(t) y(t+\tau)] \\ &\approx \frac{1}{T} \int_0^T x(t) y(t+\tau) \, dt, \end{aligned} \tag{A.9}$$

$$\begin{aligned} C_{xy}(t+\tau, t) &= \mathbb{E}[y(t) x(t+\tau)] \\ &\approx \frac{1}{T} \int_0^T x(t+\tau) y(t) \, dt. \end{aligned} \tag{A.10}$$

It is noted that the cross-correlation function is generally nonsymmetric, namely, $C_{xy}(t, t+\tau) \neq C_{xy}(t+\tau, t)$.

The cross-correlation function has the following properties:

- The cross-correlation function is bounded by the cross-correlation inequality [82]

$$|C_{xy}(\tau)|^2 \leq C_{xx}(0)C_{yy}(0) = \sigma_x^2 \sigma_y^2, \tag{A.11}$$

where $\sigma_x^2 = \mathbb{E}[x^2(t)]$ and $\sigma_y^2 = \mathbb{E}[y^2(t)]$ denote the power of $x(t)$ and $y(t)$, respectively.

- In terms of spectral representations, the cross-correlation function can be written as the inverse Fourier transform

$$C_{xy}(\tau) = \frac{1}{2\pi} \int_{-\infty}^{\infty} S_{xy}(\omega) e^{j\omega\tau}\, d\omega, \tag{A.12}$$

where $S_{xy}(\omega)$ denotes the cross-spectrum density.

- The correlation coefficient (also called normalized cross-correlation) between two random signals $x(t)$ and $y(t)$ is defined as

$$\rho_{xy} = \frac{C_{xy}(0)}{\sqrt{\text{var}[x(t)]\,\text{var}[y(t)]}}. \tag{A.13}$$

From (A.11), it follows that the correlation coefficient ρ_{xy} ranges between -1 and 1. Positive/negative ρ_{xy} indicates $x(t)$ and $y(t)$ are positively/negatively correlated; $\rho_{xy} = 0$ indicates that they are uncorrelated.

In the frequency domain, let $X(\omega)$ and $Y(\omega)$ denote the Fourier transform of $x(t)$ and $y(t)$, respectively; then the cross-spectrum of $X(\omega)$ and $Y(\omega)$ is defined as

$$S_{XY}(\omega) = \mathbb{E}[X(\omega)Y^*(\omega)], \tag{A.14}$$

where the asterisk denotes the complex conjugate. In a similar vein, the *normalized cross-spectrum* is defined as

$$\tilde{S}_{XY}(\omega) = \frac{S_{XY}(\omega)}{\sqrt{\text{var}[X(\omega)]\,\text{var}[Y(\omega)]}}, \tag{A.15}$$

and its magnitude $|\tilde{S}_{XY}(\omega)|$ is a real function between 0 and 1 that gives a measure of correlation between $x(t)$ and $y(t)$ at each frequency ω. Observe that $|\tilde{S}_{XY}(\omega)|^2$ bears some similarity to ρ_{xy}^2; however, $|\tilde{S}_{XY}(\omega)|^2$ takes into account out-of-phase relationships and can examine the variance of two signals in a selected frequency range.

- The relationship between the cross-correlation and convolution is established as

$$\int x(t)y(t+\tau)\,dt = \int x(t)y(\tau-(-t))\,d\tau \equiv x(t) \otimes y(-t). \quad \text{(A.16)}$$

If $y(t)$ is an even (possibly noncausal) function, then these two operations are essentially identical. Therefore, convolution operation is commutative (symmetric), while the cross-correlation operation is generally noncommutative (nonsymmetric).

- Let $x_1(t)$ denote the Hilbert transform of $x(t)$; then the cross-correlation function between $x_1(t)$ and $x(t)$ is defined by

$$C_{xx_1}(\tau) = \frac{1}{T}\int_0^T x(t)x_1(t+\tau)\,dt, \quad \text{(A.17)}$$

it can be shown [525] that

$$C_{xx_1}(\tau) = -\frac{1}{\pi}\int_{-\infty}^{\infty} \frac{C_{xx}(\tau')}{\tau-\tau'}\,d\tau', \quad \text{(A.18)}$$

$$C_{xx}(\tau) = \frac{1}{\pi}\int_{-\infty}^{\infty} \frac{C_{xx_1}(\tau')}{\tau-\tau'}\,d\tau', \quad \text{(A.19)}$$

and

$$C_{xx_1}(0) = 0. \quad \text{(A.20)}$$

The last property is often used for minimum direction finding.

Let $x_1(t)$ and $x_2(t)$ be two zero-mean, mutually uncorrelated real-valued signals, namely, $\mathbb{E}[x_1(t)x_2(t)] = 0$, $\mathbb{E}[x_1(t)] = 0$, and $\mathbb{E}[x_2(t)] = 0$; also let $X_1(\omega)$ and $X_2(\omega)$ denote the Fourier transforms of $x_1(t)$ and $x_2(t)$, respectively; then the following properties hold:

- $X_1(\omega)$ and $X_2(\omega)$ are uncorrelated in the sense that

$$\mathbb{E}[X_1(\omega)X_2(\omega)] = \int_{-\infty}^{\infty}\int_{-\infty}^{\infty} \mathbb{E}[x_1(t)x_2(t)]e^{-j\omega(t_1+t_2)}\,dt_1\,dt_2 = 0. \quad \text{(A.21)}$$

Likewise, $\mathbb{E}[X_1(\omega)X_2^*(\omega)] = 0$.

- If in addition, $x_1(t)$ is stationary (i.e., with constant variance), then $\mathbb{E}[X_1^2(\omega)] = 0$ for $\omega \neq 0$.
- If, in addition, $x_1(t)$ and $x_2(t)$ are both stationary (i.e., both with constant variance), then $\mathbb{E}[X_1^2(\omega)] = \mathbb{E}[X_2^2(\omega)] = \mathbb{E}[X_1(\omega)X_2(\omega)] = 0$ for $\omega \neq 0$.
- If $x_1(t)$ is temporally uncorrelated with a time-varying variance $q(t)$, namely $\mathbb{E}[x_1(t_1)x_1(t_2)] = q(t_1)\delta(t_1-t_2)$, then $X_1(\omega)$ is a stationary, correlated process with an autocorrelation function $Q(\omega)$, which is defined as the Fourier transform of $q(t)$.

A.3 DERIVATIVE STOCHASTIC PROCESSES

If $\{x(t)\}$ is a stochastic process, then its associative derivative stochastic process, denoted by $\{\dot{x}(t)\}$, is defined as [82]

$$\dot{x}(t) = \frac{dx(t)}{dt} = \lim_{\varepsilon \to 0} \left[\frac{x(t+\varepsilon) - x(t)}{\varepsilon} \right]. \tag{A.22}$$

If $\{x(t)\}$ is stationary and its autocorrelation function is defined as $C_{xx}(\tau) = \mathbb{E}[x(t)x(t+\tau)] = \mathbb{E}[x(t-\tau)x(t)]$, then the following equalities can be derived [82]:

$$\begin{aligned} C'_{xx}(\tau) &= \frac{dC_{xx}(\tau)}{d\tau} = \mathbb{E}[x(t)\dot{x}(t+\tau)] = C_{x\dot{x}}(\tau) \\ &= -\mathbb{E}[\dot{x}(t-\tau)(t)] = -C_{\dot{x}x}(\tau), \end{aligned} \tag{A.23}$$

$$C'_{xx}(\tau) = -C'_{xx}(-\tau), \tag{A.24}$$

and

$$C'_{xx}(0) = C_{x\dot{x}}(0) = -C_{\dot{x}x}(0) = 0. \tag{A.25}$$

Namely, a maximum value of the autocorrelation function $C_{xx}(\tau)$ corresponds to the zero crossing of its derivative $C'_{xx}(\tau)$; this is an important observation since finding the zero-crossing points in practice is easier than determining the location of maximum values. In addition, the above equations imply that, for stationary signals $\{x(t)\}$, $x(t)$ and $\dot{x}(t)$ are statistically uncorrelated:

$$\mathbb{E}[x(t)\dot{x}(t)] = 0. \tag{A.26}$$

Similarly, one can further define the second-order derivative random process

$$\ddot{x}(t) = \frac{d^2x(t)}{dt^2} = \lim_{\varepsilon \to 0} \left[\frac{\dot{x}(t+\varepsilon) - \dot{x}(t)}{\varepsilon} \right], \tag{A.27}$$

and correspondingly we obtain

$$\begin{aligned} C''_{xx}(\tau) &= \frac{dC'_{xx}(\tau)}{d\tau} = \frac{dC_{x\dot{x}}(\tau)}{d\tau} \\ &= C_{x\ddot{x}}(\tau) = -C_{\dot{x}\dot{x}}(\tau), \end{aligned} \tag{A.28}$$

$$C''_{xx}(\tau) = C''_{xx}(-\tau), \tag{A.29}$$

and

$$-C''_{xx}(0) = -C_{x\ddot{x}}(0) = C_{\dot{x}\dot{x}}(0) = \mathbb{E}[\dot{x}^2(t)]. \tag{A.30}$$

APPENDIX B

STOCHASTIC APPROXIMATION

As we have observed in this book, most online stochastic learning rules, in one form or another, use the following recursive computation equation:

$$\boldsymbol{\theta}(t+1) = \boldsymbol{\theta}(t) + \eta(t)h(\boldsymbol{\theta}(t), \mathbf{x}(t)), \qquad t = 0, 1, 2, \ldots, \tag{B.1}$$

where $\boldsymbol{\theta}(\cdot)$ is a sequence of vectors that are the object of interest and $\mathbf{x}(t)$ is an observation vector present at time t. Note that the vectors $\boldsymbol{\theta}(t)$ and $\mathbf{x}(t)$ may or may not have the same dimension. As time goes on, the change of parameter vector, $\Delta\boldsymbol{\theta}(t)$, will gradually be proportional to the expected value, $\langle h(\boldsymbol{\theta}(t), \mathbf{x}(t))\rangle$, which, in many cases, can be decomposed into a series of correlation terms, either Hebbian or anti-Hebbian. In fact, a large family of stochastic learning rules with the form of (B.1) can be viewed as *stochastic approximation* algorithms [514, 515, 567, 764].

In the stochastic approximation framework, it is often assumed that $\mathbf{x}(t)$ is a sample drawn from a stochastic process or a distribution function. The elements of the vector $\boldsymbol{\theta}$ are referred to as the synaptic weights, or the unknown parameters (organized in a vector form) to be learned. The scalar sequence $\eta(\cdot)$, determining the time-varying or time-invariant learning-rate parameter, is assumed to be a sequence of nonincreasing positive scalars. The *update function* $h(\cdot, \cdot)$ is a *deterministic* (either linear or nonlinear) function with certain conditions imposed on it. This function, together with the learning-rate sequence $\eta(\cdot)$, specifies the complete structure of the algorithm.

The convergence analysis of the stochastic learning algorithm with the form of (B.1) is often tackled within the stochastic approximation framework. This is often done by relating the difference equation with a deterministic, linear or nonlinear, ordinary differential equation (ODE) followed by conventional mathematical

Correlative Learning: A Basis for Brain and Adaptive Systems, by Zhe Chen, Simon Haykin, Jos J. Eggermont, and Suzanna Becker

analysis. Rearranging (B.1), we may have

$$\frac{\boldsymbol{\theta}(t+1)-\boldsymbol{\theta}(t)}{\eta} = h(\boldsymbol{\theta}(t), \mathbf{x}(t)). \tag{B.2}$$

When η is sufficiently small, (B.2) can be approximated by an ODE.

Generally, the following regular conditions are often assumed within the stochastic approximation framework:

1. The learning-rate sequence $\eta(t)$ is a decreasing sequence of positive real numbers that satisfy

$$\sum_{t=1}^{\infty} \eta(t) = \infty, \tag{B.3}$$

$$\sum_{t=1}^{\infty} \eta^p(t) < \infty \qquad (p > 1), \tag{B.4}$$

$$\lim_{t\to\infty} \eta(t) \to 0. \tag{B.5}$$

2. The sequence of parameter vector $\boldsymbol{\theta}(\cdot)$ is bounded with probability 1.
3. The update function $h(\boldsymbol{\theta}, \mathbf{x})$ is continuously differentiable with respect to $\boldsymbol{\theta}$ and $\mathbf{x}$, and its derivatives are bounded in time.
4. The limit

$$\bar{h}(\boldsymbol{\theta}) = \lim_{t\to\infty} \mathbb{E}[h(\boldsymbol{\theta}, \mathbf{x})] \tag{B.6}$$

   exists for each $\boldsymbol{\theta}$; the statistical expectation operator is taken over $\mathbf{x}$.
5. There is a locally asymptotically stable (in the Lyapunov sense) solution to the ODE

$$\frac{d\theta(t)}{dt} = \bar{h}(\boldsymbol{\theta}(t)), \tag{B.7}$$

   where t here denotes continuous time.
6. Let $\mathbf{q}_0$ denote the solution to equation (B.7) with a basin of attraction $\mathcal{B}(\mathbf{q}_0)$; then the parameter vector $\boldsymbol{\theta}(t)$ enters a compact subset $\mathcal{A}$ of the basin of attraction $\mathcal{B}(\mathbf{q}_0)$ infinitely often, with probability 1.

The above six conditions are all reasonable. Equations (B.3) and ((B.5) are necessary conditions that guarantee the convergence of the algorithm to the desired estimate regardless of its initial conditions. Equation (B.4) specifies a condition on how fast the learning-rate sequence $\eta(\cdot)$ will approach to zero; it is much less restrictive than the usual condition

$$\sum_{t=1}^{\infty} \eta^2(t) < \infty. \tag{B.8}$$

One example of the learning-rate annealing procedure satisfying condition 1 is

$$\eta(t) = \frac{\alpha + \beta}{t + \beta}, \tag{B.9}$$

where α and β are two predefined scalars. Equation (B.6) specifies the assumption that makes it possible to associate (B.1) with an ODE.

Given a recursive (online) stochastic learning rule that satisfies conditions 1–6, the following *asymptotic stability theorem* [514, 567] establishes the convergence of learning rule (B.1):

$$\lim_{t \to \infty} \boldsymbol{\theta}(t) \to \mathbf{q}_0 \quad \text{infinitely often with probability 1.}$$

Note that the above convergence analysis of stochastic approximation algorithms assumes that $\mathbf{x}(t)$ is drawn from a stationary stochastic process or a time-invariant probability distribution; if, however, this assumption is not valid, it is advisable to maintain the learning-rate parameter $\eta(t)$ as a small value to keep tracking the time-variant data.

APPENDIX C

PRIMER ON LINEAR ALGEBRA

Let **a** and **b** denote two m-length real-valued column vectors and let $\mathbf{a}^{\mathrm{T}}$ denote the transpose of the vector **a**. When the vectorial variable is complex valued, the Hermitian transpose will correspondingly replace the transpose operator wherever it appears.

Norm: The L_2 norm of vector **a** is defined as

$$\|\mathbf{a}\| = \sqrt{a_1^2 + a_2^2 + \cdots + a_m^2}. \tag{C.1}$$

Inner Product: The inner product (or dot product) between vectors **a** and **b** is defined as

$$\langle \mathbf{a}, \mathbf{b} \rangle = \mathbf{a}^{\mathrm{T}}\mathbf{b} = \sum_{i=1}^{m} a_i b_i. \tag{C.2}$$

Outer Product: The outer product between **a** and **b** defines an $m \times m$ matrix

$$\mathbf{R} = \mathbf{a}\mathbf{b}^{\mathrm{T}} \tag{C.3}$$

with components $R_{ij} = a_i b_j$.

Angle: The angle between two vectors **a** and **b**, defined as $\angle(\mathbf{a}, \mathbf{b})$, satisfies the relationship

$$\cos \angle(\mathbf{a}, \mathbf{b}) = \frac{\langle \mathbf{a}, \mathbf{b} \rangle}{\|\mathbf{a}\| \cdot \|\mathbf{b}\|}. \tag{C.4}$$

When $\cos(\mathbf{a}, \mathbf{b}) = 0$, it is said that **a** and **b** are orthogonal.

Correlative Learning: A Basis for Brain and Adaptive Systems, by Zhe Chen, Simon Haykin, Jos J. Eggermont, and Suzanna Becker

Trace: Let $\mathbf{A}$ denote an arbitrary $m \times m$ square matrix; the trace of matrix $\mathbf{A}$ is defined as the sum of its diagonal elements:

$$\text{tr}(\mathbf{A}) = \sum_{i=1}^{m} a_{ii}. \tag{C.5}$$

The trace operator relates the inner product and outer product via the following equation:

$$\text{tr}(\mathbf{a}\mathbf{a}^{\mathrm{T}}) = \mathbf{a}^{\mathrm{T}}\mathbf{a} = \|\mathbf{a}\|^2.$$

Determinant: The determinant of a square matrix $\mathbf{A}$ is defined by the *Laplacian expansion by minors*

$$\det(\mathbf{A}) = \sum_{i=1}^{m} (-1)^{i+j} a_{ij} M_{ij}, \tag{C.6}$$

where M_{ij} denotes the minor of matrix $\mathbf{A}$ that is formed by eliminating the ith row and the jth column from the matrix $\mathbf{A}$.

Frobenius Norm: The Frobenius norm of an $m \times n$ matrix $\mathbf{A}$ is defined as the square root of the sum of the absolute squares of its elements:

$$\|\mathbf{A}\|_{\mathrm{F}} = \sqrt{\sum_{i=1}^{m}\sum_{j=1}^{n} |a_{ij}|^2} = \sqrt{\text{tr}(\mathbf{A}\mathbf{A}^{\mathrm{T}})} = \sqrt{\text{tr}(\mathbf{A}^{\mathrm{T}}\mathbf{A})}. \tag{C.7}$$

Rayleigh Quotient: The Rayleigh quotient of the real symmetric matrix $\mathbf{A}$ is defined as

$$\rho(\mathbf{A}) = \frac{\mathbf{a}^{\mathrm{T}}\mathbf{A}\mathbf{a}}{\mathbf{a}^{T}\mathbf{a}} \quad \text{for } (\mathbf{a} \neq 0). \tag{C.8}$$

C.1 EIGENANALYSIS

Let $\mathbf{C}$ denote an $m \times m$ symmetric (or Hermitian), positive-definite correlation matrix and $\mathbf{v}$ be an $m \times 1$ nonzero real-valued (or complex-valued) column vector; the eigenequation is stated as

$$\mathbf{C}\mathbf{v} = \lambda\mathbf{v}, \tag{C.9}$$

or equivalently

$$(\mathbf{C} - \lambda\mathbf{I})\mathbf{v} = 0. \tag{C.10}$$

In light of the *spectral theorem*, the eigenvalue decomposition (EVD) states that

$$\mathbf{C} = \mathbf{U}\boldsymbol{\Lambda}\mathbf{U}^{\mathrm{T}} = \sum_{i=1}^{m} \lambda_i \mathbf{u}_i \mathbf{u}_i^{\mathrm{T}} \quad \text{or} \quad \mathbf{C} = \mathbf{U}\boldsymbol{\Lambda}\mathbf{U}^{\mathrm{H}} = \sum_{i=1}^{m} \lambda_i \mathbf{u}_i \mathbf{u}_i^{\mathrm{H}}, \tag{C.11}$$

where $\boldsymbol{\Lambda}$ is a diagonal matrix, with its nonnegative diagonal elements $\{\lambda_i\}$ as *eigenvalues*, and the column vectors $\mathbf{u}_i$ of the orthogonal (or unitary) matrix $\mathbf{U}$ are called the *eigenvectors*; the eigenvectors consist of a set of orthogonal basis vectors that satisfy the eigenequation

$$\mathbf{C}\mathbf{u}_i = \lambda_i \mathbf{u}_i. \tag{C.12}$$

In functional analysis, the matrix operation will be substituted by an operator. The functional analog of the eigenvector is the eigenfunction, denoted as $e(t)$, which satisfies

$$\int K(t, t')e(t')\,dt' = \lambda e(t), \tag{C.13}$$

where $K(t, t')$ is a linear integral operator which plays a similar role as the matrix $\mathbf{C}$ in (C.9). If $K(t, t')$ is translationally invariant, namely $K(t, t') = K(t - t')$, then the eigenfunctions are complex exponentials:

$$\int K(t - t') \exp(j\omega t')\,dt' = \left(\int K(\tau) \exp(-j\omega\tau)\,d\tau\right) \exp(j\omega t), \tag{C.14}$$

where we have used substitution $\tau = t - t'$ in the above equality; the eigenvalue for the eigenfunction is defined as

$$\lambda(\omega) = \int K(\tau) \exp(-j\omega\tau)\,d\tau. \tag{C.15}$$

Hence, the discrete eigenvalues in matrix analysis will turn into the continuous eigenspectrum in functional analysis. Likewise, a functional analog of expanding a vector using eigenvectors as bases, is the inverse Fourier transform, which expands a function using complex exponential eigenfunctions as the bases, and the Fourier transform is used to determine the coefficients of the expansion. This property indeed serves as the basis of spectrum analysis for discrete-time stochastic processes.

Specifically, an important property of eigenvalue in the context of spectrum analysis is stated as follows

> The eigenvalues of the correlation matrix of a discrete-time stochastic process are bounded by the minimum and maximum values of the power spectral density of the process.

Stated mathematically, let λ_i and $\mathbf{u}_i$ $(i = 1, 2, \ldots, m)$ denote, respectively, the eigenvalues of the $m \times m$ correlation matrix $\mathbf{C}$ (which is assumed to be Hermitian symmetric) of a stochastic process $x(t)$ and their associative eigenvectors. According to the eigenvalue definition, we have

$$\lambda_i = \frac{\mathbf{u}_i^{\mathrm{H}}\mathbf{C}\mathbf{u}_i}{\mathbf{u}_i^{\mathrm{H}}\mathbf{u}_i}, \tag{C.16}$$

where the numerator may be expressed in an expanded form

$$\mathbf{u}_i^{\mathrm{H}}\mathbf{C}\mathbf{u}_i = \sum_{k=1}^{m}\sum_{j=1}^{m} u_{ik}^{*} c(j-k) u_{ij}, \tag{C.17}$$

with u_{ik}^{*} being the kth element of the row vector $\mathbf{u}_i^{\mathrm{H}}$, $c(j-k)$ being the (k, j)th element of the matrix $\mathbf{C}$, and u_{ij} being the jth element of the column vector $\mathbf{u}_i$. In light of the *Wiener–Khinchin equation*, we may have

$$c(j-k) = \frac{1}{2\pi}\int_{-\pi}^{\pi} S(\omega) e^{j\omega(j-k)}\, d\omega, \tag{C.18}$$

where $S(\omega)$ is the power spectral density of the stochastic process $x(t)$. It can be proven [369] that

$$\lambda_i = \frac{\int_{-\pi}^{\pi} |U_i'(e^{j\omega})|^2 S(\omega)\, d\omega}{\int_{-\pi}^{\pi} |U_i'(e^{j\omega})|^2\, d\omega}, \tag{C.19}$$

where $U_i'(e^{j\omega})$ denotes the discrete Fourier transform of the sequence $u_{i1}^{*}, u_{i2}^{*}, \ldots, u_{im}^{*}$:

$$U_i'(e^{j\omega}) = \sum_{k=1}^{m} q_{ik}^{*} e^{-j\omega k}. \tag{C.20}$$

Let $S_{\min}$ and $S_{\max}$ denote, respectively, the absolute minimum and maximum values of the power spectral density $S(\omega)$; then it further follows that

$$S_{\min}\int_{-\pi}^{\pi} |U_i'(e^{j\omega})|^2\, d\omega \le \int_{-\pi}^{\pi} |U_i'(e^{j\omega})|^2 S(\omega)\, d\omega \le S_{\max}\int_{-\pi}^{\pi} |U_i'(e^{j\omega})|^2\, d\omega,$$

and

$$S_{\min} \le \lambda_i \le S_{\max}. \tag{C.21}$$

C.2 GENERALIZED EIGENVALUE PROBLEM

The generalized eigenvalue analysis is an extension of the conventional eigenvalue analysis. Given two square matrices **A** and **B**, the generalized eigenvalue problem is to find the pairs $\{\alpha_i, \beta_i\}$ and the vectors $\mathbf{v} \neq 0$ such that

$$\beta_i \mathbf{A}\mathbf{v}_i = \alpha_i \mathbf{B}\mathbf{v}_i, \tag{C.22}$$

where $\mathbf{v}_i$ is called the generalized eigenvector and $\lambda_i = \alpha_i/\beta_i$ is called the generalized eigenvalue.

If the determinant of the matrix $\mathbf{A} - \lambda\mathbf{B}$ does not vanish, then the matrix pair $(\mathbf{A}, \mathbf{B})$ is said to be *regular*; otherwise it is called *singular*. If the matrix pair is regular and the matrix **B** is nonsingular, then $\mathbf{v}_i$ is the eigenvector of the matrix $\mathbf{B}^{-1}\mathbf{A}$ with associated eigenvalue λ_i.

C.3 SVD AND CHOLESKY FACTORIZATION

Singular-value decomposition is an extension of EVD. Let **A** denote an $m \times n$ *arbitrary* real matrix

$$\mathbf{A} = \mathbf{U}\mathbf{S}\mathbf{V}^{\mathrm{T}}, \tag{C.23}$$

where **U** is an $m \times n$ matrix and **V** is an $n \times n$ square matrix, both of which are unitary matrices that consist of orthogonal columns such that

$$\mathbf{U}^{\mathrm{T}}\mathbf{U} = \mathbf{V}^{\mathrm{T}}\mathbf{V} = \mathbf{I}.$$

The matrix **S** is degenerate and contains a $p \times p$ [where $p = \mathrm{rank}(\mathbf{A})$] diagonal matrix with the nonzero singular values appearing in the diagonal.

Singular-value decomposition can be used to efficiently calculate the eigenvalue decomposition, especially when the dimensionality of the variable is very large compared to the total number of observations. In particular, let **A** be the $m \times n$ (assuming $m < n$) data matrix upon appropriate centering (i.e., with zero mean) and $\mathbf{C} = \mathbf{A}\mathbf{A}^{\mathrm{T}}/n$ be the $m \times m$ sample correlation matrix; provided $\mathbf{C} = \mathbf{W}\mathbf{\Lambda}\mathbf{W}^{\mathrm{T}}$ represents the EVD and $\mathbf{A} = \mathbf{U}\mathbf{S}\mathbf{V}^{\mathrm{T}}$ represents the SVD, the following relationship can be established:

$$\begin{aligned}
\mathbf{A}\mathbf{A}^{\mathrm{T}} &= \mathbf{U}\mathbf{S}\mathbf{S}^{\mathrm{T}}\mathbf{U}^{\mathrm{T}}, \\
\mathbf{A}^{\mathrm{T}}\mathbf{A} &= \mathbf{V}\mathbf{S}^{\mathrm{T}}\mathbf{S}\mathbf{V}^{\mathrm{T}}, \\
\mathbf{\Lambda} &= \mathbf{S}\mathbf{S}^{\mathrm{T}}, \\
\mathbf{W} &= \mathbf{U}.
\end{aligned}$$

If we truncate the zero entries within the $m \times n$ matrix $\mathbf{S}$ and rewrite it as a full-rank $m \times m$ diagonal matrix $\hat{\mathbf{S}}$, then we have $\boldsymbol{\Lambda} = \hat{\mathbf{S}}^2$; namely, the square of the singular value of $\mathbf{A}$ is equivalent to the eigenvalues of $\mathbf{AA}^{\mathrm{T}}$.

Similar to the generalized EVD, we can also define the generalized (or quotient) SVD. Given an $m \times p$ matrix $\mathbf{A}$ and an $n \times p$ matrix $\mathbf{B}$, the generalized SVD (GSVD) is to find two unitary matrices $\mathbf{U}$ and $\mathbf{V}$ such that

$$
\begin{aligned}
\mathbf{A} &= \mathbf{URQ}^{\mathrm{T}}, \\
\mathbf{B} &= \mathbf{VSQ}^{\mathrm{T}}, \\
\mathbf{I} &= \mathbf{R}^{\mathrm{T}}\mathbf{R} + \mathbf{S}^{\mathrm{T}}\mathbf{S}.
\end{aligned}
$$

The sizes of the matrices $\mathbf{U}$, $\mathbf{V}$, and $\mathbf{Q}$ are, respectively, $m \times m$, $n \times n$, and $p \times q$, where $q = \min\{m + n, p\}$, and the dimensionality of $\mathbf{R}$ and $\mathbf{S}$ are of $m \times q$ and $n \times q$, respectively. Let $\boldsymbol{\Sigma}_1 = \mathbf{R}^{\mathrm{T}}\mathbf{R} = \mathrm{diag}\{\alpha_1^2, \ldots, \alpha_q^2\}$ and $\boldsymbol{\Sigma}_2 = \mathbf{S}^{\mathrm{T}}\mathbf{S} = \mathrm{diag}\{\beta_1^2, \ldots, \beta_q^2\}$ denote two $q \times q$ diagonal matrices; then the values $\{\alpha_1/\beta_1, \ldots, \alpha_q/\beta_q\}$ are called the *generalized singular values* of the matrix pair $(\mathbf{A}, \mathbf{B})$.

Several additional comments are noteworthy:

- When $\mathbf{B}$ is an identity matrix, the GSVD reduces to the ordinary SVD as a special case.
- If $\mathbf{B}$ is square and nonsingular, then the GSVD of matrix pair $(\mathbf{A}, \mathbf{B})$ is equivalent to the SVD of the matrix $\mathbf{B}^{-1}\mathbf{A}$.
- If the columns of $(\mathbf{A}^{\mathrm{T}}\ \mathbf{B}^{\mathrm{T}})^{\mathrm{T}}$ are orthonormal, then the GSVD of $(\mathbf{A}, \mathbf{B})$ is equivalent to the *cosine–sine decomposition* of $(\mathbf{A}^{\mathrm{T}}\ \mathbf{B}^{\mathrm{T}})^{\mathrm{T}}$:

$$
\begin{pmatrix} \mathbf{A} \\ \mathbf{B} \end{pmatrix} = \begin{pmatrix} \mathbf{U} & \mathbf{0} \\ \mathbf{0} & \mathbf{V} \end{pmatrix} \begin{pmatrix} \mathbf{R} \\ \mathbf{S} \end{pmatrix} \mathbf{Q}^{\mathrm{T}}.
$$

Assuming that $\mathbf{C}$ is an $m \times m$ symmetric, positive-definite matrix, Cholesky factorization provides another way of matrix decomposition. Specifically, $\mathbf{C}$ can be factorized into the outer product between a lower triangular matrix $\mathbf{L}$ and its transpose, or the inner product between a upper triangular matrix $\mathbf{U}$ and its transpose, namely,

$$
\mathbf{C} = \mathbf{LL}^{\mathrm{T}} = \mathbf{U}^{\mathrm{T}}\mathbf{U}. \tag{C.24}
$$

C.4 GRAM–SCHMIDT ORTHOGONALIZATION

Gram–Schmidt orthogonalization is a procedure to obtain a set of orthogonal vectors $\{\mathbf{u}_i\}$ from any linearly independent set $\{\mathbf{x}_i\}$. Start with the first vector $\mathbf{u}_1 = \mathbf{x}_1$; then take the second vector $\mathbf{x}_2$ and subtract from it the part that lies along the direction $\mathbf{x}_1$: $\mathbf{u}_2 = \mathbf{x}_2 - \alpha\mathbf{u}_1$, where the scalar α is defined as

$$
\alpha = \frac{\langle \mathbf{x}_2, \mathbf{u}_1 \rangle}{\langle \mathbf{u}_1, \mathbf{u}_1 \rangle}. \tag{C.25}
$$

For $k = 3, 4, \ldots$, continuing the same process yields the ensuing orthogonal vectors:

$$\mathbf{u}_k = \mathbf{x}_k - \sum_{i=1}^{k-1} \frac{\langle \mathbf{x}_k, \mathbf{u}_i \rangle}{\langle \mathbf{u}_i, \mathbf{u}_i \rangle} \mathbf{u}_i. \tag{C.26}$$

C.5 PRINCIPAL CORRELATION

Given an $m \times p$ matrix $\mathbf{A}$ and an $m \times q$ matrix $\mathbf{B}$, let r be the minimum of the ranks of these two matrices. Let us define a function subcorr$\{\mathbf{A}, \mathbf{B}\} = \{c_1, c_2, \ldots, c_r\}$, where the scalars c_k are defined as follows [329]:

$$c_k = \max_{\mathbf{a} \in \mathbf{U_A}} \max_{\mathbf{b} \in \mathbf{U_B}} \mathbf{a}^{\mathrm{T}} \mathbf{b} = \mathbf{a}_k^{\mathrm{T}} \mathbf{b}_k \tag{C.27}$$

subject to

$$\|\mathbf{a}\| = \|\mathbf{b}\| = 1,$$
$$\mathbf{a}^{\mathrm{T}} \mathbf{a}_i = 0, \qquad \mathbf{b}^{\mathrm{T}} \mathbf{b}_i = 0 \qquad (i = 1, \ldots, k-1).$$

The vectors $\{\mathbf{a}_1, \ldots, \mathbf{a}_r\}$ and $\{\mathbf{b}_1, \ldots, \mathbf{b}_r\}$ are the *principal vectors* between the two subspaces spanned by $\mathbf{A}$ and $\mathbf{B}$; denoted by $\mathbf{U_A}$ and $\mathbf{U_B}$, respectively; each set of vectors represents an orthogonal basis. Note that $1 \geq c_1 \geq c_2 \geq \cdots \geq c_r \geq 0$. The angle $\theta_k = \arccos c_k$ is the *principal angle*, which represents the geometric angle between $\mathbf{a}_k$ and $\mathbf{b}_k$; the value c_k denotes the *principal correlation* between these two vectors. Several points are noteworthy:

- When matrices $\mathbf{A}$ and $\mathbf{B}$ are of the same subspace dimension, then the measure $\sin\theta_r = \sqrt{1 - c_r^2}$ is called the *distance* between the two subspaces spanned by $\mathbf{A}$ and $\mathbf{B}$.
- Minimizing the distance is equivalent to maximizing the minimum principal correlation (i.e., c_r) between $\mathbf{A}$ and $\mathbf{B}$.
- The fact that $c_r = 1$ implies $\mathbf{A}$ and $\mathbf{B}$ are in parallel subspaces, whereas $c_r = 0$ indicates at least of one basis of $\mathbf{A}$ is orthogonal to $\mathbf{B}$, or vice versa.
- If the principal correlation $c_1 = 0$, then all bases are orthogonal.

The procedure for calculating principal correlations is based on a SVD procedure, which was described in depth in [329].

APPENDIX D

PROBABILITY DENSITY AND ENTROPY ESTIMATORS

Information-theoretic learning often requires the use of the probability density function (pdf), entropy, or mutual information. In this appendix, we provide a brief overview of some efficient methods for estimating the pdf as well as the entropy function. The pdf and entropy estimators discussed here are practically useful because of their simplicity and the basis of sample statistics.

For discussion simplicity, we restrict our attention to continuous, real-valued univariate random variables, for which the estimators of pdf and its associated entropy are sought.

Definition D.1 *A real-valued Lebesgue-integrable function $p(x)$ ($x \in \mathbb{R}$) is called a pdf if it satisfies*

$$p(x) = \int_{-\infty}^{x} F(x)\, dx,$$

where $F(x)$ is a cumulative probability distribution function. A pdf is everywhere nonnegative and its integral from $-\infty$ to $+\infty$ is equal to 1; namely $0 \leq p(x) \leq 1$ *and* $\int_{-\infty}^{\infty} p(x)\, dx = 1$.

Definition D.2 *Given the pdf of a continuous random variable x, its differential Shannon entropy is defined as*

$$H(x) = \mathbb{E}[-\log p(x)] = -\int_{-\infty}^{\infty} p(x)\, \log p(x)\, dx.$$

Correlative Learning: A Basis for Brain and Adaptive Systems, by Zhe Chen, Simon Haykin, Jos J. Eggermont, and Suzanna Becker

Definition D.3 *The characteristic function of a random variable x that has a pdf $p(x)$ is defined as*

$$\varphi_x(\omega) = \int_{-\infty}^{\infty} p(x)e^{j\omega x}\,dx,$$

where $j = \sqrt{-1}$ and $\omega \in \mathbb{R}$; namely, $\varphi_x(\omega)$ is the Fourier transform of the pdf $p(x)$, except for a sign change in the exponent.

The characteristic function $\varphi_x(\omega)$ is a complex number and can be expanded in a power series in a neighborhood of $\omega = 0$ as follows:

$$\varphi_x(\omega) = 1 + \sum_{k=1}^{\infty} \frac{(j\omega)^k}{k!} m_k, \tag{D.1}$$

where m_k is the kth-order moment of the random variable x, as defined by

$$m_k = \mathbb{E}[x^k] = \int_{-\infty}^{\infty} x^k p(x)\,dx. \tag{D.2}$$

The logarithm of $\varphi_x(\omega)$ can also be expanded in terms of cumulant statistics

$$\log \varphi_x(\omega) = \sum_{k=1}^{\infty} \frac{\kappa_k}{k!}(j\omega)^k, \tag{D.3}$$

where κ_k is the kth order cumulant of the random variable x. For a random variable x with zero mean ($\kappa_1 = 0$) and unit variance ($\kappa_2 = 1$), we then obtain that

$$\log \varphi_x(\omega) = -\frac{1}{2}\omega^2 + \sum_{k=3}^{\infty} \frac{\kappa_k}{k!}(j\omega)^k. \tag{D.4}$$

The cumulant statistics can also be calculated from the moment statistics

$$\kappa_1 = m_1, \qquad \kappa_2 = m_2, \qquad \kappa_3 = m_3, \qquad \kappa_4 = m_4 - 3m^2, \ldots .$$

D.1 GRAM-CHARLIER EXPANSION

The Gram–Charlier expansion is a popular method for approximating a pdf. According to the definition, we have

$$p(x) = \mathcal{N}(x)\left(\sum_{k=0}^{\infty} c_k \mathcal{H}_k(x)\right) = \mathcal{N}(x)\left(1 + \sum_{k=3}^{\infty} c_k \mathcal{H}_k(x)\right) \tag{D.5}$$

where $\mathcal{N}(x)$ denotes the standard Gaussian pdf as $\mathcal{N}(x) = (1/\sqrt{2\pi})\exp(-x^2/2)$ and c_k denotes the expansion coefficient of the characteristic function $\varphi_x(\omega)$ that relates to the cumulant statistics

$$
\begin{aligned}
c_0 &= 1, \quad c_1 = c_2 = 0, \\
c_3 &= \frac{\kappa_3}{6}, \\
c_4 &= \frac{\kappa_4}{24}, \\
c_5 &= \frac{\kappa_5}{120}, \\
c_6 &= \frac{\kappa_6 + 10\kappa_3^2}{720}, \quad \ldots
\end{aligned}
$$

and $\mathcal{H}_k(x)$ denotes the k-order *Chebyshev–Hermite polynomial*. Some typical Hermite polynomials are

$$
\begin{aligned}
\mathcal{H}_0(x) &= 1, \\
\mathcal{H}_1(x) &= x, \\
\mathcal{H}_2(x) &= x^2 - 1, \\
\mathcal{H}_3(x) &= x^3 - 3x, \\
\mathcal{H}_4(x) &= x^4 - 6x^2 + 3, \\
\mathcal{H}_5(x) &= x^5 - 10x^3 + 15x, \\
\mathcal{H}_6(x) &= y^6 - 15x^4 + 45x^2 - 15.
\end{aligned}
$$

A recursive relation for these Hermite polynomials is

$$\mathcal{H}_{k+1}(x) = x\mathcal{H}_k(x) - k\mathcal{H}_{k-1}(x). \tag{D.6}$$

The kth-order Hermite polynomial and the nth derivative of the Gaussian pdf $\mathcal{N}(x)$ are biorthogonal, namely,

$$\int_{-\infty}^{\infty} \mathcal{H}_k(x)\mathcal{N}^{(n)}(x)\,dx = (-1)^n n!\delta_{kn}, \qquad (k, n) = 0, 1, \ldots, \tag{D.7}$$

where δ_{kn} denotes the Kronecker delta, which is equal to unity if $k = n$ and zero otherwise.

In light of the above definitions, for a random variable x, we may obtain its up-to-sixth-order Gram–Charlier expansion

$$p(x) \approx \mathcal{N}(x)\left(1 + \frac{\kappa^3}{6}\mathcal{H}_3(x) + \frac{\kappa^4}{24}\mathcal{H}_4(x) + \frac{\kappa_6 + 10\kappa_3^2}{720}\mathcal{H}_6(x)\right). \tag{D.8}$$

If $p(x)$ is symmetric with respect to the origin (which implies the odd-order moment statistics are all zeros), then the above equation is further simplified to

$$p(x) \approx \mathcal{N}(x)\left(1 + \frac{\kappa^4}{24}\mathcal{H}_4(x) + \frac{\kappa_6}{720}\mathcal{H}_6(x)\right). \tag{D.9}$$

Correspondingly, the differential entropy of x may be approximated by

$$\begin{aligned} H(x) \approx \mathcal{N}(x)&\left(1 + \frac{\kappa^4}{24}\mathcal{H}_4(x) + \frac{\kappa_6}{720}\mathcal{H}_6(x)\right) \\ &\times \left(\log \mathcal{N}(x) + \log\left(1 + \frac{\kappa^4}{24}\mathcal{H}_4(x) + \frac{\kappa_6}{720}\mathcal{H}_6(x)\right)\right). \end{aligned} \tag{D.10}$$

D.2 EDGEWORTH EXPANSION

The Edgeworth series expansion is another popular method for approximating the pdf. Without loss of generality, we assume the random variable x has zero mean and unit variance; then the Edgeworth expansion of the pdf $p(x)$ is given by [862]

$$\begin{aligned} p(x) &= \mathcal{N}(x)\left(1 + \frac{\kappa^3}{3!}\mathcal{H}_3(x) + \frac{\kappa^4}{4!}\mathcal{H}_4(x) + \frac{\kappa_5}{5!}\mathcal{H}_5(x) + \frac{\kappa_6 + 10\kappa_3^2}{6!}\mathcal{H}_6(x)\right. \\ &\left. + \frac{35\kappa_3\kappa_4}{7!}\mathcal{H}_7(x) + \frac{56\kappa_3\kappa_5 + 35\kappa_4^2}{8!}\mathcal{H}_8(x) + \frac{280\kappa_3^3}{9!}\mathcal{H}_9(x) + \cdots\right). \end{aligned} \tag{D.11}$$

The key feature of the Edgeworth expansion is that its coefficients decrease uniformly, whereas the terms in the Gram–Charlier expansion do not approach uniformly to zero from the viewpoint of numerical errors; that is, generally no term is negligible compared to a preceding term.

The Gram–Charlier and Edgeworth expansions have been widely used in the ICA literature for approximating the pdf or the marginal entropy [29, 180, 986].

D.3 ORDER STATISTICS

The entropy function can also be estimated by a spacing estimator in light of the order statistics [77]. Let $\{x^{(i)}\}_{i=1}^{\ell}$ denote the random samples of a univariate random variable x, and the order statistics of x are simply the elements of the sample rearranged in a *nondecreasing* order: $x^{(1)} \leq x^{(2)} \leq \cdots \leq x^{(\ell)}$. A spacing of order m, or m-spacing, is defined to be $x^{(i+m)} - x^{(i)}$ for $1 \leq i < i + m \leq \ell$.

The m-spacing estimator of the entropy may be defined as [622, 719, 913]

$$H(x) \approx \frac{m}{\ell - 1} \sum_{i=0}^{(\ell-1)/m-1} \log\left(\frac{\ell+1}{m}\left(x^{(m(i+1)+1)} - x^{(mi+1)}\right)\right). \tag{D.12}$$

The estimator (D.12) is known to be asymptotically consistent when the conditions $m, \ell \to \infty$ and $m/\ell \to 0$ hold [622]. In practice, only a finite number of m is selected. In the special case of $m = 1$, the 1-spacing estimator of the entropy is obtained by

$$H(x) \approx \frac{1}{\ell - 1} \sum_{i=0}^{\ell-1} \log\big[(\ell + 1)\left(x^{(i+1)} - x^{(i)}\right)\big]. \tag{D.13}$$

Miller and Fisher [622] also proposed a modified version of the m-spacing entropy estimator (that allows m-spacing overlap to reduce the variance) as follows:

$$H(x) \approx \frac{1}{\ell - m} \sum_{i=1}^{\ell-m} \log\left(\frac{\ell + 1}{m}\left(x^{(i+m)} - x^{(i)}\right)\right), \tag{D.14}$$

which is known to be asymptotically efficient.

D.4 KERNEL ESTIMATOR

Kernel smoothing is a popular statistical method for estimating both the pdf and entropy [835, 934]. Let us consider the *Parzen estimator* for a univariate random variable x given a finite set of i.i.d. samples $\{x^{(i)}\}_{i=1}^{\ell}$. Consider a simple isotropic kernel (such as the Gaussian kernel) with the form $K_h(x) = (1/h)K(x/h)$, which is the scaled version of the kernel function $K(x)$, where $h > 0$ represents the kernel bandwidth. The Parzen estimator of the pdf $p(x)$ is given by

$$p(x) = \frac{1}{C\ell} \sum_{i=1}^{\ell} K\left(\frac{x - x^{(i)}}{h}\right), \tag{D.15}$$

where $C = \int_{-\infty}^{\infty} K_h(x)\,dx$. In practice, the kernel function $K(x)$ is often chosen to be a symmetric pdf such that $C = 1$ and

$$\int xK(x)\,dx = 0 \qquad \text{and} \qquad \int x^2 K(x)\,dx < \infty.$$

It can be shown that under the limit $h \to 0$ the Gaussian kernel function converges to a Dirac delta function: $\lim_{h\to 0} K_h(x) \to \delta(x)$. The value of the scalar h controls the degree of smoothness of the pdf: the smaller is h, the less smoothing (and therefore the greater variance) is imposed; the larger is h, the greater is the bias. Choosing an optimal kernel bandwidth is the key issue for the Parzen estimator [835, 934].

When the number of samples, ℓ, is sufficiently large, the entropy can be estimated by

$$H(x) \approx -\frac{1}{\ell}\sum_{j=1}^{\ell}\log\left[\frac{1}{\ell}\sum_{i=1}^{\ell}K\left(\frac{x^{(j)}-x^{(i)}}{h}\right)\right]. \tag{D.16}$$

For applications and discussions of entropic kernel estimators in the context of ICA, see [720].

Finally, it is noteworthy that in addition to the classic Shannon entropy other definitions of the entropy, such as α-*Rényi entropy* and (nonextensive) *Tsallis entropy*, are also available in the literature. However, an in-depth exploration of these issues is beyond the scope of the current discussion; the interested reader is referred to [261–263, 382] for discussions regarding these issues. The estimators of entropy or mutual information for discrete random variables are also discussed in [700].

APPENDIX E

EXPECTATION–MAXIMIZATION ALGORITHM

The EM algorithm [211, 608] is an elegant and powerful statistical estimation procedure to tackle the incomplete (or missing) data or parameter estimation problem. Given some observation data $\boldsymbol{x}$ and a model family parameterized by $\boldsymbol{\theta}$, the goal of the EM algorithm is to find the unknown parameters $\boldsymbol{\theta}$ such that the log-likelihood $\log p(\boldsymbol{x}|\boldsymbol{\theta})$ is maximized. Put another way, the EM algorithm solves an unconstrained optimization problem with respect to the unknown parameter $\boldsymbol{\theta}$. The EM procedure consists of two alternating steps: first, the expectation (E) step, which computes an expectation of the likelihood by including the latent variables as if they were observed; second, the maximization (M) step, which computes the MLE of the parameters by maximizing the expected likelihood found in the E step. The parameters found in the M step are then used for the next E step, and the iteration process is repeated until convergence.

E.1 ALTERNATING FREE-ENERGY MAXIMIZATION

From the statistical physics viewpoint, the EM algorithm can be understood as an alternating maximization procedure of free energy [658]. Specifically, given the observed data $\boldsymbol{x}$, we can rewrite the log-likelihood in the following form:

$$\log\, p(\boldsymbol{x}|\boldsymbol{\theta}) = \log\left(\int_z p(\boldsymbol{x}, z|\boldsymbol{\theta})\, dz\right) = \max_{q \in \mathcal{P}} \mathcal{F}(q, \boldsymbol{\theta}), \tag{E.1}$$

where $\mathcal{P}$ denotes the set of all probability distributions defined on the missing variable z and $\mathcal{F}(q, \boldsymbol{\theta})$ is the so-called *free energy* that defines the *lower bound* of

Correlative Learning: A Basis for Brain and Adaptive Systems, by Zhe Chen, Simon Haykin, Jos J. Eggermont, and Suzanna Becker

the log-likelihood:

$$
\begin{aligned}
\mathcal{F}(q,\boldsymbol{\theta}) &= \mathbb{E}_{q(z)}\left[\log p(\boldsymbol{x},z|\boldsymbol{\theta})\right] + H(q(z)) \\
&= \int q(z)\log\frac{p(z|\boldsymbol{x},\boldsymbol{\theta})p(\boldsymbol{x}|\boldsymbol{\theta})}{q(z)}dz \\
&= \int q(z)\,\log\,p(\boldsymbol{x}|\boldsymbol{\theta})\,dz + \int q(z)\,\log\,\frac{p(z|\boldsymbol{x},\boldsymbol{\theta})}{q(z)}\,dz \\
&= \log\,p(\boldsymbol{x}|\boldsymbol{\theta})\int q(z)\,dz - \int q(z)\,\log\,\frac{q(z)}{p(z|\boldsymbol{x},\boldsymbol{\theta})}\,dz \\
&= \log\,p(\boldsymbol{x}|\boldsymbol{\theta}) - \mathrm{KL}\left(q(z)\|p(z|\boldsymbol{x},\boldsymbol{\theta})\right),
\end{aligned} \tag{E.2}
$$

where the first term of the right-hand side of (E.2) denotes the *energy*, whereas the second term denotes the *entropy* (which is independent of $\boldsymbol{\theta}$). The EM algorithm comprises two alternating maximization steps with respect to q and $\boldsymbol{\theta}$, respectively:

- E step: Fix $\boldsymbol{\theta}$ and find and solve $q = \arg\max_{q'\in\mathcal{P}} \mathcal{F}(q',\boldsymbol{\theta})$;
- M step: Fix q and find and solve $\boldsymbol{\theta} = \arg\max_{\boldsymbol{\theta}'} \mathcal{F}(q,\boldsymbol{\theta}')$.

The two steps are iterated until a local maximum of free energy $\mathcal{F}(q,\boldsymbol{\theta})$ is reached.

E.2 FITTING GAUSSIAN MIXTURE MODEL

Consider a d-dimensional multivariate Gaussian mixture model as follows:

$$
\begin{aligned}
p(\mathbf{x}) &= \sum_{j=1}^{K} p(j)p(\mathbf{x}|j) \\
&= \sum_{j=1}^{K} c_j \frac{1}{\sqrt{(2\pi)^d|\boldsymbol{\Sigma}_j|}} \exp\left(-\frac{1}{2}|\mathbf{x}-\boldsymbol{\mu}_j|^{\mathrm{T}}\boldsymbol{\Sigma}_j^{-1}|\mathbf{x}-\boldsymbol{\mu}_j|\right),
\end{aligned} \tag{E.3}
$$

where K denotes the number of mixtures and $(\boldsymbol{\mu}_j, \boldsymbol{\Sigma}_j)$ denotes the mean and (full) covariance matrix of the jth mixture, $p(j) \equiv c_j$ denotes the prior probability of the jth mixture and $p(\mathbf{x}|j)$ denotes the probability of $\mathbf{x}$ generated from the jth mixture.

Given ℓ observations of i.i.d. data samples $\{\mathbf{x}_i\}_{i=1}^{\ell}$, the EM algorithm for fitting a K mixture of Gaussians can be derived as follows [231]:

- E step:

$$
p_{ij} \equiv p(j|\mathbf{x}_i) = \frac{p(\mathbf{x}_i|j)c_j}{\sum_{k=1}^{K} p(\mathbf{x}_i|k)c_k} = \frac{p(\mathbf{x}_i|j)c_j}{p(\mathbf{x}_i)}. \tag{E.4}
$$

- M step:

$$c_j^{\text{new}} = \frac{1}{\ell} \sum_{i=1}^{\ell} p(j|\mathbf{x}_i) = \frac{p_j}{\ell}, \tag{E.5}$$

$$\boldsymbol{\mu}_j^{\text{new}} = \frac{\sum_{i=1}^{\ell} p(j|\mathbf{x}_i)\mathbf{x}_i}{\sum_{i=1}^{\ell} p(j|\mathbf{x}_i)} = \frac{\sum_i p_{ij}\mathbf{x}_i}{\sum_i p_{ij}} = \frac{\sum_i p_{ij}\mathbf{x}_i}{\ell c_j^{\text{new}}}, \tag{E.6}$$

$$\boldsymbol{\Sigma}_j^{\text{new}} = \frac{\sum_{i=1}^{\ell} p_{ij}(\mathbf{x}_i - \boldsymbol{\mu}_j^{\text{new}})(\mathbf{x}_i - \boldsymbol{\mu}_j^{\text{new}})^{\text{T}}}{\ell c_j^{\text{new}}}. \tag{E.7}$$

The computational complexity of the above EM procedure is $\mathcal{O}(d\ell + K\ell^2)$.

Let $\boldsymbol{\theta} = \{c_j, \boldsymbol{\mu}_j, \boldsymbol{\Sigma}_j\}_{j=1}^K$; then the log-likelihood of the observed data $\{\mathbf{x}_i\}_{i=1}^{\ell}$ is calculated as

$$\mathcal{L} = \log \prod_{i=1}^{\ell} p(\mathbf{x}_i|\boldsymbol{\theta}) = \sum_{i=1}^{\ell} \log p(\mathbf{x}_i|\boldsymbol{\theta}). \tag{E.8}$$

Repeating the E and M steps alternatingly will produce a monotonically increasing likelihood or log-likelihood sequence until a local maximum or saddle point is approached. The convergence analysis of the EM algorithm for the Gaussian mixture model is referred to [981].

BIBLIOGRAPHY

1. L. F. Abbott and P. Dayan. The effect of correlated activity on the accuracy of a population code. *Neural Computation*, 11:91–101, 1999.
2. L. F. Abbott and W. G. Regehr. Synaptic computation. *Nature*, 431:796–803, 2004.
3. M. Abeles. *Local Cortical Circuits: An Electrophysiological Study*. Springer, Berlin, 1982.
4. M. Abeles. *Corticonics: Neural Circuits of the Cerebral Cortex*. Cambridge University Press, Cambridge, 1991.
5. M. Abeles, G. Hayon, and D. Lehmann. Modeling compositionality by dynamic binding of synfire chains. *Journal of Computational Neuroscience*, 17:179–201, 2004.
6. D. H. Ackley, G. E. Hinton, and T. J. Sejnowski. A learning algorithm for Boltzmann machines. *Cognitive Science*, 9:147–169, 1985.
7. T. Adali, T. Kim, and V. Calhoun. Independent component analysis by complex nonlinearities. In *Proceedings of IEEE ICASSP'04*, pp. 525–528, Montreal, Canada, 2004, IEEE Press, Piscataway, NJ.
8. A. Aertsen, M. Erb, and G. Palm. Dynamics of functional coupling in the cerebral cortex: An attempt at a model-based interpretation. *Physica D*, 75:103–128, 1994.
9. N. C. Aggelopoulos, L. Franco, and E. T. Rolls. Object perception in natural scenes: Encoding by inferior temporal cortex simultaneously recorded neurons. *Journal of Neurophysiology*, 93:1342–1357, 2005.
10. E. Ahissar, M. Abeles, M. Ahissar, S. Haidarliu, and E. Vaadia. Hebbian-like functional plasticity in the auditory cortex of the behaving monkey. *Neuropharmacology*, 37:633–655, 1998.
11. E. Ahissar, E. Vaadia, M. Ahissar, H. Bergman, A. Arieli, and M. Abeles. Dependence of cortical plasticity on correlated activity of single neurons and on behavioral context. *Science*, 257:1412–1415, 1992.
12. N. Ahmed and S. Vijayendra. An algorithm for line enhancement. *Proceedings of the IEEE*, 70:1459–1460, 1982.
13. J. S. Albus. A theory of cerebellar function. *Mathematical Biosciences*, 10:25–61, 1971.
14. J. S. Albus. *Brain, Behavior, and Robotics*. Byte Books, Petersborough, NH, 1981.
15. K. D. Alloway, M. Zhang, S. H. Dick, and S. A. Roy. Pervasive synchronization of local neural networks in the secondary somatosensory cortex of cats during focal cutaneous stimulation. *Experimental Brain Research*, 147:227–242, 2002.
16. J-M. Alonso, W. M. Usrey, and R. C. Reid. Precisely correlated firing in cells of the lateral geniculate nucleus. *Nature*, 383:815–819, 1996.

17. J. Alspector, R. B. Allen, V. Hu, and S. Satyanarayana. Stochastic learning networks and their electronic implementation. In D. Z. Anderson, Ed., *Advances in Neural Information Processing Systems*, pp. 9–21. American Institute of Physics, New York, 1988.

18. S. Amari. Theory of adaptive pattern classifiers. *IEEE Transactions on Electronic Computers*, 16:299–307, 1967.

19. S. Amari. Learning patterns and pattern sequences by self-organizing nets of threshold elements. *IEEE Transactions on Computers*, 21:1197–1206, 1972.

20. S. Amari. Neural theory of association and concept-formation. *Biological Cybernetics*, 26:175–185, 1977.

21. S. Amari. Topographic organization of nerve fields. *Bulletin of Mathematical Biology*, 42:339–364, 1980.

22. S. Amari. Mathematical analysis of the Alopex process for determination of visual receptive fields. *Neuroscience Letters*, Suppl. 6:S119, 1981.

23. S. Amari. Field theory of self-organizing neural nets. *IEEE Transactions on Systems, Man, and Cybernetics*, 13:741–748, 1983.

24. S. Amari. Mathematical foundations of neurocomputing. *Proceedings of the IEEE*, 78: 1443–1463, 1990.

25. S. Amari. Natural gradient works efficiently in learning. *Neural Computation*, 10: 251–276, 1998.

26. S. Amari. Natural gradient learning for over- and under-complete bases in ICA. *Neural Computation*, 11:1875–1883, 1999.

27. S. Amari, T. Chen, and A. Cichocki. Stability analysis of adaptive blind source separation. *Neural Networks*, 10(8):1345–1351, 1997.

28. S. Amari, T. Chen, and A. Cichocki. Nonholonomic orthogonal learning algorithms for blind source separation. *Neural Computation*, 12:1463–1484, 2000.

29. S. Amari, A. Cichocki, and H. H. Yang. A new learning algorithm for blind signal separation. In D. S. Touretzky, M. C. Mozer, and M. E. Hasselmo, Eds., *Advances in Neural Information Processing Systems*, Vol. 8, pp. 757–763. MIT Press, Cambridge, MA, 1996.

30. S. Amari and K. Maginu. Statistical neurodynamics of associative memory. *Neural Networks*, 1(1):63–73, 1988.

31. S. Amari and H. Nagaoka. *The Methods of Information Geometry*. AMS and Oxford University Press, New York, 2000.

32. S. Amari and A. Takeuchi. Mathematical theory on formation of category detecting nerve cells. *Biological Cybernetics*, 29:127–136, 1978.

33. J. A. Anderson. A memory storage model utilizing spatial correlation functions. *Kybernetik*, 5(3):113–119, 1969.

34. J. A. Anderson. A simple neural network generating an interactive memory. *Mathematical Biosciences*, 14:197–220, 1972.

35. J. A. Anderson. Cognitive and psychological computation with neural models. *IEEE Transactions on Systems, Man, and Cybernetics*, 13:799–815, 1983.

36. J. A. Anderson. What hebb synapses build. In W. B. Levy, J. A. Anderson, and S. Lehmkuhle, Eds., *Synaptic Modification, Neuron Selectivity, and Nervous System Organization*, pp. 153–173. Erlbaum, Hillsdale, NJ, 1985.

37. J. A. Anderson. *An Introduction to Neural Networks.* MIT Press, Cambridge, MA, 1995.

38. J. A. Anderson, M. T. Gately, P. A. Penz, and D. R. Collins. Radar signal categorization using a neural network. *Proceedings of the IEEE*, 78:1646–1657, 1990.

39. J. A. Anderson and E. Rosenfeld, Eds. *Neurocomputing: Foundations of Research.* MIT Press, Cambridge, MA, 1988.

40. J. A. Anderson, J. W. Silverstein, S. A. Ritz, and R. S. Jones. Distinctive features, categorical perception, and probability learning: Some applications of a neural model. *Psychological Review*, 84:413–451, 1977.

41. M. J. Anderson and E. Tzanakou. Auditory stimulus optimization with feedback from fuzzy clustering of neuronal responses. *IEEE Transactions on Information Technology in Biomedicine*, 6(2):159–169, 2002.

42. J. Anemüller, T. J. Sejnowski, and S. Makeig. Complex independent component analysis of frequency-domain electroencephalographic data. *Neural Networks*, 16:1311–1323, 2003.

43. S. Araki, R. Mukai, S. Makino, T. Nishikawa, and H. Saruwatari. The fundamental limitation of frequency domain blind source separation for convolutive mixtures of speech. *IEEE Transactions on Speech and Audio Processing*, 11(2):109–115, 2003.

44. S. R. Arnott, C. L. Grady, S. J. Hevenor, S. Graham, and C. Alain. The functional organization of auditory working memory as revealed by fMRI. *Journal of Cognitive Neuroscience*, 17(5):819–831, 2005.

45. N. Aronszajn. Theory of reproducing kernels. *Transactions of American Mathematical Society*, 68:337–404, 1950.

46. F. Asano, S. Ikeda, M. Ogawa, H. Asoh, and N. Kitawaki. Combined approach of array processing and independent component analysis for blind separation of acoustic signals. *IEEE Transactions on Audio and Speech Processing*, 11(3):204–215, 2003.

47. J. J. Atick and A. N. Redlich. Towards a theory of early visual processing. *Neural Computation*, 2:308–320, 1990.

48. J. J. Atick and A. N. Redlich. What does the retina know about natural scenes? *Neural Computation*, 4:196–210, 1992.

49. H. Attias. Independent factor analysis. *Neural Computation*, 11:803–851, 1999.

50. M. Atzori, S. Lei, D. I. Evans, P. O. Kanold, E. Phillips-Tansey, O. McIntyre, and C. J. McBain. Differential synaptic processing separates stationary from transient inputs to the auditory cortex. *Nature Neuroscience*, 4:1230–1237, 2001.

51. F. Bach and M. I. Jordan. Predictive low-rank decomposition for kernel methods. In *Proceedings of the 22nd International Conference on Machine Learning (ICML'2005)*, Proceedings was self-published but ACM include it in the ACM digital Library. pp. 33–40, Bonn, Germany, 2005.

52. F. R. Bach and M. I. Jordan. Kernel independent component analysis. *Journal of Machine Learning Research*, 3:1–48, 2002.

53. W. Bair, E. Zohary, and W. T. Newsome. Correlated firing in macaque visual area MT: Time scales and relationship to behavior. *Journal of Neuroscience*, 21(5): 1676–1697, 2001.

54. P. Baldi and K. Hornik. Neural networks and principal component analysis: Learning from examples without local minimum. *Neural Networks*, 1:53–58, 1989.

55. D. H. Ballard. Cortical connections and parallel processing: Structure and function. *Behavior and Brain Sciences*, 9:67–119, 1986.

56. S. Bao, V. T. Chan, and M. M. Merzenich. Cortical remodelling induced by activity of ventral tegmental dopamine neurons. *Nature*, 412:79–83, 2001.

57. S. Bao, V. T. Chan, L. Zhang, and M. M. Merzenich. Suppression of cortical representation through background conditioning. *Proceedings of the National Academy of Sciences, USA*, 100:1405–1408, 2003.

58. H. B. Barlow. Possible principles underlying the transformation of sensory messages. In W. Rosenblith, Ed., *Sensory Communication*, pp. 217–234. MIT Press, Cambridge, MA, 1961.

59. H. B. Barlow. Single units and sensation: A neuron doctrine for perceptual psychology? *Perception*, 1:371–394, 1972.

60. H. B. Barlow. Unsupervised learning. *Neural Computation*, 1:295–311, 1989.

61. H. B. Barlow and P. Földiák. Adaptation and decorrelation in the cortex. In R. M. Durin, C. Miall, and G. J. Mitchison, Eds., *The Computing Neuron*, pp. 54–72. Addison-Wesley, Wokingham, England, 1989.

62. H. B. Barlow, T. P. Kaushal, and G. J. Mitchison. Finding minimum entropy codes. *Neural Computation*, 1:412–423, 1989.

63. C. A. Barnes, B. L. McNaughton, S. J. Y. Mizumori, and B. W. Leonard. Comparison of spatial and temporal characteristics of neuronal activity in sequential stages of hippocampal processing. *Progress in Brain Research*, 83:287–300, 1990.

64. A. G. Barto, R. S. Sutton, and C. W. Anderson. Neuronlike adaptive elements that can solve difficult learning control-problems. *IEEE Transactions on Systems, Man, and Cybernetics*, 13(5):834–846, 1983.

65. G. Baudat and F. Anouar. Generalized discriminant analysis using a kernel approach. *Neural Computation*, 12:2385–2404, 2000.

66. M. F. Bear, L. N. Cooper, and F. F. Ebner. A physiological basis for a theory of synapse modification. *Science*, 237:42–47, 1987.

67. S. Becker. Unsupervised learning procedures for neural networks. *International Journal of Neural Systems*, 2:17–33, 1991.

68. S. Becker. Unsupervised learning with global objective functions. In M. A. Arbib, Ed., *Handbook of Brain Theory and Neural Networks*, pp. 997–1001. MIT Press, Cambridge, MA, 1995.

69. S. Becker. Mutual information maximization: Models of cortical self-organization. *Network: Computation in Neural Systems*, 7:7–31, 1996.

70. S. Becker. Implicit learning in 3D object recognition: The importance of temporal context. *Neural Computation*, 10:347–374, 1999.

71. S. Becker. A computational principle for hippocampal learning and neurogenesis. *Hippocampus*, 15(6):722–738, 2005.

72. S. Becker. Modeling the mind: From circuits to systems. In S. Haykin, J. C. Principe, T. J. Sejnowski, and J. McWhirter, Eds., *New Directions in Statistical Signal Processing: From Systems to Brain*, pp. 1–21. MIT Press, Cambridge, MA, 2006.

73. S. Becker and I. C. Bruce. Neural coding in the auditory periphery: Insights from physiology and modeling lead to a novel hearing compensation algorithm. Paper presented at the Workshop in Neural Information Coding, Les Houches, France, 2002.

74. S. Becker and G. E. Hinton. A self-organizing neural network that discovers surfaces in random-dot stereograms. *Nature*, 355:161–163, January 1992.

75. S. Becker and M. D. Plumbley. Unsupervised neural network learning procedures for feature extraction and classification. *International Journal of Applied Intelligence*, 6(3):185–205, 1996.

76. S. Becker and R. Zemel. Unsupervised learning with global objective functions. In M. A. Arbib, Ed., *Handbook of Brain Theory and Neural Networks*, 2nd ed., pp. 1183–1187. MIT Press, Cambridge, MA, 2005.

77. J. Beirlant, E. J. Dudewicz, L. Györfi, and E. C. van der Meulen. Nonparametric entropy estimation: An overview. *International Journal of Mathematical Statistical Sciences*, 6(1):17–39, 1997.

78. A. Bell and T. J. Sejnowski. An information-maximization approach to blind separation and blind deconvolution. *Neural Computation*, 7:1129–1159, 1995.

79. A. Bell and T. J. Sejnowski. The "independent components" of natural scenes are edge filters. *Vision Research*, 37(3):3327–3338, 1997.

80. C. C. Bell, V. Z. Han, Y. Sugawara, and K. Grant. Synaptic plasticity in a cerebellum-like structure depends on temporal order. *Nature*, 387:278–281, 1997.

81. A. Belouchrani, K. Abed-Meraim, J.-F. Cardoso, and E. Moulines. A blind source separation technique based on second order statistics. *IEEE Transactions on Signal Processing*, 45(2):434–444, 1988.

82. J. S. Bendat and A. G. Piersol. *Random Data: Analysis and Measurement Procedures*, 2nd ed. Wiley, New York, 1986.

83. N. Benvenuto and F. Piazza. On the complex backpropagation algorithm. *IEEE Transactions on Signal Processing*, 40(4):967–969, 1992.

84. G. S. Berns, P. Dayan, and T. J. Sejnowski. A corrrelational model for the development of disparity selectivity in visual cortex that depends on prenatal and postnatal phases. *Proceedings of the National Academy of Sciences, USA*, 90:8277–8281, 1993.

85. D. P. Bertsekas and J. N. Tsitsiklis. *Neuro-Dynamic Programming*. Athena Scientific, Belmont, MA, 1996.

86. R. L. Beurle. Properties of a mass of cells capable of regenerating pulses. *Philosophical Transactions of the Royal Society of London, B*, 240:55–94, 1956.

87. G-Q. Bi and M. Poo. Synaptic modifications in cultured hippocampal neurons: Dependence on spike timing, synaptic strength, and postsynaptic cell type. *Journal of Neuroscience*, 18:10464–10472, 1998.

88. G-Q. Bi and M. Poo. Distributed synaptic modification in neural networks induced by patterned simulation. *Nature*, 401:792–796, 1999.

89. G-Q. Bi and M. Poo. Synaptic modification of correlated activity: Hebb's postulate revisited. *Annual Review of Neuroscience*, 24:139–166, 2001.

90. A. Bia. Alopex-B: A new, simple, but yet faster version of the Alopex training algorithm. *International Journal of Neural Systems*, 11(6):497–507, 2001.

91. W. Bialek, F. Rieke, R. de Ruyter van Steveninck, and D. Warland. Reading a neural code. *Science*, 252:1854–1857, 1991.

92. E. Bienenstock. A model of neocortex. *Network: Computation in Neural Systems*, 6: 179–224, 1995.

93. E. Bienenstock, L. N. Cooper, and P. W. Munro. Theory for the development of neuron selectivity: Orientation specificity and binocular interaction in visual cortex. *Journal of Neuroscience*, 2:32–48, 1982.

94. E. Bingham and A. Hyvarinen. A fast fixed-point algorithm for independent component analysis of complex-valued signals. *International Journal of Neural Systems*, 10(1):1–8, 2000.

95. N. Birbaumer, W. Lutzenberger, P. Montoya, W. Larbig, K. Unertl, S. Topfner, W. Grodd, E. Taub, and H. Flor. Effects of regional anesthesia on phantom limb pain are mirrored in changes in cortical reorganization. *Journal of Neuroscience*, 17:5503–5508, 1997.

96. F. Black and M. Scholes. The pricing of options and corporate liabilities. *Journal of Political Economy*, 81:637–659, 1973.

97. B. S. Blais, N. Intrator, H. Shouval, and L. N. Cooper. Receptive field formation in natural scene environments: Comparison of single cell learning rules. *Neural Computation*, 10:1797–1813, 1998.

98. B. H. Bland and L. V. Colom. Extrinsic and intrinsic properties underlying oscillation and synchrony in limbic cortex. *Progress in Neurobiology*, 41:157–208, 1993.

99. T. Blaschke, P. Berkes, and L. Wiskott. What is the relation between slow feature analysis and independent component analysis? *Neural Computation*, 18(10): 2495–2508, 2006.

100. T. V. P. Bliss and T. Lomo. Long-lasting potentiation of synaptic transmission in the dendate area of anaesthetized rabbit following stimulation of the prefrant path. *Journal of Physiology*, 232:551–556, 1973.

101. J. Bondy, S. Becker, I. Bruce, L. Trainor, and S. Haykin. A novel signal-processing strategy for hearing-aid design: Neurocompensation. *Signal Processing*, 84:1239–1253, 2004.

102. J. Bondy, I. Bruce, R. Dong, S. Becker, and S. Haykin. Modeling intelligibility of hearing-aid compression circuits. In *Proceedings of the 37th Asilomar Conference on Signals, Systems, and Computers*, pp. 720–724, 2003, IEEE Press Pacific Grove, CA.

103. B. H. Bonham, S. W. Cheung, B. Godey, and C. E. Schreiner. Spatial organization of frequency response areas and rate/level functions in the developing A1. *Journal of Neurophysiology*, 91:841–854, 2004.

104. V. S. Borkar. Stochastic approximation with two time scales. *Systems and Control Letters*, 29:291–294, 1997.

105. R. J. C. Bosman, W. A. van Leeuwen, and B. Wemmenhove. Combining Hebbian and reinforcement learning in minibrain model. *Neural Networks*, 17:29–36, 2004.

106. H. R. Bourne and R. Nicoll. Molecular machines integrate coincident synaptic signals. *Cell*, 72:841–854, 1993.

107. O. Bousquet, K. Balakrishnan, and V. Honavar. Is the hippocampus a Kalman filter. Technical Report, 97-11, Department of Computer Science, Iowa State University, July 1997.

108. O. Bousquet, K. Balakrishnan, and V. Honavar. Is the hippocampus a Kalman filter? In *Proc. Pacific Symposium on Biocomputing*, pp. 657–668, 1998.

109. E. S. Boyden, A. Katoh, and J. L. Raymond. Cerebellum-dependent learning: The role of multiple plasticity mechanisms. *Annual Review of Neuroscience*, 27:581–609, 2004.

110. V. Braitenberg. Thoughts on the cerebral cortex. *Journal of Theoretical Biology*, 46(2):421–447, 1974.

111. N. Brenner, W. Bialek, and R. de Ruyter van Steveninck. Adaptive rescaling maximizes information transmission. *Neuron*, 26:695–702, 2000.

112. T. Briegel and V. Tresp. Fisher scoring and a mixture of modes approach for approximate inference and learning in nonlinear state space models. In M. Kearns, S. Solla, and D. Cohn, Eds., *Advances in Neural Information Processing Systems*, Vol. 11, pp. 403–409. MIT Press, Cambridge, MA, 1999.

113. D. R. Brillinger. An introduction to polyspectra. *Annals of Mathematical Statistics*, 36:1351–1374, 1965.

114. D. R. Brillinger. Statistical inference for stationary point processes. In M. L. Puri, Ed., *Stochastic Processes and Related Topics*, pp. 55–99. Academic, New York, 1975.

115. R. W. Brockett. Dynamical systems that sort lists, diagonalize matrices, and solve linear programming problems. *Linear Algebra and Applications*, 146:79–91, 1991.

116. C. D. Brody and J. J. Hopfield. Simple networks for spike-timing-based computation, with application to olfactory processing. *Neuron*, 37:843–852, 2003.

117. M. Brosch and C. E. Schreiner. Correlations between neural discharges are related to receptive field properties in cat primary auditory cortex. *European Journal of Neuroscience*, 11:3517–3530, 1999.

118. E. N. Brown, R. E. Kass, and K. P. Mitra. Multiple neural spike train data analysis: State-of-the-art and future challenges. *Nature Neuroscience*, 7(5):456–461, 2004.

119. G. J. Brown and D. L Wang. Modelling the perceptual segregation of concurrent vowels with a network of neural oscillation. *Neural Networks*, 10(9):1547–1558, 1997.

120. M. Brown, D. R. Irvine, and V. N. Park. Perceptual learning on an auditory frequency discrimination task by cats: Association with changes in primary auditory cortex. *Cerebral Cortex*, 14(9):952–965, 2004.

121. T. H. Brown, P. F. Chapman, E. W. Kairiss, and C. L. Keenan. Long-term synaptic potentiation. *Science*, 242:724–728, 1988.

122. T. H. Brown, E. W. Kairiss, and C. L. Keenan. Hebbian synapses: Biophysical mechanisms and algorithms. *Annual Review of Neuroscience*, 13:475–511, 1990.

123. I. C. Bruce, M. B. Sachs, and E. Young. An auditory-periphery model of the effects of acoustic trauma on auditory nerve responses. *Journal of the Acoustical Society of America*, 113(1):369–388, 2003.

124. R. M. Bruno and B. Sakmann. Cortex is driven by weak but synchronously active thalamocortical synapses. *Science*, 312:1622–1627, 2006.

125. D. V. Buonomano and M. M. Merzenich. Cortical plasticity: From synapses to maps. *Annual Review of Neuroscience*, 21:149–186, 1998.

126. J. J. Bussgang. Cross-correlation functions of amplitude-distored Gaussian signals. Technical Report 216, MIT Research Laboratory of Electronics, 1952.

127. D. A. Butts, M. B. Feller, C. J. Shatz, and D. S. Rokhsar. Retinal waves are governed by collective network properties. *Journal of Neuroscience*, 19:3580–3593, 1999.

128. G. Buzsáki. Theta rhythm of navigation: Link between path integration and landmark navigation, episodic and semantic memory. *Hippocampus*, 15:827–840, 2005.

129. G. Buzsáki, Z. Horvath, R. Urioste, J. Hetke, and K. Wise. High-frequency network oscillation in the hippocampus. *Science*, 256:1025–1027, 1992.

130. G. Buzsáki and A. Kandel. Somadendritic backpropagation of action potentials in cortical pyramidal cells of the awake rat. *Journal of Neurophysiology*, 79:1587–1591, 1998.

131. W. Byrne, A. Parkinson, and P. Newall. Hearing aid gain and frequency response requirements for the severely/profoundly hearing impaired. *Ear and Hearing*, 11:40–49, 1990.

132. E. R. Caianiello. Outline of a theory of thought-processes and thinking machines. *Journal of Theoretical Biology*, 1:204–235, 1961.

133. M. B. Calford. Dynamic representational plasticity in sensory cortex. *Neuroscience*, 111(4):709–738, 2002.

134. M. B. Calford and R. Tweedale. Immediate and chronic changes in responses of somatosensory cortex in adult flying-fox after digit amputation. *Nature*, 332:446–448, 1988.

135. M. B. Calford, C. Wang, V. Taglianetti, W. J. Waleszczyk, W. Burke, and B. Dreher. Plasticity in adult cat visual cortex (area 17) following circumscribed monocular lesions of all retinal layers. *Journal of Physiology*, 524:587–602, 2000.

136. M. B. Calford, L. L. Wright, A. B. Metha, and V. Taglianetti. Topographic plasticity in primary visual cortex is mediated by local corticocortical connections. *Journal of Neuroscience*, 23:6434–6442, 2003.

137. V. Calhoun and T. Adali. Complex Infomax: Convergence and approximation of Infomax with complex nonlinearities. In *Proceedings of IEEE Neural Networks for Signal Processing (NNSP'02)*, pp. 307–316, Martigny, Swizerland, 2002, IEEE Press Piscataway, NJ.

138. V. D. Calhoun, T. Adali, G. D. Pearlson, P. C. M. van Zijl, and J. J. Pekar. Independent component analysis of fMRI data in the complex domain. *Magnetic Resonance in Medicine*, 48:180–192, 2002.

139. J. L. Cantero, M. Atienza, R. Stickgold, M. J. Kahana, J. R. Madsen, and B. Kocsis. Sleep-dependent theta oscillations in the human hippocampus and neocortex. *Journal of Neuroscience*, 23:10897–10903, 2003.

140. J. B. Caplan, J. R. Madsen, A. Schulze-Bonhage, R. Aschenbrenner-Scheibe, E. L. Newman, and M. J. Kahana. Human theta oscillations related to sensorimotor integration and spatial learning. *Journal of Neuroscience*, 23:4726–4736, 2003.

141. O. Cappé, E. Moulines, and T. Rydén. *Inference in Hidden Markov Models*. Springer, Berlin, 2005.

142. J.-F. Cardoso. Super-symmetric decomposition of the fourth-order cumulant tensors: Blind identification of more sources than sensors. In *Proceedings of IEEE ICASSP'91*, pp. 3109–3112, 1991, IEEE Press Piscataway, NJ.

143. J.-F. Cardoso. An efficient technique for the blind separation of complex sources. In *Proc. Higher-Order Statistics (HOS'93)*, pp. 275–279, South Lake Tahoe, CA, 1993.

144. J.-F. Cardoso. Infomax and maximum likelihood for blind source separation. *IEEE Signal Processing Letters*, 4(4):112–114, April 1997.

145. J.-F. Cardoso. Blind signal separation: Statistical principles. *Proceedings of the IEEE*, 86(10):2029–2025, October 1998.

146. J.-F. Cardoso. High-order contrasts for independent component analysis. *Neural Computation*, 11(1):157–192, 1999.

147. J-F. Cardoso. Entropic contrasts for souce separation: Geometry and stability. In S. Haykin, Ed., *Unsupervised Adaptive Filtering*, Vol. I, pp. 139–190. Wiley, New York, 2000.

148. J.-F. Cardoso and B. Laheld. Equivariant adaptive source separation. *IEEE Transactions on Signal Processing*, 44(12):3017–3030, December 1996.

149. J.-F. Cardoso and A. Solouminac. Blind beamforming for non-Gaussian signals. *IEE Proceedings of Vision, Image and Signal Processing*, 140(6):362–370, December 1993.

150. G. Carpenter and S. Grossberg. The ART of adaptive pattern recognition by a self-organizing neural networks. *Computer*, 21(3):77–88, March 1980.

151. C. E. Carr and M. Konishi. A circuit for detection of interaural time differences in the brain stem of the barn owl. *Journal of Neuroscience*, 10:3227–3246, 1990.

152. G. C. Carter. Coherence and time delay estimation. *Proceedings of the IEEE*, 75:236–255, 1987.

153. M. V. Chafee and P. S. Goldman-Rakic. Matching patterns of activity in primate prefrontal area 8a and parietal area 7ip neurons during a spatial working memory task. *Journal of Neurophysiology*, 79(6):2919–2940, 1998.

154. S. V. Chakravarthy and J. Ghosh. A complex-valued associative memory for storing patterns as oscillatory states. *Biological Cybernetics*, 75(3):229–238, 1996.

155. J.-P. Changeux and T. Heidmann. Allosteric receptors and molecular models of learning. In G. M. Edelman, W. E. Gall, and W. D. Cowan, Eds., *Synaptic Function*, pp. 549–601. Wiley, New York, 1987.

156. T.-P. Chen, S. Amari, and Q. Lin. A unified algorithm for principal and minor components extraction. *Neural Networks*, 11(3):385–390, 1998.

157. Y. Chen and C. Hou. High resolution adaptive bearing estimation using a complex-weighted neural network. In *Proceedings of ICASSP'92*, pp. 317–320, 1992, IEEE Press Piscataway, NJ.

158. Z. Chen. Bayesian filtering: From Kalman filters to particle filters, and beyond. Technical Report, Adaptive Systems Lab, McMaster University. Available: http://soma.crl.mcmaster.ca/~zhechen/download/ieee_bayesian.ps, Feburary 2003.

159. Z. Chen. Stochastic correlative firing figure-ground segregation. *Biological Cybernetics*, 92(3):192–198, 2005.

160. Z. Chen, S. Becker, J. Bondy, I. Bruce, and S. Haykin. A novel model-based hearing compensation design using a gradient-free optimization method. *Neural Computation*, 17(12):2648–2671, 2005.

161. Z. Chen, S. L. Gay, and S. Haykin. Proportionate adaptation: New paradigms in adaptive filters. In S. Haykin and B. Widrow, Eds., *Least Mean Squared Filters*, pp. 293–334. Wiley, New York, 2003.

162. Z. Chen and S. Haykin. On different facets of regularization theory. *Neural Computation*, 14(12):2791–2846, 2002.

163. Z. Chen, S. Haykin, and S. Becker. Sampling-based ALOPEX algorithms for neural networks and optimization. Technical Report, Adaptive Systems Lab, McMaster University, Available: http://soma.crl.mcmaster.ca/~zhechen/download/TR_alopex.pdf, June 2003.

164. Z. Chen and J. Ma. Contrast functions for non-circular and circular sources separation in complex-valued ICA. In *Proceedings of Int. Joint Conf. Neural Networks (IJCNN'06)*, pp. 1192–1199, Vancouver, Canada, 2006.

165. Z. X. Chen, J. W. Shuai, J. C. Zheng, R. T. Liu, and B. X. Wu. The storage capacity of the complex phasor neural network. *Physica A*, 225(2):157–163, 1996.

166. E. C. Cherry. Some experiments on the recognition of speech, with one and two ears. *Journal of the Acoustical of Society of America*, 25:975–979, 1953.

167. J. J. Chrobak and G. Buzsáki. Selective activation of deep layer (V–VI) retrohippocampal cortical-neurons during hippocampal sharp waves in the behaving rat. *Journal of Neuroscience*, 14:6160–6170, 1994.

168. J. J. Chrobak and G. Buzsáki. High-frequency oscillations in the output networks of the hippocampal-entorhinal axis of the freely behaving rat. *Journal of Neuroscience*, 16(9):3056–3066, 1996.

169. J. J. Chrobak and G. Buzsáki. Gamma oscillations in the entorhinal cortex of the freely behaving rat. *Journal of Neuroscience*, 18(1):388–398, 1998.

170. J. J. Chrobak, A. Lorincz, and G. Buzsáki. Physiological patterns in the hippocampo-entorhinal cortex system. *Hippocampus*, 10(4):457–465, 2000.

171. P. S. Churchland and T. J. Sejnowski. *The Computational Brain*. MIT Press, Cambridge, MA, 1992.

172. A. Cichocki and S. Amari. *Adaptive Blind Signal and Image Processing*. Wiley, New York, 2002.

173. A. Cichocki, W. Kasprzak, and S. Amari. Multi-layer neural networks with a local adaptive learning rule for blind separation of source signals. In *Proceedings of International Symposium on Nonlinear Theory Applications*, pp. 61–65, Las Vegas, NV, 1995.

174. S. A. Clark, T. Allard, W. M. Jenkins, and M. M. Merzenich. Receptive fields in the body-surface map in adult cortex defined by temporally correlated inputs. *Nature*, 332:444–445, 1988.

175. J. D. Cohen, W. M. Perlstein, T. S. Braver, L. E. Nystrom, D. C. Noll, J. Jonides, and E. E. Smith. Temporal dynamics of brain activation during a working memory task. *Nature*, 386:604–608, 1997.

176. L. Cohen. Time-frequency distribution—-a review. *Proceedings of the IEEE*, 77(7): 941–981, July 1989.

177. L. Cohen. *Time-Frequency Analysis*. Prentice-Hall, Englewood Cliffs, NJ, 1995.

178. M. A. Cohen and S. Grossberg. Absolute stability of global pattern formation and parallel memory storage by competitive neural networks. *IEEE Transactions on Systems, Man, and Cybernetics*, 13(3):815–826, 1983.

179. Y. E. Cohen and E. I. Knudsen. Maps versus clusters: Different representations of auditory space in the midbrain and forebrain. *Trends in Neuroscience*, 22(3):128–135, 1999.

180. P. Comon. Independent component analysis, a new concept? *Signal Processing*, 36:287–314, 1994.

181. P. Comon. Contrast for multichannel blind deconvolution. *IEEE Signal Processing Letters*, 3(7):209–211, 1996.

182. I. Constantin, C. Richard, R. Lengelle, and L. Soufflet. Regularized kernel-based Wiener filtering: Application to magnetoencephalographic signals denoising. In

Proceedings of ICASSP'2005, pp. 289–292, Philadelphia, PA, 2005, IEEE Press Piscataway, NJ.

183. J. E. Cook. Correlated activity in the CNS: A role on every timescale? *Trends in Neuroscience*, 14:397–401, 1991.

184. M. Cooke. *Modelling Auditory Processing and Organization.* Cambridge University Press, Cambridge, 1993.

185. L. N. Cooper. A possible organization of animal memory and learning. In B. Lundqvist and S. Lundqvist, Eds., *Collective Properties of Physical Systems*, pp. 252–264. Academic, New York, 1973.

186. L. N. Cooper, N. Intrator, B. S. Blais, and H. Z. Shouval. *Theory of Cortical Plasticity.* World Scientific, Singapore, 2004.

187. C. Cortes and V. Vapnik. Support-vector networks. *Machine Learning*, 20:273–297, 1995.

188. S. M. Courtney, L. G. Ungerleider, K. Keil, and J. V. Haxby. Transient and sustained activity in a distributed neural system for human working memory. *Nature*, 386:608–611, 1997.

189. T. M. Cover and J. A. Thomas. *Elements of Information Theory.* Wiley, New York, 1991.

190. J. D. Cowan. Statistical mechanics of neural nets. In E. R. Caianiello, Ed., *Neural Networks*, pp. 181–188. Springer, Berlin, 1968.

191. D. R. Cox and V. Isham. *Point Processes.* Chapman and Hall, London, 1980.

192. D. R. Cox and P. A. W. Lewis. *The Statistical Analysis of Series of Events.* Chapman and Hall, London, 1966.

193. F. Crick. Function of the thalamic reticular complex: The searchlight hypothesis. *Proceedings of the National Academy of Sciences, USA*, 81:4586–4590, 1984.

194. S. J. Cruikshank and N. M. Weinberger. Receptive-field plasticity in the adult auditory cortex induced by Hebbian covariance. *Journal of Neuroscience*, 16:861–875, 1996.

195. Y. Dan and M. Poo. Spike timing-dependent plasticity of neural circuits. *Neuron*, 44:23–30, 2004.

196. C. Darian-Smith and C. D. Gilbert. Axonal sprouting accompanies functional reorganization in adult cat striate cortex. *Nature*, 368:737–740, 1994.

197. A. Das and C. D. Gilbert. Receptive field expansion in adult visual cortex is linked to dynamic changes in strength of cortical connections. *Journal of Neurophysiology*, 74:779–792, 1995.

198. T. J. Dasey and E. M. Tzanakou. Detection of multiple sclerosis with visual evoked potentials—An unsupervised computational intelligence system. *IEEE Transactions on Information Technology in Biomedicine*, 4(3):216–224, 2000.

199. J. G. Daugman. Uncertainty relations for resolution in space, spatial frequency, and orientation optimized by two-dimensional visual cortical filters. *Journal of the Optical Society of America, A*, 2:1160–1169, 1985.

200. P. Dayan. Arbitrary elastic topologies and ocular dominance. *Neural Computation*, 5:392–401, 1993.

201. P. Dayan and L. F. Abbott. *Theoretical Neuroscience: Computational and Mathematical Modeling of Neural Systems.* MIT Press, Cambridge, MA, 2001.

202. P. Dayan and B. W. Balleine. Reward, motivation and reinforcement learning. *Neuron*, 36:285–298, 2002.

203. S. A. Deadwyler and R. E. Hapson. The significance of neural ensemble coding during behavior and cognition. *Annual Review of Neuroscience*, 20:217–244, 1997.

204. S. Debener, C. S. Herrmann, C. Kranczioch, D. Gembris, and A. K. Engel. Top-down attentional processing enhances auditory evoked gamma band activity. *Neuroreport*, 14(5):683–686, 2003.

205. R. C. deCharms and M. M. Merzenich. Primary cortical representation of sounds by the coordination of action-potential timing. *Nature*, 381:610–613, 1996.

206. R. C. deCharms and A. Zador. Neural representation and the cortical code. *Annual Review of Neuroscience*, 23:613–647, 2000.

207. G. Deco and D. Obradovic. *An Information-Theoretic Approach to Neural Computing*. Springer-Verlag, Berlin, 1996.

208. J. F. G. deFreitas. Bayesian methods for neural networks. Ph.D. thesis, Engineering Department, Cambridge University, 1999.

209. J. F. G. deFreitas, M. Niranjan, A. H. Gee, and A. Doucet. Sequential Monte Carlo methods to train neural network models. *Neural Computation*, 12(4):955–993, 2000.

210. T. DelSole and P. Chang. Predictable component analysis, canonical correlation analysis, and autoregressive models. *Journal of the Atmospheric Sciences*, 60(2):409–416, 2003.

211. A. Dempster, N. Laird, and D. Rubin. Maximum likelihood from incomplete data via the EM algorithm (with discussions). *Journal of the Royal Statistical Society, Series B*, 39:1–38, 1977.

212. R. Descartes. *Traité de l'homme*. 1664. Translated by J. Cottingham et al. *The Philosophical Writings of Descartes*, Vol. 1, pp. 99–108. Cambrige University Press, 1985.

213. A. Destexhe, D. Contreras, and M. Steriade. Cortically-induced coherence of a thalamic-generated oscillation. *Neuroscience*, 92(2):427–443, 1999.

214. E. A. DeYoe and D. C. Van Essen. Concurrent processing streams in monkey visual cortex. *Trends in Neurosciences*, 11:219–226, 1988.

215. K. Diamantaras and S. Kung. Cross-correlation neural networks models. *IEEE Transactions on Signal Processing*, 42(11):3218–3223, 1994.

216. K. Diamantaras and S. Kung. *Principal Component Neural Networks: Theory and Applications*. Wiley, New York, 1996.

217. D. M. Diamond and N. M. Weinberger. Role of context in the expression of learning-induced plasticity of single neurons in auditory cortex. *Behavior Neuroscience*, 103(3):471–494, 1989.

218. Z. Ding and Y. Li, Eds. *Blind Equalization and Identification*. Marcel Dekker, New York, 2001.

219. T. J. Dodd and C. J. Harris. Identification of nonlinear time series via kernels. *International Journal of Systems Science*, 33(9):737–750, 2002.

220. M. Dominguez, S. Becker, I. Bruce, and H. Read. A spiking neuron model of cortical correlates of sensorineural hearing loss: Spontaneous firing, synchrony, and tinnitus. *Neural Computation*, 18(12):2942–2958, 2006.

221. R. Dong. Perceptual binaural speech enhancement in noisy environments. Master's thesis, Department of Electrical and Computer Engineering, McMaster University, 2005.

222. R. Dony and S. Haykin. Neural network approaches to image compression. *Proceedings of the IEEE*, 83(2):288–303, 1995.

223. G. Dornhege, B. Blankertz, M. Krauledat, F. Losch, G. Curio, and K.-R. Müller. Combined optimization of spatial and temporal filters in improving brain-computer interface. *IEEE Transactions on Biomedical Engineering*, 53(11):2274–2281, 2006.

224. G. Dornhege, J. del R. Millán, T. Hinterberger, D. McFarland, and K.-R. Müller., Eds. *Towards Brain-Computer Interfacing*. MIT Press, Cambridge, MA, 2007.

225. A. Doucet, N. de Freitas, and N. Gordon, Eds. *Sequential Monte Carlo Methods in Practice*. Springer, New York, 2001.

226. S. C. Douglas. Fixed-point fastICA algorithms for the blind separation of complex-valued signal mixtures. In *Proceedings of the 39th Asilomar Conference on Signals, Systems, and Computers*, pp. 1320–1325, 2005.

227. S. C. Douglas and A. Cichocki. Neural networks for blind decorrelation of signals. *IEEE Transactions on Signal Processing*, 45(11):2849–2842, November 1997.

228. B. Dreher, W. Burke, and M. B. Calford. Cortical plasticity revealed by circumscribed retinal lesions or artificial scotomas. *Progress of Brain Research*, 134:217–246, 2001.

229. P. J. Drew and L. F. Abbott. Extending the effects of spike-timing-dependent plasticity to behavioral timescales. *Proceedings of the National Academy of Sciences, USA*, 103:8876–8881, 2006.

230. S. Duane, A. D. Kennedy, B. J. Pendleton, and D. Roweth. Hybrid Monte Carlo. *Physics Letter B*, 55:2774–2777, 1987.

231. R. O. Duda, P. E. Hart, and D. G. Stork. *Pattern Classification*, 2nd ed. Wiley, New York, 2001.

232. R. Durbin and G. Mitchison. A dimension reduction framework for understanding cortical maps. *Nature*, 343:644–647, 1990.

233. R. Durbin, R. Szeliski, and A. Yuille. An analysis of the elastic net approach to the traveling salesman problem. *Neural Computation*, 1:348–358, 1989.

234. R. Durbin and D. Willshaw. An analogue approach to the traveling salesman problem using an elastic net method. *Nature*, 326:689–691, 1987.

235. R. Eckhorn, R. Bauer, W. Jordan, M. Brosch, W. Kruse, M. Munk, and H. J. Reitboeck. Coherent oscillations: A mechanism of feature linking in the visual cortex? *Biological Cybernetics*, 60:121–130, 1988.

236. J.-P. Eckmann, S. O. Kamphorst, and D. Ruelle. Recurrence plots of dynamical systems. *Europhysics Letters*, 4:973–977, 1987.

237. J. M. Edeline, P. Pham, and N. M. Weinberger. Rapid development of learning-induced receptive field plasticity in the auditory cortex. *Behavior Neuroscience*, 107(4):539–551, 1993.

238. G. M. Edelman. Group selection and phasic reentrant signaling: A theory of higher brain function. In G. M. Edelman and V. B. Mountcastle, Eds., *The Mindful Brain*, pp. 51–100. MIT Press, Cambridge, MA, 1978.

239. G. M. Edelman. *Neural Darwinism: The Theory of Neuronal Group Selection.* Basic Books, New York, 1987.

240. G. M. Edelman. Building a picture of the brain. *Annals of New York Academy of Sciences*, 882:68–89, 1999.

241. J. J. Eggermont. *The Correlative Brain: Theory and Experiment in Neural Interaction.* Springer-Verlag, New York, 1990.

242. J. J. Eggermont. Neural interaction in cat primary auditory cortex: Dependence on recording depth, electrode separation and age. *Journal of Neurophysiology*, 68:1216–1228, 1992.

243. J. J. Eggermont. Functional aspects of synchrony and correlation in the auditory nervous system. *Concepts in Neuroscience*, 4(2):105–129, 1993.

244. J. J. Eggermont. Neural interaction in cat primary auditory cortex II: Effects of sound stimulation. *Journal of Neurophysiology*, 71:246–270, 1994.

245. J. J. Eggermont. Differential maturation rates for response parameters in cat primary auditory cortex. *Auditory Neuroscience*, 2:309–327, 1996.

246. J. J. Eggermont. The magnitude and phase of temporal modulation transfer functions in cat primary auditory cortex. *Journal of Neuroscience*, 19(7):2780–2788, 1999.

247. J. J. Eggermont. Sound induced correlation of neural activity between and within three auditory cortical areas. *Journal of Neurophysiology*, 83:2708–2722, 2000.

248. J. J. Eggermont. Between sound and perception: Reviewing the search for a neural code. *Hearing Research*, 157:1–42, 2001.

249. J. J. Eggermont. Temporal modulation transfer functions in cat primary auditory cortex: Separating stimulus effects from neural mechanisms. *Journal of Neurophysiology*, 87(1):305–321, 2002.

250. J. J. Eggermont. Properties of correlated neural activity clusters in cat auditory cortex resemble those of neural assemblies. *Journal of Neurophysiology*, 96(2):746–764, 2006.

251. J. J. Eggermont and H. Komiya. Moderate noise trauma in juvenile cats results in profound cortical topographic map changes in adulthood. *Hearing Research*, 142:89–101, 2000.

252. J. J. Eggermont and J. E. Mossop. Azimuth coding in primary auditory cortex of the cat I: Spike synchrony vs. spike count representations. *Journal of Neurophysiology*, 80:2133–2150, 1998.

253. J. J. Eggermont and L. E. Roberts. The neuroscience of tinnitus. *Trends in Neuroscience*, 27(11):678–682, 2004.

254. J. J. Eggermont and G. M. Smith. Synchrony between single-unit activity and local field potentials in relation to periodicity coding in primary auditory cortex. *Journal of Neurophysiology*, 73(1):227–245, 1995.

255. H. Eichenbaum and J. L. Davis, Eds. *Neuronal Ensembles: Strategies for Recording and Decoding*. Wiley-Liss, New York, 1998.

256. A. D. Ekstrom, M. J. Kahana, J. B. Caplan, T. A. Fields, E. A. Isham, E. L. Newman, and I. Fried. Cellular networks underlying human spatial navigation. *Nature*, 425:184–187, 2003.

257. M. Elhilali. Neural basis and computational strategies for auditory processing. Ph.D. thesis, Department of Electrical and Computer Engineering, University of Maryland, 2004.

258. P. Elias. Predictive coding I, II. *IRE Transactions on Information Theory*, 1:16–33, March 1955.

259. A. K. Engel, P. König, and W. Singer. Direct physiological evidence for scene segmentation by temporal coding. *Proceedings of the National Academy of Sciences, USA*, 88:9136–9140, 1991.

260. A. K. Engel, A. K. Kreiter, P. König, and W. Singer. Synchronization of oscillatory neuronal responses between striate and extrastriate visual cortical areas of the cat. *Proceedings of the National Academy of Sciences, USA*, 88:6048–6052, 1991.

261. D. Erdogmus, K. E. Hild II, and J. C. Principe. Blind source separation using Renyi's alpha-marginal entropies. *Neurocomputing*, 49(1):25–38, 2002.

262. D. Erdogmus, K. E. Hild II, and J. C. Principe. On-line entropy manipulation: Stochastic information gradient. *IEEE Signal Processing Letters*, 10(8):242–245, 2003.

263. D. Erdogmus and J. C. Principe. From linear adaptive filtering to nonlinear information processing: The design and analysis of information processing systems. *IEEE Signal Processing Magazine*, 23(6):14–33, November 2006.

264. D. Erdogmus and J. C. Principe. *Information Theoretic Learning*. Wiley, New York, 2007.

265. J. Eriksson and V. Koivunen. Complex random vectors and ICA models: Identifiability, uniqueness and separability. *IEEE Transactions on Information Theory*, 52(3):1017–1029, March 2006.

266. J. Eriksson, A-M. Seppola, and V. Koivunen. Complex ICA for circular and non-circular sources. In *Proceedings of the 13th European Signal Processing Conference (EUSIPCO'2005)*, Antalya, Turkey, 2005.

267. T. Evgeniou, M. Pontil, and T. Poggio. Regularization networks and support vector machines. *Advances in Computational Mathematics*, 13(1):1–50, 2000.

268. U. T. Eysel. Functional reconnections without new axonal growth in a partially denervated visual relay nucleus. *Nature*, 299:442–444, 1982.

269. U. T. Eysel, G. Schweigart, T. Mittmann, D. Eyding, Y. Qu, F. Vandesande, G. Orban, and L. Arckens. Reorganization in the visual cortex after retinal and cortical damage. *Restorative Neurology and Neuroscience*, 15:153–164, 1999.

270. B. M. Faggin, K. T. Nguyen, and M. A. Nicolelis. Immediate and simultaneous sensory reorganization at cortical and subcortical levels of the somatosensory system. *Proceedings of the National Academy of Sciences, USA*, 94:9428–9433, 1997.

271. M. S. Falconbridge, R. L. Stamps, and D. R. Badcock. A simple Hebbian/anti-Hebbian network learns the sparse, independent components of natural images. *Neural Computation*, 18(2):415–429, 2006.

272. B. G. Farley and W. A. Clark. Simulation of self-organizing systems by digital computer. *IRE Transactions on Information Theory*, 4:76–84, 1954.

273. L. Feldkamp and G. V. Puskorius. A signal processing framework based on dynamic neural networks with applications to problems in adaptation, filtering and classification. *Proceedings of the IEEE*, 86(11):2259–2277, 1998.

274. D. J. Felleman and D. C. Van Essen. Distributed hierarchical processing in the primate cerebral cortex. *Cerebral Cortex*, 1:1–47, 1991.

275. J-M. Fellous, P. Tiesinga, P. J. Thomas, and T. J. Sejnowski. Discovering spike patterns in neuronal responses. *Journal of Neuroscience*, 24:2989–3001, 2004.

276. D. J. Field. Relations between the statistics of natural images and the response properties of cortical cells. *Journal of the Optical Society of America, A*, 4(12):2379–2394, 1987.

277. D. J. Field. What is the goal of sensory coding? *Neural Computation*, 6:559–601, 1994.

278. S. Fiori. Blind separation of circularly distributed source signals by the neural extended APEX algorithm. *Neurocomputing*, 34(1–4):239–252, 2000.

279. S. Fiori. Neural minor component analysis approach to robust constrained beamforming. *IEE Proceedings of Vision, Image and Signal Processing*, 150(4):205–218, August 2003.

280. S. Fiori. Nonlinear complex-valued extensions of Hebbian learning: An essay. *Neural Computation*, 17:779–838, 2005.

281. R. Fletcher. *Practical Methods of Optimization*, 2nd ed., Wiley, New York, 2000.

282. H. Flor, T. Elbert, S. Knecht, C. Wienbruch, C. Pantev, N. Birbaumer, W. Larbig, and Taub E. Phantom-limb pain as a perceptual correlate of cortical reorganization following arm amputation. *Nature*, 375:482–484, 1995.

283. P. Földiák. Adaptive network for optimal linear feature extraction. In *Proceedings of IJCNN'89*, pp. 401–405, Washington, DC, 1989, IEEE Press Piscataway, NJ.

284. P. Földiák. Forming sparse representations by local anti-Hebbian learning. *Biological Cybernetics*, 64:165–170, 1990.

285. P. Földiák. Learning invariance from transformation sequence. *Neural Computation*, 3:194–200, 1991.

286. P. Földiák and M. Young. Sparse coding in the primate cortex. In M. A. Arbib, ed., *Handbook of Brain Theory and Neural Networks*, pp. 895–898. MIT Press, Cambridge, MA, 1995.

287. D. J. Foster and M. A. Wilson. Reverse replay of behavioural sequences in hippocampal place cells during the awake state. *Nature*, 440:680–683, 2006.

288. M. O. Franz and B. Schölkopf. Implicit Wiener series for higher-order image analysis. In L. K. Saul, Y. Weiss, and L. Bottou, Eds., *Advances in Neural Information Processing Systems*, Vol. 17, pp. 465–472. MIT Press, Cambridge, MA, 2005.

289. W. J. Freeman. Simulation of chaotic EEG patterns with a dynamic model of the olfactory system. *Biological Cybernetics*, 56:139–150, 1987.

290. W. J. Freeman, Y. Yao, and B. Burke. Central pattern generating and recognizing in olfactory bulb: A correlation learning rule. *Neural Networks*, 1:277–288, 1988.

291. J. H. Freidman. Exploratory projection pursuit. *Journal of the American Statistical Association*, 82:249–266, 1987.

292. S. Freud. A project for a scientific psychology. In E. Jones, ed., *The Standard Edition of the Complete Psychological Works of Sigmund Freud*, Vol. 1, pp. 295–397. Hogarth London, 1966.

293. P. Fries, J. H. Reynolds, A. E. Rorie, and R. Desimone. Modulation of oscillatory neuronal synchronization by selective visual attention. *Science*, 291:1560–1563, 2001.

294. U. Frisch. *Turbulence: The Legacy of A. N. Kolmogorov*. Cambridge University Press, Cambridge, 1995.

295. J. Fritz, M. Elhilali, and S. Shamma. Active listening: Task-dependent plasticity of spectrotemporal receptive fields in primary auditory cortex. *Hearing Research*, 206:159–176, 2005.

296. B. Fritzke. Some competitive learning methods. Techical Report, Institute of Neural Computation, Ruhr-Universität Bochum, April 1997.

297. R. C. Froemke and Y. Dan. Spike-timing-dependent synaptic modification induced by natural spike trains. *Nature*, 416:433–438, 2002.

298. R. C. Froemke, M. Poo, and Y. Dan. Spike-timing-dependent synaptic plasticity depends on dendritic location. *Nature*, 434:221–225, 2005.

299. S. Frurukawa, L. Xu, and J. C. Middlebrooks. Coding of sound-source location by ensembles of cortical neurons. *Journal of Neuroscience*, 20:1216–1228, 2000.

300. M. Fujita. Adaptive filter model of the cerebellum. *Biological Cybernetics*, 45: 195–206, 1982.

301. O. Fujita. Trial-and-error correlation learning. *IEEE Transactions on Neural Networks*, 4(4):720–722, 1993.

302. K. Fukushima. Cognitron: A self-organizing multilayered neural network. *Biological Cybernetics*, 20:121–136, 1975.

303. C. Fyfe. *Hebbian Learning and Negative Feedback Networks*. Springer, Berlin, 2005.

304. D. Gabor. A new microscopic principle. *Nature*, 161:777, 1948.

305. S. Gais and J. Born. Low acetylcholine during slow-wave sleep is critical for declarative memory consolidation. *Proceedings of the National Academy of Sciences, USA*, 101:2140–2144, 2004.

306. W. J. Gao, D. E. Newman, A. B. Wormington, and S. Pallas. Development of inhibitory circuitry in visual and auditory cortex of postnatal ferrets: Immunocytochemical localization of GABAergic neurons. *Journal of Comparative Neurology*, 409:261–273, 1999.

307. W. A. Gardner. *Statistical Spectral Analysis: A Nonprobabilistic Theory*. Prentice-Hall, Englewood Cliffs, NJ, 1987.

308. W. A. Gardner. *Introduction to Random Processes*. McGraw-Hill, New York, 1989.

309. W. A. Gardner, Ed. *Cyclostationarity in Communications and Signal Processing*. IEEE Press, New York, 1994.

310. W. A. Gardner and L. E. Franks. Characteristics of cyclostationary random signal processes. *IEEE Transactions on Information Theory*, 21(1):4–14, 1975.

311. N. D. Gaubitch and P. A. Naylor. The complex multichannel LMS algorithm for adaptive blind system identification. In *Proceedings of International Workshop on Acoustic Echo and Noise Control (IWAENC'06)*, Paris, France, 2006.

312. D. D. Gehr, H. Komiya, and J. J. Eggermont. Neuronal responses of cat primary auditory cortex to natural and altered species-specific calls. *Hearing Research*, 150:27–42, 2000.

313. S. Geman, E. Bienenstock, and R. Doursat. Neural networks and the bias-variance dilemma. *Neural Computation*, 4:1–58, 1992.

314. M. G. Genton. Classes of kernels for machine learning: A statistics perspective. *Journal of Machine Learning Research*, 2:299–312, 2001.

315. A. P. Geogopoulos, A. B. Schwartz, and R. E. Kettner. Neuronal population coding of movement direction. *Science*, 233:1416–1419, 1986.

316. G. L. Gerstein and K. L. Kirkland. Neural assemblies: Technical issues, analysis, and modeling. *Neural Networks*, 14:589–598, 2001.

317. W. Gerstner. Coding properties of spiking neurons: Reverse and cross-correlations. *Neural Networks*, 14:559–610, 2001.

318. W. Gerstner, R. Kempter, J. L. van Hemmen, and H. Wagner. A neuronal learning rule for sub-millisecond temporal coding. *Nature*, 383:76–81, 1996.

319. W. Gerstner and W. M. Kistler. Mathematical formulations of Hebbian learning. *Biological Cybernetics*, 87:404–415, 2002.

320. W. Gerstner and W. M. Kistler. *Spiking Neuron Models: Single Neurons, Populations, Plasticity*. Cambridge University Press, Cambridge, 2002.

321. R. R. Gharieb and A. Cichocki. Noise reduction in brain evoked potentials based on third-order correlations. *IEEE Transactions on Biomedical Engineering*, 48(5): 501–512, 2001.

322. Z. Gil, B. W. Conners, and Y. Amitai. Differential regulation of neocortical synapses by neuromodulators and activity. *Neuron*, 19:679–686, 1997.

323. C. D. Gilbert. Adult cortical dynamics. *Physiological Review*, 78(2):467–485, 1998.

324. M. Girolami and C. Fyfe. An extended exploratory projection pursuit network with linear and nonlinear anti-Hebbian lateral connections applied to the cocktail party problem. *Neural Networks*, 10(9):1607–1618, 1997.

325. F. Girosi, M. Jones, and T. Poggio. Regularization theory and neural network architecture. *Neural Computation*, 7:219–269, 1995.

326. R. Gnanadesikan. *Methods for Statistical Data Analysis of Multivariate Observations*. Wiley, New York, 1977.

327. D. N. Godard. Self-recovering equallization and carrier tracking in two-dimensional data communication systems. *IEEE Transactions on Communications*, 28(11):1867–1875, 1980.

328. S. L. Goh and D. P. Mandic. A complex-valued RTRL algorithm for recurrent neural networks. *Neural Computation*, 16:2699–2713, 2004.

329. G. H. Golub and C. F. Van Loan. *Matrix Computations*, 3rd ed. Johns Hopkins University Press, Baltimore, MD, 1996.

330. G. J. Goodhill. Topology and ocular dominance: A model exploring positive correlations. *Biological Cybernetics*, 69:109–118, 1993.

331. G. J. Goodhill and D. J. Willshaw. Application of the elastic net algorithm to the formation of ocular dominance stripes. *Network: Computation in Neural Systems*, 1:41–59, 1990.

332. N. Gordon, D. Salmond, and A. F. M. Smith. Novel approach to nonlinear/non-gaussian Bayesian state estimation. *IEE Proceedings of Vision, Image and Signal Processing*, 140:107–113, 1993.

333. L. A. Grande, G. A. Kinney, G. L. Miracle, and W. J. Spain. Dynamic influences on coincidence detection in neocortical pyramidal neurons. *Journal of Neuroscience*, 24:1839–1851, 2004.

334. C. M. Gray. Synchronous oscillations in neuronal systems: Mechanisms and functions. *Journal of Computational Neuroscience*, 1:11–38, 1994.

335. C. M. Gray. The temporal correlation hypothesis of visual feature integration: Still alive and well. *Neuron*, 24:31–47, 1999.

336. C. M. Gray, P. König, A. K. Engel, and W. Singer. Oscillatory responses in cat visual cortex exhibit intercolumnar synchronization which reflects global stimulus properties. *Nature*, 338:334–337, 1989.

337. C. M. Gray and W. Singer. Stimulus-specific neuronal oscillations in orientation columns of cat visual cortex. *Proceedings of the National Academy of Sciences, USA*, 86:1698–1702, 1989.

338. M. S. Grewal and A. P. Andrews. *Kalman Filtering: Theory and Practice*. Prentice-Hall, Englewood Cliffs, NJ, 1993.

339. J. S. Griffith. *Mathematical Neurobiology*. Academic, London, 1971.

340. D. Grimes and R. P. N. Rao. Bilinear sparse coding for invariant vision. *Neural Computation*, 17:47–73, 2005.

341. J. Gross, F. Schmitz, I. Schnitzler, K. Kessler, K. Shapiro, B. Hommel, and A. Schnitzler. Modulation of long-range neural synchrony reflects temporal limitations of visual attention in humans. *Proceedings of the National Academy of Sciences, USA*, 101:13050–13055, 2004.

342. S. Grossberg. Adaptive pattern classification and universal recoding: I. parallel development and coding of neural feature detectors. *Biological Cybernetics*, 23:121–134, 1976.

343. S. Grossberg. Competitive learning: From interactive activation to adaptive resonance. *Cognitive Science*, 11:23–63, 1987.

344. S. Grossberg. Birth of a learning law. *INNS/ENNS/JNNS Newsletter*, 21:1–4, 1998.

345. B. Grothe. New roles for synaptic inhibition in sound localization. *Nature Review Neuroscience*, 4:540–550, 2003.

346. S. Guderian and E. Duzel. Induced theta oscillations mediate large-scale synchrony with mediotemporal areas during recollection in humans. *Hippocampus*, 15(7):901–912, 2005.

347. F. Gustafsson, Ed. *Adaptive Filtering and Change Detection*. Wiley, New York, 2000.

348. S. L. Hahn. *Hilbert Transforms in Signal Processing*. Artech House, London, 1996.

349. P. J. B. Hancock, L. S. Smith, and W. A. Phillips. A biologically supported error-correcting learning rule. *Neural Computation*, 3:201–212, 1991.

350. A. I. Hanna and D. P. Mandic. A general fully adaptive normalised gradient descent learning algorithm for complex-valued nonlinear adaptive filters. *IEEE Transactions on Signal Processing*, 51(10):2540–2549, 2003.

351. T. Hara and A. Hirose. Plastic mine detecting radar system using complex-valued self-organizing map that deals with multiple-frequency interferometric images. *Neural Networks*, 17:1201–1210, 2004.

352. D. R. Hardoon, S. Szedmak, and J. Shawe-Taylor. Canonical correlation analysis: An overview with application to learning methods. *Neural Computation*, 16(12): 2639–2664, 2004.

353. H. H. Harman. *Modern Factor Analysis*, 3rd ed. University of Chicago Press, Chicago, IL, 1976.

354. E. Harth, T. Kalogeropoulos, and A. S. Pandya. A universal optimization network. In *Proc. Symposium on Maturing Technology and Emerging Horizons in Biomedical Engineering*, pp. 97–107, 1988.

355. E. Harth and E. Tzanakou. Alopex: A stochastic method for determining visual receptive fields. *Vision Research*, 14:1475–1482, 1974.

356. E. Harth, K. P. Unnikrishnan, and A. S. Pandya. The inversion of sensory processing by feedback pathways: A model of visual cognitive functions. *Science*, 237:184–187, 1987.

357. M. E. Hasselmo, C. Bodelon, and B. P. Wyble. A proposed function for hippocampal theta rhythm: Separate phases of encoding and retrieval enhance reversal of prior learning. *Neural Computation*, 14(4):793–817, 2002.

358. M. E. Hasselmo and E. Schnell. Laminar selectivity of the cholinergic suppression of synaptic transmission in rat hippocampal region CA1: Computational modeling and brain slice physiology. *Journal of Neuroscience*, 14(6):3898–3914, 1994.

359. M. E. Hasselmo, B. P. Wyble, and G. V. Wallenstein. Encoding and retrieval of episodic memories: Role of cholinergic and GABAergic modulation in the hippocampus. *Hippocampus*, 6(6):693–708, 1996.

360. N. G. Hatsopoulos, L. Paninski, and J. P. Donoghue. Sequential movement representation based on correlated neuronal activity. *Experimental Brain Research*, 149:478–486, 2003.

361. S. Haykin, Ed. *Nonlinear Methods of Spectrum Analysis*, 2nd Ed. Springer-Verlag, Berlin, 1983.

362. S. Haykin, Ed. *Advances in Spectrum Analysis and Array Processing*, Vols. I and II. Prentice-Hall, Englewoods Cliff, NJ, 1991.

363. S. Haykin, Ed. *Blind Deconvolution*. Prentice-Hall, Englewoods Cliff, NJ, 1994.

364. S. Haykin. *Neural Networks: A Comprehensive Foundation*, 2nd Ed. Prentice-Hall, Upper Saddle River, NJ, 1999.

365. S. Haykin, Ed. *Unsupervised Adaptive Filtering*, Vols. I and II. Wiley, New York, 2000.

366. S. Haykin. *Communications Systems*, 4th ed. Wiley, New York, 2001.

367. S. Haykin, Ed. *Kalman Filtering and Neural Networks*. Wiley, New York, 2001.

368. S. Haykin. Signal processing: Where physics and mathematics meet. *IEEE Signal Processing Magazine*, 18(4):6–7, July 2001.

369. S. Haykin. *Adaptive Filter Theory*, 4th ed. Prentice-Hall, Upper Saddle River, NJ, 2002.

370. S. Haykin. Kalman filtering and its neural implications. In M. A. Arbib, Ed., *Handbook of Brain Theory and Neural Networks*, 2nd ed., pp. 590–594. MIT Press, Cambridge, MA, 2002.

371. S. Haykin and J. A. Cadzow. Special issue on spectral estimation. *Proceedings of the IEEE*, 70(9), September 1992.

372. S. Haykin and Z. Chen. The cocktail party problem. *Neural Computation*, 17(9): 1875–1902, 2005.

373. S. Haykin and Z. Chen. The machine cocktail party problem. In S. Haykin, J. C. Principe, T. J. Sejnowski, and J. McWhirter, Eds., *New Directions in Statistical Signal Processing: From Systems to Brain*, pp. 51–75. MIT Press, Cambridge, MA, 2006.

374. S. Haykin, Z. Chen, and S. Becker. Stochastic correlative learning algorithms. *IEEE Transactions on Signal Processing*, 52(8):2200–2209, August 2004.

375. S. Haykin and D. J. Thomson. Signal detection in a nonstatonary environment reformulated as an adaptive pattern classification problem. *Proceedings of the IEEE*, 86(10):2325–2344, November 1998.

376. S. Haykin and B. Widrow, Eds. *Least-Mean-Square Adaptive Filters*. Wiley, New York, 2003.

377. D. Hebb. *Organization of Behavior: A Neuropsychological Theory*. Wiley, New York, 1949.

378. R. Hecht-Nielsen. *Neurocomputing*. Addison-Wesley, Redwood City, CA, 1990.

379. M. Heerema and W. A. van Leeuwen. Derivation of Hebb's rule. *Journal of Physics A*, 32:263–286, 1999.

380. J. A. Henry, K. C. Dennis, and M. A. Schechter. General review of tinnitus: Prevalence, mechanisms, effects, and management. *Journal of Speech, Language, and Hearing Research*, 48(5):1204–1235, 2005.

381. J. Hertz, A. Krogh, and R. Palmer. *Introduction to the Theory of Neural Computation*. Addison-Wesley, Reading, MA, 1991.

382. K. E. Hild II, D. Erdogmus, and J. C. Principe. An analysis of entropy estimators for blind source separation. *Signal Processing*, 86(1):182–194, 2005.

383. G. E. Hinton. Deterministic Boltzmann learning performs steepest descent in weight-space. *Neural Computation*, 1:143–150, 1989.

384. G. E. Hinton. Training products of experts by minimizing contrastive divergence. Technical Report, GCNU TR 2000-004, Gatsby Computational Neuroscience Unit, University College London, 2000.

385. G. E. Hinton. Training products of experts by minimizing contrastive divergence. *Neural Computation*, 14:1771–1800, 2002.

386. G. E. Hinton and A. Brown. Spiking Boltzmann machines. In S. Solla, T. Leen, and K.-R. Müller, Eds., *Advances in Neural Information Processing Systems*, Vol. 12, pp. 122–128. MIT Press, Cambridge, MA, 2000.

387. G. E. Hinton, P. Dayan, R. Frey, and R. Neal. The "wake-sleep" algorithm for unsupervised neural networks. *Science*, 268:1158–1161, May 1995.

388. G. E. Hinton, S. Osindero, and Y-W. Teh. A fast learning algorithm for deep belief nets. *Neural Computation*, 18(7):1527–1554, 2006.

389. G. E. Hinton and T. Sejnowski, Eds. *Unsupervised Learning: Foundations of Neural Computation*. MIT Press, Cambridge, MA, 1999.

390. G. E. Hinton and T. J. Sejnowski. Optimal perceptual learning. In *Proceedings of IEEE Conference on Computer Vision and Pattern Recognition*, pp. 448–453, Washington, DC, 1983.

391. G. E. Hinton and T. J. Sejnowski. Learning and relearning in Boltzmann machines. In D. Rumelhart and J. McClelland, Eds., *Parallel Distributed Processing: Explorations in the Microstructure Cognition*, Vol. 1, pp. 282–317. MIT Press, Cambridge, MA, 1986.

392. A. Hirose, Ed. *Complex-Valued Neural Networks: Theories and Applications*. World Scientific, Singapore, 2003.

393. A. Hirose. *Complex-Valued Neural Networks*. Springer, Berlin, 2006.

394. J. A. Hirsch and C. D. Gilbert. Long-term changes in synaptic strength along specific intrinsic pathways in the cat visual cortex. *Journal of Physiology*, 461:247–262, 1993.

395. A. L. Hodgkin and A. F. Huxley. A quantitative description of membrane current and its application to conduction and excitation in nerve. *Journal of Physiology*, 117:500–544, 1952.

396. P. M. Hofman, J. G. A. van Riswick, and A. J. van Opstal. Relearning sound localization with new ears. *Nature Neuroscience*, 1(5):417–421, 1998.

397. A. O. Holcombe and P. Cavanagh. Early binding of feature pairs for visual perception. *Nature Neuroscience*, 4(2):127–128, 2001.

398. C. Holscher, R. Anwyl, and M. J. Rowan. Stimulation on the positive phase of hippocampal theta rhythm induces long-term potentiation that can be depotentiated

by stimulation on the negative phase in area CA1 in vivo. *Journal of Neuroscience*, 17:6470–6477, 1997.

399. J. J. Hopfield. Neural networks and physical systems with emergent collective computational abilities. *Proceedings of the National Academy of Sciences, USA*, 79:2554–2558, July 1982.

400. J. J. Hopfield. Transforming neural computations and representing time. *Proceedings of the National Academy of Sciences, USA*, 93:15440–15444, December 1996.

401. J. J. Hopfield and C. D. Brody. Learning rules and network repair in spike-timing-based computation networks. *Proceedings of the National Academy of Sciences, USA*, 101(1):337–342, 2004.

402. J. J. Hopfield and D. W. Tank. Neural computation of decisions in optimization problems. *Biological Cybernetics*, 52:141–152, 1985.

403. J. D. Horel. Complex principal component analysis: Theory and example. *Journal of Climate and Applied Meteorology*, 23:1660–1673, 1984.

404. R. A. Horn and C. R. Johnson. *Matrix Analysis*. Cambridge University Press, Cambridge, 1985.

405. H. Hotelling. Relation between two sets of variates. *Biometrika*, 28:322–377, 1936.

406. J. C. Houk, J. T. Buckingham, and A. G Barto. Models of the cerebellum and motor learning. *Behavioral and Brain Sciences*, 19(3):368–383, 1996.

407. M. W. Howard, D. S. Rizzuto, J. B. Caplan, J. R. Madsen, J. Lisman, R. Aschenbrenner-Scheibe, A. Schulze-Bonhage, and M. J. Kahana. Gamma oscillations correlate with working memory load in humans. *Cerebral Cortex*, 13:1369–1374, 2003.

408. P. O. Hoyer. Non-negative sparse coding. In *Proceedings of IEEE Workshop on Neural Networks for Signal Processing (NNSP'02)*, pp 557–565, Martigny, Switzerland, 2002.

409. P. O. Hoyer. Non-negative matrix factorization with sparseness constraints. *Journal of Machine Learning Research*, 5:1457–1469, 2004.

410. P. O. Hoyer and A. Hyvärinen. Independent component analysis applied to feature extraction from colour and stereo images. *Network: Computation in Neural Systems*, 11:191–210, 2000.

411. C. Y. Hsieh, S. J. Cruikshank, and R. Metherate. Differential modulation of auditory thalamocortical and intracortical synaptic transmission by cholinergic agonist. *Brain Research*, 880:51–64, 2000.

412. W. W. Hsieh. Nonlinear canonical correlation analysis by neural networks. *Neural Networks*, 13(10):1095–1105, 2000.

413. N. E. Huang and S. P. Shen, Eds. *Hilbert-Huang Transform and Its Applications*. World Scientific, Singapore, 2005.

414. N. E. Huang, Z. Shen, S. R. Long, M. C. Wu, H. H. Shih, Q. Zheng, N-C. Yen, C. C. Tung, and H. L. Liu. The empirical mode decomposition and the Hilbert spectrum for nonlinear and non-stationary time series analysis. *Proceedings of Royal Society of London, A*, 454:903–995, 1998.

415. Y. A. Huang and J. Benesty. Adaptive multi-channel least mean square and Newton algorithms for blind channel identification. *Signal Processing*, 82:1127–1138, 2002.

416. D. H. Hubel and T. N. Wiesel. *Brain and Visual Perception*. Oxford University Press, New York, 2004.

417. P. T. Huerta and J. E. Lisman. Heightened synaptic plasticity of hippocampal CA1 neurons during a cholinergically induced rhythmic state. *Nature*, 364:723–725, 1993.

418. P. T. Huerta and J. E. Lisman. Bidirectional synaptic plasticity induced by a single burst during cholinergic theta-oscillation in CA1 in-vitro. *Neuron*, 15(5):1053–1063, 1995.

419. J. M. Hupé, A. C. James, B. R. Payne, S. G. Lomber, P. Girard, and J. Bullier. Cortical feedback improves discrimination between figure and background by V1, V2 and V3 neurons. *Nature*, 394:784–787, 1998.

420. J. M. Hutchinson. A radial basis function approach to financial time series analsyis. Ph.D. thesis, Department of Electrical Engineering and Computer Science, Massachusetts Institute of Technology, 1994.

421. J. M. Hutchinson, A. W. Lo, and T. Poggio. A nonparametric approach to pricing and hedging derivative securities via learning networks. *Journal of Finance*, 49(3):851–889, 1994.

422. J. Huxter, N. Burgess, and J. O'Keefe. Independent rate and temporal coding in hippocampal pyramidal cells. *Nature*, 425:828–832, 2003.

423. J. M. Hyman, B. P. Wyble, V. Goyal, C. A. Rossi, and M. E. Hasselmo. Stimulation in hippocampal region CA1 in behaving rats yields long-term potentiation when delivered to the peak of theta and long-term depression when delivered to the trough. *Journal of Neuroscience*, 23:11725–11731, 2003.

424. J. M. Hyman, E. A. Zilli, A. M. Paley, and M. E. Hasselmo. Medial prefrontal cortex cells show dynamic modulation with the hippocampal theta rhythm dependent on behavior. *Hippocampus*, 15(6):739–749, 2005.

425. A. Hyvärinen. Complexity pursuit: Separating interesting components from time series. *Neural Computation*, 13:883–898, 2001.

426. A. Hyvärinen and P. O. Hoyer. A two-layer sparse coding model learns simple and complex cell receptive fields and topography from natural images. *Vision Research*, 41(8):2413–2423, 2001.

427. A. Hyvärinen, P. O. Hoyer, and M. Inki. Topographic independent component analysis. *Neural Computation*, 13(7):1527–1558, 2001.

428. A. Hyvärinen, J. Karhunen, and E. Oja. *Independent Component Analysis*. Wiley, New York, 2001.

429. S. Ikeda, S. Amari, and H. Nakahara. Convergence of the wake-sleep algorithm. In M. S. Kearns, S. A. Solla, and D. A. Cohn, Eds., *Advances in Neural Information Processing Systems*, Vol. 11, pp. 239–245. MIT Press, Cambridge, MA, 1999.

430. N. Intrator and L. N. Cooper. Objective function formulation of the BCM theory. *Neural Networks*, 5:3–17, 1993.

431. D. R. Irvine, R. Rajan, and S. Smith. Effects of restricted cochlear lesions in adult cats on the frequency organization of the inferior colliculus. *Journal of Comparative Neurology*, 467(3):354–374, 2003.

432. M. Ito, Ed. *The Crebellum and Neural Control*. Raven, New York, 1984.

433. M. Ito. Long-term depression. *Annual Review of Neuroscience*, 12:85–102, 1989.

434. E. M. Izhikevich. *Dynamical Systems in Neuroscience: The Geometry of Excitability and Bursting*. MIT Press, Cambridge, MA, 2006.

435. E. M. Izhikevich, J. A. Gally, and G. M. Edelman. Spike-timing dynamics of neuronal groups. *Cerebral Cortex*, 14(8):933–944, 2004.

436. W. James. *Psychology (Briefer Course)*. Holt, New York, 1890.

437. J. Janakiraman and K. P. Unnikrishnan. A feedback model of visual attention. In *Proceedings of the International Joint Conference on Neural Networks (IJCNN'92)*, pp. 541–546, 1992.

438. S. Jankowski, A. Lozowski, and J. M. Zurada. Complex-valued multistate neural associative memory. *IEEE Transactions on Neural Networks*, 7(6):1491–1496, 1996.

439. D. C. Javitt, M. Steinschneider, C. E. Schroeder, and J. C. Arezzo. Role of cortical N-methyl-D-aspartate receptors in auditory sensory memory and mismatch negativity generation: Implications for schizophrenia. *Proceedings of the National Academy of Sciences, USA*, 93:11962–11967, 1996.

440. A. H. Jazwinski. *Stochastic Processes and Filtering Theory*. Academic, New York, 1970.

441. L. A. Jeffress. A place theory of sound localization. *Journal of Comparative and Physiological Psychology*, 41:35–39, 1948.

442. P. Jezzard, P. M. Matthews, and S. M. Smith, Eds. *Functional MRI: An Introduction to Methods*. Oxford University Press, New York, 2001.

443. G. Johansson. Visual perception of biological motion and a model for its analysis. *Perception and Psychophysics*, 14(2):201–211, 1973.

444. E. R. John. Switchboard versus statistical theories of learning and memory. *Science*, 177:850–864, 1972.

445. D. H. Johnson and N. Y. Kiang. Analysis of discharges recorded simultaneously from pairs of auditory nerve fibers. *Biophysics Journal*, 16:719–734, 1976.

446. R. Johnson Jr., P. Schniter, T. J. Endres, J. D. Behm, D. R. Brown, and R. A. Casas. Blind equalization using the constant modulus criterion: A review. *Proceedings of the IEEE*, 86(10):1927–1950, 1998.

447. I. T. Jolliffe. *Principal Component Analysis*, 2nd edn., Springer, New York, 2002.

448. E. G. Jones. Cortical and subcortical contributions to activity-dependent plasticity in primate somatosensory cortex. *Annual Review of Neuroscience*, 23, 2000.

449. M. I. Jordan. Computational aspects of motor control and motor learning. In H. Heuer and S. Keele, Eds., *Handbook of Perception and Action: Motor Skills*. Academic, New York, 1996.

450. K. G. Jöreskog. Some contributions to maximum likelihood factor analysis. *Psychometrika*, 32:443–482, 1967.

451. P. X. Joris, P. H. Smith, and T. C. T. Yin. Coincident detection in the auditory system: 50 years after Jeffress. *Neuron*, 21:1235–1238, December 1998.

452. S. Julier and J. Uhlmann. Unscented filtering and nonlinear estimation. *Proceedings of the IEEE*, 92(3):401–422, 2004.

453. M. W. Jung and B. L. McNaughton. Spatial selectivity of unit activity in the hippocampal granular layer. *Hippocampus*, 3(2):165–182, 1993.

454. C. Jutten and J. Herault. Blind separation of sources, part I–III. *Signal Processing*, 24:1–29, 1991.

455. T. Kailath. Correlation detection of signals perturbed by a random channel. *IRE Transactions on Information Theory*, 6(3):361–366, June 1960.

456. T. Kailath. RKHS approach to detection and estimation problems—Part I: Deterministic signals in Gaussian noise. *IEEE Transactions on Information Theory*, 17(5):530–549, 1971.

457. T. Kailath. A view of three decades of linear filtering theory. *IEEE Transactions on Information Theory*, 20(2):146–181, March 1974.

458. T. Kailath and V. Poor. Detection of stochastic processes. *IEEE Transactions on Information Theory*, 44(6):2230–2259, October 1998.

459. T. Kailath, A. H. Sayed, and B. Hassibi. *Linear Estimation*. Prentice-Hall, Englewood Cliffs, NJ, 2000.

460. S. Káli and P. Dayan. Off-line replay maintains declarative memories in a model of hippocampal-neocortical interactions. *Nature Neuroscience*, 7(3):286–294, 2004.

461. R. E. Kalman. A new approach to linear filtering and prediction problems. *Transactions of the ASME, Journal of Basic Engineering*, 82:35–45, March 1960.

462. R. E. Kalman and R. S. Bucy. New results in linear filtering and prediction theory. *Transactions of the ASME, Journal of Basic Engineering*, 83:95–107, December 1961.

463. J. A. Kaltenbach, J. Zhang, and P. Finlayson. Tinnitus as a plastic phenomenon and its possible neural underpinnings in the dorsal cochlear nucleus. *Hearing Research*, 206:200–226, 2005.

464. M. R. Kamke, M. Brown, and D. R. Irvine. Plasticity in the tonotopic organization of the medial geniculate body in adult cats following restricted unilateral cochlear lesions. *Journal of Computational Neurology*, 459:355–367, 2003.

465. H. J. Kappen and F. B. Rodriguez. Efficient learning in Boltzmann machine using linear response theory. *Neural Computation*, 10:1137–1156, 1998.

466. J. Karhunen and J. Jourtensalo. Representation and separation of signals using nonlinear PCA type learning. *Neural Networks*, 7:113–127, 1994.

467. S. Kaur, R. Lazar, and R. Metherate. Intracortical pathways determine breadth of subthreshold frequency receptive fields in primary auditory cortex. *Journal of Neurophysiology*, 91:2551–2567, 2004.

468. M. Kawato. Cerebellum and motor control. In M. A. Arbib, Ed., *Handbook of Brain Theory and Neural Networks*, 2nd ed. pp. 190–195. MIT Press, Cambridge, MA, 2002.

469. J. Kay. Feature discovery under contextual supervision using mutual information. In *Proceedings of the International Joint Conference on Neural Networks (IJCNN'92)*, Vol. IV, pp. 79–84, 1992.

470. J. Kay and W. A. Phillips. Activation functions, computational goals, and learning rules for local processors with contextual guidance. *Neural Computation*, 9:895–910, 1997.

471. S. M. Kay. *Fundamentals of Statistical Signal Processing*, Vol. II: *Detection Theory*. Prentice-Hall, Upper Saddle River, NJ, 1998.

472. S. R. Kelso, A. H. Ganong, and T. H. Brown. Hebbian synapses in hippocampus.1 *Proceedings of the National Academy of Sciences, USA*, 83:5326–5330, 1986.

473. J. Kettenring. Canonical analysis of several sets of variables. *Biometrika*, 58:433–451, 1971.

474. A. Ya. Khinchin. Korrelationstheorie der statistischen stochcstischen prozesse. *Mathematischen Annalen*, 109:604–615, 1934.

475. M. P. Kilgard and M. M. Merzenich. Cortical map reorganization enabled by nucleus basalis activity. *Science*, 279:1714–1718, 1998.

476. M. P. Kilgard and M. M. Merzenich. Plasticity of temporal information processing in the primary auditory cortex. *Nature Neuroscience*, 1:727–731, 1998.

477. M. P. Kilgard and M. M. Merzenich. Order-sensitive plasticity in adult primary auditory cortex. *Proceedings of the National Academy of Sciences, USA*, 99:3205–3209, 2002.

478. M. P. Kilgard, P. K. Pandya, J. Vazquez, A. Gehi, C. E. Schreiner, and M. M. Merzenich. Sensory input directs spatial and temporal plasticity in primary auditory cortex. *Journal of Neurophysiology*, 86:326–338, 2001.

479. K. I. Kim, M. O. Franz, and B. Schölkopf. Iterative kernel principal component analysis for image modeling. *IEEE Transactions on Pattern Analysis and Machine Intelligence*, 27(9):1351–1365, 2005.

480. T. Kim and T. Adali. Approximation by fully complex multilayer perceptrons. *Neural Computation*, 15(7):1641–1666, 2003.

481. R. R. Kimpo, F. E. Theunnissen, and A. J. Doupe. Propagation of correlated activity through multiple stages of a neural circuit. *Journal of Neuroscience*, 23:5760–5761, 2003.

482. F. Kimura, M. Fukuada, and T. Tusomoto. Acetylcholine suppresses the spread of excitation in the visual cortex revealed by optical recording: Possible differential effect depending on the source of input. *European Journal of Neuroscience*, 11:3597–3609, 1999.

483. S. Kirkpatrick, C. D. Gelatt, Jr., and M. P. Vecchi. Optimization by simulated annealing. *Science*, 220:671–680, May 1983.

484. W. M. Kistler. Spike-timing dependent synaptic plasticity: A phenomenonological framework. *Biological Cybernetics*, 87:416–427, 2002.

485. W. M. Kistler and W. Gerstner. Stable propagation of activity pulses in populations of spiking neurons. *Neural Computation*, 14:987–997, 2002.

486. D. J. Klein, J. Z. Simon, D. A. Depireux, and S. A. Shamma. Stimulus-invariant processing and spectrotemporal reverse correlation in primary auditory cortex. *Journal of Computational Neuroscience*, 20:111–136, 2006.

487. R. Klein. Donald O. Hebb. In R. A. Wilson and F. C. Keil, Eds., *MIT Encyclopedia of Cognitive Science*, pp. 366–367. MIT Press, Cambridge, MA, 1999.

488. A. Klopf. A drive-reinforcement model of single neuron function: An alternative to the Hebbian neuronal model. In J. S. Denker, Ed., *Neural Networks for Computing: AIP Conference Proceedings*, pp. 265–270. American Institute of Physics, New York, 1986.

489. C. H. Knapp and G. C. Carter. The generalized correlation method for estimation of time delay. *IEEE Transactions on Acoustics, Speech, and Signal Processing*, 24(4):320–327, 1976.

490. D. Knill and W. Richard, Eds. *Perception as Bayesian Inference*. Cambridge University Press, Cambridge, 1995.

491. E. I. Knudsen. Eary auditory experience aligns the auditory map of space in the optic tectum of the barn owl. *Science*, 222:939–942, 1983.

492. E. I. Knudsen and M. Konishi. A neural map of auditory space in the owl. *Science*, 200:795–797, 1978.

493. C. Koch. Computation and the single neuron. *Nature*, 385:207–210, January 1997.

494. J. J. Koenderink. Geometrical structures determined by the functional order in nervous nets. *Biological Cybernetics*, 50:43–50, 1984.

495. J. J. Koenderink. Simultaneous order in nervous nets from a functional standpoint. *Biological Cybernetics*, 50:35–41, 1984.

496. T. Kohonen. Correlation matrix memories. *IEEE Transactions on Computers*, 21:353–359, 1972.

497. T. Kohonen. The self-organized formation of topologically correct feature maps. *Biological Cybernetics*, 43:59–69, 1982.

498. T. Kohonen. *Self-organization and Associative Memory*. Springer, Berlin, 1984.

499. T. Kohonen. *Self-organizing Maps*, 3rd ed. Springer, Berlin, 2001.

500. T. Kohonen and E. Oja. Fast adaptive formation of orthogonalizing filters and associative memory in recurrent networks of neuron-like elements. *Biological Cybernetics*, 21:85–95, 1976.

501. P. König and A. K. Engel. Correlated firing in sensory-motor systems. *Current Opinions in Neurobiology*, 5:511–519, 1995.

502. P. König, A. K. Engel, and W. Singer. Relation between oscillatory activity and long-range synchronization in cat visual cortex. *Proceedings of the National Academy of Sciences, USA*, 92:290–294, 1995.

503. P. König, A. K. Engel, and W. Singer. Integrator or coincidence detector? The role of the cortical neuron revisited. *Trends in Neuroscience*, 19:130–137, 1996.

504. M. Konishi. Deciphering the brain's codes. *Neural Computation*, 3(1):1–18, 1991.

505. B. Kosko. Differential Hebbian learning. In J. S. Denker, Ed., *Neural Networks for Computing: AIP Conference Proceedings*, pp. 277–288. American Institute of Physics, New York, 1986.

506. B. Kosko. Bidirectional associative memories. *IEEE Transactions on Systems, Man, and Cybernetics*, 18:49–60, 1988.

507. S. G. Krantz. *Function Theory of Several Complex Variables*. AMS Chelsea Publishing, Providence, RI, 1992.

508. I. Kreitschmann-Andermahr, T. Rosburg, U. Demme, E. Gaser, H. Nowak, and H. Sauer. Effect of ketamine on the neuromagnetic mismatch field in healthy humans. *Brain Research. Cognitive Brain Research*, 12(1):109–116, 2001.

509. S. Kullback. *Information Theory and Statistics*. Wiley, New York, 1959.

510. B. V. K. Kumar, D. P. Casasent, and A. Mahalanobis. Correlation filters for target detection in a Markov model. *Applied Optics*, 28(15):3112–3119, August 1989.

511. S-I. Kung, K. I. Diamantaras, and J. S. Taur. Adaptive principal component extraction (APEX) and applications. *IEEE Transactions on Signal Processing*, 42(5):1202–1217, 1994.

512. Y. Kuroe. A model of complex-valued associative memories and its dynamics. In A. Hirose, Ed., *Complex-Valued Neural Networks: Theories and Applications*, pp. 57–79. World Scientific, Singapore, 2003.

513. Y. Kuroe and Y. Taniguchi. Models of self-correlation type complex-valued associative memories and their dynamics. In W. Duch, J. Kacprzyk, E. Oja, and S. Zadrozny, editors, *Proc. ICANN'05 (Lecture Notes in Computer Science 3696)*, pp. 185–192. Springer, Berlin, 2005.

514. H. J. Kushner and D. S. Clark. *Stochastic Approximation Method for Constrained and Unconstrained Systems*. Springer-Verlag, Berlin, 1978.

515. H. J. Kushner and G. G. Yin. *Stochastic Approximation Algorithms and Applications*. Springer-Verlag, New York, 1997.

516. M. Kuss and T. Graepel. The geometry of kernel canonical correlation analysis. Technical Report No. 108, Max-Planck Institute for Biological Cybernetics, May 2003.

517. H. Kwon and N. M. Nasrabadi. Kernel spectral matched filter for hyperspectral imagery. *International Journal of Computer Vision*, 71(2):127–141, 2007.

518. P. J. A. Lago, A. P. Rocha, and N. B. Jones. Covariance density estimation for autoregressive spectral density of point processes. *Biological Cybernetics*, 61:195–203, 1989.

519. P. L. Lai and C. Fyfe. A neural network implementation of canonical correlation analysis. *Neural Networks*, 12:1391–1397, 1999.

520. P. L. Lai and C. Fyfe. Kernel and nonlinear canonical correlation analysis. *International Journal of Neural Systems*, 10(5):365–377, 2000.

521. V. A. F. Lamme and H. Spekreijse. Neuronal synchrony does not represent texture segregation. *Nature*, 396:362–366, 1998.

522. I. Lampl, I. Reichova, and D. Ferster. Synchronous membrane potential fluctuations in neurons of the cat visual cortex. *Neuron*, 22(2):361–374, 1999.

523. G. Lanckriet, L. El Ghaoui, C. Ghattacharyya, and M. I. Jordan. A robust minimax approach to classification. *Journal of Machine Learning Research*, 3:555–582, 2002.

524. K. J. Lang and M. J. Witbrock. Learning to tell two spirals apart. In D. Touretzky, G. E. Hinton, and T. J. Sejnowski, Eds., *Proceedings of the 1988 Connectionist Models Summer School*, pp. 52–59. Morgan Kaufmann, San Mateo, CA, 1989.

525. F. H. Lange. *Correlation Techniques: Foundations and Applications of Correlation Analysis in Modern Communications, Measurement and Control*. Van Nostrand, Princeton, NJ, 1967.

526. J. Larson and G. Lynch. Induction of synaptic potentiation in hippocampus by patterned stimulation involves two events. *Science*, 232:985–988, 1986.

527. M. Laubach, J. Wessberg, and M. A. Nicolelis. Cortical ensemble activity increasingly predicts behaviour outcomes during learning of a motor task. *Nature*, 405:567–571, June 2000.

528. S. B. Laughlin. *Coding Efficiency and the Metabolic Cost of Sensory and Neural Information: Information Theory and the Brain*. Cambridge University Press, Cambridge, 1999.

529. S. B. Laughlin. Energy as a constraint on the coding and processing of sensory information. *Current Opinion in Neurobiology*, 11(4):475–480, 2001.

530. S. B. Laughlin and T. J. Sejnowski. Communication in neuronal networks. *Science*, 301:1870–1874, 2003.

531. G. Laurent and H. Davidowitz. Encoding of oscillatory information with oscillating neural assemblies. *Science*, 265:1872–1875, September 1994.

532. C. C. Law and L. N. Cooper. Formation of receptive fields in realistic visual environment according to the Bienestock, Cooper and Munro (BCM) theory. *Proceedings of the National Academy of Sciences, USA*, 91:7797–7801, 1994.

533. Y. LeCun. Une procedure d'apprentissage pour reseau a seuil asymmetrique (a learning scheme for asymmetric threshold networks). In *Proceedings of Cognitiva'85*, pp. 599–604, Paris, France, 1985.

534. Y. LeCun, L. Bottou, Y. Bengio, and P. Haffner. Gradient-based learning applied to document recognition. *Proceedings of the IEEE*, 86(11):2278–2323, November 1998.

535. A. K. Lee and M. A. Wilson. Memory of sequential experience in the hippocampus during slow wave sleep. *Neuron*, 36(6):1183–1194, 2002.

536. C. C. Lee, K. Imaizumi, C. E. Schreiner, and J. A. Winer. Concurrent tonotopic processing streams in auditory cortex. *Cerebral Cortex*, 14:441–451, 2004.

537. C. C. Lee, C. E. Schreiner, K. Imaizumi, and J. A. Winer. Tonotopic and heterotopic projection systems in physiologically defined auditory cortex. *Neuroscience*, 128:871–887, 2004.

538. D. D. Lee and H. S. Seung. Learning the parts of objects by nonnegative matrix factorization. *Nature*, 401(21):788–791, October 1999.

539. D. D. Lee and H. S. Seung. Algorithms for nonnegative matrix factorization. In S. Solla, T. Leen, and K.-R. Müller, Eds., *Advances in Neural Information Processing Systems*, Vol. 12, pp. 556–562. Cambridge, MA, 2000.

540. D.-L. Lee and W. J. Wang. A multivalued bidirectional associative memory operating on a complex domain. *Neural Networks*, 11(9):1623–1635, 1998.

541. T. S. Lee and D. Mumford. Hierarchical Bayesian inference in the visual cortex. *Journal of the Optical Society of America, A*, 20(7):1434–1448, 2003.

542. T.-W. Lee, M. Girolami, and T. J. Sejnowski. Independent component analysis using an extended Infomax algorithm for mixed subgaussian and supergaussian sources. *Neural Computation*, 11:417–441, 1999.

543. T. K. Leen. Dynamics of learning in linear feature discovery networks. *Network: Computation in Neural Systems*, 2:85–105, 1991.

544. S. Lemm, B. Blankertz, G. Curio, and K.-R. Müller. Spatio-spectral filters for robust classification of single trial EEG. *IEEE Transactions on Biomedical Engineering*, 52(9):1541–1548, 2005.

545. H. Leung and S. Haykin. The complex backpropagation algorithm. *IEEE Transactions on Signal Processing*, 33(9):2101–2104, 1991.

546. D. S. Levine. *Introduction to Neural and Cognitive Modeling*, 2nd ed. Erlbaum, Mahwah, NJ, 2000.

547. N. Levinson and R. Redheffer. *Complex Variables*. Holden-Day, San Francisco, CA, 1970.

548. W. B. Levy, J. A. Anderson, and S. Lehmkuhle, Eds. *Synaptic Modification, Neuron Selectivity, and Nervous System Organization*. Erlbaum, Hillsdale, NJ, 1985.

549. W. B. Levy and R. A. Baxter. Energy efficient neural codes. *Neural Computation*, 8:531–543, 1996.

550. W. B. Levy, C. M. Colbert, and N. L. Desmond. Elemental adaptive processes of neurons and synapses: A statistical/computational perspective. In M. Gluck and D. Rumelhart, Eds., *Neuroscience and Connectionist Models*, pp. 187–235. Erlbaum, Hillsdale, NJ, 1990.

551. M. S. Lewicki. A review of methods for spike sorting: The detection and classification of neural action potentials. *Network: Computation in Neural Systems*, 9:R53–R78, 1998.

552. M. S. Lewicki. Efficient coding of natural sounds. *Nature Neuroscience*, 5(4):356–363, 2002.

553. M. S. Lewicki and B. Olshausen. A probabilistic framework for the adaptation and comparison of image codes. *Journal of the Optical Society of America, A*, 16(7):1587–1601, 1996.

554. S. Z. Li, X. Hou, H. Zhang, and Q. Cheng. Learning spatially localized, parts-based representation. In *Proceedings of IEEE Computer Vision and Pattern Recognition (CVPR'01)*, pp. 207–210, 2001, IEEE Computer Society Press, New York.

555. T. M. Liggett. *Interacting Particle Systems*. Springer-Verlag, New York, 1985.

556. L. Lin, R. Osan, and J. Z. Tsien. Organizing principles of real-time memory encoding: Neural clique assemblies and universal neural codes. *Trends in Neuroscience*, 29:48–57, 2006.

557. R. Linsker. From basic network principles to neural architecture: Emergence of orientation columns. *Proceedings of the National Academy of Sciences, USA*, 83:8779–8783, 1986.

558. R. Linsker. From basic network principles to neural architecture: Emergence of orientation-selective cells. *Proceedings of the National Academy of Sciences, USA*, 83:8390–8394, 1986.

559. R. Linsker. From basic network principles to neural architecture: Emergence of spatial opponent cells. *Proceedings of the National Academy of Sciences, USA*, 83:7508–7512, 1986.

560. R. Linsker. Self-organization in a perceptual network. *Computer*, 21:105–117, March 1988.

561. R. Linsker. How to generate ordered maps by maximizing the mutual information between input and output signals. *Neural Computation*, 1:402–411, 1989.

562. R. Linsker. Local synaptic rules suffice to maximize mutual information in a linear network. *Neural Computation*, 4:691–702, 1992.

563. R. Linsker. A local learning rule that enables information maximization for arbitrary input distributions. *Neural Computation*, 9:1661–1665, 1997.

564. J. E. Lisman, J. M. Fellous, and X. J. Wang. A role for NMDA-receptor channels in working memory. *Nature Neuroscience*, 1:273–275, 1998.

565. W. Liu, P. P. Pokharel, and J. C. Principe. Correntropy: A localized similarity measure. In *Proceedings of IJCNN'06*, pp. 4919–4924, Vancouver, Canada, 2006, IEEE Press Piscataway, NJ.

566. W. Liu, P. P. Pokharel, and J. C. Principe. Correntropy: Properties and applications in non-Gaussian signal processing. *IEEE Transactions on Signal Processing*, 2007 (in press).

567. L. Ljung. Analysis of recursive stochastic algorithms. *IEEE Transactions on Automatic Control*, 22:551–574, 1977.

568. L. Ljung. *System Indentification: Theory for the User*, 2nd ed. Prentice-Hall, Englewood Cliffs, NJ, 1999.

569. M. Loève. *Probability Theory*, 3rd ed. Van Nostrand, New York, 1963.

570. H. C. Longuet-Higgins. Holographic model of temporal recall. *Nature*, 217:104, 1968.

571. A. Lörincz and G. Buzsáki. Two-phase computational model training long-term memories in the entorhinal-hippocampal region. *Annals of New York Academy of Sciences*, 911:83–111, 2000.

572. K. Louie and M. A. Wilson. Temporally structured replay of awake hippocampal ensemble activity during rapid eye movement sleep. *Neuron*, 29(1):145–156, 2001.

573. S. Lowel and W. Singer. Selection of intrinsic horizontal connections in the visual cortex by correlated neuronal activity. *Science*, 255:209–212, 1992.

574. T. Lu and X. Wang. Information content of auditory cortical responses to time-varying acoustic stimuli. *Journal of Neurophysiology*, 91:301–313, 2004.

575. R. W. Lucky. Techniques for adaptive equalization of digital communication systems. *Bell Systems of Technical Journal*, 45:255–286, 1966.

576. J. S. Lund, Q. Wu, and J. B. Levitt. Visual cortex cell types and connections. In M. A. Arbib, Ed., *Handbook of Brain Theory and Neural Networks*, pp. 1016–1021. MIT Press, Cambridge, MA, 1995.

577. F-L. Luo and R. Unbehauen. *Applied Neural Networks for Signal Processing*. Cambridge University Press, Cambridge, 1997.

578. F-L. Luo, R. Unbehauen, and A. Cichocki. A minor component analysis algorithm. *Neural Networks*, 10(2):291–297, March 1997.

579. D. J. C. Mackay. Introduction to Monte Carlo methods. In M. I. Jordan, Ed., *Learning in Graphical Models*, pp. 175–204. Kluwer Academic, Norwell, MA, 1998.

580. D. J. C. MacKay. *Information Theory, Inference, and Learning Algorithms*. Cambridge University Press, Cambridge, 2003.

581. D. J. C. MacKay and K. D. Miller. Analysis of Linsker's simulations of Hebbian rules. *Neural Computation*, 2:173–183, 1990.

582. K. MacLeod, A. Bäcker, and G. Laurent. Who reads temporal information contained across synchronized and oscillatory spike trains. *Nature*, 395:693–696, 1998.

583. K. MacLeod and G. Laurent. Distinct mechanisms for synchronization and temporal patterning of odor-encoding neural assemblies. *Science*, 274:976–979, 1996.

584. J. C. Magee and D. Johnston. A synaptically controlled, associative signal for Hebbian plasticity in hippocampal neurons. *Science*, 275:209–213, 1997.

585. J. Makhoul. Spectral linear prediction: Properties and applications. *IEEE Transactions on Acoustics, Speech, and Signal Processing*, 23:283–296, 1975.

586. R. C. Malenka and R. A. Nicoll. Long-term potentiation—A decade of progress? *Science*, 285:1870–1874, 1999.

587. G. Mallat, S. Papanicolaou, and Z. Zhang. Adaptive covariance estimation of locally stationary process. *Annals of Statistics*, 26(1):1–47, 1998.

588. K. V. Mardia. *Statistics of Directional Data*. Academic, London, 1972.

589. H. Markram, J. Lübke, M. Frotscher, and B. Sakmann. Regulation of synaptic plasticity by coincidence of postsynaptic APs and EPSPs. *Science*, 275:213–215, January 1997.

590. D. Marr. A theory of cerebellar cortex. *Journal of Physiology*, 202:437–470, 1969.

591. D. Marr. Simple memory: A theory for archicortex. *Philosophical Transactions of the Royal Society of London, B*, 262:23–81, 1971.

592. L. Martignon, G. Deco, K. Laskey, M. Diamond, W. Freiwald, and E. Vaadia. Neural coding: Higher-order temporal patterns in the neurostatistics of cell assemblies. *Neural Computation*, 12(11):2621–2653, 2000.

593. S. J. Martin, L. de Hoz, and R. G. M. Morris. Retrograde amnesia: Neither partial nor complete hippocampal lesions in rats result in preferential sparing of remote spatial memory, even after reminding. *Neuropsychologia*, 43(4):609–624, 2005.

594. W. Martin and P. Flandrin. Wigner-Ville spectral analysis of non-stationary processes. *IEEE Transactions on Acoustics, Speech, and Signal Processing*, 33: 1461–1470, 1985.

595. T. M. Martinetz. Competitive Hebbian learning rule forms perfectly topology preserving maps. In *Proceedings of the International Conference on Artificial Neural Networks (ICANN'93)*, pp. 427–434. Springer, 1993.

596. T. M. Martinetz, S. G. Berkovich, and K. Schulten. "Neural gas" network for vector quantization and its application to time-series prediction. *IEEE Transactions on Neural Networks*, 4(4), 558–568 1993.

597. T. M. Martinetz and K. Schulten. Topology representing networks. *Neural Networks*, 7(3), 507–522 1994.

598. S. Martinkauppi, P. Rama, H. J. Aronen, A. Korvenoja, and S. Carlson. Working memory of auditory localization. *Cerebral Cortex*, 10(9):889–898, 2000.

599. N. Masuda and S. Amari. Modeling memory transfer and savings in cerebellar motor learning. In Y. Weiss, B. Schölkopf, and J. Platt, Eds., *Advances in Neural Information Processing Systems*, Vol. 18, pp. 859–866. MIT Press, Cambridge, MA, 2006.

600. N. Matsumura, H. Nishijo, R. Tamura, S. Eifuku, S. Endo, and T. Ono. Spatial- and task-dependent neuronal responses during real and virtual translocation in the monkey hippocampal formation. *Journal of Neuroscience*, 19(6):2381–2393, 1999.

601. E. M. Maynard, N. G. Hatsopoulos, C. L. Ojakangas, B. D. Acuna, J. N. Sanes, R. A. Normann, and J. P. Donoghue. Neuronal interactions improve cortical population coding of movement direction. *Journal of Neuroscience*, 19:8083–8093, 1999.

602. R. A. Mazzoni, P. ad Anderson, and M. I. Jordan. A more biologically plausible learning rule for neural networks. *Proceedings of the National Academy of Sciences, USA*, 88:4433–4437, 1991.

603. C. J. McAdams and J. H. R. Maunsell. Effects of attention on the reliability of individual neurons in monkey visual cortex. *Neuron*, 23:765–773, 1999.

604. J. L. McClelland and N. H. Goddard. Considerations arising from a complementary learning systems perspective on hippocampus and neocortex. *Hippocampus*, 6:654–665, 1996.

605. J. L. McClelland, B. L. McNaughton, and R. C. O'Reilly. Why there are complementary learning systems in the hippocampus and neocortex: Insights from the successes and failures of connectionist models of learning and memory. *Psychological Review*, 102(3):419–457, 1995.

606. W. McCulloch and W. Pitts. A logical calculus of the ideas immanent in nervous activity. *Bulletin of Mathematical Biophysics*, 5:115–133, 1943.

607. H. J. McDermott, M. Lech, M. S. Kornblum, and D. R. Irvine. Loudness perception and frequency discrimination in subjects with steeply sloping hearing loss: Possible correlates of neural plasticity. *Journal of the Acoustical Society of America*, 104(4):2314–2325, 1998.

608. G. L. McLachlan and T. Krishnan. *The EM Algorithm and Extensions*. Wiley, New York, 1997.

609. B. L. McNaughton, B. Leonard, and L. Chen. Cortical-hippocampal interactions and cognitive mapping: A hypothesis based on reintegration of the parietal and inferotemporal pathways for visual processing. *Psychobiology*, 17:230–235, 1989.

610. B. L. McNaughton and R. G. M. Morris. Hippocampal synaptic enhancement and information storage within a distributed memory system. *Trends in Neurosciences*, 10:408–415, 1987.

611. J. McQueen. Some methods for classification and analysis of multivariate observations. In *Proceedings of the Fifth Berkeley Symposium on Mathematical Statistics and Probability*, pp. 281–297. University of California Press, Berkely, CA, 1967.

612. M. Meister and M. J. Berry II. The neuronal code of the retina. *Neuron*, 22:435–450, 1999.

613. M. Meister, R. O. Wong, D. A. Baylor, and C. J. Shatz. Synchronous bursts of action potentials in ganglion cells of the developing mammalian retina. *Science*, 252:939–943, 1991.

614. R. J. Meleca, J. A. Kaltenbach, and P. R. Falzarano. Changes in the tonotopic map of the dorsal cochlear nucleus in hamsters with hair cell loss and radial nerve bundle degeneration. *Brain Research*, 750:201–213, 1997.

615. M. A. Meredith and B. E. Stein. Visual, auditory, and somatosensory convergence on cells in superior colliculus results in multisensory integration. *Journal of Neurophysiology*, 56:640–662, 1986.

616. M. M. Merzenich, J. H. Kaas, J. T. Wall, M. Sur, R. J. Nelson, and D. J. Felleman. Progression of change following median nerve section in the cortical representation of the hand in areas 3b and 1 in adult owl and squirrel monkeys. *Neuroscience*, 10:639–665, 1983.

617. N. Metropolis, A. W. Rosenbluth, M. N. Rosenbluth, A. H. Teller, and E. Teller. Equations of state calculations by fast computing machines. *Journal of Chemical Physics*, 21:1087–1091, March 1953.

618. R. B. Michaels and B. R. Upadhyaya. A complex valued neural network local learning laws. In C. H. Dagli, Ed., *Intelligent Engineering Systems through Artificial Neural Networks*, pp. 101–109. *American Society of Mechanical Engineers*, New York, 1999.

619. J. C. Middlebrooks, A. E. Clock, L. Xu, and D. M. Green. A panoramic code for sound location by cortical neurons. *Science*, 264:842–844, 1994.

620. R. Miikkulainen, J. A. Bednar, Y. Choe, and J. Sirosh. *Computational Maps in the Visual Cortex*. Springer, Berlin, 2005.

621. S. Mika, G. Rätsch, J. Weston, B. Schölkopf, and K.-R. Müller. Fisher discriminant analysis with kernels. In *Proceedings of IEEE Workshop on Neural Networks for Signal Processing (NNSP'99)*, pp. 41–48, 1999, IEEE Press Piscataway, NJ.

622. E. G. Miller and J. W. Fisher III. ICA using spacings estimates of entropy. *Journal of Machine Learning Research*, 4:1271–1295, 2003.

623. K. Miller. *Complex Stochastic Processes*. Addison-Wesley, Reading, MA, 1974.

624. K. D. Miller. Correlation-based models of neural development. In M. Gluck and D. Rumelhart, Eds., *Neuroscience and Connectionist Theory*, pp. 267–353. Erlbaum, Hilsdale, NJ, 1990.

625. K. D. Miller. Equivalence of a sprouting-and-retraction model and correlation-based plasticity models of neural development. *Neural Computation*, 10:529–547, 1998.

626. K. D. Miller, J. B. Keller, and M. P. Stryker. Ocular dominance column development: Analysis and simulation. *Science*, 245:605–615, 1989.

627. K. D. Miller and D. J. C. MacKay. The role of constraints in Hebbian learning. *Neural Computation*, 6:100–126, 1994.

628. P. M. Milner. The mind and Donald O. Hebb. *Scientific American*, 268:124–129, 1986.

629. M. Minsky. Steps towards artificial intelligence. *Proceedings of the IRE*, 49:8–30, 1961.

630. M. Minsky and S. Pappert. *Perceptrons*, expanded from 1969 edn. MIT Press, Cambridge, MA, 1988.

631. M. Mishkin, L. G. Ungerleider, and K. A. Macko. Object vision and spatial vision: Two cortical pathways. *Trends in Neurosciences*, 6:414–417, 1983.

632. G. Mitchison. Removing time variation with the anti-Hebbian differential synapse. *Neural Computation*, 3:312–320, 1991.

633. L. Molgedey and H. G. Schuster. Separation of a mixture of independent signals using time delayed correlations. *Physical Review Letters*, 72(23):3634–3637, 1994.

634. J. Moran and R. Desimone. Selective attention gates visual processing in the extrastriate cortex. *Science*, 229:782–784, 1985.

635. E. Moreau and O. Macchi. High order contrast for self-adaptive source separation. *International Journal of Adaptive Control and Signal Processing*, 10:19–46, 1996.

636. S. M. Morton and A. J. Bastian. Prism adaptation during walking generalizes to reaching and requires the cerebellum. *Journal of Neurophysiology*, 92:2497–2509, 2004.

637. M. Moscovitch. Multiple dissociations of function in amnesia. In L. S. Cermak, Ed., *Human Memory and Amnesia*, pp. 337–370. Erlbaum, Hillsdale, NJ, 1982.

638. M. Moscovitch, L. Nadel, G. Winocur, A. Gilboa, and R. S. Rosenbaum. The cognitive neuroscience of remote episodic, semantic and spatial memory. *Current Opinion in Neurobiology*, 16(2):179–190, 2006.

639. J. R. Movellan. Contrastive Hebbian learning in the continuous Hopfield model. In D. S. Touretzky, G. E. Hinton, and T. J. Sejnowski, Eds., *Proceedings of the 1989 Connectionist Models Summer School*, pp. 10–17. Morgan Kaufman, San Mateo, CA, 1990.

640. M. C. Mozer, R. S. Zemel, M. Behrmann, and C. K. I. Williams. Learning to segment images using dynamic feature binding. *Neural Computation*, 4:650–665, 1992.

641. M. K. Müezzinoğlu, C. Güzeliş, and J. M. Zurada. A new design method for the complex-valued multistate Hopfield associative memory. *IEEE Transactions on Neural Networks*, 14(4):891–899, July 2003.

642. W. Muhlnickel, T. Elbert, E. Taub, and H. Flor. Reorganization of auditory cortex in tinnitus. *Proceedings of the National Academy of Sciences, USA*, 95:10340–10343, 1998.

643. D. Mumford. On the computational architecture of the neocortex: I. the role of thalamo-cortical loop. *Biological Cybernetics*, 65:135–145, 1991.

644. D. Mumford. Thalamus. In M. Arbib, Ed., *The Handbook of Brain Theory and Neural Networks*, pp. 981–984. MIT Press, Cambridge, MA, 1995.

645. N. Murata, S. Ikeda, and A. Ziehe. An approach to blind source separation based on temporal structure of speech signals. *Neurocomputing*, 41(1):1–24, 2001.

646. R. Näätänen, A. W. Gaillard, and S. Mäntysalo. Early selective attention effect on evoked potentials reinterpreted. *Acta Psychology*, 42(4):313–329, 1978.

647. J.-P. Nadal and N. Parga. Nonlinear neurons in the low noise limit: A factorial code maximises information transfer. *Network: Computation in Neural Systems*, 5:561–581, 1994.

648. Z. Nadasdy, H. Hirase, A. Czurko, J. Csicsvari, and G. Buzsaki. Replay and time compression of recurring spike sequences in the hippocampus. *Journal of Neuroscience*, 19:9497–9507, 1999.

649. H. Nakahara, S. Amari, and O. Hikosaka. Self-organization in the basal ganglia with modulation of reinforcement signals. *Neural Computation*, 14:819–844, 2002.

650. H. Nakahara, H. Itoh, R. Kawagoe, Y. Takikawa, and O. Hikosaka. Dopamine neurons can represent context-dependent prediction error. *Neuron*, 41:269–280, 2004.

651. K. Nakano. Associatron—A model of associative memory. *IEEE Transactions on Systems, Man, and Cybernetics*, 2(3):380–388, 1972.

652. K. Nakazawa, M. C. Quirk, R. A. Chitwood, M. Watanabe, M. F. Yeckel, L. D. Sun, A. Kato, C. A. Carr, D. Johnston, M. A. Wilson, and S. Tonegawa. Requirement for hippocampal CA3 NMDA receptors in associative memory recall. *Science*, 297:211–218, 2002.

653. M. Á. Carreira-Perpi nán and G. J. Goodhill. Influence of lateral connections on the structure of cortical maps. *Journal of Neurophysiology*, 92:2947–2959, 2004.

654. A. K. Nandi and V. Zarzoso. Fourth-order cumulant based blind source separation. *IEEE Signal Processing Letters*, 3(12):312–314, 1996.

655. V. H. Nascimento and A. H. Sayed. On the learning mechanism of adaptive filters. *IEEE Transactions on Signal Processing*, 48(6):1609–1625, June 2000.

656. N. M. Nasrabadi and H. Kwon. Kernel spectral matched filter for hyperspectral target detection. In *Proc. ICASSP'05*, Vol. 4, pp. 665–668, Philadelphia, PA, 2005, IEEE Press Piscataway, NJ.

657. R. Neal and P. Dayan. Factor analysis using delta-rule wake-sleep learning. *Neural Computation*, 9:1781–1803, 1997.

658. R. M. Neal and G. E. Hinton. A view of the EM algorithm that justifies incremental, sparse, and other variants. In M. I. Jordan, Ed., *Learning in Graphical Models*, pp. 355–368. Kluwer Academic, Norwell, MA, 1998.

659. M. A. Nicolelis, A. Ghazanfar, C. R. Stambaugh, L. M. Oliveira, M. Laubach, J. K. Chapin, R. J. Nelson, and J. H. Kaas. Simultaneous encoding of tactile information by three primate cortical areas. *Nature Neuroscience*, 1:621–630, 1998.

660. C. L. Nikias and A. P. Petropulu. *Higher-Order Spectra Analysis. A Nonlinear Signal Processing Framework*. Prentice-Hall, Englewoods Cliff, NJ, 1993.

661. T. Nitta. Orthogonal decision boundaries and generalization of complex-valued neural networks. In A. Hirose, Ed., *Complex-Valued Neural Networks: Theories and Applications*, pp. 7–28. World Scientific, Singapore, 2003.

662. T. Nitta. Orthogonality of decision boundaries in complex-valued neural networks. *Neural Computation*, 16:73–97, 2004.

663. H. Noda, S. Manohar, and W. R. Adey. Correlated firing of hippocampal neuron pairs in sleep and wakefulness. *Experimental Neurology*, 24(2):232–247, 1969.

664. A. J. Noest. Discrete-state phasor neural network. *Physical Review A*, 38(4): 2196–2199, 1988.

665. A. J. Noreña and J. J. Eggermont. Comparison between local field potentials and unit cluster activity in primary auditory cortex and anterior auditory field in the cat. *Hearing Research*, 166:202–213, 2002.

666. A. J. Noreña and J. J. Eggermont. Changes in spontaneous neural activity immediately after an acoustic trauma: Implications for neural correlates of tinnitus. *Hearing Research*, 183:137–153, 2003.

667. A. J. Noreña and J. J. Eggermont. Enriched acoustic environment after noise trauma reduces hearing loss and prevents cortical map reorganization. *Journal of Neuroscience*, 25:699–705, 2005.

668. A. J. Noreña and J. J. Eggermont. Enriched acoustic environment after noise trauma abolishes neural signs of tinnitus. *Neuroreport*, 17:559–563, 2006.

669. A. J. Noreña, B. Gourévitch, N. Aizawa, and J. J. Eggermont. Spectrally enhanced acoustic environment disrupts frequency representation in cat auditory cortex. *Nature Neuroscience*, 9(7):932–939, 2006.

670. A. J. Noreña, M. Tomita, and J. J. Eggermont. Neural changes in cat auditory cortex after a transient pure-tone trauma. *Journal of Neurophysiology*, 90:2387–2401, 2003.

671. M. Norgaard. *Neural Network Based System Identification Toolbox: For Use with MATLAB*. MathWorks, Natick, MA, 2000.

672. S. J. Nowlan. Maximum likelihood competitive learning. In D. Touretzky, Ed., *Advances in Neural Information Processing Systems*, Vol. 2, pp. 574–582. Morgan Kaufmann, San Mateo, CA, 1990.

673. K. Obermayer, H. Ritter, and K. Schulten. A principle for the formation of the spatial structure of cortical feature maps. *Proceedings of the National Academy of Sciences, USA*, 87:8345–8349, 1990.

674. K. Obermayer and T. J. Sejnowski, Eds., *Self-Organization Map Formation: Foundations of Neural Computation*. MIT Press, Cambridge, MA, 2001.

675. K. Obermayer, T. J. Sejnowski, and G. G. Blasdel. Neural pattern formation via a competitive Hebbian mechanism. *Behavioural Brain Research*, 66:161–167, 1995.

676. E. Oja. A simplified neuron model as a principal component analyzer. *Journal of Mathematical Biology*, 15:267–273, 1982.

677. E. Oja. Neural networks, principal components, and subspaces. *International Journal of Neural Systems*, 1:61–68, 1989.

678. E. Oja. Principal components, minor components, and linear neural networks. *Neural Networks*, 5:927–936, 1992.

679. E. Oja and J. Karhunen. A stochastic approximation of the eigenvectors and eigenvalues of the expectation of a random matrix. *Journal of Mathematical Analysis and Applications*, 106:69–84, 1985.

680. E. Oja, H. Ogawa, and J. Wangviwattana. Learning in nonlinear constrained Hebbian network. In T. Kohonen, Ed., *Artificial Neural Networks*, pp. 385–390. North-Holland, Amsterdam, 1991.

681. J. O'Keefe and N. Burgess. Dual phase and rate coding in hippocampal place cells: Theoretical significance and relationship to entorhinal grid cells. *Hippocampus*, 15(7):853–866, 2005.

682. J. O'Keefe and L. Nadel. *The Hippocampus as a Cognitive Map*. Clarendon, London, 1978.

683. J. O'Keefe and M. L. Recce. Phase relationship between hippocampal place units and the EEG theta rhythm. *Hippocampus*, 3(3):317–330, 1993.

684. B. A. Olshausen. Sparse codes and spikes. In R. P. N. Rao, B. A. Olshausen, and M. S. Lewicki, Eds., *Probabilistic Models of the Brain: Perception and Neural Function*, pp. 257–272. MIT Press, Cambridge, MA, 2002.

685. B. A. Olshausen and D. J. Field. Emergence of simple-cell receptive field properties by learning a sparse code for natural images. *Nature*, 381:607–609, 1996.

686. B. A. Olshausen and D. J. Field. Natural image statistics and efficient coding. *Network*, 7(2):333–340, 1996.

687. B. A. Olshausen and D. J. Field. Sparse coding with an overcomplete basis set: A strategy employed by V1? *Vision Research*, 37:3311–3325, 1997.

688. J. C. O'Neill and W. J. Williams. A function of time, frequency, lag, and doppler. *IEEE Transactions on Signal Processing*, 47(3):789–799, March 1999.

689. T. Ono, K. Nakamura, H. Nishijo, and S. Eifuku. Monkey hippocampal neurons related to spatial and nonspatial functions. *Journal of Neurophysiology*, 70(4):1516–1529, 1993.

690. M. W. Oram and D. I. Perrett. Modeling visual recognition from neurobiological constraints. *Neural Networks*, 7:945–972, 1994.

691. R. C. O'Reilly. Six principles for biologically based computational models of cortical cognition. *Trends in Cognitive Sciences*, 2(11):455–462, 1998.

692. R. C. O'Reilly and J. L. McClelland. Hippocampal conjunctive encoding, storage, and recall: Avoiding a tradeoff. *Hippocampus*, 4:661–682, 1994.

693. R. C. O'Reilly and J. W. Rudy. Conjunctive representations in learning and memory: Principles of cortical and hippocampal function. *Psychological Review*, 108:311–345, 2001.

694. F. P. Ottes, J. A. M. van Gisbergen, and J. J. Eggermont. Visuomotorfields of the superior colliculus: A quantitative model. *Vision Research*, 26:857–873, 1986.

695. C. Paciorek. Nonstationary Gaussian processes for regression and spatial modelling. Ph.D. thesis, Department of of Statistics, Carnegie Mellon University, Pittsburgh, PA, 2003.

696. G. Palm. On representation and approximation of nonlinear systems. *Biological Cybernetics*, 31:119–124, 1978.

697. G. Palm and T. Poggio. Stochastic identification methods for nonlinear systems: An extension of Wiener theory. *SIAM Journal of Applied Mathematics*, 34(3):524–534, 1978.

698. F. Palmeri, J. Zhu, and C. Chang. Anti-Hebbian learning in topologically constrained linear networks: A tutorial. *IEEE Transactions on Neural Networks*, 4(5):746–761, 1993.

699. A. S. Pandya, E. Sen, and S. Hsu. Buffer allocation optimization in ATM switching networks using ALOPEX algorithm. *Neurocomputing*, 24:1–11, 1999.

700. L. Paninski. Estimation of entropy and mutual information. *Neural Computation*, 15:1191–1253, 2003.

701. C. Papageorgiou, F. Girosi, and T. Poggio. Sparse correlation kernel analysis and reconstruction. AI Memo 1635, Massachusetts Institute of Technology, Cambridge, MA, 1998.

702. A. Papoulis and S. U. Pillai. *Probability, Random Variables and Stochastic Processes*, 4th ed. McGraw-Hill, New York, 2002.

703. D. Parker. Learning-logic: Casting the cortex of the human brain in silicon. MIT Center for Computational Research in Economics and Management Science, 1985, Cambridge, MA.

704. L. Parra and P. Sajda. Blind source separation via generalized eigenvalue decomposition. *Journal of Machine Learning Research*, 4:1261–1269, 2003.

705. L. C. Parra. Symplectic nonlinear component analysis. In G. Tesauro, D. S. Touretzky, and T. K. Leen, Eds., *Advances in Neural Information Processing Systems*, Vol. 8, pp. 437–443. MIT Press, Cambridge, MA, 1996.

706. E. Parzen. An approach to time series analysis. *Annals of Mathematical Statistics*, 32:951–989, 1961.

707. E. Parzen. *Stochastic Processes*. Holden-Day, San Francisco, CA, 1962.

708. E. Parzen. *Time Series Analysis Papers*. Holden-Day, San Francisco, CA, 1967.

709. G. S. Patel, S. Becker, and R. Racine. Learning shape and motion from image sequences. In S. Haykin, Ed., *Kalman Filtering and Neural Networks*, pp. 69–81. Wiley, New York, 2001.

710. M. G. Paulin. Neural representations of moving systems. *International Review of Neurobiology*, 41:515–533, 1997.

711. C. Pavlides, Y. J. Greenstein, M. Grudman, and J. Winson. Long-term potentiation in the dentate gyrus is induced preferentially on the positive phase of theta-rhythm. *Brain Research*, 439:383–387, 1988.

712. C. Pavlides and J. Winson. Influences of hippocampal place cell firing in the awake state on the activity of these cells during subsequent sleep episodes. *Journal of Neuroscience*, 9(8):2907–2918, 1989.

713. B. A. Pearlmutter, G. E. Hinton and J. S. Denker Eds., G-maximization: An unsupervised learning procedure. In *AIP Conference Proceedings on Neural Networks for Computing*, pp. 333–338. American Institute of Physics, New York, 1986.

714. R. S. Petersen, S. Panzeri, and M. E. Diamond. Population coding of stimulus location in rat somatosensory cortex. *Neuron*, 32:503–514, 2001.

715. C. Peterson and J. R. Anderson. A mean field theory learning algorithm for neural networks. *Complex Systems*, 1:995–1019, 1987.

716. A. Pezeshki, M. R. Azimi-Sadjadi, and L. L. Scharf. A network for recursive extraction of canonical coordinates. *Neural Networks*, 16:801–808, 2003.

717. R. Pfeifer and C. Scheier. *Understanding Intelligence*. MIT Press, Cambridge, MA, 1999.

718. G. Pfurtscheller and C. Neuper. Motor imagery and direct brain-computer communication. *Proceedings of the IEEE*, 89(7):1123–1134, 2001.

719. D. T. Pham. Blind separation of instantaneous mixture of sources based on order statistics. *IEEE Transactions on Signal Processing*, 48(2):363–375, 2000.

720. D. T. Pham. Fast algorithm for estimating mutual information, entropies and score functions. In *Proceedings of the Fourth International Symposium on Independent Component Analysis and Blind Signal Separation (ICA'2003)*, Self-published online proceedings, pp. 17–22, Nara, Japan, 2003.

721. D. T. Pham and F. Vrins. Local minima of information-theoretic contrasts in blind source separation. *IEEE Signal Processing Letters*, 12(11):788–791, 2005.

722. W. A. Phillips, D. Floreano, and J. Kay. Contextually guided unsupervised learning using local multivariate binary processors. *Neural Networks*, 11(1):117–140, 1998.

723. B. Picinbono. On circularity. *IEEE Transactions on Signal Processing*, 42(12): 3473–3482, December 1994.

724. B. Picinbono. Second-order complex random vectors and normal distributions. *IEEE Transactions on Signal Processing*, 44(10):2637–2640, October 1996.

725. B. Picinbono and P. Bondon. Second-order statistics of complex signals. *IEEE Transactions on Signal Processing*, 45(2):411–419, 1997.

726. C. Piepenbrock and K. Obermayer. The effect of intracortical competition on the formation of topographic maps of Hebbian learning. *Biological Cybernetics*, 82(4):345–353, 2000.

727. A. Pikovsky, M. Rosenblum, and J. Kurths. *Synchronization—A Universal Concept in Nonlinear Sciences*. Cambridge University Press, Cambridge, 2001.

728. M. D. Plumbley. A Hebbian/anti-Hebbian network which optimizes information capacity. In J. Taylor, Ed., *Proceedings of the Artificial Neural Networks*, pp. 86–90, Brighton, UK, 1993, Elseiver. Amesterdam.

729. B. Póczos and A. Lörincz. Kalman-filtering using local interactions. Department of Information Systems, Eötvös Loránd University, Hungary, February 2003.

730. T. Poggio. On optimal nonlinear associative recall. *Biological Cybernetics*, 19:201–209, 1975.

731. T. Poggio and F. Girosi. Networks for approximation and learning. *Proceedings of the IEEE*, 78(10):1481–1497, 1990.

732. T. Poggio and F. Girosi. A sparse representation for function approximation. *Neural Computation*, 10:1445–1454, 1998.

733. P. P. Pokharel, J-W. Xu, D. Erdogmus, and J. C. Principe. A closed form solution for a nonlinear Wiener filter. In *Proceedings of IEEE ICASSP'06*, pp. 720–723, Toulouse, France, 2006.

734. D. B. Polley, E. E. Steinberg, and M. M. Merzenich. Perceptual learning directs auditory cortical map reorganization through top-down influences. *Journal of Neuroscience*, 26:4970–4982, 2006.

735. A. Pouget, P. Dayan, and R. Zemel. Information processing with population codes. *Nature Review Neuroscience*, 1:125–132, 2000.

736. A. Pouget, P. Dayan, and R. Zemel. Inference and computation with population codes. *Annual Review of Neuroscience*, 26:381–410, 2003.

737. J. C. Principe, N. R. Euliano, and W. C. Lefebvre. *Neural and Adaptive Systems: Fundamentals through Simulations*. Wiley, New York, 2000.

738. J. C. Principe, D. Xu, and J. W. Fisher. Information-theoretic learning. In S. Haykin, Ed., *Unsupervised Adaptive Filtering*, Vol. I, pp. 265–319. Wiley, New York, 2000.

739. N. Qian. On the momentum term in gradient descent learning algorithms. *Neural Networks*, 12:145–151, 1999.

740. G. J. Quirk, R. U. Muller, and J. L. Kubie. The firing of hippocampal place cells in the dark depends on the rat's recent experience. *Journal of Neuroscience*, 10:2008–2017, 1990.

741. R. J. Racine, C. A. Chapman, C. Trepel, G. C. Teskey, and N. W. Milgram. Post-activation potentiation in the neocortex: IV. Multiple sessions required for induction

of long-term potentiation in the chronic preparation. *Brain Research*, 702:87–93, 1995.

742. M. R. Raghuveer. Bispectrum estimation: Digital processing framework. *Proceedings of the IEEE*, 75:869–891, 1987.

743. R. Rajan and D. R. Irvine. Absence of plasticity of frequency map in dorsal cochlear nucleus of adult cats after unilateral partial cochlear lesions. *Journal of Comparative Neurology*, 399:35–46, 1998.

744. R. Rajan, D. R. Irvine, L. Z. Wise, and P. Heil. Effect of unilateral partial cochlear lesions in adult cats on the representation of lesioned and unlesioned cochleas in primary auditory cortex. *Journal of Comparative Neurology*, 338:17–49, 1993.

745. V. S. Ramachandran, C. Armel, C. Foster, and R. Stoddard. Object recognition can drive motion perception. *Nature*, 395:852–853, 1998.

746. V. S. Ramachandran, D. Rogersramachandran, and S. Cobb. Touching the phantom limb. *Nature*, 377:489–490, 1995.

747. S. Ramón y Cajal. *Histologie du systéme nerveux de l'homme et des vertebras*, Vols. 1 and 2. Maloine, Paris, 1909 and 1911.

748. H. Ramoser, J. Müller-Gerking, and G. Pfurtscheller. Optimal spatial filtering of single trial EEG during imagined hand movement. *IEEE Transactions on Neural Systems and Rehabilitation Engineering*, 8(4):441–446, 2000.

749. R. P. Rao. An optimal estimation approach to visual perception and learning. *Vision Research*, 39:1963–1989, 1999.

750. R. P. Rao and D. Ballard. Dynamic model of visual recognition predicts neural response properties in the visual cortex. *Neural Computation*, 9:721–763, 1997.

751. R. P. Rao and D. Ballard. Predictive coding in the visual cortex: A functional interpretation of some extra-classical receptive-field effects. *Nature Neuroscience*, 2:79–87, 1999.

752. R. P. Rao and T. J. Sejnowski. Spike-timing-dependent Hebbian plasticity as temporal difference learning. *Neural Computation*, 13:2221–2237, 2001.

753. R. P. Rao and T. J. Sejnowski. Self-organizing neural systems based on predictive learning. *Philosophical Transactions on the Royal Society of London, A*, 361:1149–1175, 2003.

754. R. P. N. Rao, B. A. Olshausen, and M. S. Lewicki, Eds., *Probabilistic Models of the Brain: Perception and Neural Function.* MIT Press, Cambridge, MA, 2002.

755. C. E. Rasmussen and C. K. I. Williams. *Gaussian Processes for Machine Learning*. MIT Press, Cambridge, MA, 2006.

756. S. S. P. Rattan and W. W. Hsieh. Complex-valued neural networks for nonlinear complex principal component analysis. *Neural Networks*, 18:61–69, 2005.

757. G. H. Recanzone, M. M. Merzenich, W. M. Jenkins, K. A. Grajski, and H. R. Dinse. Topographic reorganization of the hand representation in cortical area 3b owl moneky trained in a frequency-discrimination task. *Journal of Neurophysiology*, 67:1031–1056, 1992.

758. R. Rescorla and A. Wagner. A theory of Pavlovian conditioning: Variations in the effectiveness of reinforcement and nonreinforcement. In A. Black and W. Prokasy, Eds., *Classical Conditioning II: Current Research and Theory*, pp. 64–99. Appleton-Century-Crofts, New York, 1972.

759. A. D. Reyes. Synchrony-dependent propagation of firing rate in iteratively constructed networks in vitro. *Nature Neuroscience*, 6:593–599, 2003.

760. A. Riehle, S. Grün, M. Diesmann, and A. Aertsen. Spike synchronization and rate modulation differentially involved in motor cortical functions. *Science*, 278:1950–1953, 1997.

761. F. Rieke, D. Warland, R. van Steveninck, and W. Bialek. *Spikes: Exploring the Neural Code*. MIT Press, Cambridge, MA, 1996.

762. H. Ritter, T. Martinetz, and K. Schulten. *Neural Computation and Self-organizing Maps: An Introduction*. Addison-Wesley, Reading, MA, 1992.

763. D. S. Rizzuto, J. R. Madsen, E. B. Bromfield, A. Schulze-Bonhage, and M. J. Kahana. Human neocortical oscillations exhibit theta phase differences between encoding and retrieval. *NeuroImage*, 31(3):1352–1358, 2006.

764. H. Robbins and S. Monro. A stochastic approximation model. *Annals of Mathematical Statistics*, 22:400–407, 1951.

765. P. D. Roberts. Computational consequence of temporally asymmetric learning rules: I. Differential Hebbian learning. *Journal of Computational Neuroscience*, 7:235–246, 1999.

766. P. D. Roberts and C. C. Bell. Spike-timing dependent synaptic plasticity: Mechanisms and implications. *Biological Cybernetics*, 87:392–403, 2002.

767. D. Robertson and D. R. Irvine. Plasticity of frequency organization in auditory cortex of guinea pigs with partial unilateral deafness. *Journal of Computational Neurology*, 282:456–471, 1989.

768. E. Rodriguez, N. George, J. P. Lachaux, J. Martinerie, B. Renault, and F. J. Varela. Perception's shadow: Long-distance synchronization of human brain activity. *Nature*, 397:430–433, 1999.

769. E. T. Rolls. Functions of neural networks in the hippocampus and neocortex in memory. In J. H. Byrne and W. O. Berry, Eds., *Neural Models of Plasticity: Theoretical and Empirical Approaches*, pp. 240–265. Academic New York, 1989.

770. E. T. Rolls, L. Franco, N. C. Aggelopoulos, and S. Reece. An information theoretic approach to the contributions of the firing rates and the correlations between the firing of neurons. *Journal of Neurophysiology*, 89:2810–2822, 2003.

771. E. T. Rolls and T. Milward. A model of invariant object recognition in the visual system: Learning rules, activation functions, lateral inhibition and information-based performance measures. *Neural Computation*, 12:2547–2572, 2000.

772. E. T. Rolls and S. M. Stringer. Invariant object recognition in the visual system with error correction and temporal difference learning. *Network*, 12:111–129, 2001.

773. F. Rosenblatt. *Principles of Neurodynamics*. Spartan Books, Washington, DC, 1962.

774. A. L. Roskies. The binding problem. *Neuron*, 24:7–9, 1999.

775. Y. Rossetti, G. Rode, L. Pisella, A. Farne, L. Li, D. Boisson, and M. T. Perenin. Prism adaptation to a rightward optical deviation rehabilitates left hemispatial neglect. *Nature*, 395:166–169, 1998.

776. S. A. Roy and K. D. Alloway. Coincidence detection or temporal integration? What the neurons in somatosensory cortex are doing. *Journal of Neuroscience*, 21:2462–2473, 2001.

777. D. B. Rubin and D. T. Thayer. EM algorithms for ML factor analysis. *Psychometrika*, 47(1):69–76, 1982.

778. M. Rucci, G. Tononi, and G. M. Edelman. Registration of neural maps through value-dependent learning: Modeling the alignment of auditory and visual maps in the barn owl's optic tectum. *Journal of Neuroscience*, 17:334–352, 1997.

779. M. Rudolph and A. Destexhe. Tuning neocortical pyramidal neurons between integrators and coincidence detectors. *Journal of Computational Neuroscience*, 14:239–251, 2003.

780. D. E. Rumelhart, G. E. Hinton, and R. J. Williams. Learning internal representations by propagating error. *Nature*, 323:533–536, October 1986.

781. D. E. Rumelhart and J. L. McClelland, Eds. *Parallel Distributed Processing: Explorations in the Microstructure of Cognition*, Vols. I and II. MIT Press, Cambridge, MA, 1986.

782. D. E. Rumelhart and D. Zipser. Feature discovery by competitive learning. *Cognitive Science*, 9:75–112, 1985.

783. M. Sakurai. Synaptic modification of parallel fibre-Purkinje cell transmission in *in vitro* guinea-pig cerebellar slices. *Journal of Physiology*, 394:463–480, 1987.

784. M. Salami, C. Itami, T. Tsumoto, and F. Kimura. Change of conduction velocity by regional myelination yields constant latency irrespective of distance between thalamus and cortex. *Proceedings of the National Academy of Sciences, USA*, 100:6174–6179, 2003.

785. E. Salinas and L. F. Abbott. Vector reconstruction from firing rates. *Journal of Computational Neuroscience*, 1:89–108, 1994.

786. E. Salinas and T. J. Sejnowski. Correlated neuronal activity: High- and low-level views. In J. Feng, Ed., *Computational Neuroscience*, pp. 341–373. Chapman & Hall/CRC Press, 2004, Boca Raton, FL.

787. J. M. Samonds, J. D. Allison, H. A. Brown, and A. B. Bonds. Cooperation between area 17 neuron pairs enhance discrimination of orientation. *Journal of Neuroscience*, 23:2416–2425, 2003.

788. J. M. Samonds, J. D. Allison, H. A. Brown, and A. B. Bonds. Cooperative synchronized assemblies enhance orientation discrimination. *Proceedings of the National Academy of Sciences, USA*, 101:6722–6727, 2004.

789. T. E. Sanger. Optimal unsupervised learning in a single-layer linear feedforward neural network. *Neural Networks*, 2(6):459–473, 1989.

790. I. Santamaría, P. Pokharel, and J. C. Principe. Generalized correlation function: Definition, properties and application to blind equalization. *IEEE Transactions on Signal Processing*, 54(6):2187–2197, 2006.

791. P. S. Sastry, M. Magesh, and K. P. Unnikrishnan. Two timescale analysis of Alopex algorithm for optimization. *Neural Computation*, 14:2729–2750, 2002.

792. Y. Sato. Two extensional applications of the zero-forcing equalization. *IEEE Transactions on Communications*, 23:684–687, 1975.

793. A. H. Sayed. *Fundamentals of Adaptive Filtering*. Wiley, New York, 2003.

794. R. Schaette and R. Kempter. Development of tinnitus-related neuronal hyperactivity through homeostatic plasticity after hearing loss: A computational model. *European Journal of Neuroscience*, 23:3124–3138, 2006.

795. R. Schneggenburger and E. Neher. Intracellular calcium dependence of transmitter release rates at a fast central synapse. *Nature*, 406:889–893, 2000.

796. M. J. Schnitzer and M. Meister. Multineuronal firing patterns in the signal from eye to brain. *Neuron*, 37:499–511, 2003.

797. J. W. Schnupp, T. M. Hall, R. F. Kokelaar, and B. Ahmed. Plasticity of temporal pattern codes for vocalization stimuli in primary auditory cortex. *Journal of Neuroscience*, 26:4785–4795, 2006.

798. B. Schölkopf. The kernel trick for distances. In T. Leen, T. Dietterich, and V. Tresp, Eds., *Advances in Neural Information Processing Systems,* Vol. 13, pp. 301–307. MIT Press, Cambridge, MA, 2001.

799. B. Schölkopf and A. Smola. *Learning with Kernels: Support Vector Machines, Regularization, Optimization and Beyond.* MIT Press, Cambridge, MA, 2002.

800. B. Schölkopf, A. Smola, and K.-R. Müller. Nonlinear component analysis as a kernel eigenvalue problem. *Neural Computation*, 10:1299–1319, 1998.

801. B. Schölkopf, K. Tsuda, and J-P. Vert, Eds. *Kernel Methods in Computational Biology.* MIT Press, Cambridge, MA, 2004.

802. N. N. Schraudolph and T. J. Sejnowski. Competitive anti-Hebbian learning of invariants. In J. Moody, S. J. Hanson, and R. P. Lippmann, Eds., *Advances in Neural Information Processing Systems,* Vol. 4, pp. 1017–1024. Morgan Kaufmann, San Mateo, CA, 1992.

803. S. Schuett, T. Bonhoeffer, and M. Hubener. Pairing-induced changes of orientation maps in cat visual cortex. *Neuron*, 32:325–337, 2001.

804. K. Schulten and M. Zeller. Topology representing maps and brain function. *Nova Acta Leopoldina*, 72:133–157, 1996.

805. W. Schultz, P. Dayan, and P. R. Montague. A neural substrate of prediction and reward. *Science*, 275:499–544, 1997.

806. W. Schultz and A. Dickinson. Neuronal coding of prediction errors. *Annual Review of Neuroscience*, 23:473–500, 2000.

807. E. L. Schwartz. Afferent geometry in the primate visual cortex and the generation of neuronal trigger features. *Biological Cybernetics*, 28:1–14, 1977.

808. E. L. Schwartz. Spatial mapping in the primate sensory projection: Analytic structure and relevance to perception. *Biological Cybernetics*, 25:181–194, 1977.

809. E. L. Schwartz. Computational anatomy and functional architecture of striate cortex: A spatial mapping approach to perceptual coding. *Vision Research*, 20:644–669, 1980.

810. E. L. Schwartz. Anatomical and physiological correlates of visual computation from striate cortex to infero-temporal cortex. *IEEE Transactions on Systems, Man, and Cybernetics*, 14:257–271, 1984.

811. O. Schwartz and E. Simoncelli. Natural sound statistics and divisive normalization in the auditory system. In T. Leen, T. Dietterich, and V. Tresp, Eds., *Advances in Neural Information Processing Systems,* Vol. 13, pp. 166–172. MIT Press, Cambridge, MA, 2001.

812. M. Seeger. Gaussian processes for machine learning. *International Journal of Neural Systems*, 14(2):69–106, 2004.

813. T. Seidenbecher, T. R. Laxmi, O. Stork, and H. C. Pape. Amygdalar and hippocampal theta rhythm synchronization during fear memory retrieval. *Science*, 301:846–850, 2003.

814. T. J. Sejnowski. Statistical constraints on synaptic plasticity. *Journal of Theoretical Biology*, 69:385–389, 1977.

815. T. J. Sejnowski. Storing covariance with nonlinearly interacting neurons. *Journal of Mathematical Biology*, 4:303–321, 1977.

816. T. J. Sejnowski. The book of Hebb. *Neuron*, 24:773–776, 1999.

817. T. J. Sejnowski, S. Chattarji, and P. Stanton. Induction of synaptic plasticity by Hebbian covariance in the hippocampus. In R. Durbin, C. Miall, and G. Mitchison, Eds., *The Computing Neuron*, pp. 105–124. Addison-Wesley, Reading, MA, 1989.

818. T. J. Sejnowski and G. Tesauro. The Hebb rule for synaptic plasticity: Algorithms and implementations. In J. H. Byrne and W. O. Berry, Eds., *Neural Models of Plasticity*, pp. 94–103. Academic, San Diego, CA, 1989.

819. W. Senn, I. Segev, and M. Tsodyks. Reading neuronal synchrony with depressing synapses. *Neural Computation*, 10:815–819, 1998.

820. M. N. Shadlen and J. Movshon. Synchrony unbound: A critical evaluation of the temporal binding hypothesis. *Neuron*, 24:67–77, 1999.

821. S. Shah and P. S. Sastry. New algorithms for learning and pruning oblique decision trees. *IEEE Transactions on Systems, Man, and Cybernetics, Part C*, 29:494–505, November 1999.

822. S. A. Shamma. On the role of space and time in auditory processing. *Trends in Cognitive Sciences*, 5(8):340–348, 2001.

823. C. E. Shannon. A mathematical theory of communication. *Bell Systems Technical Journal*, 27:379–423, 623–656, 1948.

824. R. V. Shannon, F-G. Zeng, and J. Wygonski. Speech recognition with altered spectral distribution of envelope cues. *Journal of the Acoustical Society of America*, 104:2467–2476, 1998.

825. R. M. Shapley and J. D. Victor. The contrast gain conrol of the cat retina. *Vision Research*, 19:431–434, 1979.

826. C. J. Shatz. Emergence of order in visual system development. *Proceedings of the National Academy of Sciences, USA*, 93:602–608, 1996.

827. J. Shawe-Taylor and N. Cristianini. *Kernel Methods for Pattern Analysis*. Cambridge University Press, Cambridge, 2004.

828. M. Sherman and C. Koch. The control of retinogeniculate transmission in the mammalian LGN. *Experimental Brain Research*, 63:1–20, 1986.

829. C. S. Sherrington. The central nervous system. In M. Foster, Ed., *A Text Book of Physiology*, 7th ed. Macmillan, London, 1897.

830. H. Shouval, B. Blais, and L. N. Cooper. Formation of direction selectivity in natural scene environments. *Neural Computation*, 12:1057–1066, 2000.

831. L. Shpigelman, Y. Singer, R. Paz, and E. Vaadia. Spikernels: Predicting arm movements by embedding population spike rate patterns in inner-product spaces. *Neural Computation*, 17(3):671–690, 2005.

832. O. Shriki, H. Sompolinsky, and D. Lee. An information maximization approach to overcomplete and recurrent representations. In T. Leen, T. Dietterich, and V. Tresp, Eds., *Advances in Neural Information Processing Systems*, Vol. 13, pp. 612–618. MIT Press, Cambridge, MA, 2001.

833. A. Sillito, H. Jones, G. Gerstein, and D. West. Feature-linked synchronization of thalamic relay cell firing induced by feedback from the visual cortex. *Nature*, 369:479–482, 1994.

834. F. M. Silva and L. B. Almeida. A distributed decorrelation algorithm. In *Proc. ICANN'91*, pp. 943–948, Espoo, Finland, 1991, Elsevier, Amsterdam.

835. B. W. Silverman. *Density Estimation for Statistics and Data Analysis*. Chapman and Hall, London, 1986.

836. W. Singer. Synchronization of cortical activity and its putative role in information processing and learning. *Annual Review of Physiology*, 55:349–374, 1993.

837. W. Singer. Synchronization of neuronal responses as a putative binding mechanism. In M. A. Arbib, Ed., *Handbook of Brain Theory and Neural Networks*, pp. 960–964. MIT Press, Cambridge, MA, 1995.

838. W. Singer and C. M. Gray. Visual feature integration and the temporal correlation hypothesis. *Annual Review of Neuroscience*, 18:555–586, 1995.

839. J. Sjöberg, Q. Zhang, L. Ljung, A. Benveniste, B. Delyon, P. Glorennec, H. Hjalmarsson, and A. Juditsky. Nonlinear black-box modeling in system identification: A unified overview. *Automatica*, 31(12):1691–1724, 1995.

840. W. E. Skaggs, B. L. McNaughton, M. A. Wilson, and C. A. Barnes. Theta phase precession in hippocampal neuronal populations and the compression of temporal sequences. *Hippocampus*, 6(2):149–172, 1996.

841. R. L. Snyder, D. G. Sinex, J. D. McGee, and E. W. Walsh. Acute spiral ganglion lesions change the tuning and tonotopic organization of cat inferior colliculus neurons. *Hearing Research*, 147:200–220, 2000.

842. S. Song and L. F. Abbott. Cortical development and remapping through spike-timing dependent plasticity. *Neuron*, 32(2):339–350, 2001.

843. S. Song, K. D. Miller, and L. F. Abbott. Competitive Hebbian learning through spike-timing-dependent synaptic plasticity. *Nature Neuroscience*, 3:919–926, 2000.

844. H. Spencer. *The Principle of Psychology*, 3rd ed. D. Appleton and Company, New York, 1855.

845. O. Sporns, J. A. Gally, G. N. Reeke, Jr., and G. M. Edelman. Reentrant signaling among simulated neuronal groups leads to coherency in their oscillatory activity. *Proceedings of the National Academy of Sciences, USA*, 86:7265–7269, 1989.

846. M. W. Spratling and M. H. Johnson. Dendritic inhibition enhances neural coding properties. *Cerebral Cortex*, 11:1144–1149, 2001.

847. M. W. Spratling and M. H. Johnson. Pre-integration lateral inhibition enhances unsupervised learning. *Neural Computation*, 14(9):2157–2179, 2002.

848. L. R. Squire, R. E. Clark, and B. J. Knowlton. Retrograde amnesia. *Hippocampus*, 11(1):50–55, 2001.

849. G. B. Stanley and R. M. Webber. A point process analysis of sensory encoding. *Journal of Computational Neuroscience*, 15:321–333, 2003.

850. P. Stanton and T. J. Sejnowski. Associative long-term depression in the hippocampus: Induction of synaptic plasticity by Hebbian covariance. *Nature*, 339:215–218, 1989.

851. K. Steinbuch. Die lernmatrix. *Kybernetik*, 1:36–45, 1961.

852. K. Steinbuch. *Automat und Mensch*, 3rd ed., Springer-Verlag, Hidelberg, 1965.

853. K. Steinbuch and U. A. W. Piske. Learning matrices and their applications. *IEEE Transactions on Electronic Computers*, 12:846–862, 1963.

854. P. N. Steinmetz, A. Roy, P. J. Fitzgerald, S. S. Hsiao, K. O. Johnson, and E. Niebur. Attention modulates synchronized neuronal firing in primate somatosensory cortex. *Nature*, 404:187–190, 2000.

855. G. S. Stent. A physiological mechanism of Hebb's postulate of learning. *Proceedings of the National Academy of Sciences, USA*, 70:997–1001, 1973.

856. M. Steriade. *The Intact and Sliced Brain*. MIT Press, Cambridge, MA, 2001.

857. J. V. Stone. Learning perceptually salient visual parameters using spatiotemporal smoothness constraints. *Neural Computation*, 8:1463–1492, 1996.

858. J. V. Stone. Object recognition: View-specificity and motion-specificity. *Vision Research*, 39:4032–4044, 1999.

859. J. V. Stone. Blind source separation using temporal predictability. *Neural Computation*, 13:1559–1574, 2001.

860. J. V. Stone. *Independent Component Analysis: A Tutorial Introduction*. MIT Press, Cambridge, MA, 2004.

861. S. M. Stringer and E. T. Rolls. Invariant object recognition in the visual system with novel views of 3D objects. *Neural Computation*, 14(11):2585–2596, 2002.

862. A. Stuart and J. Ord. *Kendall's Advanced Theory of Statistics*, Vol. 1: *Distribution Theory*. Edward Arnold, London, 1994.

863. G. J. Stuart and B. Sakmann. Active propagation of somatic action potentials into neocortical pyramidal cell dendrites. *Nature*, 367:69–72, 1994.

864. E. Sussman, W. Ritter, and H. G. Vaughan. Attention affects the organization of auditory input associated with the mismatch negativity system. *Brain Research*, 789:130–138, 1998.

865. R. S. Sutton. Learning to predict by the methods of temporal differences. *Machine Learning*, 3:9–44, 1988.

866. R. S. Sutton and A. G. Barto. Toward a modern theory of adaptive networks: Expectation and prediction. *Psychology Review*, 88:135–170, 1981.

867. R. S. Sutton and A. G. Barto. Time-derivative models of Pavlovian reinforcement. In M. Gabriel and J. Moore, Eds., *Learning and Computational Neuroscience: Foundations of Adaptive Networks*, pp. 497–537. MIT Press, Cambridge, MA, 1990.

868. R. S. Sutton and A. G. Barto. *Reinforcement Learning: An Introduction*. MIT Press, Cambridge, MA, 1998.

869. N. V. Swindale. The development of topography in the visual cortex: A review of models. *Network: Computation in Neural Systems*, 7(2):161–247, 1996.

870. I. Szita and A. Lörincz. Kalman filter control embedded into the reinforcement learning framework. *Neural Computation*, 16(3):491–499, 2004.

871. C. Tallon-Baudry and O. Bertrand. Oscillatory gamma activity in humans and its role in object representation. *Trends in Cognitive Scicences*, 3(4):151–162, 1999.

872. C. Tallon-Baudry, O. Bertrand, M. A. Henaff, J. Isnard, and C. Fischer. Attention modulates gamma-band oscillations differently in the human lateral occipital cortex and fusiform gyrus. *Cerebral Cortex*, 15(5):654–662, 2005.

873. A. Y. Tan, L. I. Zhang, M. M. Merzenich, and C. E. Schreiner. Tone-evoked excitatory and inhibitory synaptic conductances of primary auditory cortex neurons. *Journal of Neurophysiology*, 92:630–643, 2004.

874. S. Tanaka. Theory of ocular dominance column formation: Mathematical basis and computer simulation. *Biological Cybernetics*, 64(4):263–272, 1991.

875. T. Tanaka. Generalized weighted rules for principal components tracking. *IEEE Transactions on Signal Processing*, 53(4):1243–1253, 2005.

876. J. G. Taylor and S. Coombes. Learning higher order correlations. *Neural Networks*, 6:423–427, 1993.

877. J. G. Taylor and M. D. Plumbley. Information theory and neural networks. In J. G. Taylor, Ed., *Mathematical Properties of Neural Networks*, pp. 307–337. Elsevier, 1993.

878. Y. W. Teh and G. E. Hinton. Rate-coded restricted Boltzmann machines for face recognition. In T. Leen, T. Dietterich, and V. Tresp, Eds., *Advances in Neural Information Processing Systems*, Vol. 13, pp. 908–914. MIT Press, Cambridge, MA, 2001.

879. A. Thiele and G. Stoner. Neuronal synchrony does not correlate with motion coherence in cortical area MT. *Nature*, 421:366–370, 2003.

880. A. M. Thomson and J. Deuchars. Synaptic interactions in neocortical local circuits: Dual intracellular recordings in vitro. *Cerebral Cortex*, 7:510–522, 1997.

881. D. J. Thomson. Spectrum estimation and harmonic analysis. *Proceedings of the IEEE*, 70:1055–1096, September 1982.

882. E. L. Thorndike. *Human Nature and the Social Order*. Macmillan, New York, 1940.

883. S. Thorpe, D. Fize, and C. Marlot. Speed of processing in the human visual system. *Nature*, 381:520–522, 1996.

884. H. Tiitinen, J. Sinkkonen, K. Reinikainen, K. Alho, J. Lavikainen, and R. Naatanen. Selective attention enhances the auditory 40-Hz transient-response in humans. *Nature*, 364:59–60, 1993.

885. E. Todorov and M. I. Jordan. Optimal feedback control theory as a theory of motor coordination. *Nature Neuroscience*, 5:1226–1235, 2002.

886. M. Tomita and J. J. Eggermont. Cross-correlation and joint spectro-temporal receptive field properties in auditory cortex. *Journal of Neurophysiology*, 93:378–392, 2005.

887. L. Tong, V. Soon, Y. F. Huang, and R. Liu. Indeterminacy and identifiability of blind identification. *IEEE Transactions on Circuits and Systems*, 38:499–509, 1991.

888. J. R. Treichler and B. G. Agee. A new approach to multipath correction of constant modulus signals. *IEEE Transactions on Acoustics, Speech, and Signal Processing*, 31(2):459–472, 1983.

889. A. M. Treisman. Features and objects—The 14th Bartlett memorial lecture. *Quarterly Journal of Experimental Psychology, Section A—Human Experimental Psychology*, 40(2):201–237, 1988.

890. A. M. Treisman. The binding problem. *Current Opinion in Neurobiology*, 6:171–178, 1996.

891. A. M. Treisman and G. Gelade. A feature integration theory of attention. *Cognitive Psychology*, 12:97–136, 1980.

892. A. Treves and E. T. Rolls. Computational constraints suggest the need for two distinct input systems to the hippocampal CA3 network. *Hippocampus*, 2:189–200, 1992.

893. D. Y. Ts'o, C. D. Gilbert, and T. N. Wiesel. Relationships between horizontal interactions and functional architecture in cat striate cortex as revealed by cross-correlation analysis. *Journal of Neuroscience*, 6:1160–1170, 1986.

894. M. Tuck and A. Pentland. Eigenfaces for recognition. *Journal of Cognitive Neuroscience*, 3(1):71–86, 1991.

895. A. Turing. The chemical basis of morphogenesis. *Philosophical Transactions of the Royal Society of London, B*, 237:5–72, 1952.

896. G. G. Turrigiano. Homeostatic plasticity in neuronal networks: The more things change, the more they stay the same. *Trends in Neuroscience*, 22(5):221–227, 1999.

897. G. G. Turrigiano and S. B. Nelson. Hebb and homeostasis in neuronal plasticity. *Current Opinion in Neurobiology*, 10:358–364.

898. E. Tzanakou. When a feature detector becomes a feature generator. *IEEE Engineering in Medicine and Biology Magazine*, 9(1):44–46, 1990.

899. E. Tzanakou. *Supervised and Unsupervised Pattern Recognition: Feature Extraction and Computational Intelligence*. CRC Press, Roca Raton, FL, 2000.

900. E. Tzanakou, R. Michalak, and E. Harth. The Alopex process: Visual receptive fields by response feedback. *Biological Cybernetics*, 35:161–174, 1979.

901. K. P. Unnikrishnan and K. P. Venugopal. Learning in connectionist networks using the Alopex algorithm. In *Proceedings of the IJCNN*, Vol. 1, pp. 926–931, 1992, IEEE Press Piscataway, NJ.

902. K. P. Unnikrishnan and K. P. Venugopal. Alopex: A correlation-based learning algorithm for feedforward and recurrent neural networks. *Neural Computation*, 6(3):469–490, 1994.

903. W. M. Usrey and R. C. Reid. Synchronous activity in the nervous system. *Annual Review of Physiology*, 61:435–456, 1999.

904. E. Vaadia, Y. Gottlieb, and M. Abeles. Single-unit activity related to sensorimotor association in auditory cortex. *Journal of Neurophysiology*, 48(5):1201–1213, 1982.

905. A. van den Bos. Complex gradient and Hessian. *IEE Proceedings of Vision, Image and Signal Processing*, 141(6):380–383, 1994.

906. R. van der Merwe, J. F. G. de Freitas, A. Doucet, and E. Wan. The unscented particle filter. TR-30, Cambridege University Engineering Department, August 2000.

907. J. H. van Hateren and D. L. Ruderman. Independent component analysis of natural image sequences yields spatio-temporal filters similar to simple cells in primary visual cortex. *Proceedings of the Royal Society of London, B*, 265:2315–2320, 1998.

908. J. H. van Hateren and A. van der Schaaf. Independent component filters of natural images compared with simple cells in primary visual cortex. *Proceedings of the Royal Society of London, B*, 265:359–366, 1998.

909. H. L. Van Trees. *Detection, Estimation, and Modulation*. Wiley, New York, 1968.

910. B. D. van Veen and K. M. Buckley. Beamforming: A versatile approach to spatial filtering. *IEEE ASSP Magazine*, 5(2):4–24, 1988.

911. B. D. van Veen and K. M. Buckley. Beamforming techniques for spatial filtering. In V. K. Madisetti and D. B. Williams, Eds., *Digital Signal Processing Handbook*. CRC Press, Boca Raton, FL, 1997.

912. V. Vapnik. *Statistical Learning Theory*. Wiley, New York, 1998.

913. O. Vasicek. A test for normality based on sample entropy. *Journal of the Royal Statistical Society, Series B*, 38(1):54–59, 1976.

914. K. P. Venugopal, A. S. Pandya, and R. Sundhakar. A recurrent network controller and learning algorithm for the on-line learning control of autonomous underwater vehicles. *Neural Networks*, 7(5):833–846, 1994.

915. B. V. K. Vijayakumar, A. Mahalanobis, and R. D. Juday. *Correlation Pattern Recognition*. Cambridge University Press, Cambridge, 2005.

916. A. E. P. Villa, B. Hyland, I. V. Tetko, and A. Najam. Dynamical cell assemblies in the rat auditory cortex in a reaction-time task. *BioSystems*, 48:269–277, 1998.

917. W. E. Vinje and J. L. Gallant. Sparse coding and decorrelation in primary visual cortex during natural vision. *Science*, 287:1273–1276, 2000.

918. V. Virsu, B. B. Lee, and O. D. Creutzfeldt. Dark adaptation and receptive field organisation of cells in the cat lateral geniculate nucleus. *Experimental Brain Research*, 27(1):35–50, 1977.

919. T. Voegtlin. Recursive principal component analysis. *Neural Networks*, 18:1051–1063, 2005.

920. T. P. Vogels and L. F. Abbott. Signal propagation and logic gating in networks of integrate-and-fire neurons. *Journal of Neuroscience*, 25:10786–10795, 2005.

921. C. von der Malsburg. Self-organization of orientation sensitive cells in the striate cortex. *Kybernetick*, 14:85–100, 1973.

922. C. von der Malsburg. The correlation theory of brain function. Internal Report 81-2, Department of Neurobiology, Max-Planck-Institute for Biophysical Chemistry, 1981.

923. C. von der Malsburg. Am I thinking assemblies. In G. Palm and A. Aertsen, Eds., *Brain Theory*, pp. 161–176. Springer, Berlin, 1986.

924. C. von der Malsburg. Binding in models of perception and brain function. *Current Opinion in Neurobiology*, 5:520–526, 1995.

925. C. von der Malsburg. Dynamic link architecture. In M. Arbib, Ed., *The Handbook of Brain Theory and Neural Networks*, pp. 329–331. MIT Press, Cambridge, MA, 1995.

926. C. von der Malsburg. The what and why of binding: The modeler's perspective. *Neuron*, 24:95–104, 1999.

927. C. von der Malsburg and W. Schneider. A neural cocktail-party processor. *Biological Cybernetics*, 54:29–40, 1986.

928. C. von der Malsburg and W. Schneider. Sensory segmentation with coupled neural oscillators. *Biological Cybernetics*, 67:233–242, 1992.

929. C. von der Malsburg and W. Singer. Principles of cortical network organization. In P. Rakic and W. Singer, Eds., *Neurobiology of Neocortex*, pp. 69–99. Wiley, New York, 1988.

930. G. Wahba. *Spline Models for Observational Data*. SIAM, Philadephia, PA, 1990.

931. J. T. Wall, J. Xu, and X. Wang. Human brain plasticity: An emerging view of the multiple substrates and mechanisms that cause cortical changes and related sensory dysfunctions after injuries of sensory inputs from the body. *Brain Research Reviews*, 39:181–215, 2002.

932. M. N. Wallace, L. M. Kitzes, and E. G. Jones. Intrinsic inter- and intralaminar connections and their relationship to the tonotopic map in cat primary auditory cortex. *Experimental Brain Research*, 86:527–544, 1991.

933. E. Wan and R. van der Merwe. The unscented Kalman filter. In S. Haykin, Ed., *Kalman Filtering and Neural Networks*, pp. 221–280. Wiley, New York, 2001.

934. M. P. Wand and M. C. Jones. *Kernel Smoothing*. Chapman & Hall, London, 1995.

935. D. L. Wang. Primitive auditory segregation based on oscillatory correlation. *Cognitive Science*, 20(3):409–456, 1996.

936. D. L. Wang. The time dimension for scene analysis. *IEEE Transactions on Neural Networks*, 16(6):1401–1426, 2005.

937. D. L. Wang and G. J. Brown. Separation of speech from interfering sounds based on oscillatory correlation. *IEEE Transactions on Neural Networks*, 10(3):684–697, 1999.

938. D. L. Wang, J. Buhmann, and C. von der Malsburg. Pattern segmentation in associative memory. *Neural Computation*, 2:94–106, 1990.

939. L. Wang and J. Karhunen. A unified neural bigradient algorithm for robust PCA and MCA. *International Journal of Neural Systems*, 7:53–67, 1996.

940. Y. Wang, P. Berg, and M. Scherg. Common spatial subspace decomposition applied to analysis of brain responses under multiple task conditions: A simulation study. *Clinical Neurophysiology*, 110:604–614, 1999.

941. Y. Washizawa and Y. Yamashita. Non-linear Wiener filter in a reproducing kernel Hilbert space. In *Proceedings of IEEE Conference on Pattern Recognition*, pp. 967–970, Hong Kong, China, 2006, IEEE Press, Piscataway, NJ.

942. C. Watkins. Learning from delayed rewards. Ph.D. thesis, King's College, Cambridge University, UK, 1989.

943. A. Webb, Ed. *Statistical Pattern Recognition*. Oxford University Press, New York, 1999.

944. C. Weber and S. Wermter. Image segmentation by complex-valued units. In W. Duch, J. Kacprzyk, E. Oja, and S. Zadrozny, Eds., *Proc. ICANN'05 (Lecture Notes in Computer Science 3696)*, pp. 519–524. Springer, Berlin, 2005.

945. N. M. Weinberger. Physiological memory in primary auditory cortex: Characteristics and mechanisms. *Neurobiology of Learning and Memory*, 70:226–251, 1998.

946. E. Weinstein, M. Feder, and A. V. Oppenheim. Multi-channel signal separation by decorrelation. *IEEE Transactions on Signal and Audio Processing*, 1:405–413, 1993.

947. M. Weliky and L. C. Katz. Disruption of orientation tuning in visual cortex by artificially correlated neuronal activity. *Nature*, 386:680–685, 1997.

948. P. Werbos. Beyond regression: New tools for prediction and analysis in behavioral sciences. Ph.D. thesis, Harvard University, Cambridge, MA, 1974.

949. J. Wessberg, C. R. Stambaugh, J. D. Kralik, P. D. Beck, M. Laubach, J. K. Chapin, J. Kim, J. Biggs, M. A. Sirinivasan, and M. A. Nicolelis. Real-time prediction of hand trajectory by ensembles of cortical neurons in primates. *Nature*, 408:361–365, 2000.

950. R. H. White. Competitive Hebbian learning: Algorithm and demonstrations. *Neural Networks*, 5(2):261–275, 1992.

951. B. Widrow and M. E. Hoff, Jr. Adaptive switch circuits. In *IRE WESCON Convention Record*, pp. 96–104, 1960.

952. B. Widrow, J. McCool, and M. Ball. The complex LMS algorithm. *Proceedings of the IEEE*, 63:719–720, 1975.

953. B. Widrow and S. D. Stearns. *Adaptive Signal Processing*. Prentice-Hall, Englewood Cliffs, NJ, 1985.

954. N. Wiener. Generalized harmonic analysis. *Acta Mathematica*, 55:117–258, 1930.

955. N. Wiener. *Cybernetics: Or Control and Communications in the Animal and the Machine*. Wiley, New York, 1948.

956. N. Wiener. *Time Series Analysis*. MIT Press, Cambridge, MA, 1948.

957. N. Wiener. *Extrapolation, Interpolation and Smoothing of Time Series*. MIT Press, Cambridge, MA, 1949.

958. T. N. Wiesel and D. H. Hubel. Ordered arrangement of orientation columns in monkey lacking visual experiences. *Journal of Comparative Neurology*, 158:307–318, 1974.

959. C. K. I. Williams. Computation with infinite neural networks. *Neural Computation*, 10:1203–1216, 1998.

960. C. K. I. Williams. Prediction with Gaussian processes: From linear regression to linear prediction and beyond. In M. I. Jordan, ed., *Learning in Graphical Models*, pp. 599–621. Kluwer Academic, Norwell, MA, 1998.

961. D. J. Willshaw, O. P. Buneman, and H. C. Longuet-Higgins. Nonholographic associative memory. *Nature*, 222:960–962, 1969.

962. D. J. Willshaw and P. Dayan. Optimal plasticity from matrix memories: What goes up must come down. *Neural Computation*, 2:85–93, 1990.

963. D. J. Willshaw and C. von der Malsburg. How patterned neural connections can be set up by self-organization. *Proceedings of the Royal Society of London, B*, 194:431–445, 1976.

964. M. A. Wilson and B. L. McNaughton. Reactivation of hippocampal ensemble memories during sleep. *Science*, 265:676–679, 1994.

965. J. Winson. Interspecies differences in the occurrence of theta. *Behavioral Biology*, 7:479–487, 1972.

966. L. Wiskott and T. J. Sejnowski. Slow feature analysis: Unsupervised learning of invariances. *Neural Computation*, 14(4):715–770, 2002.

967. T. Wolansky, E. A. Clement, S. R. Peters, M. A. Palczak, and C. T. Dickson. Hippocampal slow oscillation: A novel EEG state and its coordination with ongoing neocortical activity. *Journal of Neuroscience*, 26:6213–6229, 2006.

968. L. Wolf and A. Shashua. Kernel principal angles for classification machines with application to image sequence interpretation. In *Proceedings of IEEE Computer Vision and Pattern Recognition (CVPR'03)*, pp. 635–640, 2003, IEEE Computer Society Press, New York.

969. L. Wolf and A. Shashua. Learning over sets using kernel principal angles. *Journal of Machine Learning Research*, 4:913–931, 2003.

970. J. M. Wolfe. Visual search. In H. Pashler, Ed., *Attention*, pp. 13–74. Psychology Press, Hove, East Sussex, England, 1998.

971. D. M. Wolpert and Z. Ghahramani. Computational principles of movement neuroscience. *Nature Neuroscience*, 3:1212–1217, 2000.

972. D. M. Wolpert, Z. Ghahramani, and M. I. Jordan. An internal model for sensorimotor integration. *Science*, 269:1880–1882, September 1995.

973. R. O. Wong, M. Meister, and C. J. Shatz. Transient period of correlated bursting activity during development of the mammalian retina. *Neuron*, 11:923–938, 1993.

974. R. Wooding. The multivariate distribution of complex normal variables. *Biometrika*, 43:212–215, 1956.

975. F. Wörgötter and B. Porr. Temporal sequence learning, prediction, and control: A review of different models and their relation to biological mechanisms. *Neural Computation*, 17:245–319, 2005.

976. W. Wu, Y. Gao, E. Bienenstock, J. P. Donoghue, and M. J. Black. Bayesian population coding of motor cortical activity using a Kalman filter. *Neural Computation*, 18:80–118, 2005.

977. X. Xie and H. S. Seung. Spike-based learning rules and stabilization of persistent neural activity. In S. A. Solla, T. K. Leen, and K.-R. Müller, Eds., *Advances in Neural Information Processing Systems*, Vol. 12, pp. 199–208. MIT Press, Cambridge, MA, 2000.

978. X. Xie and H. S. Seung. Equivalence of backpropagation and contrastive Hebbian learning in a layered network. *Neural Computation*, 15:441–454, 2003.

979. G. Xu, H. Liu, L. Tong, and T. Kailath. A least-squares approach to blind channel identification. *IEEE Transactions on Signal Processing*, 43(12):2982–2993, 1995.

980. L. Xu. Least mean square error recognition principle for self organizing neural networks. *Neural Networks*, 6:627–648, 1993.

981. L. Xu and M. I. Jordan. On convergence properties of the EM algorithm for Gaussian mixtures. *Neural Computation*, 8:129–151, 1996.

982. L. Xu, E. Oja, and C. Y. Suen. Modified Hebbian learning for curve and surface fitting. *Neural Networks*, 5:441–457, 1992.

983. Y. Xu, J.-Y. Yang, and J. Yang. A reformulative kernel Fisher discriminant analysis. *Pattern Recognition*, 37:1299–1302, 2004.

984. M. Yamada and M. Azimi-Sadjadi. Kernel Wiener filter using canonical correlation analysis framework. In *Proceedings of IEEE 13th Workshop on Statistical Signal Processing*, pp. 769–774, Bordeaux, France, 2005, IEEE Press, Piscataway, NJ.

985. M. Yamada and M. Azimi-Sadjadi. Kernel Wiener filter with distance constraint. In *Proc. IEEE ICASSP'06*, pp. 596–599, Toulouse, France, 2006, IEEE Press, Piscataway, NJ.

986. H. H. Yang and S. Amari. On-line learning algorithms for blind separation—Maximum entropy and minimum mutual information. *Neural Computation*, 9:1457–1482, 1997.

987. J. Yang, Z. Jin, J.-Y. Yang, D. Zhang, and A. F. Frangi. Essence of kernel Fisher discriminant analysis: KPCA plus LDA. *Pattern Recognition*, 37:2097–2100, 2004.

988. H. Yao and Y. Dan. Stimulus timing-dependent plasticity in cortical processing of orientation. *Neuron*, 32:315–323, 2001.

989. D. Yellin and E. Weinstein. Multichannel signal separation: Methods and analysis. *IEEE Transactions on Signal Processing*, 44:106–118, 1996.

990. J. Z. Young. The evolution of the nervous system and of the relationship of organism and environment. In G. R. de Beer, Ed., *Evolution: Essays on Aspects of Evolutionary Biology, Presented to Professor E. S. Goodrich on His 70th Birthday*, pp. 179–204. Clarendon, Oxford, 1938.

991. M. P. Young and S. Yamane. Sparse population coding of faces in the inferotemporal cortex. *Science*, 256(2):1327–1330, 1992.

992. A. J. Yu and P. Dayan. Uncertainty, neuromodulation, and attention. *Neuron*, 46:681–692, 2005.

993. A. L. Yuille. Generalized deformable models, statistical physics, and matching problems. *Neural Computation*, 2(1):1–24, 1990.

994. A. L. Yuille and N. M. Grzywacz. A winner-take-all mechanism based on presynaptic inhibition. *Neural Computation*, 1:334–347, 1989.

995. A. L. Yuille, D. M. Kammen, and D. Cohen. Quadrature and the development of orientation selective cortical cells by Hebb rules. *Biological Cybernetics*, 61:183–194, 1988.

996. R. S. Zemel and G. E. Hinton. Discovering viewpoint-invariant relationships that characterize objects. In R. P. Lippmann, J. E. Moody, and D. S. Touretzky, Eds., *Advances in Neural Information Processing Systems*, Vol. 3, pp. 299–305. Morgan Kaufmann, San Mateo, CA, 1991.

997. R. S. Zemel, C. K. I. Williams, and M. C. Mozer. Lending direction to neural networks. *Neural Networks*, 8(4):503–512, 1995.

998. L. I. Zhang, S. Bao, and M. M. Merzenich. Disruption of primary auditory cortex by synchronous auditory inputs during a critical period. *Proceedings of the National Academy of Sciences, USA*, 99:2309–2314, 2002.

999. Y. Zhang. Complex-valued generalized Hebbian algorithm and its applications to sensor array signal processing. In A. Hirose, Ed., *Complex-Valued Neural Networks: Theories and Applications*, pp. 227–250. World Scientific, Singapore, 2003.

1000. Y. Zhang and Y. Ma. CGHA for principal component extraction in the complex domain. *IEEE Transactions on Neural Networks*, 8(5):1031–1036, 1997.

1001. W. Zheng. Class-incremental generalized discriminant analysis. *Neural Computation*, 18:979–1006, 2006.

1002. E. Zohary, M. N. Shadlen, and W. T. Newsome. Correlated neuronal discharge rate and its implications for psychophysical performance. *Nature*, 370:140–143, 1994.

INDEX

Correlative Learning: A Basis for Brain and Adaptive Systems, by Zhe Chen, Simon Haykin, Jos J. Eggermont, and Suzanna Becker